Synchronizing 5G Mobile Networks

Dennis Hagarty, Shahid Ajmeri, Anshul Tanwar

T0111863

Cisco Press

221 River St.

Hoboken, NJ 07030 USA

Synchronizing 5G Mobile Networks

Dennis Hagarty, Shahid Ajmeri, Anshul Tanwar

Copyright©2021 Cisco Systems, Inc.

Cisco Press logo is a trademark of Cisco Systems, Inc.

Published by:
Cisco Press
221 River St.
Hoboken, NJ 07030 USA

1 2021

Library of Congress Control Number: 2021903473

ISBN-13: 978-0-13-683625-4
ISBN-10: 0-13-683625-9

Warning and Disclaimer

Trademark Acknowledgments

All terms mentioned in this book that are known to be trademarks or service marks have been appropriately capitalized. Cisco Press or Cisco Systems, Inc., cannot attest to the accuracy of this information. Use of a term in this book should not be regarded as affecting the validity of any trademark or service mark.

Special Sales

For information about buying this title in bulk quantities, or for special sales opportunities (which may include electronic versions; custom cover designs; and content particular to your business, training goals, marketing focus, or branding interests), please contact our corporate sales department at corpsales@pearsoned.com or (800) 382-3419.

For government sales inquiries, please contact governmentsales@pearsoned.com.

For questions about sales outside the U.S., please contact intlcs@pearson.com.

Feedback Information

At Cisco Press, our goal is to create in-depth technical books of the highest quality and value. Each book is crafted with care and precision, undergoing rigorous development that involves the unique expertise of members from the professional technical community.

Readers' feedback is a natural continuation of this process. If you have any comments regarding how we could improve the quality of this book, or otherwise alter it to better suit your needs, you can contact us through email at feedback@ciscopress.com. Please make sure to include the book title and ISBN in your message.

We greatly appreciate your assistance.

Editor-in-Chief: Mark Taub

Alliances Manager, Cisco Press: Arezou Gol

Director, ITP Product Management: Brett Bartow

Executive Editor: James Manly

Managing Editor: Sandra Schroeder

Development Editor: Christopher A. Cleveland

Project Editor: Mandie Frank

Copy Editor: Bill McManus

Technical Editors: Mike Gilson, Peter Meyer

Editorial Assistant: Cindy Teeters

Designer: Chuti Prasertsith

Composition: codeMantra

Indexer: Ken Johnson

Proofreader: Charlotte Kughen

Americas Headquarters
Cisco Systems, Inc.
San Jose, CA

Asia Pacific Headquarters
Cisco Systems (USA) Pte. Ltd.
Singapore

Europe Headquarters
Cisco Systems International BV Amsterdam,
The Netherlands

Cisco has more than 200 offices worldwide. Addresses, phone numbers, and fax numbers are listed on the Cisco Website at **www.cisco.com/go/offices.**

Cisco and the Cisco logo are trademarks or registered trademarks of Cisco and/or its affiliates in the U.S. and other countries. To view a list of Cisco trademarks, go to this URL: www.cisco.com/go/trademarks. Third party trademarks mentioned are the property of their respective owners. The use of the word partner does not imply a partnership relationship between Cisco and any other company. (1110R)

Credits

Figure Number	Credit Attribution
Figure 1-1	zechal/123RF
Figure 1-3A	Brett Holmes/Shutterstock
Figure 1-3B	Petr Vaclavek/Shutterstock
Figure 1-3C	LongQuattro/Shutterstock
Figure 3-1	Aliaksei Tarasau/Shutterstock
Figure 3-14	U.S. Coast Guard Navigation Center, United States Department of Homeland Security
Figure 5-19	© ITU 2008
Figure 8-2	Based on Figure 8-5 of ITU-T Recommendation G.803
Figure 8-3	Based on Figure 7-1 of ITU-T Recommendation G.8271.1
Figure 8-4	Based on Figure 1 of ITU-T Recommendation G.8271.2
Figure 9-14	© ITU 2008
Figure 9-18	Based on Figure III.2 of ITU-T Recommendation G.8273.2
Figure 11-2a	Wikipedia Commons
Figure 11-2b	Wikipedia Commons
Figure 12-9	Calnex Solutions plc © 2021
Figure 12-10	Calnex Solutions plc © 2021
Figure 12-24	Calnex Solutions plc © 2021

About the Authors

Dennis Hagarty is an experienced technical specialist in the fields of information technology and telecommunications. He has led presales, consulting, and engineering efforts for major utilities, corporations, and service providers in Australasia and Europe. Having worked in numerous technical areas, Dennis has concentrated on the mobile space for almost 30 years and has specialized in timing and synchronization for the last 12 years. In his current role, Dennis is the Cisco communications interface between engineering, field sales teams, and the global Cisco customer community for all matters related to 5G timing and synchronization. This mandate sees him talking with many large service providers, including most of the world's tier 1 mobile operators.

Shahid Ajmeri is a senior product manager at Cisco with responsibility for leading its 5G transport and mobile edge architecture strategy. He has more than 20 years of experience in the service provider industry, focusing on various technologies ranging from 3G/4G to 5G mobile networks, mobile edge computing, telco data center, service provider security, time synchronization, and end-to-end network design. Shahid has been instrumental in driving network transformation projects and architecting next-generation networks for customers across the globe. He currently works across disciplines, bringing together engineering, standards development organizations, and customers to develop and translate product requirements from industry and standard-setting bodies to the market.

Anshul Tanwar is a principal engineer at Cisco Systems, where he is known as a technologist with a combination of R&D expertise and business sensibility. During his tenure of more than 20 years at Cisco, Anshul has architected many routing and switching products used by large tier 1 mobile and Metro Ethernet service providers across the world. He has led the SyncE and PTP architecture definition and implementation for multiple access and pre-aggregation routers in Cisco. In his most recent role, Anshul was responsible for defining the deployment architecture of phase timing synchronization for one of the world's largest service provider LTE/LTE-A networks. He is also a co-inventor on three patents, including one covering synchronization.

About the Technical Reviewers

Mike Gilson is a senior manager in BT and is responsible for the Synchronization and Timing Platforms. He has played a major role in BT's synchronization strategy from 1988 to present day and currently leads a team responsible for timing-related research, timing design, development, and the delivery into the operational estate. Through the years, he has represented BT on various national and international working groups, standards committees, and regulatory bodies. He currently contributes to ITU, is on the steering committee for the ITSF/WSTS forums, and several UK advisory groups. He has authored or co-authored books, papers, conference presentations, and produced standards contributions on many aspects of synchronization. Mike joined BT in 1983 working in advanced transmission, before moving onto timing. He has a BA (Hons) in Business from University of East Anglia and is a Member of the IET.

Peter Meyer is a systems engineering manager at Microchip Technology's timing and communications business unit. He has worked in network synchronization and telecommunications for over 20 years at Mitel, Zarlink, Microsemi, and Microchip in a variety of roles, including applications engineering and system architecture. Peter has covered the transition of network synchronization distribution from T1/E1/PDH through SONET/SDH and now via Ethernet/IP using SyncE and PTP/IEEE1588. He represents Microchip as a timing expert at ITU-T, IETF, IEEE, and other standards development organizations.

Dedications

Dennis Hagarty: I dedicate this book to my incredibly supportive parents, Willa and Bill, for their steadfast belief in the value of a good education and their support no matter where my learning and development took me. To my wife, Dr. Ingrid Slembek, for her unwavering support and assistance through the long process of writing and editing. To my co-authors, Anshul and Shahid, who worked tirelessly to make this book the best and most comprehensive it could be. Finally, to Johann Sebastian Bach (1685–1750), whose genius is clearly recognized in the *Brandenburg Concertos,* the various recordings of which kept me company during the lonely hours on this journey.

Shahid Ajmeri: I dedicate this book to my entire family, who have always been encouraging and supportive during my work on this book. To my wonderful wife, Radhika, for inspiring and encouraging me to write this book. Without her support, it would have been impossible to finish this book. To my co-authors, Dennis and Anshul, for their patience during our many regular review calls and technology discussions. Finally, to the many mentors, co-workers, and friends who, over the last 20-odd years, have helped me—this book reflects of all those learnings.

Anshul Tanwar: I must start by thanking my wife, Soma, for patiently putting up with me and always encouraging me to keep writing. To my wonderful son, Aryan, and beautiful daughter, Anika, for their unwavering belief in me. To my mother, Aruna, for infusing the ever-lasting positivity in me. To my co-authors, Dennis and Shahid, for reviewing, editing, and keeping me focused on the book. I have a learned a lot from our formal and informal discussions.

Acknowledgments

Writing a book takes an immense amount of patience, discipline, and, of course, time. We would like to especially acknowledge the tremendous support we received from staff inside Cisco—especially our management team and colleagues.

We owe a huge debt of gratitude to the reviewers, Peter and Mike, for their amazing efforts and insights that led to the improvement of our text and correction of our frequent mistakes and misunderstandings. The amount of time and dedication that it took to understand and scrutinize our draft material is truly impressive.

Similarly, we would like to thank contributors from various companies, most especially the team at Calnex Solutions, and especially their CEO, Tommy Cook, who cheerfully agreed to write the foreword. We are also grateful for their permission to reuse photos of their equipment to help clarify the text, especially in Chapter 12.

We would like to extend our appreciation to James Manly, Cisco Press Executive Editor, for his patience with changing deadlines and flexibility required to fit this book around our day jobs. We would also like to thank Chris Cleveland, the development editor, for his solid guidance throughout. Their assistance through the entire process has made writing this book an interesting and rewarding experience.

Lastly, we would like to thank the many standards development organizations, technologists, and mobile experts that continue to contribute to the fields of both mobile communications and time synchronization—especially 5G mobile. Some of these professionals design and produce hardware, others build and code software, and many also make valuable contributions to standards development organizations. Without their hard work in striving for the best of the best, we would not have a book to write.

Contents at a Glance

Contents

Icons Used in This Book

Frequency signal

Set-top box (STB)

Voice gateway

Time/phase
timing signal

Handset/user
equipment (UE)

Time source

Cell site

Router

Home router
for DSL

Radio unit

Satellites

CMTS/DOCSIS
head end device

Power substation

Content acquirer

Central mobile core

Foreword

The distribution of synchronization across telecommunications networks has been a requirement since the emergence of digital networks back in the 1970s. Initially, frequency synchronization was required for the transfer of voice calls. Over the years, multiple generations of equipment standards increased the requirement for frequency synchronization. In the relatively recent past, the requirement was expanded to include time or, more precisely, phase synchronization. This need is principally driven by mobile base stations requiring phase alignment with other base stations to support overlapping radio footprints.

One of the main issues when dealing with network synchronization is that when synchronization goes wrong, it initially doesn't look like a synchronization problem! I started my engineering career as a digital designer, and one of the first lessons I learned when doing printed circuit board (PCB) layout design was to always do the clock distribution plans first and make them as robust as possible. This was, and still is, best practice, because on PCB assemblies, when clocking issues occur, they manifest as logic design issues, not as clocking issues, making these types of problems very challenging to debug.

Networks are just the same. Synchronization issues in networks lead to sporadic outages and/or data loss events, often hours or days apart, which can appear to be traffic loading/management issues and thus are very challenging to track down. This is why network architects have always taken great care to design quality into their synchronization networks to ensure the synchronization is distributed robustly by design, not by trial and error.

With the evolution of mobile networks, the requirement for frequency synchronization broadened to include phase synchronization. The challenges faced by designers of equipment and networks for frequency distribution are also present in phase, but with an additional set of challenges that come on top. An additional factor is that recent international standards from the ITU Telecommunication Standardization Sector (ITU-T) specify tighter performance requirements for the transfer of phase. Combined with the increasing criticality of accurate phase synchronization to the future mobile network (due to the forecasted far-higher numbers of radio stations resulting in increased overlapping broadcast footprints), and the use of time/phase in several emerging applications, it has never been more important to ensure that quality and performance are robustly designed in at every stage of the life cycle.

Many other industry sectors are now looking at the need to include time as a core element of their architecture; for example, factory automation, audio/video systems, financial networks utilizing machine trading, and so on. Although the accuracy requirements will vary between applications, several application-specific implementation challenges will arise.

Core to all of these applications is an understanding of the dynamics of time/phase transfer. Understanding these principles, the challenges involved, and the effects that cause problems is vital. Most importantly, if you are a designer, understanding how to mitigate against problems that impede reliable time/phase delivery is critical. Once you develop this knowledge for one application, you will be able to use these skills to design synchronization networks that meet the specific needs of any application that requires time/phase.

As you can see from looking at the table of contents of this book, there are many dimensions and areas of knowledge involved in developing a broad and deep understanding of all aspects of synchronization. If deep and comprehensive understanding is your objective, you will find this book essential reading. However, not everyone needs, wants, or has the time to understand the topic of synchronization to this depth. The book is structured to meet your needs whether you have a need for specific knowledge or seek to be an expert. Whatever your objective, I'm sure you will find this book essential literature to support your goal. Enjoy!

--Tommy Cook

Founder and CEO of Calnex Solutions, Plc

Introduction

Maintaining high-quality synchronization is very important for many forms of communication, and with mobile networks it is an especially critical precondition for good performance. Synchronization will have a dramatic (negative) effect on the efficiency, reliability, and capacity of the network, if the timing distribution network is not properly designed, implemented, and managed.

The stringent clock accuracy and precision requirements in these networks require that network engineers have a deep knowledge and understanding of synchronization protocols, their behavior, and their deployment needs.

Synchronization standards are also evolving to address a wide range of application needs using real-time network techniques. Factory automation, audio/video systems, synchronizing wireless sensors, and the Internet of things are some of the use cases where synchronization is extensively used, in addition to 4G/5G mobile and radio systems.

Motivation for Writing This Book

Timing and synchronization are not easy concepts to grasp, and they are becoming increasingly complex and nuanced as newer technologies become available. The learning curve is steep, and much of the existing material on the topic is dispersed across many resources and in a variety of formats. When viewed from outside, the subject of timing appears as a castle with a high wall and a broad moat.

It is no surprise, then, that timing tends to frighten off the newcomer, as at first its complexity is daunting. Recent research confirms our belief that the engineer starting out in the field is not well served by any educational material currently available—a situation which this work aims to rectify.

This book collects, arranges, and consolidates the necessary knowledge into a format that makes it easier to digest. The aim is to provide in-depth information on timing and synchronization—at the basic as well as advanced level. This book includes topics such as timing standards and protocols, clock design, operational and testing aspects, solution design, and deployment trade-offs.

Although 5G mobile is the primary focus of this book, it is written to be relevant to other industries and use cases. This is because the need for a timing solution is becoming important in an increasing number of scenarios—the mobile network being only one very specific example. Many of the concepts and principles apply equally to these other use cases.

The goal is for this book to be educational and informative. Our collective years of experience in both developing timing products and helping customers understand how to implement it is now available to the reader.

Who Should Read This Book?

The primary audience of this book is engineers and network architects who need to understand the field—as specified by international standards—and apply that knowledge to design and deploy a working synchronization solution. Because the book covers a broad spectrum of topics, it is also well suited for anybody who is involved in selecting equipment or designing timing solutions.

This book is written to be suitable for any level of technical expertise, including the following:

- Transport design engineers and radio engineers looking to design and implement mobile networks

- Test engineers preparing to validate clock equipment or certifying a synchronization solution in a production network

- Networking consultants interested in understanding the time synchronization technology evolution that affects their mobile service provider customers

- Students preparing for a career in the mobile service provider or private 5G network fields

- Chief technology officers seeking a greater understanding of the value that time synchronization brings to mobile networks

Throughout the book, you will see practical examples of how an engineer might architect a solution to deliver timing to the required level of accuracy. The authors have designed this book so that even a network engineer with almost no knowledge of the topic can easily understand it.

How This Book Is Organized

This book starts with the fundamental concepts and builds toward the level of knowledge needed to implement a timing solution. The basic ideas are laid out in advance, so that you gain confidence and familiarity as you progress through the chapters. So, for those new to the field, the recommended approach is to progress sequentially through the chapters to get the full benefit of the book.

Depending on the level of technical depth that you require, you may use this book as a targeted reference if you are interested in a specific area. The content is organized so that you can move between sections or chapters to cover just the material that interests you.

This book is ideal for anybody who is looking for an educational resource built with the following methods:

- Written in an educational style that is very readable for most of the content, because the goal is to first understand the concept and leave the complexity and details to those who need it

- Progresses through the chapters from easy overview, through basic concepts, to more advanced material

- Limits the heavy technical treatment and mathematical equations to smaller sections and only where necessary

- Comprehensively covers the topic at both a technical and a product-feature level, including equipment design, selection, and testing

- Includes the complete field of mobile timing, including packet transport, satellite systems, radio fronthaul networks, network timing redundancy, and more

- Helps network and transport engineers, radio engineers, and executives understand how to validate their selection and design for 5G timing solutions

- Benefits anybody selecting or designing timing solutions, especially for those using the PTP telecom profiles

- Covers the latest standards and functionality from both the standards development organizations and industry

- Compares and contrasts different approaches to implementation—objectively and with vendor neutrality. There are only two mentions of any Cisco product or router models in this book.

From the very first chapters, the book introduces concepts from mobile communications to give you background to the key technology trends and decision points facing today's mobile operators. It also addresses the implementation and characterization of timing for several deployment methods and use cases using best-in-class design practices.

The book is divided into the following four parts, each of which contains two to four chapters:

Part I, "Fundamentals of Synchronization and Timing," introduces synchronization requirements and timing concepts for communications networks. It includes the following chapters:

- **Chapter 1**, "Introduction to Synchronization and Timing," covers the fundamental concepts of what timing is, what type of synchronization exists, why it is needed, and how it can be sourced and transported.

- **Chapter 2**, "Usage of Synchronization and Timing," covers the uses of timing, where it is applied, and what sort of industries need time—although with particular emphasis on telecommunications.

- **Chapter 3**, "Synchronization and Timing Concepts," starts with the specifics of synchronous networks and the foundational aspect of the topic: clocks, timing reference sources, the time distribution network, and the application use or consumption of the timing signal. It also covers the GNSS satellite systems in some detail and describes various methods of transporting synchronization information and timing signals.

Part II, "SDOs, Clocks, and Timing Protocols," introduces the relevant standards development organizations, explains clocks in detail, and discusses various timing protocols. It includes the following chapters:

- **Chapter 4**, "Standards Development Organizations," takes a slightly different tack and introduces the amazing standards-setting community that underpins much of what makes the world's information and communications systems work. Of course, there is the 3GPP, the IEEE, and the ITU-T, but also later entrants such as the CPRI and the O-RAN Alliance, defining the modern radio access network (RAN).

- **Chapter 5**, "Clocks, Time Error, and Noise," is where the material starts to get more intense, covering the details of clocks, clock signals, and components of clocks. This is where many of the metrics for timing are first explored in topics such as TE, TIE, MTIE, and TDEV along with an explanation of jitter and wander. This sets up the foundation knowledge for a detailed examination of distributing frequency using physical methods.

- **Chapter 6**, "Physical Frequency Synchronization," provides comprehensive coverage of the topic of frequency synchronization, mainly using older TDM techniques such as E1/T1/SDH/SONET and the newer SyncE and eSyncE. It includes discussion of building a clock hierarchy using clocks of different quality or stratum levels—such as BITS/SSU devices. For this to work, there needs to be a traceability mechanism to support the transport of clock quality information—a mechanism that also helps to avoid timing loops, which this chapter then covers in a separate section.

- **Chapter 7**, "Precision Time Protocol," is all about PTP and the details of many of its characteristics and features contained first in IEEE 1588-2008 and then 1588-2019. The chapter provides details on the PTP datasets, the messages and how they are encapsulated, the BMCA, the clock types, and the mathematics used to recover frequency and time/phase. It ends with coverage of the different PTP profiles, including the ITU-T telecom profiles, and PTP security.

Part III, "Standards, Recommendations, and Deployment Considerations," provides an overview of the various ITU-T recommendations and discusses the factors involved in distributing time over both physical and packet-based networks. It includes the following chapters:

- **Chapter 8**, "ITU-T Timing Recommendations," is a comprehensive overview of the ITU-T and its process for making recommendations. The rest of the chapter contains a list of the many recommendations used to outline the architecture, metrics, network budgets, and clock performance and define the behavior of timing solutions. There are recommendations covering both physical (TDM and SyncE) and packet methods to transport timing signals and quality-level information.

- **Chapter 9**, "PTP Deployment Considerations," is where the theory meets the practicalities of design and deployment. This chapter contains many valuable lessons in the areas of design trade-offs and deployment concerns when putting together a timing solution. It includes more details on timing metrics and end-to-end time error budgeting, as well as options for holdover and redundancy. The information in this chapter is for more general synchronization use cases, while Chapter 11 covers some of the same subject matter but is much more specifically focused on the mobile network.

Part IV, "Timing Requirements, Solutions, and Testing," discusses various 4G/5G mobile network requirements, time synchronization deployment options, and validation methodologies. It includes following chapters:

- **Chapter 10**, "Mobile Timing Requirements," is where most of the mobile information is located, including timing requirements for LTE and 5G, evolution of the 5G architecture, and the features behind the 5G New Radio and the synchronization that it now demands. This chapter covers many of the major features, services, and radio optimization techniques now available in the radio access network (RAN). It also covers the different timing requirements for both the distributed and centralized RAN and covers the special challenges in the disaggregated and virtualized 5G RAN.

- **Chapter 11**, "5G Timing Solutions," has some echoes of Chapter 9 but with more detailed description of the deployment considerations for a timing solution targeted for mobile. It examines the trade-offs in synchronizing various types of cell sites, different topologies, radio types, and generations of mobile technology. It examines both the frequency-only case with the 4G mobile network and the additional time/phase requirements in both the 5G mobile backhaul and fronthaul/midhaul deployment.

  This chapter also examines the various options for the design of a timing solution based on the individual circumstances that the operator is faced with. This includes where the access network is owned by third parties, where GNSS is unavailable, and when autonomous operation of the timing solution (no reliance on GNSS) is required.

- **Chapter 12**, "Operating and Verifying Timing Solutions," covers the operational aspects of deploying and managing a timing solution. It introduces the hardware design techniques for clocks, explains how to define specifications for timing equipment, and provides extended coverage of testing timing performance. Lastly, it covers the operational aspects of configuration, management, and monitoring of a timing solution, including troubleshooting and field testing.

Note that the early chapters in the book may define a few technical terms somewhat loosely. Subsequent chapters unlock additional technical details and offer more exacting treatment of the topics. As the details are revealed, the terminology becomes more precisely defined.

As an example, at the beginning, the authors may use everyday terms—such as clocking, synchronization, and timing—somewhat interchangeably. Some sections talk about carrying time or transporting synchronization, clocking a radio, or timing a network device, when in fact, some methods of synchronization do not deal with actual time at all!

The book does not give recommendations on which of these technologies should be deployed, nor does it provide a transition plan for a mobile operator. Each mobile operator can evaluate the technologies and make decisions based on their own criteria and circumstances. However, it does cover the details of the pros and cons behind each approach, allowing the engineer to make better-informed decisions.

We hope you enjoy reading and using this book as much as we enjoyed writing it!

Chapter 1

Introduction to Synchronization and Timing

In this chapter, you will learn the following:

- **Overview of time synchronization:** Covers the basic concepts behind time, beginning with some history and coming forward to the present day.
- **What is synchronization and why is it needed?:** Introduces the different types of synchronization, namely frequency, phase, and time of day.
- **What is time?:** Covers the concepts of time, the currently used timescales, such as TAI and UTC, and modern sources of time.
- **How can GPS provide timing and synchronization?:** Provides a basic overview of the satellite systems that are widely used to distribute time.
- **Accuracy versus precision versus stability:** Covers these three concepts, vital to the discussion of the quality with which time is kept, transported, and used.

This chapter begins with a look at the historical record to help you understand how the world defines time today and why it is defined that way. Although seemingly a very simple concept, the implementation and practical applications of time are very involved. A few searches on Wikipedia on the subject of time quickly leads down rabbit holes of details, so this text will stick to the highlights.

After an introduction to time, the focus moves to synchronization, addressing why you might need it and what happens when it goes wrong. The term "synchronization" covers several different types (frequency, phase, and time), so it is important to understand which type of synchronization is required for any specific use case, why that application might need it, and how you accomplish it.

Overview of Time Synchronization

In medieval times, before anybody had any form of accurate timepiece, a village might use a clocktower (see Figure 1-1) or the clock at the local church as a time reference for the surrounding community. Time was announced to villagers by bells, commonly on the quarter hour (one chime for each quarter), and then a much deeper set of "bongs" on the hour—one for each hour from 1–12.

Figure 1-1   *Zytturm (Clock Tower) in Zug, Switzerland*

Therefore, every community had its own local version of time, as proclaimed to all the people within earshot of the bells. However, the clocks in neighboring villages could be significantly different from each other, and there was no easy means to synchronize them.

Commonly, the position of the sun was used as the basis for a timepiece (the *sundial*); however, this meant that every location had its own "local" version of time (since, at any one time, the sun is in a different position in the sky for every place on the earth). The position of the sun at specific times, such as noon (when the sun is highest in the sky), was used to help set the time on the local community clock. But since noon is quite dependent on position (particularly longitude), the clocks across a country would vary quite significantly (moving from west to east, the clocks would show the time to be increasingly later and later).

With the arrival of faster transport like railways, this situation became untenable, as every station on the railway line needed to agree on a common time so that standardized timetables would become meaningful. It was this need which led to the adoption of standardized times and eventually to Greenwich Mean Time (GMT) and the system of time zones used today. With the arrival of more accurate mechanical timepieces and particularly communications via the telegraph, the ability to synchronize clocks across a wider area also became possible.

GMT was established in 1884 at the International Meridian Conference held in Washington, DC, where the meridian passing through Greenwich (near London) was adopted as the initial or prime meridian for longitude and timekeeping. Given that there are 24 hours in a day and 360 degrees of longitude around the earth, it is obvious that the world could have 24 time zones with the average single hour time zone covering longitude of 15 degrees (360 / 24).

With this scheme, various governments around the world could adopt one or more of the time zones (or even quarter- and half-hour time zones) that worked for their situation. The result was that all clocks within a time zone were aligned to the same time reference rather than being locally defined. That way, local communities could have a common understanding of scheduled events like train timetables. Larger countries could adopt more than one time zone so that the local time did not diverge too far from the daily movement of the sun.

What Is Synchronization and Why Is It Needed?

Time synchronization is simply a mechanism to allow coordination and alignment among

- Processes, such as robots in a factory working on a single widget.

- Clocks that are running independently of each other, such as wall clocks across a railway network (including changes for summertime or daylight saving time).

- Computers that should process information in the correct time order, such as executing trades on a stock trading system.

- Multiple streams of information that should be presented in a strict sequence or coordinated with each other—a good example is audio and video in a TV signal.

- Events observed at multiple locations. Observers can determine the location of an event (for example, tracking storms via lightning strikes) if several lightning detectors can all record the detection time with closely aligned time clocks.

- Sensors monitoring real-world processes. If an accurate timestamp is logged against an event, operators can determine the correct sequence in a chain of events leading up to a critical situation (such as monitoring the stability of the power grid).

Accurate timing is also a critical attribute of a modern mobile network. These networks require a well-designed and implemented synchronization system to maximize their efficiency, reliability, and capacity. Not getting that right means that the mobile subscribers will likely suffer dropped calls, interrupted data usage, and a generally poor user experience. At the same time, operators will suffer network instability, inefficient usage of the radio spectrum, and unhappy customers. Refer to Chapter 10, "Mobile Timing Requirements," for more information on mobile standards and requirements.

Modern 5G networks are created using very sophisticated radio technologies designed to increase data rates and reliability, enhance the subscribers' experience, and maximize the utilization of the spectrum (which is normally purchased at very high cost). The techniques used in the radios rely on coordination between transmitting and receiving equipment placed quite far apart from one other. This includes the large, wide-area macro cells; localized small cells; and the mobile user equipment (such as your phone), which all must work together. This effective coordination relies on tight synchronization between these various components of the 5G radio system, as depicted in Figure 1-2.

Synchronization

Figure 1-2 *Radio Synchronization in a 5G Network*

The successful deployment of synchronization makes use of a combination of techniques to source and carry timing. Therefore, the secret to a successful synchronization solution is the selection of the technologies with the flexibility to support numerous network topologies and designs. Typically, this combines a network of strategically located time sources—most likely receivers of a satellite navigation system such as the Global Positioning System (GPS)—and a well-designed transport network to carry the timing out to where it is needed.

Frequency Versus Phase Versus Time Synchronization

As stated previously, many people loosely use the terms clocking, synchronization, and timing interchangeably, but it is time to get more specific about the different forms of synchronization. Figure 1-3 illustrates a way to demonstrate the difference, using music as an analogy.

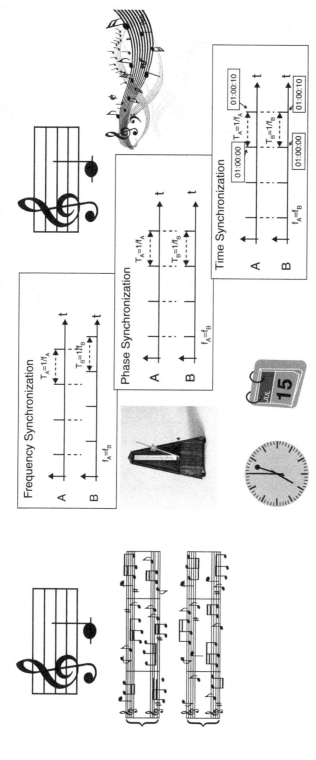

Figure 1-3 *Frequency Versus Phase Versus Time Synchronization*

According to this analogy, *frequency synchronization* is the process of ensuring that two vibrating sources of sound in an instrument are playing the same note. An example would be where a musician plays the note "A above middle C" on an instrument (sound waves carrying A above middle C, or A_4, oscillate at 440.00 Hz). Playing the same note on another instrument such as a piano should generate a sound wave at that same frequency.

On the other hand, *phase synchronization* is concerned with ensuring that two or more separate processes execute a planned action only when it is supposed to happen. Another way to think of this would be when the second hand of a clock actually "ticks" over. Two clocks can be running at the same speed (frequency synchronized), but if they do not tick over the second at the same time, their phase is not aligned. Processes that are phase synchronized (or *phase aligned*) understand time in a way that guarantees the correct sequence is followed, and simultaneous events occur at the same instant.

In the music portion of this analogy, a metronome or orchestral conductor would be responsible for ensuring that the musicians play to the correct beat or tempo and that the same passage of music is simultaneously played by each musician—the choir needs to be "singing off the same sheet." If the musicians do not play the notes at the correct time or tempo, then the music will immediately become jarring and unmelodic.

When two processes are phase synchronized, the term *phase accuracy* describes how far away the two processes are from each other in their understanding of when an event should happen. Phase accuracy is generally expressed in fractions of a second, so you might say, for example, that two clocks are phase aligned within 100 milliseconds (100 ms) when one is up to 100 ms faster than the other.

Note that neither frequency synchronization nor phase synchronization deals with the actual time of the day. The final term, *time synchronization*, is about providing a date and time that people can agree on. More correctly, it is phase synchronization combined with a count of the passage of time since a defined starting point (called an *epoch*). If everyone agrees with the starting point of time (epoch) and the rate at which time passes (frequency), then everybody is synchronized in date and time.

Here our music analogy breaks down somewhat. Let's assume that a concert is scheduled to start at 7 p.m. (19:00) on July 15, and so anybody who wants to be in the audience to hear the event should be present by that time. But this requires that the current date and time are agreed upon so that everybody understands when that concert will begin. Therefore, time synchronization is useful for communicating an absolute point in time by giving it a value; or by recording an event and being able to place it in sequence with other events (building a timeline).

So, the evening news starts daily at 10 p.m. (22:00:00) local time and Brexit occurred at 31-Jan-2020 23:00:00 GMT. Of course, for this scheme to work, all our clocks must be time synchronized to an agreed time reference. You are probably very familiar with time synchronization from experience with everyday mobile phones, computers, and smart watches, as they are automatically time synchronized from the network, unlike the older wristwatch.

Frequency Synchronization

As previously stated, frequency synchronization is simply the capability to make two things vibrate or oscillate at the same frequency or rate. This is what the piano tuner does, and why all the players in the orchestra adjust their instruments before a concert—they are "frequency synchronizing" their instruments. Those instruments that are correctly frequency synchronized are said to be "in tune," while a timing expert would say they are "frequency synchronized" or *syntonized* (the correct technical term).

If you are synchronizing the frequency of machines such as computers, radios, or even clocks, you are basically doing the same thing as a piano tuner. Each device, such as the quartz wristwatch, your home computer, or your mobile phone, contains an oscillator (possibly more than one). The silicon engineer designs this oscillator to run at a very precise frequency, known as the *nominal frequency*. This nominal frequency will be designed for a target application while meeting other requirements like power consumption, so the nominal frequency might vary—for example, 20.00 MHz in a network router or 32768 Hz in a quartz watch.

The first thing to realize is that physics comes into play and therefore a device will never naturally oscillate at exactly its correct nominal frequency. Out of the box, it will be either fast or slow, even if only by a miniscule amount. The amount the oscillator can be expected to deviate is part of the specification of the oscillator (*frequency accuracy*) and can be expressed in terms of parts per million (ppm) or parts per billion (ppb). The better the oscillator, the lower that number will be—a typical value for a 20-MHz oscillator in a Cisco router might be ±4.6 ppm, which means the measured frequency could be anywhere between 19,999,908 and 20,000,092 Hz.

Furthermore, the actual output frequency can dynamically change because of many factors, so an engineer cannot simply measure the actual output frequency of an oscillator and somehow correct for it. As you would expect, it will change based on temperature, but it might surprise you to know that as the device gets older, the frequency will change slowly over time (a process called *aging*) and might change by several ppb per day. Oscillators will be discussed in more depth in Chapter 5, "Clocks, Time Error, and Noise."

Frequency synchronization is the process whereby one uses another, more accurate, and stable source of frequency to make the oscillator tick over much more closely to its nominal design frequency. As an example, the accuracy of an oscillator might improve from, say, ±4.6 ppm to ±16 ppb (now between 19,999,999.68 and 20,000,000.32 Hz). The frequency will still drift, but much less than a standalone oscillator running without the assistance of a more accurate source from elsewhere.

Phase Synchronization

As previously stated, phase synchronization can be described as the process of aligning the tick of the second hand on our clocks. In normal life, if all our clocks did not tick over the second at near the same instant, nobody would notice. However, many processes require a much more accurate alignment.

To illustrate, imagine two industrial-type robots, neither of which has vision capability, passing objects to each other in a factory assembly process. One of the robots is the sender (of the object) and one is the receiver. If you were to program the robot pair, then the steps, along with the time to start that step, might look something like the following.

The timeline for the sender robot, with time in seconds, might be

Step 1.   Start + 0.000: Move robot arm to position x, y, z

Step 2.   Start + 0.350: Check arm is in position x, y, z

Step 3.   Start + 0.400: Open fingers to release object

Step 4.   Start + 0.600: Close fingers

Step 5.   Start + 0.650: Move robot arm to position 0, 0, 0

Step 6.   Start + 0.850: Retrieve another object

Step 7.   Start + 1.000: Restart process

The timeline for the receiver robot, with time in seconds, might be

Step 1.   Start + 0.000: Move robot arm to position x, y, z – 100 (100 mm below the other robot)

Step 2.   Start + 0.350: Check arm is in position x, y, z – 100

Step 3.   Start + 0.400: Open fingers to catch object

Step 4.   Start + 0.600: Close fingers

Step 5.   Start + 0.650: Move robot arm to position 0, 0, 0

Step 6.   Start + 0.850: Release object

Step 7.   Start + 1.000: Restart process

For this coordination to work between the two robots, they must clearly agree when the time of 0.000 starts (regardless of the time of day); otherwise one robot will be dropping objects on the floor because the other robot is not ready to catch them. Once these two robots agree on that zero time, they are phase synchronized. Now when you see video of robots working together to assemble a car in an automobile factory, you'll understand that they need phase synchronization to ensure not only that they get the job done but that they don't crash into each other.

Note   You might detect a slight flaw in this scenario, because the sender robot is dropping the object at 0.400 and the receiver robot is opening its hand at the same time. If our robots are only phase synchronized to a low degree of precision (for example, only 50 ms), the sender robot could be 50 ms ahead of the receiver in time. Therefore, the sender robot could drop the object up to 50 ms before the receiver has its hand in place, and the object could drop on the floor. The solution would be either to reduce the allowable time difference between the robots or to put an extended wait time in the program in case one robot gets too far ahead of the other.

The resulting slowdown in production would be money wasted. So, the closer the alignment, the faster the process can be made to go, as you can remove extra wait times. You will see that the same issue arises in communications systems in Chapter 2, "Usage of Synchronization and Timing."

You can see that there are many processes in our modern world where groups of devices, although separated by distance, need to agree on phase and to be phase synchronized or phase aligned to a high degree of accuracy. This alignment accuracy is commonly expressed in terms of the phase offset between the two devices and represented by the Greek letter theta (θ).

For the robot scenario to work, the robots might need to be phase aligned to within about 5 ms to eliminate the risk of objects falling on the floor. This degree of alignment might not seem too difficult, because the robots are very close to each other and connected to a high-speed local-area network (LAN) infrastructure. In a modern 5G radio network, the problem is much harder to solve because the operator must phase align radio equipment, distributed around the whole country, to within a few *microseconds* (μs) of any neighboring equipment.

Time of Day Synchronization

You can easily find examples of time synchronization in daily life. A parent might let their teenagers go off alone in the shopping mall, if they promise to meet up again at a specific time. Before letting them wander off, the parent typically asks, "What time do you have now?" which is a crude method to ensure their "watches are synchronized." Of course, they all carry mobile phones these days, so setting meeting times is almost a thing of the past, and additionally, all our phones are automatically time synchronized for us.

However, once you set your watch to the correct time, the watch must rely on its oscillator to keep the time as best it can. For the 32768-Hz oscillator in a quartz watch, this means that every time the crystal undergoes 32768 oscillations, a counter will tick over one second and move the second hand by one graduation on the watch face. Therefore, the better the frequency accuracy of the oscillator (the closer it is to exactly 32768 Hz), the more accurate the watch will keep time. At the very high end, oscillators with exceptional frequency accuracy are called *atomic clocks*.

The concept of time drift typically refers to the accuracy of the time reported by a device and how much it has moved over a fixed time period—for example, the accuracy of a quartz wristwatch might be expressed as 1.5 seconds per day. This means that one month after the hypothetical parent and teens in a shopping mall coordinate the time on their watches, they could expect the watches to disagree with each other by up to 45 seconds or so.

If a set of devices need to keep more accurate time over an extended period, they need a mechanism to synchronize time automatically to compensate for the inaccuracy in the frequency of the oscillator. Various mechanisms exist that do this, and although you might not notice, it occurs frequently all around you, most commonly in your laptop or mobile phone. Other clocks, especially those used for public purposes (for example, railway clocks), are synchronized by a radio signal.

Accuracy comes at a cost. A highly accurate oscillator (such as an atomic clock) can be set once and it will maintain very accurate time for an extended period; however, this type of oscillator is very expensive to buy and maintain. A far more cost-effective solution is to put cheaper "good enough" oscillators in our devices and use some form of time synchronization to periodically reset them to the correct time. This compromise will be revisited later, as it is an important factor in building a synchronization network: inherently better clocks versus better correction mechanisms.

Now it should be apparent that to align time accurately, you need a combination of all forms of synchronization: frequency, phase, and time. The oscillators must spin at the correct frequency, the clocks must tick over at the correct moment, and the date on the watch face must correctly display what everybody accepts to be the correct time. Which raises the next question. What exactly is the time everybody agrees on?

What Is Time?

The basic unit of time that most directly affects us all is the day, which is controlled by the rotation of the earth and its orbit around the sun. Even with something seemingly immutable as the day, there is nothing absolute about time. Rather, it is an agreement among us all about what time is—just as everybody accepts what the meter or kilogram is. This agreement covers how we define both larger periods of time (the calendar) and smaller periods of time (such as the second).

Originally, *metrologists*, the people who develop standards for our units of measure, defined the second (before 1956) to be 1/86,400 the value of a *mean solar day*, and developed what is known as *universal time* (UT). The main problem with tying a definition of the second to the earth's rotation is that the length of the earth's day can vary (which is why the metrologists were specific about the mean solar day). But even that day slows down and speeds up, only to a miniscule degree, but basing the second on such a varying phenomenon is too inaccurate for some purposes.

After some time, because metrologists knew the year was more stable than the day, they changed the definition of the second to one based on a fraction of a year (1/31,556,925.9747). Of course, it is no surprise that the year is not really a stable measure of time either, as the earth's orbit around the sun is slowly changing and, anyway, observations of the sun can be somewhat difficult and inaccurate. These days, universal time is determined by observations of distant celestial bodies (technically, there are three forms of UT known as UT0, UT1, and UT2, but that is beyond the scope of what is needed here to explain time).

By the 1950s, scientists developed atomic clocks, which were more stable than the earth's rotation (or orbit), and so in 1967 the metrologists changed the definition of the second to "the duration of 9,192,631,770 periods of the radiation corresponding to the transition between the two hyperfine levels of the ground state of the cesium 133 atom." Since 1967, cesium atomic clocks have been the standard for defining the second, although there are new generations of extremely accurate optical clocks which might take over at some point.

Overseeing and managing the definitions of these units is an organization based in Paris, the *Bureau International des Poids et Mesures* (BIPM), known in English as the International Bureau of Weights and Measures. The BIPM website (https://www.bipm.org) is full of very detailed information on metrology and International System of Units (SI) units, including the second, so that is an excellent source should you desire to better understand the definitions of SI units.

Note Effective May 2019, the BIPM changed the definition of the second. It is now described as follows:

The second, symbol s, is the SI unit of time. It is defined by taking the fixed numerical value of the caesium [cesium] frequency, the unperturbed ground-state hyperfine transition frequency of the cesium 133 atom, to be 9 192 631 770 when expressed in the unit Hz, which is equal to s^{-1}.

So, that was a quick history of how the world decided to use atomic clocks as our source of the length of the second. The next question to address is, "Which is the precise second of the day/hour that you are currently experiencing?"

What Is TAI?

International Atomic Time, or from the French, *Temps Atomique International* (TAI), is the world's time as determined by the weighted average of a worldwide network of atomic clocks (over 400 in 80 national laboratories). The scientists in these laboratories use numerous techniques to measure very precisely the time difference between clocks, allowing each lab to compare the time of its clock with that of the other labs' clocks. BIPM publishes these comparisons monthly in its *Circular T* (more on that in the following section).

TAI is what is known as a *monotonic timescale*, meaning that since it was aligned with universal time on 1 January 1958 00:00:00 (its "time zero," known as its *epoch*), it has never jumped time (as in some form of leap event) and continues to count forward one second at a time. It is uniform and very stable, meaning that it is not exactly in step with the slightly irregular rotation and orbit of the earth. For this reason, another timescale beyond TAI was needed, one which more closely aligns with our day and years (although the differences are small).

What Is UTC?

To have a timescale that reflects the realities of our world (look at what happened to the calendar before the leap year problem was corrected), scientists developed UTC, or *Coordinated Universal Time*. UTC is identical to TAI except that it is a *discontinuous timescale* because it occasionally has *leap seconds* inserted to ensure it reflects changes in the earth's rotation and therefore aligns with our days and years. Leap seconds are extra seconds added or deducted from TAI to keep UTC within 0.9 second of the earth's orbit—a process managed by the International Earth Rotation and Reference Systems (IERS) service, which is responsible for monitoring the earth's rotation.

IERS announces, about 5 months ahead of time (in a document called *Bulletin C*), the dates of any upcoming leap second. Usually, IERS announces that it will insert an extra second (as the earth's rotation is generally slowing down) on 30 June or 31 December. As of 1 January 2017 UTC, there have been 37 leap seconds, which means that UTC time is 37 seconds behind TAI. In the normal course of events, there is a single leap second added every few years.

Interestingly, adding or subtracting these extra seconds means that the last minute on the day of the leap event could have either 58, 59, 61, or 62 seconds. So, assuming the addition of a single leap second, at the forecast time of the insertion, a clock displaying UTC can read 23:59:60 for one added leap second (or 23:59:61 for two). But another clock displaying TAI would still only have 60 seconds in that minute, meaning that for one added leap second, UTC would fall one more second behind TAI.

It is this adjusted UTC time that is the basis for the dates and times used in our daily life and is a replacement for the more commonly quoted GMT. Technically, there is only one UTC across the world, and the time from every atomic clock is only a local approximation of UTC, known as UTC(k) (where k is a name of the laboratory that manages the clock).

The interesting fact about UTC is that only an approximate value for it is known in real time. The single "official" UTC is determined by post-processing masses of data collected from atomic clocks a couple of weeks after the end of the month. Once a month, *Circular T* from the BIPM gives us a readout of how closely aligned all the atomic clocks participating in the scheme are to "real" UTC. To quote from the BIPM website:

> BIPM Circular T *is the monthly publication of the Time Department that provides traceability to Coordinated Universal Time (UTC) for the local realizations UTC(k) maintained by national institutes.* Circular T *provides the values of the differences [UTC – UTC(k)] every five days, for about 80 institutes regularly contributing clock and clock comparison data to the BIPM.*

So, *Circular T* announces what "real" UTC was and then gives information on how accurate UTC(k) on each individual atomic clock was compared to the newly calculated UTC.

The following is a sample of a few lines of a *Circular T* report (some date columns and headings have been removed to fit across the page):

```
Date 2019/20 0h UTC        DEC  7   DEC 12   DEC 17
Laboratory k                        [UTC-UTC(k)]/ns


AUS  (Sydney)             -369.4   -381.2   -378.4
SU   (Moskva)                2.6      2.4      2.6
USNO (Washington DC)         0.6      0.9      0.4
```

This snippet of *Circular T* shows the difference between three versions of UTC(k) and "real" UTC. The three versions are from laboratories based in Sydney, Moscow, and Washington, DC. The three values show the difference (in nanoseconds) between UTC and UTC estimated by that laboratory on three separate dates in December 2019.

Note the last line; that version of UTC, namely UTC(USNO), is shown to be within a nanosecond of UTC, which is a good thing, because UTC(USNO) is the time that is distributed by the GPS constellation (USNO is the United States Naval Observatory).

How Can GPS Provide Timing and Synchronization?

A very powerful tool in the distribution of time information (frequency, phase, and time) is the *global navigation satellite system* (GNSS), the most well-known of which is GPS. Many documents on timing refer to GPS specifically, but it is important to know that there are numerous similar systems, and so this book uses the generic term, GNSS, except when specifically referring to any one system. Table 1-1 shows the most common GNSSs, including some regional systems.

Table 1-1   *Major Global Navigation Satellite Systems*

Name	Country
GPS	United States of America
GLONASS	Russian Federation
Galileo	European Union
Beidou	People's Republic of China
INRSS	Republic of India
QZSS	Japan

Conceptually, GNSSs are quite simple, although there is a lot of complexity behind their success that is outside the scope of this book. You can think of a GNSS as a network of atomic clocks flying around in orbit, which is absolutely what they are (for GPS, at 20,200 km above the earth). Transmitters on the satellites use the onboard clocks to broadcast timing signals toward the earth such that receivers below can determine time, location, and speed.

Controlling these satellites are ground and monitoring stations that ensure the satellites are in the correct position, that they are transmitting the correct data, and that their clocks are accurately timed. The timescale used by GPS is based on the UTC(USNO), which as you read previously is the version of UTC from the U.S. Naval Observatory and is aligned to within a nanosecond or so of the "real" UTC. Consequently, there is little difficulty in using a suitable receiver to recover a very accurate representation of UTC time from GPS.

The GPS receiver in your mobile phone, which commonly now also contains a Galileo receiver, simultaneously locks onto GPS signals from several satellites and quickly determines position and speed. It would also be possible to get UTC(USNO) time from GPS, although most phones get their time synchronization from the cellular network or the Internet.

However, there are more specialized receivers that are not constructed to determine location or help you navigate around the world but are designed specifically to deliver an accurate source of time. To recover accurate time information, these devices only need a connection to an antenna located outside (because GNSS signals will not penetrate walls). Normally, these receivers also have up to three physical connectors that allow cables to carry timing information away to any device that needs it. The three signals consist of

- Frequency (some sort of sine or square wave)

- Pulse (representing the start of the second or phase)

- Time of day (a string of characters that conveys the UTC date and time)

These three signals can then be used to provide frequency, phase, and time synchronization to any nearby equipment. Because GPS is a worldwide system, and it consists of many satellites (currently 31) that are in overlapping orbits, the coverage is global and will work anywhere on the earth. Because the antenna for the receiver is placed outside or on the roof, there is almost nothing to get in the way between the receiver and the transmitter (except in dense urban canyons). Consequently, everybody takes for granted that GPS is extremely reliable and "always there" because it has worked so well for decades, and many users of time and navigation rely on it, including countless mobile operators.

The only major disadvantage to these GNSSs is that the signal received on the ground is very weak and therefore quite susceptible to interference and jamming. Because the information transmitted (at least for civilian users) is not encrypted, there is also an increasing danger from bad actors spoofing the signals, thereby fooling the receivers into accepting fake time data.

Many experts fear that everybody is far too complacent in thinking that GPS will always be there, when in fact it is quite vulnerable. This topic is revisited when discussing the options for deploying synchronization across a wide-area network. Chapter 11, "5G Timing Solutions," covers the issues regarding the resilience of the GNSS systems in much more detail.

There is a lot more information on GNSS systems in Chapter 3, "Synchronization and Timing Concepts," and an interesting background paper for using GNSS for timing is found in the ITU-T technical report *GSTR-GNSS*, referenced at the end of the chapter.

Accuracy Versus Precision Versus Stability

Accuracy is defined as how close a measured value is to the actual (or accepted) value. For example, accuracy would indicate how close a clock has stayed to real UTC time or how close an oscillator has stayed to its nominal frequency. As shown previously, time accuracy can be expressed in terms of a time error over a period (1 second per day) or a frequency error compared to the nominal frequency (20 Hz for a 20-MHz oscillator, or 1 ppm).

Precision is defined as how close the values measuring some quantity are to each other—in other words, how repeatable and reproduceable the measured values are. Another word for precision might be *stability* in that it indicates how well a measurement continues to reveal the same value—no matter whether that value is accurate (meaning reflecting the real value).

Imagine that a 20-MHz oscillator is measured to be oscillating at 20.200000 MHz every time you measure it. Because it is about 1% fast, it is hardly an accurate oscillator, and if used in a clock, the resulting time will not be accurate either, because you would expect it to be about 1% (or 864 seconds per day) too fast. Although not accurate, it is precise in that its performance is consistent or *stable*.

If something like an oscillator is stable, it is a good candidate as a frequency source because it can be relied on to deliver the same output over time. Stability only indicates whether the frequency (or any other characteristic) stays the same and does not drift. The behavior of cesium atoms is very stable, which is why metrologists use it as a source of the second and to keep UTC time.

Taking this oscillator that is always spinning at 20.2 MHz, it would be possible to build a very accurate clock out of it, since the designer could compensate for the 1% excessive frequency. One does this by either slowing it down (for example, by lowering the input voltage) or counting 20.2 million cycles for a second rather than 20.0. Designing a clock around this oscillator and using compensation for the inaccurate (but stable) frequency would result in a very accurate clock.

When a stable clock behaves the same all day and every day, then it should be possible to compensate for any inaccuracy and make the device both accurate and stable. In contrast, a clock that fluctuates within a range of ±5 minutes of the correct time every day is unstable, and although on occasion it might have the correct time, it could never be made accurate. As they say, even a stopped (analog) watch shows the correct time twice per day.

Summary

This chapter covered the basic concepts of timing and synchronization. The next step is to learn why timing and synchronization is important and which use cases it can be applied to. Chapter 2 starts with some historical background before dealing with more modern use cases. It will introduce timing and synchronization applied to real-world problems.

References in This Chapter

Bizouard, C. "Information on UTC – TAI." *International Earth Rotation and Reference Systems Service (IERS)*, Bulletin C 61, 2021. https://datacenter.iers.org/data/latestVersion/16_BULLETIN_C16.txt

IEEE Standards Association. Annex B, "Timescales and Epochs in PTP," from "IEEE Standard for Precision Clock Synchronization Protocol for Networked Measurement and Control Systems. *IEEE Std. 1588:2008*. https://standards.ieee.org/standard/1588-2008.html

Sobel, D. *Longitude: The True Story of a Lone Genius Who Solved the Greatest Scientific Problem of His Time.* New York: Walker, 1995.

The 13th Conférence Générale des Poids et Mesures. "Resolution 1 of the 13th Conférence Générale des Poids et Mesures (CGPM)." *The International Bureau of Weights and Measures (BIPM)*, 1967. https://www.bipm.org/en/CGPM/db/13/1/

The International Bureau of Weights and Measures Time Department. "Circular T." *The International Bureau of Weights and Measures (BIPM)*. https://www.bipm.org/jsp/en/TimeFtp.jsp?TypePub=Circular-T

GSTR-GNSS. "Considerations on the use of GNSS as a primary time reference in telecommunications " *ITU-T Technical Report*, 2020-02. https://handle.itu.int/11.1002/pub/815052de-en

Chapter 1 Acronyms Key

The following table expands the key acronyms used in this chapter.

Term	Value
5G	5th generation (mobile telecommunications system)
BIPM	Bureau International des Poids et Mesures
CGPM	Conférence Générale des Poids et Mesures
DST	daylight saving time
GLONASS	GLObalnaya NAvigazionnaya Sputnikovaya Sistema (Global Navigation Satellite System)
GMT	Greenwich Mean Time

Term	Value
GNSS	global navigation satellite system
GPS	Global Positioning System
IERS	International Earth Rotation and Reference Systems
INRSS	Indian Regional Navigation Satellite System
LAN	local-area network
μs	microsecond or 1×10^{-6} second
NIST	National Institute of Standards and Technology (USA)
ns	nanosecond or 1×10^{-9} second
ppb	parts per billion
ppm	parts per million
QZSS	Quasi-Zenith Satellite System
SI	International System of Units
TAI	Temps Atomique International
USNO	United States Naval Observatory
UT	universal time
UTC	Coordinated Universal Time
UTC(k)	Coordinated Universal Time at laboratory k

Further Reading

Refer to the following recommended sources for further information about the topics covered in this chapter.

European GNSS Service Centre (for the Galileo GNSS system): https://www.gsc-europa.eu/

GPS: The Global Positioning System: https://www.gps.gov/systems/

Higgins, K., D. Miner, C.N. Smith, and D.B. Sullivan (2004), *A Walk Through Time* (version 1.2.1): https://physics.nist.gov/time

International System of Units (SI): https://www.bipm.org/en/measurement-units/

National Institute of Standards and Technology (NIST) Time and Frequency Division: https://www.nist.gov/pml/time-and-frequency-division/

Resilient Navigation and Timing Foundation: https://rntfnd.org/

U.S. Coast Guard Navigation Center GPS Constellation: https://www.navcen.uscg.gov/?Do=constellationStatus

Chapter 2

Usage of Synchronization and Timing

In this chapter, you will learn the following:

- **Use of synchronization in telecommunications:** Provides an overview of the use of synchronization in telecommunications and especially in mobile communications. This discussion is somewhat historical but explains the position the industry finds itself in today.

- **Use of time synchronization in finance, business, and enterprise:** Gives some examples for the scenarios where time synchronization is a necessary tool in the corporate and enterprise space.

- **Industrial uses of time—power industry:** Covers an example for the application of synchronization to an industrial segment, namely the power ("smart grid") industry.

Chapter 1, "Introduction to Synchronization and Timing," introduced some basic concepts of time, including the different forms of synchronization and how they are applied. This chapter covers more of the practical uses of time synchronization, including some historical background, before moving on to pre-5G mobile use cases that are already deployed. Chapter 10, "Mobile Timing Requirements," covers 5G mobile in more depth.

The standards for synchronization were defined starting several decades ago, and they are still applicable to many widely deployed use cases of today. For example, the synchronous Ethernet (SyncE) standards can be directly traced back to standards published on traditional legacy networks like synchronous digital hierarchy (SDH) from the ITU-T and synchronous optical network (SONET) from ANSI. If your application area is circuit emulation, where one replaces time-division multiplexing (TDM) networks with IP/MPLS packet technology, then these "legacy" TDM standards are still current and relevant. Chapter 4, "Standards Development Organizations," shares much more information on the standards organizations and their roles.

Use of Synchronization in Telecommunications

Synchronous communications arose with the adoption of TDM digital circuits (such as E1 and T1) being used to interconnect equipment over long distances. To improve the throughput and quality of these circuits, engineers developed and adopted synchronous forms of communication. Synchronization was also introduced to allow channels (such as individual voice connections) to be multiplexed onto a single communications medium. Therefore, network engineers needed to have a good understanding of clocking because they configured routers to use such circuits as an everyday task. Compare this to modern network engineers, who might never have had to configure an E1/T1 circuit in their career, and so are not as familiar with the continued importance of synchronous communications.

Efficient operation of digital transport continues to be the most common usage of frequency synchronization today in the telecommunications industry, such as in legacy TDM circuits like those based on E1/T1, or SDH and SONET. However, frequency (and sometimes phase) timing is widely deployed in other data transports such as cable, passive optical networking (PON), and mobile, as well as in modern packet networks such as Ethernet (so-called synchronous Ethernet) and optical technologies.

Legacy Synchronization Networks

This section addresses the fundamental ideas and history of legacy networks only to provide the basic concepts that are needed to understand the discussion of the standardization behind SyncE. This section also addresses the levels of accuracy used to source and transport frequency.

Frequency synchronization provided benefits to engineers designing the then-new optical networks such as SDH and SONET, as it allowed the new networks to transmit more data with less overhead than available in previous schemes such as plesiochronous digital hierarchy (PDH). It was also clearly superior to other methods, such as those that used dedicated Start and Stop bits to indicate the beginning and end of a data byte or other techniques such as bit-stuffing. The difference here was that synchronization was used to implement the transport itself, rather than being a requirement for some application using the transport (such as telephone voice channels).

Synchronous networks (such as SDH and SONET) send data throughout a network with the physical layer of the network synchronized to a common frequency. For this to work, engineers need to provide a source of accurate frequency to all elements across the whole network because oscillators need help to stay accurately aligned. This allows a data receiver in a router to read the incoming data stream at the same frequency as the transmitter used when sending it.

Engineers achieve this alignment by using the data circuit itself as a source of clock (*line clocking*), or they built a dedicated, independent network using specialized timing devices with stable oscillators as an external frequency source for each of the routers. Figure 2-1 gives a conceptual view of how the frequency is distributed across a network to align every device. In the interests of clarity, references to the ITU-T and ANSI (and now ATIS) standards that define each of these components has been omitted.

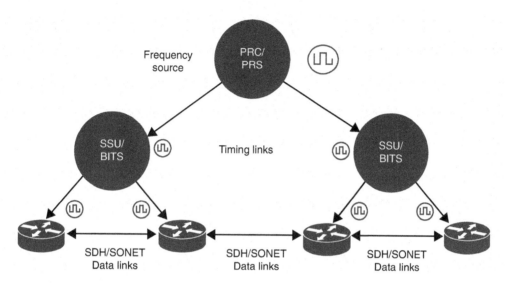

Figure 2-1   *Frequency Distribution Network*

You must also understand that SDH and SONET use different terminology for equivalent items in each system. As an example, note that the standards refer to the source of frequency (for example, an atomic clock) as a primary reference clock (PRC) in SDH and as a primary reference source (PRS) in SONET. Similarly, an intermediate node is known as a synchronization supply unit (SSU) in SDH and a building integrated timing supply (BITS) in SONET. For clarity, where items from both systems are being discussed, this book uses the SDH term before the slash and the SONET after it (for example, SSU/BITS for an intermediate timing node).

The PRC/PRS sources of frequency are either atomic clocks or a device combining a *global navigation satellite system* (GNSS) receiver with a stable oscillator. PRC/PRS devices based on atomic clocks can keep frequency accurately over a long period of time because they are stable sources of frequency (based on fundamental physics of the cesium atom). GNSS receivers normally include a good-quality (stable) oscillator to be able to keep an accurate frequency whenever it temporarily loses the signal from space (a process known as *holdover*).

A typical SSU/BITS device also includes hardware that allows it to maintain a clock signal with higher accuracy for a holdover period should it ever lose its connection to the PRC/PRS source. Engineers typically implement the SSU/BITS device with a piece of hardware called a *digital phase-locked loop* (DPLL) driven by a very stable rubidium or high-quality quartz oscillator. So, the ability of the SSU/BITS device to maintain frequency during holdover is better than the normal SDH/SONET transport node, but not as good as a PRC/PRS.

Chapter 6, "Physical Frequency Synchronization," and Chapter 7, "Precision Time Protocol," explore sources of frequency and time in more detail.

If the frequency on either end of a data link is not accurately aligned with the other, then the receiving device will have a lower or higher frequency than the transmitting device. When this happens, the receiver cannot read and store the data in its buffer at the same rate as the transmitter sends it. Eventually the receiver either will be waiting with an empty buffer for data that has not yet arrived (known as an *underflow*) or will not be able to store it quickly enough in its buffer and will lose data (known as an *overflow*). These events lead to loss of quality on the data circuits and show up in what network engineers know as *slips*, as shown in Figure 2-2.

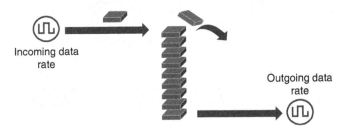

Incoming data rate

Outgoing data rate

Figure 2-2   *Slips in a TDM Network*

The slip rate is proportional to the frequency differences between transmitter and receiver, so if one measures the differences between the signals, it is possible to calculate the expected number of slips over a time period. Highly stable, closely aligned frequency sources might only result in a slip once every several months, whereas network devices running without any accurate source of frequency may generate multiple slips per second.

Closely aligning the frequency on every node in the network is paramount for a quality network. Slips are detrimental to the quality of the user experience, resulting in problems such as clicks in audio streams, poor-quality video, freezes, and the like. Although the use of TDM is increasingly declining, building a source of frequency and distributing it throughout a network is still very relevant to many use cases in modern communications.

Legacy Mobile Synchronization—Frequency

In earlier generations of mobile telecoms (approximately up to 4G/LTE), the radio equipment at the cell tower only required frequency synchronization. Fortunately, for early 2G systems, this was quite easy to provide because the cell tower was connected to the core mobile network by TDM circuits (such as E1 or T1). Because the backhaul circuits were synchronous, the radio could recover a stable source of frequency using the signal from the incoming line.

Figure 2-3 shows how the frequency synchronization of those networks worked.

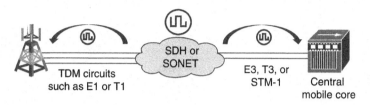

Figure 2-3 *Frequency Synchronization Using TDM Circuits*

Because the frequency of the circuit backhauling the traffic to/from the cell tower was coming from accurate sources in the SDH/SONET network, it was an accurate source of frequency for the equipment in the cell tower. Frequency had one additional important use besides the need to synchronize the backhaul circuits. The cell site equipment required frequency synchronization because its inbuilt oscillators were the source for the frequency being transmitted by the radio. This meant that if the oscillator in the radio drifted off its frequency, the radio was transmitting on the wrong frequency and was either

- interfering with neighboring cells

  or

- transmitting its signal outside its licensed radio bands

As a result, frequency synchronization was used to "tune" the radios so that they transmitted on the exact frequency that the mobile operator intended. So, hardware engineers designed the radios to use the oscillators that were being assisted by the frequency recovered from the external source.

Radios for 3G and later generations evolved over time from carrying mobile traffic using a TDM transport to carrying mobile traffic using packet-based methods (for example, Ethernet, IP, and IP/MPLS). But because the transmission of data in packet networks is not synchronous, the Ethernet circuits do not require frequency synchronization. This created a problem because these links to the cell site no longer (by default) provided a source of frequency synchronization for the radio equipment. Some operators solved this problem by retaining a single E1/T1 circuit connected to the radio purely for the purposes of providing it a source of frequency.

It was this deployment of packet-based methods that made SyncE popular because it added the capability for the packet network to carry frequency synchronization in the Ethernet's electrical and optical layers. Figure 2-4 shows how the frequency synchronization of those networks was achieved using either a single TDM circuit or enabling SyncE (only one of the two options was required).

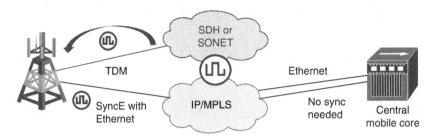

Figure 2-4   *Frequency Synchronization Using Packet Backhaul*

Enabling Ethernet to propagate frequency hop-by-hop through a network required hardware support, so it was not something that could be enabled only by software changes. But as equipment manufacturers increasingly incorporated SyncE support into mobile backhaul routers, operators moved to decommission any costly TDM links and provided mobile synchronization via the Ethernet connections (this process is largely complete for many operators).

The section "Circuit Emulation" later in this chapter as well as Chapter 7 and Chapter 9, "PTP Deployment Considerations," describe how a packet-based solution based on the precision time protocol (PTP) can address the problem of frequency synchronization across the IP/MPLS backhaul network instead of using SyncE. This approach makes sense where the operator is not able to deploy SyncE-enabled equipment but needs another method.

Today, many Long Term Evolution (LTE) networks require only frequency synchronization, although it depends on the services and radio techniques that the operator decides to deploy. Many operators decided not to switch on these advanced services because their networks were unprepared to provide the phase synchronization those services need. With each subsequent updated release of the mobile standards, that option is increasingly unavailable, and phase synchronization is becoming mandatory.

Legacy Mobile Synchronization—Phase

In the late 1980s, regulators, manufacturers, and operators were working on the design and deployment of 2G networks to replace existing analogue systems. In Europe at this time, the European Telecommunications Standards Institute (ETSI) produced the Global System for Mobile communications (GSM) mobile standard based on time-division multiple access (TDMA) radio techniques. Most of the world eventually adopted GSM as the standardized digital mobile system, but some countries, most notably the United States, selected a different system more compatible with their existing analogue systems. The U.S. operators adopted several competing systems, but one was based on code-division multiple access (CDMA) radio technology and marketed as cdmaOne.

This early split in the mobile world over 2G technologies persisted into later decisions concerning 3G standards. While most of the world evolved their GSM systems to a 3G system developed by the 3GPP standards organization, called the Universal Mobile Telecommunications System (UMTS), operators in the United States chose another system based on CDMA. This CDMA system was much more compatible with the U.S. operators' existing 2G networks at the signaling layer and so made sense for interoperability and migration. This system, designed by a collaboration called the 3rd Generation Partnership Project 2 (3GPP2), was known as CDMA2000 and was also adopted by operators across several other countries such as Korea, China, Japan, and Canada.

While CDMA radio has some advantages, it requires phase synchronization to ensure correct operation, especially for transfer of handsets from the radio in one cell site to the next (a process called *handover*). This presented operators with the challenge of how to achieve very accurate phase synchronization with cell towers spread all around the huge landmass of North America.

The simple answer was to use the Global Positioning System (GPS). Most of the cell towers then were large transmitter towers, sometimes called *macro cells*. Because the frequency of the original 2G spectrum was quite low (~800–850 MHz), these radios gave very good coverage over a wide area and great reception inside buildings—as low frequencies tend not to be readily absorbed. This combination of long range and good propagation properties allowed operators quite a deal of flexibility in positioning their radio towers.

Because of the open locations, all that was required was for the operator to put a GPS antenna on top of the tower and connect a GPS receiver to the CMDA radio. This was a widely implemented approach in North America at the time. When later generations of mobile required phase synchronization, operators outside the U.S. were somewhat hesitant about adopting a timing solution (GPS) seen to be controlled by the U.S. government. This reluctance to overly rely on GPS gave operators and standards organizations an incentive to develop alternative systems to distribute phase and frequency.

One other factor determined whether phase was required for a mobile network. There are two techniques (see Figure 2-5) for separating the full-duplex channels (one uplink and one downlink) for a radio system. One technique (and historically the most common) was frequency division duplex (FDD), whereby the uplink (transmit from handset) and downlink (receive from tower) used different bands—frequency is used to divide the duplex channels. This system needed only the normal frequency synchronization, although some applications running on top of FDD radios may need phase synchronization.

Figure 2-5 *FDD vs. TDD Radios*

The other option is known as time division duplex (TDD), whereby the uplink and downlink channels share the same band, but both ends of the link take turns either transmitting or receiving—time is used to divide the duplex channels. Because TDD involves time, it is no surprise that this system requires phase synchronization. TDD is the predominant radio type for 5G deployments.

Cable and PON

There are several transport technologies that require phase synchronization to efficiently function. Examples include the hybrid fiber coax (HFC) cable and PON systems used to provide consumer broadband services to residential households. The design of these systems reflects the fact that most broadband subscribers download much more traffic than they upload, and so the traffic flows are inherently asymmetric.

As a result of this requirement, these systems commonly transmit data in the downstream direction differently than they transmit data in the upstream direction. For example, the broadband router might broadcast the same downstream signal to every receiving set-top box (STB) but implement a unicast transmission technique in the upstream direction. Because the consumer's STB is sharing the cable or fiber with numerous other subscribers, it requires exclusive access to the cable while transmitting. Therefore, there needs to be arbitration between competing upstream transmitters to ensure that they don't transmit over the top of each other's signals.

While details vary somewhat depending on the specific implementation of the transport, in shared media broadband, the STB must bid for permission to use a timeslot (or similar resource) to allow it to transmit. When it transmits, all other devices on that piece of media must remain in a receive-only state. To ensure the STB transmits without interfering with its neighbors, three prerequisites must be met:

- All the devices on a shared piece of media must be phase aligned so that transmission begins and ends at the correct moment after the headend has granted the STB a transmission timeslot.

- There must be a *guard band* or *guard time* between the transmissions from different STBs to ensure transmissions do not talk over the top of each other if there is a small error in phase alignment.

- The distance from the headend to the STB must be measured (a process known as *ranging*) to help calculate the time that the signal takes from the transmitter to the headend. Once the STB knows the distance to the headend receiver, it will transmit its data earlier to compensate for the time delay before the headend receives it (in fiber, this advance might be around 5 ns per meter of distance).

Figure 2-6 shows the situation where two STBs are transmitting at the correct time so that the signal arrives at the headend correctly aligned, even when allowing for the longer distance to set-top box #1 compared to set-top boxes #2 and #3.

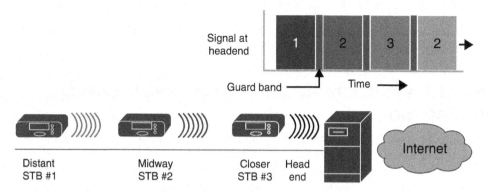

Figure 2-6   *Correctly Synchronized Signals on a Shared Media*

A side-effect of a system being tightly aligned with accurate range measurements is that the parameters of the length of transmission and the width of the guard band can be adjusted to increase the effective throughput and responsiveness on the shared cable/media. As one example, the more accurate the phase alignment, the smaller the guard band needs to be, because there is a smaller window for a transmission to be early or late.

Figure 2-7 shows what might happen when there are large phase errors and range errors on the shared media.

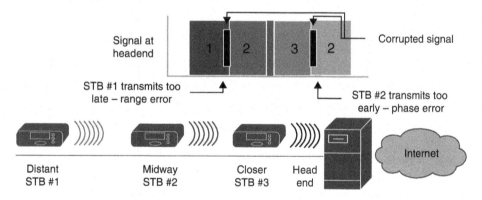

Figure 2-7   *Incorrectly Synchronized Signals on a Shared Media*

The receiver on the headend is going to struggle to decipher the corrupted transmissions on the cable because the signals are not arriving at the headend receiver at the correct time. In one case, the range to STB #1 has been measured incorrectly (too low a value) and so the STB does not advance its transmissions enough to compensate for the delay on the cable, resulting in its transmissions arriving too late—at the time STB #2 has already started transmitting.

Similarly, STB #2 has a phase alignment error that causes it to start transmitting too soon, talking over the tail end of the transmission from STB #3. One quick solution to this scenario would be to increase the width of the guard band, but that is simply consuming time on the cable that could be spent transmitting user data—and therefore a waste of money.

Use of Time Synchronization in Finance, Business, and Enterprise

Synchronization, especially of time, has always been an important part of normal operations in IT systems and network elements. You are likely aware of the Network Time Protocol (NTP) that has been used for decades to coordinate the time on machines across a network. NTP uses a two-way time protocol to transfer Coordinated Universal Time (UTC) across a hierarchy of machines, as shown in Figure 2-8. The routers in the figure connected to the source of time (for example, GNSS or an atomic clock) are known as stratum 1, and every hop further away increases the stratum number to indicate a reduction in the likely accuracy of the time on that node.

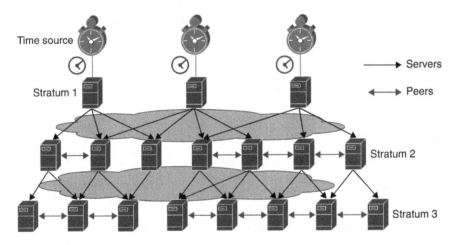

Figure 2-8   *NTP Time Distribution Network Hierarchy*

Besides being used for business processing (for example, logging the time a subscriber made a telephone call for billing purposes), NTP also has valuable operational uses, such as event correlation, fault isolation, and network-wide troubleshooting. The network engineer relies on NTP's ability to distribute time with moderate accuracy (approximately 10–100 ms) to reconstruct faults across the network (for example, from syslog event notifications) and ascertain the order in which a sequence of events occurred.

However, as machines and networks get faster and more complex, it becomes necessary for time synchronization to move from the millisecond to the microsecond level. A case in point is the 2014/65/EU directive for the finance industry that the European Union released as a response to the financial industry crisis in 2008. This directive is commonly known as the "Markets in Financial Instruments Directive," or MiFID II.

The aim of MiFID II is to regulate trading in financial instruments by operators of trading systems, particularly those known as high-frequency traders. Imagine that a trader is in a favored location with very low-latency connections to the exchange trading systems and an insurance company issues a very large buy order for a security. It might be technically possible for the high-frequency trader to preempt that transaction by first buying the securities cheaply and then offering them for sale to the insurance company at a premium—making a handy, but illegal, profit.

One way to catch traders doing this is to accurately timestamp every business transaction, which is one reason why the European Union is mandating accurate time synchronization in MiFID II. Large traders are required to timestamp all business and trading transactions to the millisecond, and they must be able to *trace* the source of that timestamp (for example, UTC sourced from GPS) and document the accuracy of the source at that point of time. For further information, refer to Commission Delegated Regulation (EU) 2017/574 of 7 June 2016, which is the portion of the directive concerning time accuracy.

NTP is normally not implemented well enough to deliver that level of accuracy because many servers and routers are only software implementations of the protocol, which limits the degree of time accuracy it can distribute. To achieve microsecond-level or better precision, time transfer needs a more accurate implementation, such as that used in PTP. PTP is a much more accurate implementation of time transfer because it is deployed with built-in hardware support and accurate timestamping. Much more is to be said on PTP in the following chapters of the book, with a detailed description in Chapter 7.

Another common use case is in the Operations, Administration, and Maintenance (OAM) functions for network management. Besides link utilization, one piece of information that a network manager is interested in is link latency. A complete snapshot of the latency for every link in the network would allow the network manager to use advanced routing protocols such as segment routing (SR) and Flex-Algo to route packets based on the lowest-latency path.

There are mechanisms to determine the link delay using packets with timestamps. Basically, the router sends out a packet with a timestamp and measures the time it takes to be returned. This gives an estimate of the round-trip time for that packet, known as the *two-way delay measurement* (2DM). From that, the *one-way delay measurement* (1DM) between two nodes can be estimated by assuming that the forward delay is equal to the reverse delay and halving the 2DM value.

The preceding assumption may be erroneous, as many conditions might arise (congestion for one) where a packet in one direction may traverse a link faster than in the other direction. This difference between the transit time of a packet in the forward and reverse directions is known as *asymmetry*, and you will read a lot more about it in later chapters. So, asymmetry invalidates our assumption and therefore results in our one-way delay estimate being inaccurate.

A more accurate way to measure 1DM is to send a timestamped packet to a target node. The receiver/listener timestamps that packet on receipt and reports the timestamp to the sender. For this method to deliver an accurate 1DM value, we require that the time on the sender be synchronized with the time on the receiver. Let's look at an example.

Imagine that the sender sends a packet out and timestamps it as 14.100100 seconds past 12:30. Now, this is a fast and short link, and, say, 10 µs later the receiver sees the packet and timestamps it. If the receiver's clock is accurately aligned with the sender, then the correct timestamp would be 14.100110 (14.100100 + 0.000010). However, what if the clock on the receiver is 50 µs slower than the clock on the sender? Then the receiver would timestamp it with 14.100060, which would show the packet arrived 40 µs before it was sent!

Therefore, the basic rule with delay and latency measurement is that the 2DM does not need time synchronization because the transmitted and received timestamps are from the same device with the same clock. However, the 1DM measurement method requires very accurate time synchronization between the two timestamping devices. The lower the expected delay, the more closely the clocks must be synchronized in order to obtain an accurate measure.

If you want to measure a link with a delay of tens of microseconds, then the time accuracy must be in microseconds (or less) or nonsensical results are possible and inaccurate results likely.

The answer to this problem lies in solutions combining GNSS time sources and a well-designed PTP network that can guarantee time synchronization to the level of microseconds.

Circuit Emulation

Although these days most network engineers are used to dealing with optical and Ethernet transport systems, the reality is that many devices around the world are still interconnected using "legacy" TDM circuits such as those based on SDH or SONET. Not only end-users and large enterprises, but major service providers still provide a large number of data circuits based on TDM-type technologies, with buildings literally full of equipment to support it.

The problem with service providers using TDM technology is that much of this equipment is beyond its useful life and no longer supported or produced by the equipment manufacturers. However, transitioning their customer base across to a different transport technology would be a disruptive and expensive migration project. Not only that, but much of the equipment in their customer sites may only support E1/T1 type TDM circuits. Examples are voice circuits on a PABX, trunked digital radios for emergency services, and signaling equipment on a railway track. Changing to non-TDM data circuits would require customers to migrate their equipment as well.

The solution is to emulate the traditional TDM circuit using packet technology and IP/MPLS. From the view of an end user, the legacy edge equipment remains in service, although it is now connected across the core IP/MPLS network. The underlying E1/T1 circuit is terminated on a special router and the transport is replaced with a packet-based solution, as shown in Figure 2-9. This technique is known as *circuit emulation* (CEM).

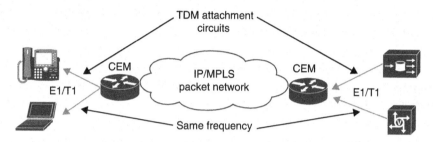

Figure 2-9   *Circuit Emulation Replacing TDM Circuits*

Circuit emulation uses a packet router with specialized TDM interface modules and a dedicated processor on those cards to perform the packetization and depacketization steps between TDM and packet. The legacy connection, known as an *attachment circuit* (AC), is connected to the TDM interface module and the TDM circuit comes up normally. The onboard processor reads the incoming data stream and every few milliseconds sends that data off in an IP/MPLS packet to the remote CEM router. At the other end, the data from the IP/MPLS packet is pushed into a framer and played out down the AC at that end. Of course, this is a duplex process, so transmission and reception both happen simultaneously at each end of the circuit.

From the preceding discussion, you know that TDM circuits need frequency synchronization to ensure that the receiver and transmitter are processing bits at the same rate. Now that we emulate the formerly synchronous circuit, the packet network breaks the chain of frequency between the transmitter and receiver. Therefore, we need another mechanism to ensure that the AC at each end is running at the same frequency.

There are two solutions (see Figure 2-10) to introduce frequency synchronization into the IP/MPLS packet-switched network. One uses SyncE, and the other uses a profile of PTP specifically designed to replace the TDM frequency timing with packet-based frequency timing. These two solutions have been widely deployed, either separately or in combination.

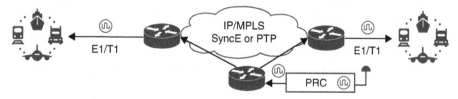

Figure 2-10   *Timing Solutions for Circuit Emulation*

Not only business, government, and enterprise customers have already taken steps to replace all SDH/SONET infrastructure with packet-based CEM routers, but also data service providers. Their data circuit customers never even realize that the T1/E1 circuit they have purchased has been migrated to packet transport inside the core network of the service provider (SP). Their TDM link only reaches to the nearest point of presence (POP) or local exchange, where it is converted to use a packet transport for delivery to the remote location over the SP's IP/MPLS network.

We have not yet mentioned another two techniques to carry clocking over the packet network, known as *adaptive clock recovery* (ACR) and *differential clock recovery* (DCR). Chapter 9 covers these techniques in more detail within the context of packet-based distribution of frequency.

Audiovisual

One example of synchronization that is commonly overlooked is that used in audiovisual production and distribution, at least until it fails. The most important aspect is the synchronization between the visual stream and the audio channels. Human beings are very quick to notice when the speech being heard does not align with the moving lips or sounds are not aligned with actions.

Figure 2-11 illustrates the importance of aligning the arrival time of audio and video and how much divergence the viewer can detect. As the two signals begin to diverge, an increasing percentage of viewers quickly notice the disconnect. When the audio is only 40 ms early, already 20% of the viewers notice—any earlier and it rapidly becomes unacceptable, distracting, and annoying.

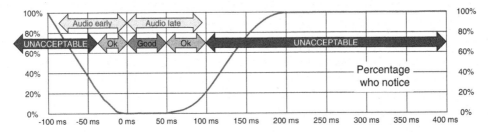

Figure 2-11   *Timing Solutions for Circuit Emulation*

The engineer needs to provide the synchronization between audio and video across the whole range of the creation process: content capture (cameras and so on), editing, distribution, and broadcast. Similarly, for packet-based A/V distribution to the subscriber—for example, with IPTV and digital video broadcasting (DVB)—the synchronization of audio and video needs to be maintained to the viewing screen (whether it be home, theatres, theme parks, or stadiums).

One of the tools used to maintain synchronization is based upon another PTP profile known as ST2059-2 from the Society of Motion Picture and Television Engineers (SMPTE). Also commonly used in the timing role for A/V is another profile of PTP known by the Institute of Electrical and Electronics Engineers (IEEE) standards number

802.1AS. There is also another profile from the Audio Engineering Society (AES), known as AES67, which is also used for audio processing. More information on the SMPTE and AES organizations and profiles is provided in Chapter 4 and Chapter 7.

Industrial Uses of Time—Power Industry

In addition to the previously mentioned examples, there are many other uses of synchronization, such as in industrial automation and mechanical processes. The use of multiple robots and machines in a factory assembly line is one common example of time synchronization, and the first standardized version of PTP was designed for industrial cases.

It is worth outlining another use case as a practical example to demonstrate that distributing timing across a wide area is not only a 5G or even a mobile problem. In fact, many of the same issues that arise when designing timing in mobile backhaul networks are present in the power industry, so there are obviously lessons we can share. Generally, many of the tools and deployment lessons used in one industry are directly applicable to other industries.

In the power industry there is an increasing focus on improving the robustness and stability of the power distribution network, or "the grid," especially as it needs to cope with the challenges of intermittent power sources such as that from renewable sources. You might hear the term "smart grid" to describe any one of several programs to increase the "intelligence" of the grid through digitalization.

One of these efforts to improve the grid in the United States is the North American SynchroPhasor Initiative (NASPI), which requires the placement of monitoring devices called *phasor measurement units* (PMUs) around the grid. These devices measure *synchrophasors*—precise measurements of the magnitude and phase of the sine waves carrying the electrical power. These measurements are taken at high speed (> 30 measurements per second) and each measurement is timestamped according to a common time reference, allowing correlation of these measurements across the whole grid.

Synchronization of the PMU equipment is generally done using an industry-specific profile of PTP, called the "Power Profile," with a local GPS receiver as the source of the common time reference. There are numerous applications in the power substation that use synchronization, with a range of required accuracies. Some of these applications, such as supervisory control and data acquisition (SCADA), are similarly deployed by other industries, such as monitoring the power grid in an electric railway track network.

The current issue facing grid operators is to how to be prepared in the case of an outage of the local GPS system, such as that caused by a jamming event. The solution is to carry a backup source of time from a remote location over the data transport network (see Figure 2-12). The obvious choice for the solution is again PTP (although the Power Profiles are not optimal for this task, other PTP profiles are designed for just this application—more on that later when addressing PTP in Chapter 7).

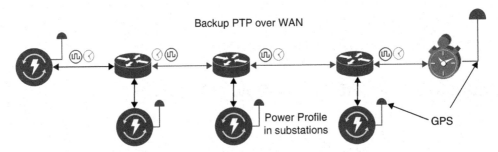

Figure 2-12 *Backup Timing Across the WAN*

Although this example highlights a program from North America, there are similar projects under way across the globe. More information on the current and future methods of synchronization in the power grid is provided in Chapter 3, "Synchronization and Timing Concepts," and Chapter 9.

Mobile operators have similar problems, whereby a cell radio is normally synchronized by a GNSS receiver but is vulnerable to being jammed and spoofed. Therefore, operators are rolling out similar solutions to provide a backup in the case of some localized GNSS outage.

Summary

This chapter examined several practical use cases that require some combination of the synchronization of frequency, phase, and time. It demonstrated that this field has a lot of application outside the 5G space, which is currently the center of attention. Also, by examining the practical application of the techniques and methods, you gained a much clearer understanding of the technology beyond using an abstract treatment involving mathematics.

References in This Chapter

European Commission. "Markets in Financial Instruments (MiFID II) – Directive 2014/65/EU." European Commission, 2018. https://ec.europa.eu/info/law/markets-financial-instruments-mifid-ii-directive-2014-65-eu_en

Ferrant, J. et al, *Synchronous Ethernet and IEEE 1588 in Telecoms: Next Generation Synchronization Networks.* John Wiley & Sons, 2013.

Juncker, J. "Commission Delegated Regulation (EU) 2017/574." *Official Journal of the European Union*, 2017. http://data.europa.eu/eli/reg_del/2017/574/oj

McNamara, J. *Technical Aspects of Data Communications.* Digital Press, 3rd Edition, 1988.

Mizrahi, T., N. Sprecher, E. Bellagamba, and Y. Weingarten. RFC 7267, "An Overview of Operations, Administration, and Maintenance (OAM) Tools." *Internet Engineering Task Force (IETF) Informational*, RFC 7267, 2014. https://tools.ietf.org/html/rfc7276

Mouly, M. and M. Pautet. *The GSM System for Mobile Communications.* Bay Foreign Language Books, 1992.

SMPTE. "SMPTE Profile for Use of IEEE-1588 Precision Time Protocol in Professional Broadcast Applications." *SMPTE ST 2059-2:2015*, 2015. https://ieeexplore.ieee.org/stamp/stamp.jsp?tp=&arnumber=7291608

The International Telecommunication Union Telecommunication Standardization Sector (ITU-T). "G.8271: Time and phase synchronization aspects of telecommunication networks." *ITU-T Recommendation*, Appendix VII, 2020. https://www.itu.int/rec/T-REC-G.8271-202003-I/en

Chapter 2 Acronyms Key

The following table expands the key acronyms used in this chapter.

Term	Value
1DM	one-way delay measurement
2DM	two-way delay measurement
3GPP	3rd Generation Partnership Project (UMTS)
3GPP2	3rd Generation Partnership Project 2 (CDMA2000)
5G	5th generation (mobile telecommunications system)
AC	attachment circuit
ACR	adaptive clock recovery
AES	Audio Engineering Society
ANSI	American National Standards Institute
ATIS	Alliance for Telecommunications Industry Solutions
BITS	building integrated timing supply (SONET)
CDMA	code-division multiple access
CEM	circuit emulation
DAB	digital audio broadcasting (digital radio)
DCR	differential clock recovery
DPLL	digital phase-locked loop
DVB	digital video broadcasting (digital TV)

Term	Value
ETSI	European Telecommunications Standards Institute
EU	European Union
FDD	frequency division duplex
Flex-Algo	flexible algorithm (part of segment routing)
GNSS	global navigation satellite system
GPS	Global Positioning System
GSM	Global System for Mobile communications
HFC	hybrid fiber coax (cable system)
IEEE	Institute of Electrical and Electronics Engineers
IP	Internet protocol
IP/MPLS	Internet protocol/Multiprotocol Label Switching
IPTV	Internet protocol television
IT	information technology
ITU	International Telecommunication Union
ITU-T	ITU Telecommunication Standardization Sector
LTE	Long Term Evolution (mobile communications standard)
MiFID II	Markets in Financial Instruments Directive (2014/65/EU)
MPLS	Multiprotocol Label Switching
NASPI	North American Synchrophasor Initiative
NTP	Network Time Protocol
OAM	Operations, Administration, and Maintenance
PDH	plesiochronous digital hierarchy
PMU	phasor measurement unit
PON	passive optical networking (optical broadband)
POP	point of presence
PRC	primary reference clock (SDH)
PRS	primary reference source (SONET)
PTP	precision time protocol
SCADA	supervisory control and data acquisition
SDH	synchronous digital hierarchy (optical transport technology)
SMPTE	Society of Motion Picture and Television Engineers

Term	Value
SONET	synchronous optical network (optical transport technology)
SP	service provider (for telecommunications services)
SR	segment routing
SSU	synchronization supply unit (SDH)
STB	set-top box
SyncE	synchronous Ethernet—a set of ITU-T standards
TDD	time division duplex
TDM	time-division multiplexing
TDMA	time-division multiple access
UMTS	Universal Mobile Telecommunications System
UTC	Coordinated Universal Time
WAN	wide-area network

Chapter 3

Synchronization and Timing Concepts

In this chapter, you will learn the following:

- **Synchronous networks overview:** Presents an overview of synchronous networks and a comparison of asynchronous and packet networks. Also defines frequency and phase synchronization as well as jitter and wander.
- **Clocks:** Provides details on oscillators, clock modes, and clock types, as well as clock stratum levels and quality.
- **Sources of frequency, phase, and time:** Explains sources of timing, with special focus on satellite systems and particularly the Global Positioning System (GPS). Also covers the short-range physical signals used to carry time.
- **Timing distribution network:** Describes different methods used to carry time and synchronization as well as the transport of clock quality and traceability information.
- **Consumer of time and sync:** Covers what factors the engineer needs to consider when determining what the end application needs for its timing and synchronous requirements.

As described in Chapter 2, "Usage of Synchronization and Timing," synchronization is a process of aligning two or more clocks to the same frequency or time, usually by aligning each of them to a common reference clock. Synchronization enables services such as

- Positioning, navigation, and timing (PNT)
- Transmitters broadcasting on single frequency networks (SFN)
- Control and command systems for Internet of things (IoT) or cloud-based applications
- Coordinated transmission and interference control in a mobile network
- Phased array antenna applications such as beamforming

This chapter focuses on numerous concepts required to understand synchronization and the components needed to build a timing distribution network. This chapter covers the definition of a clock, the different types of clocks and sources of clock signals, as well as the different operating modes of the clock. You will learn about the various clock types in the hierarchy of the network chain; methods to trace the reference clock; and how to utilize different timing reference interfaces for accurate timing distribution and measurement. This chapter also covers the difference between frequency and phase synchronization and various time distribution techniques.

Synchronous Networks Overview

Network synchronization involves distributing time and frequency, or synchronizing clocks across different geographies, by using a network. Any number of different network transport technologies, such as those based on fiber, cable, or even radio, can be used to carry timing information either as a physical signal or as data. The main goal is to synchronize all the clocks connected to the network to a single, common frequency and/or time reference.

This problem of aligning distributed clocks is widely studied, and various methods are used to maintain accurate synchronization. Figure 3-1 illustrates combining technologies to achieve this. In this case, the reference is initially distributed via a satellite-based system and then transported at ground level using some form of network.

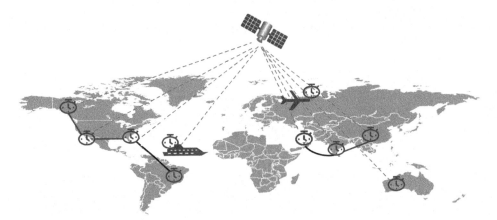

Figure 3-1 *Network Synchronization to a Common Reference Clock*

Synchronization can also be a fundamental requirement of many technologies used to transport data across a telecommunications network. In synchronous digital communications, networks require that frames arrive at the receiving node at the correct instant, so that the allocated time slot is filled and processed correctly without dropping data.

As data frames are exchanged between nodes connected to a synchronous network, it is important that clocks running at these nodes are closely synchronized. If not, eventually data would start to queue up at the slowest point and data would be lost or dropped.

But, on the other hand, there are other transmission technologies that use asynchronous methods.

Asynchronous Networks

In asynchronous networks, each node maintains its own clock, and there is no overriding control mechanism that coordinates the operation of the clocks in the network elements. Communications over an asynchronous link tend to be more byte- or packet-orientated rather than frame-orientated. These networks use methods (such as start and stop bits or a preamble) to delineate the transmission of a data segment. They also use mechanisms like flow control to stop the circuit being overwhelmed with data.

As an example, older communications standards such as RS-232 use a stop and start bit to indicate the beginning and end of a byte of data. This method was widely used in the days of dial-up modems and was frequently used for low-speed connections over copper wires.

Some later communication technologies, such as many types of Ethernet, use a preamble that allows the receiving circuit to align its receiver to the frequency of the incoming bit pattern. As defined in IEEE 802.3-2018 standard, clause 3.2.1:

> The Preamble field is a 7-octet field that is used to allow the (Physical Signaling Sublayer) PLS circuitry to reach its steady-state synchronization with the received packet's timing.

This mechanism allows the receiving circuitry to be aligned with the incoming signal; however, there are other requirements for synchronization that are important for such a link. For example, a circuit might need to have identical transmission speeds in both directions to reduce crosstalk or interference between the two data streams on adjacent copper wires.

To achieve this, forms of Ethernet using copper wires can select one end of a circuit to be a master with the other end becoming a slave. The transmitter on the master end will use its local *free running* oscillator to transmit data to the slave. The receiver at the slave end then uses the preamble to align its reception frequency with that of the incoming signal (it *recovers* the frequency). The slave uses that frequency, learned from the incoming signal, as the transmission frequency from the slave back to the master.

Note   Use of the terms "master" and "slave" is ONLY in association with the official terminology used in industry specifications and standards, and in no way diminishes Pearson's commitment to promoting diversity, equity, and inclusion, and challenging, countering, and/or combating bias and stereotyping in the global population of the learners we serve.

So, the links themselves are frequency aligned, without any centralized notion of clocking in the node. For details, see IEEE 802.3 clause 28, "Physical Layer link signaling for Auto-Negotiation on twisted pair", clause 40.4, "Physical Medium Attachment (PMA) sublayer" (for 1000BASE-T); and clause 40.4.2.6, "Clock Recovery function" (for 1000BASE-T).

For Ethernet, the requirements for frequency accuracy are quite loose, commonly only requiring accuracy to within ±100 ppm. Different Ethernet technologies and speeds can use different frequencies—as an example, the 1 Gigabit Ethernet for unshielded copper twisted-pair (1000BASE-T) has a symbol rate of 125 MHz.

But remember that every link is negotiating its own frequency based on the free-running oscillator of the port selected as master for that link. And so, without a central overriding clock as a source, a multiport switch could end up in a topology where every port is a slave and is running at a slightly different clock frequency (although within ±100 ppm of each other). Effectively, every port on the switch becomes a point-to-point synchronous link.

Conceptually, this is what happens with an Ethernet switch that is not using synchronous Ethernet (SyncE)—every port negotiates its own transmission and reception speed with the port on the neighboring switch. This means that all the oscillators used for transmission are free running (not aligned with any reference frequency signal).

SyncE corrects this by connecting the transmitter on every port to a common oscillator for the device and allows that oscillator to be frequency synchronized to a reference clock. The reference signal comes either over a SyncE link or via an external source of frequency, such as a sine wave signal fed into the switch's building integrated timing supply (BITS) port. See the section on synchronous Ethernet in Chapter 6, "Physical Frequency Synchronization," for more details on SyncE.

Synchronous Networks

In synchronous networks, all clocks are synchronized to the common clock reference. There are different approaches to synchronize the clocks in the network; they are broadly classified into two categories—centralized and decentralized control methods.

As shown in Figure 3-2 part a, in the centralized control method, the engineer assembles the clocks into a hierarchy based on a master-slave relationship between the network clock and the remote master clock. In this approach, the ultimate master clock (known in some cases as the grandmaster) is the definitive reference clock for every node in the network.

Alternatively, in a decentralized control method (Figure 3-2 part b), each clock synchronizes with its peers to determine the best quality. For example, in a mutual synchronization approach, there is no master clock concept, and every clock contributes to maintaining the network timing.

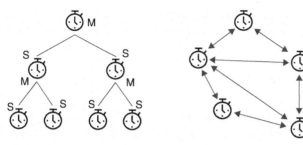

a) Master-Slave Synchronization   b) Mutual Synchronization

Figure 3-2 *Synchronous Clocks*

As covered in Chapter 2, in a synchronous network, frequency synchronization is what allows two connected systems to run at the common synchronized frequency (or bit rate). Legacy time-division multiplexing (TDM) networks use frequency synchronization to ensure that the rate of transmission of the bits is aligned to avoid frame loss or slips between nodes. Building an efficient yet affordable transport network is based on distributing frequency from a few expensive devices (with high-quality components such as oscillators) to masses of devices made using more affordable components. The section "Timing Distribution Network" later in this chapter provides further details.

Defining Frequency

Anything that rotates, swings, or ticks with a regular cycle time can be used to define a time interval. If the motion is regular and the interval of repetition remains stable, it forms the basis for building a clock. There are device movements or other phenomena that occur naturally at a specific frequency, and a device (or electrical part) that does this is called a *resonator*. In electronics, when an energy supply circuit is added to a resonator, it continues its cyclical behavior indefinitely, and that device becomes an *oscillator*.

If the length of time that one oscillation takes (the *period*) is known and remains stable, a device can accurately measure the length of a time interval by simply counting the ticks. This is the reason that timing engineers say that a clock is simply an oscillator plus a counter. By simply attaching a circuit to count and record the number of rotations, swings, or oscillations, one can build a clock and use it to measure time.

It follows that for measurements of very small time intervals, higher frequencies of oscillation are needed. An electronic circuit generates those higher frequencies by multiplying the stable frequency output by the oscillator by some factor. Take the previous example of Ethernet to illustrate. To source the 125-MHz signal to transmit data, a circuit multiplies the output of a 20-MHz quartz oscillator by 6.25, yielding 125 MHz. Another way to think of it is that for every 4 complete cycles of input from the oscillator, there are 25 complete cycles of output (4 × 6.25).

Obviously, a device used to measure a time interval must cycle at a much higher rate than the time interval being measured, and the more accuracy required, the higher the rate. To do otherwise is the equivalent of measuring the size of microbes with a meter-long ruler. So, when engineers need to accurately measure time down to millionths or billionths of a second, they need to use an electronic circuit that cycles at least billions of times per second.

Frequency is defined as the number of repetitions of the regularly occurring event over a unit of time. The standard unit for frequency is the *Hertz* (Hz), defined as the number of events or cycles per second. The frequency of rapidly occurring signals is measured in multiples of Hz, where 1 kHz is one thousand or 10^3 events per second. Similarly, 1 MHz equals one million or 10^6 events per second, while 1 GHz equals one billion or 10^9 events per second.

While frequency is the number of cycles per interval of time, the *period* is the length of time (the duration) for the motion to complete one cycle. One is the reciprocal of the other—a pendulum swinging back and forth twice every second has a frequency of 2 Hz but a period of 0.5 seconds. Mathematically, we describe it as

$$Frequency\ (f) = \frac{1}{Period\ (T)} \ or \ f = \frac{1}{T}$$

Oscillators with periods of longer than one second therefore have frequencies of less than 1 Hz. For example, a circuit that cycles once every 10 seconds is 0.1 Hz, once every 1000 seconds is 1 mHz, and once per day is 1/86,400 Hz (11.574 µHz).

The *oscillator frequency*, the rate at which the clock advances, is a characteristic based on the physical design and natural characteristics of the component. Because it is impossible to design oscillators that all have the exact same characteristics, every clock will oscillate at a different natural frequency—the difference might be small, but it will be there (and will very likely change over time as well).

You can buy desktop equipment to measure frequency, what is known as a *frequency counter*. Frequency counters measure the number of cycles of oscillation, or pulses per second in a periodic electronic signal (in Hertz). Frequency counters with today's technology can count frequencies above 1×10^{11} Hz, which allows accurate measure of very small time intervals. These devices are commonly found in lab environments for testing time accuracy.

Oscillators based on cesium-133 atoms are widely used as primary reference clocks (PRC) to provide a very stable and accurate source of frequency. To make a real "clock" out of it, you would simply add a counter that increments by 1 after it detects 9,192,631,770 cycles of cesium radiation. This counter would then increment almost exactly once per second, an *almost* perfect ticking clock (there are no perfect ones).

Oscillators based on quartz crystals (see Chapter 5, "Clocks, Time Error, and Noise"), being more affordable, are widely used to provide a frequency signal within telecommunications networks. Although quartz is an adequate oscillator capable of maintaining a somewhat stable frequency, it is not accurate enough for digital communications and nowhere near as stable as a cesium clock.

It is the distribution of the frequency reference signal from the cesium PRC to all the devices in the network that allows the embedded quartz oscillator in each network node to accurately deliver the frequency for error-free communications.

Defining Phase Synchronization

As introduced in Chapter 1, "Introduction to Synchronization and Timing" phase synchronization is when separate, but regularly occurring, events happen at the same instant or point of time. One topic covered then was the music analogy where a metronome or an orchestra conductor is used to keep musical time. In fact, they have two purposes: one is a frequency task, to make sure music is played at the correct rate or tempo, expressed in beats per minute, and the second is to provide a regular "tick" that might signal the start of a new chord in a bar of music, or a wave of the baton to start a new movement. In this analogy, these mechanisms are aligning every participant in the performance to a single agreed source of phase. Imagine an orchestra trying to play a piece of music with two different conductors doing their own thing!

Frequency or tempo is a characteristic that can be observed and measured, as you can quickly determine if the musician is playing a piece too fast or too slow by independent observation and comparing the rate at which the music is played to the tempo written in the music score. It might take a reference frequency source to determine if something is happening too slowly or too quickly, but it can be independently determined by checking the rate against a time source. Another example could be a nurse taking the pulse of a patient by checking the frequency of heartbeat against a wristwatch.

Phase, on the other hand, cannot stand alone; it is always relative to a reference clock or a reference timescale. This means that phase alignment is based upon following the lead given by the source of the phase, and you cannot measure it as a standalone metric, only relative to the source. Continuing with the analogy, the phase signaled by the conductor is only relevant to the orchestra playing, and another orchestra playing elsewhere has their own phase.

The important factor for phase is to measure how closely aligned the local phase is to the phase from the source of phase or some independent reference. So, you could say that machine A has a +100 ns phase offset to machine B, or machine B has a −250-ns phase offset compared to Coordinated Universal Time (UTC) received from GPS. But it makes no sense to say that machine A has a +200-ns phase offset (although UTC might be assumed).

So, phase is expressed as a relative offset, although when measured to a common time source like UTC, it is often treated as absolute. This difference between relative and absolute is because in some circumstances the radio in one cell site might need to have close phase alignment with its neighboring radios (relative phase offset). But when all the cell sites in one mobile network needs to be aligned with those in another mobile network, then the easiest way to make this happen is to have every radio closely phase aligned with an absolute time source such as UTC. Achieving this alignment to an absolute phase means they are also relatively phase aligned at the same time. This concept will be revisited when talking about phase synchronization in the 5G mobile network in Chapter 10 "Mobile Timing Requirements."

In the case shown in Figure 3-3, clock signals C_a, C_b, C_c, and C_d all have the same frequency, while C_e and C_f have higher frequencies. The signals are also mixed with a phase pulse, which in the case of C_a, C_b, C_c, C_d, and C_f is generated on the rising edge of every second square wave. You could imagine that the pulse could be the signal to indicate the "tick" of the second, for example. The signals C_a and C_b are phase aligned with each other, and C_d and C_e are also phase aligned, although C_e has double the frequency of C_d (the pulse is only generated on every fourth rising edge for C_e).

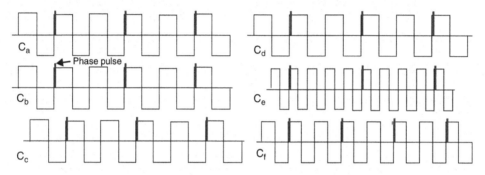

Figure 3-3 *Frequency and Phase Alignment*

The signal C_c has a phase offset from C_a and C_b, while the phase output of C_f is completely unaligned because the frequency is too high, and the pulse is still being generated every second wave. Note that the devices generating the signal C_d and C_e are frequency aligned, although they have different frequencies; it is just that the signal C_e is oscillating at twice the frequency of the signal in C_d.

The point of Figure 3-3 is to demonstrate that frequency and phase are also dependent on each other, and that once phase alignment has occurred, maintaining a very close frequency alignment will keep the phase also closely aligned. This is a topic that will be revisited in the discussion about maintaining good phase alignment during holdover (a good place to start is the section "Holdover Performance" in Chapter 5).

As shown in Figure 3-3, to determine phase alignment, an arbitrary repeatable point must be selected to measure the phase or as a point to determine the phase from an oscillator's frequency signal. This point is known as the *significant instant* and usually lies somewhere on the rising or leading edge of a wave or pulse—the important factor is that it must be a repeatable point. Specific examples could include the instant at which a signal crosses the zero baseline or reaches some value, such as the midpoint of the leading edge.

Many early standards documents relating to distribution of timing signals through the physical layer define many timing concepts in terms of significant instants. But the concept is not difficult; it is simply the exact point of time when the receiver determines that the signal has "arrived" or has been "detected." An analogy might be the postmarking of a letter sent via the post office; the significant instant of posting a letter is that instant when the mail sorter cancels the stamp.

Figure 3-4 shows examples of significant instants from a pair of ideal signal types.

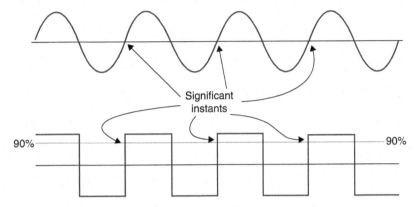

Figure 3-4   *Significant Instants in Sine and Square Wave*

When two signals have their significant instants occur at the same time, they are phase aligned.

Note that the square wave (which might have a maximum voltage of 5 V) can never go from –5 V to +5 V instantly, as the electronics to do that do not exist. There will always be a tiny time window for the signal to lose the negative voltage and acquire the new positive voltage. The times taken for the signal to rise from the low level to the high level and back again are called the *rise time* and the *fall time* respectively. So, the leading and trailing edges are never vertical—the leading edge leans slightly to the right and the trailing edge leans slightly to the left.

When the phase is not perfectly aligned, then there is a measurable phase error known as *phase offset*, as shown in Figure 3-5.

Expressing phase offset mathematically, assume that the time reported by a clock C_A at some ideal time t is $C_A(t)$. The difference between the ideal clock, which might be UTC (or an approximation of it), and the time given by the clock C_A is called the *clock offset* and usually is represented by the lowercase Greek letter theta as follows:

$$\theta = C_A(t)$$

Although this could be viewed as an absolute clock offset because it is compared to a universally accepted timescale, really it is just another relative offset compared to UTC. The offset between any two network clocks, a *relative offset*, is defined as

$$\theta_{AB} = C_A(t) - C_B(t)$$

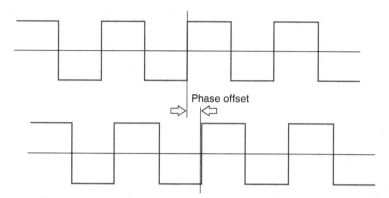

Figure 3-5   *Phase Offset*

Synchronization with Packets

In packet-based networks, the timing information is conveyed, from a packet master clock to a packet slave clock, by the periodic exchange of data between nodes. The time information is represented in the form of timestamps and, where required, exchanged using packets. The precision time protocol (PTP) calls packets that are timestamped on transmission and reception *event messages.*

Once the newly generated packet is timestamped at the transmitting end, the packet starts to "age," and unnecessary delay adds to this aging. The time between when the packet is timestamped on departure and the time recorded when it is received is time critical and must be minimized. The unavoidable delay as a result of the transmission distance is fine, because a two-way time protocol can estimate this, but having the packets sit in a buffer queue waiting for a transmission timeslot only increases the inaccuracy.

Figure 3-6 illustrates some of the basic characteristics of timing distributed by packet. First, there is the payload that contains the timing information. To allow the packet to pass over the communications path, the payload is wrapped inside a header and footer (denoted respectively as H and F in Figure 3-6). Because the delay of every packet on these types of networks is nondeterministic, time-critical packets sent out periodically will vary in their transit time—what is known as packet delay variation (PDV). PDV is a significant problem in deployment of packet-based timing distribution because of its effect on the accuracy of the recovered clock. Chapter 9, "PTP Deployment Considerations," covers strategies to manage PDV in more detail.

One more factor to determine is at what precise time the packet "departs" or "arrives." This is determined by agreeing that one precise location of the message is the significant instant. The timestamp should represent the time at which this significant instant passes the timing reference point of the clock. By convention, the location is agreed to be the end of the start-of-frame delimiter, but it can be defined differently in any given packet technology, provided the definition is consistently applied.

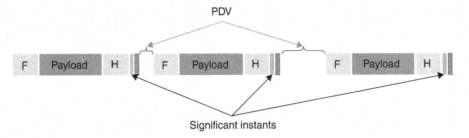

Figure 3-6 *Packet Timing Signal*

> **Note**   For further background reading on timing based on packet networks, significant instants, and more, see Chapter 6, "Description of packet timing concepts," of ITU-T G.8260.

Jitter and Wander

When the timing signal (technically, the significant instant of that signal) shows short-term, high-rate variations from the ideal position, it is called *jitter*. Alternatively, slow, long-term time variations are known as *wander*. So, the question becomes what is considered to be "short term."

According to the ITU-T (in G.810), the division between jitter and wander for physical time distribution is at 10 Hz, so variations that occur at a frequency at or above 10 Hz are referred to as jitter, and those below 10 Hz are called wander. Packet-based methods such as PTP might use a different cutoff (0.1 Hz), but just know that there will be a formal definition.

Figure 3-7 illustrates the jitter effect. Being short-term variations in the signal, jitter can be filtered out, or mathematically smoothed, so that the input timing signal is followed only for longer-term trends and each short-term variation is not immediately used to update the state of the local oscillator. Obviously, if these short-term offsets are too large, it creates a problem in sampling, decoding, and recovering time information.

Wander, on the other hand, is a longer-term effect and so cannot easily be ignored. A slave clock receiving a timing signal from a master clock cannot decide to "second guess" the accuracy of the data from the master clock and ignore (within limits) the input. That is a very dangerous assumption to make, as the whole point of a timing network is to transport a more accurate clock to the slaves with poorer accuracy. The only exception is if the input is completely out of the range of expected values and beyond the *tolerance* of the slave clock.

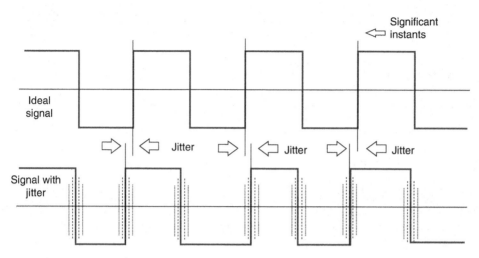

Figure 3-7   *Phase Fluctuations—Jitter*

Figure 3-8 gives an illustration of wander in a timing signal, showing the lower signal first lagging on the upper signal before running ahead of it (indicated by arrows). Note that because time is increasing to the right on the horizontal axis, the significant instant moved to the right is behind and not in front of the other signal.

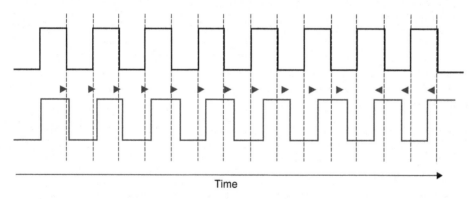

Figure 3-8   *Phase Fluctuations—Wander*

Every time-sensitive application has its own defined minimum requirement around phase accuracy for smooth operation. You can define the requirement in several ways:

■ The maximum allowed offset from a reference clock

■ An offset between master and slave clocks

■ A relative offset between two slave clocks

Therefore, it is important to track these variations and use mechanisms to minimize or partially remove errors from the timing signal. To this end, the standards define performance parameters for various measures of wander and jitter in the network equipment, which can include the following:

- Tolerance to input errors from an upstream signal

- Limits on the error generated at the output of the clock

- The amount of error transferred through the equipment

The ITU-T standards specify equipment performance for both physical-layer and packet-based timing distribution (refer to Chapter 8, "ITU-T Timing Recommendations," for more details). Of course, other standards organizations also perform similar functions.

Chapter 5 provides further coverage of jitter, wander, and noise tolerance, including many of the aspects of clock error and *noise*.

As you read previously, slight physical differences between oscillators as well as environmental variations like temperature, or airflow, mean that even identical network devices quickly diverge in synchronization. To maintain the accurate time offset between network devices, a mechanism is required to continually compare phase and time against a reference and correct any offsets. But to achieve that, the network device needs to know that it is connected to a good-quality source of frequency, phase, and time. And for that reason, it needs to be able to trace the quality of the clock signals back to its source.

Clock Quality Traceability

Both physical-layer and packet-layer synchronization techniques use several methods to transfer the clock across the network; however, it is understood that the clock accuracy degrades when the clock is transferred over the network nodes and links. For example, in a packet method, there are several factors that degrade the clock accuracy on a slave clock, including

- How accurately the machines timestamp the packets

- Variations in the packet latency and packet delay (PDV)

- Asymmetry between the forward and reverse path

- Oscillator behavior and clock recovery accuracy (including filtering) on the slave node

Obviously, this means that the more network nodes involved in the transport of time, the worse the result is at the end of the chain. The timing standards define sets of performance metrics for timing characteristics so that engineers can predict what level of accuracy can be expected after transportation across the network. This separates equipment into different categories of performance and places limits on the topology that each equipment type can support, while still delivering an acceptable result.

Additionally, ensuring the shortest path to the best reference is a fundamental goal of good timing network design. Because the timing signal degrades as it crosses every link or node, then being able to determine the path length (and dynamically recalculate it following a network rearrangement) is also valuable. And yet another reason that can lead to further significant divergence between two clocks in a network is that they are not connected to the best source of time. This could happen due to reasons such as

- Nodes are receiving a clock signal from a time source of dubious accuracy.

- The packet master is using a reference clock that is no longer a valid source of frequency, phase, and time.

- The slave selects a master clock that is not the best available.

- A node is synchronized with a clock that has lost connection to its reference and does not know it.

The conclusion from these possible issues is that slave clocks not only need a timing signal but also need to understand how good that signal is and whether it comes from a good source. The notion of traceability was developed to solve this problem. In metrology, *traceability* is defined as a situation where "the [measured timing] result can be related to a reference through a documented unbroken chain of calibrations, each contributing to the measurement uncertainty."

Therefore, just about all timing distribution mechanisms, both physical and packet, have defined a method of clock traceability to distribute information about the quality of the signal they are offering. Clock traceability allows the network elements to always select the best clock available to them and know when a valid source of time disappears. This mechanism allows the clock distribution network to optimally configure itself by continuously discovering and connecting to its best sources of synchronization.

No matter what the method of time distribution, there will always be some mechanism used to allow any clock receiving a timing signal to determine the quality and accuracy of that signal. For frequency, SDH/SONET signaled this information in some overhead bits in the data frame, what are called the *synchronization status message* (SSM) bits.

Figure 3-9 illustrates the use of this mechanism in frequency distribution using an SDH network as an example. The network element receives information from the connected devices about the source of timing they are traceable to. In the figure, the quality level (QL) is shown being distributed along with the timing signal. One of the best sources of frequency in an SDH network is the PRC, while the SSU-A is the next level down.

The SSU-A QL indicates traceability to a *primary level synchronization supply unit* (SSU), which is a level of accuracy below that of a PRC (see the section "Clock Traceability" in Chapter 6 for more details).

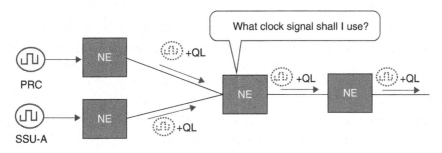

Figure 3-9 *Selecting the More Accurate Source of Frequency*

Even in the small number of timing or frequency signals that contain no quality information (for example the 2-MHz or 10-MHz sine waves), there will be some mechanism to signal to the receiver of the signal when the source of time is invalid. In these cases, the upstream reference would simply *squelch* the signal (as in, turn it off), which indicates to the downstream slave that it should try to find a new source of frequency.

Table 3-1 gives an overview of the common distribution methods and their mechanism to signal clock quality.

Table 3-1 *Clock Quality Traceability Methods*

Type	Method	Explanation
SDH/SONET	SSM bits	SSM bits indicating the quality of the clock. The synchronous digital hierarchy (SDH) and synchronous optical network (SONET) use different values for these bits and different levels of clock quality.
Ethernet SyncE	ESMC	The Ethernet synchronization message channel (ESMC) sends packets to indicate the quality of source for SyncE.
Enhanced SyncE	eESMC	The enhanced Ethernet synchronization messaging channel (eESMC) includes additional information in ESMC to support enhanced SyncE (eSyncE).
PTP	clockClass	PTP sends *Announce* messages with clock quality information (including *clockClass*) that indicates the quality of the time source at the grandmaster clock (see Chapter 7, "Precision Time Protocol").
GPS signals	Health bits	GPS satellites transmit whether their timing data is valid or not.
Global navigation satellite system (GNSS) receiver	Time of day	Most GNSS receivers have some method to signal an estimate of the clock quality. This can be included in the *time of day* messages between the receiver and the packet grandmaster. See Annex A of ITU-T G.8271 for one approach.

One thing to recognize is the problem of suffering from *timing loops* in the synchronization network. It is possible to build and configure a network where the timing information generates a loop (for example, where A gives time to B that gives time to C that gives time back to A). Some clock traceability methods, such as ESMC, have mechanisms to help mitigate such problems, although it is still possible to construct such a topology.

Chapter 6 covers the details of the specific traceability methods used by SDH/SONET, while Chapter 7 covers the methods used for packets. Chapter 6 also discusses frequency timing loops in detail, while Chapter 9 examines ways to mitigate them in an operational network.

Clocks

As mentioned previously, a clock is an oscillator plus a counter. Now it is time to revisit many aspects of the clock in further detail.

A free-running oscillator inside a device generates a frequency signal close to its nominal frequency (assume within a few parts per million of 10 MHz). This makes it a known and somewhat stable source of frequency. Then, to make a timepiece from this frequency source, you need to add a counter, some device that can count those oscillations. For every 10,000,000 cycles of the (10 MHz) oscillator, this counter increments by one (a second). So now, the result is a device that can measure periods of time much like a stopwatch. But just as with a stopwatch, although it accurately measures time intervals, it has no concept of phase and is unaware of date and time.

The specific instant (or cycle) that triggers the counter to increment the second represents the phase of the clock. So, if the counter is set to trigger that increment at the same instant as another source of phase/time, it becomes phase aligned to that source. Then it resumes counting for another 10,000,000 cycles and ticks over the second once more.

Then you could add a circuit to the clock that can generate a signal (for example, a pulse on a copper wire) to indicate the exact instant when the counter increments the second counter. This pulse is called a 1 pulse per second (1PPS) signal. The device is now a source of phase. Of course, because the free-running 10-MHz oscillator is never doing exactly 10 MHz, it will start to diverge without either a source of frequency, or constant adjustment to correct any phase offset.

Furthermore, if you calibrate the second counter to some timescale like International Atomic Time (TAI) (so that the second counter is some large number representing the seconds since an agreed epoch), the device can track the current time based on TAI. Then it is a simple conversion to UTC (by considering leap seconds) and to local time (by including information on time zone and daylight savings time changes).

Hence, in its simplest form, a clock really is a device consisting of an oscillator and a counter.

Clocks are normally characterized by a series of attributes, such as

- **Accuracy:** How far it deviates from the actual correct time
- **Stability:** How the clock stays stable over time and produces repeatable results
- **Aging:** How much the oscillator in the clock changes its frequency over time

Each of those quality factors is inherited from the characteristics of the oscillator inside it, meaning that the oscillator is the main component that underwrites the performance of the clock.

Oscillators

Quartz oscillators are widely used in watches, clocks, and electronic equipment such as counters, signal generators, oscilloscopes, computers, and many other devices. Although quartz oscillators are not the most stable oscillators available, they are "good enough" and extremely cost effective, explaining their popularity. Of course, environmental conditions such as electrical and magnetic fields, pressure, and vibration reduce or change the inherent stability of the quartz crystal. Therefore, hardware designers give a lot of consideration to oscillator design to try and mitigate the environmental effects and improve the frequency stability of the device.

The most significant factor affecting the oscillator is temperature, and there are designs to mitigate its effect. A *temperature-compensated crystal oscillator* (TCXO) includes a temperature sensor that compensates the frequency change incurred by temperature change. Similarly, the *oven-controlled crystal oscillator* (OCXO) encloses the crystal in a temperature-controlled chamber to reduce temperature change (it is not cooled but heated to a stable but higher temperature than any reasonably expected ambient temperature).

Rubidium oscillators are the cheapest form of the class of oscillators referred to as *atomic oscillators*. They make use of the resonance of the rubidium atom at around 6,834,682,611 Hz and use this signal to align a well-designed quartz oscillator. Compared to other atomic devices, such as those based on cesium, rubidium oscillators are smaller, more reliable, and less expensive.

The stability of a rubidium oscillator is much better than a quartz oscillator, which is very helpful property when the clock loses its reference signal. Rubidium oscillators are also quick to settle and reach a stable operation after only a short warm-up period. This stability, combined with reasonable cost, means that operators deploy them as an affordable option to increase holdover performance (see the following section on "Holdover Mode"). The main shortcoming of rubidium oscillators is that they are somewhat temperature sensitive, so they can only be deployed where there are reasonable environmental controls (although, to be fair, atomic clocks are hardly ruggedized devices).

Cesium oscillators use a beam of excited cesium atoms, which are then selected by magnetic fields. The device transmits a microwave signal to resonate with the cesium atoms and cause them to alter their energy state. Using a detector to determine the number of altered atoms, the device "tunes" the frequency of that signal to maximize the number of altered atoms, which increases the closer the microwave approaches the 9,192,631,770-Hz resonance frequency. This is all done in combination with a high-quality quartz oscillator.

A well-designed cesium oscillator is the most accurate reference available, with almost no deviation over time (see the upcoming discussion of *aging*), although other environmental factors (motion, magnetic fields, and so on) can cause small frequency shifts.

It requires no adjustment and is quick to warm up and settle to a stable output. Cesium oscillators are used by standards organizations to define the SI second.

A hydrogen maser is a resonator that uses hydrogen atoms that have had their energy levels elevated to certain specific states. Inside a special cavity, the atoms lose their energy and release microwaves with a resonant frequency of 1,420,405,751.786 Hz. This signal is then used to discipline a well-designed quartz oscillator. The process is maintained by a continuous feed of more hydrogen atoms. Maser is an acronym for microwave amplification by stimulated emission of radiation (it is the microwave equivalent of a laser).

For short-term measurement, hydrogen maser clocks are more accurate than cesium clocks, but as time increases, clocks based on hydrogen masers are not as accurate because they suffer a significant frequency shift through a process known as *aging*. Of the several factors that affect almost all oscillators, aging is one of the most important to consider. Aging is simply the amount by which the natural frequency of the oscillator changes over a defined period. Characterizing this can be difficult as it has many dependences; for example, aging can be accelerated by elevating the temperature. For quartz, the rate of aging can start out quite high upon initial manufacture but decrease markedly over the weeks following being placed into service. For this reason, some oscillators are intentionally "aged" before being placed in a device.

Aging is expressed as a value per year (or some other timeframe), so that an aging value of 1×10^{-9} / day indicates that the oscillator will change frequency by up to 1 ppb per day.

Table 3-2 show the most common oscillator types and their characteristics. Stability in this table shows just one of the ways to measure it—over quite a short measurement interval. In metrology, the observation time is represented by the lowercase Greek letter tau (τ).

Table 3-2 *Oscillator Types*

Value	TCXO	OCXO	Rubidium	Cesium Beam	Hydrogen Maser
Technology	Quartz	Quartz	Rubidium	Cesium	Hydrogen
Stability(τ = 1 sec)	1×10^{-8} to 1×10^{-9}	1×10^{-11} to 1×10^{-12}	1×10^{-11}	5×10^{-12} to 1×10^{-13}	1×10^{-12}
Aging / year	5×10^{-7}	5×10^{-9}	5×10^{-10}	None	1×10^{-13}

This table contains typical short-duration values at a constant temperature based on data from datasheets from several vendors. Depending on the criteria and method of testing, and specific fabrication process, some of the values may vary significantly for any technology. The table is included merely to indicate the relative values between classes of oscillators using a short-duration baseline.

Generally, atomic oscillators are more stable over longer observation intervals and quartz over shorter time intervals, meaning that an atomic clock running alone typically only offers better stability over the longer term. As a rule of thumb, a cesium oscillator has better stability for observation intervals of more than 24 hours, while a rubidium oscillator provides better stability for periods of 5 minutes to 24 hours. A good-quality quartz oscillator will generally perform better than both atomic oscillators for short observation intervals, at least for less than approximately 5 minutes.

Clocks based on atomic oscillators generally contain a good quartz oscillator, typically an OCXO, for short-term performance, which is disciplined over the long term by the atomic component. Pairing an OCXO with an atomic clock combines good long-term and short-term stability, making them great reference clocks.

Clock Modes

During the normal operation of a clock, it goes through different modes of operation. Figure 3-10 illustrates a state diagram with some of the typical transitions (based on ITU-T G.781 and G.810) that a clock can pass through. This figure also shows the transitions between states and an example of a triggering event that might cause such a transition.

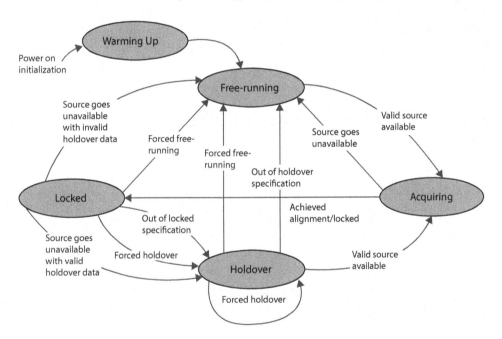

Figure 3-10 *Clock Mode States and Transitions*

Where an event is shown as "forced," it means that the event was triggered by an external event such as operator intervening and executing a control function. Some clock types might not contain every one of these states, as you will see when we discuss specific clock types, such as frequency clocks, in more detail in the following chapters.

The following sections explains the different states of the clock as depicted in Figure 3-10 and outline what those states mean and the clock behavior during the time they are in that state.

Warming Up

Oscillator characteristics such as accuracy, frequency stability, temperature sensitivity, and other characteristics are defined when the oscillator is in a steady state of operation.

Before it reaches that steady state, it is in a state known as warm-up, where the temperature, airflow, and power supply voltage have not yet stabilized.

During this time, a standalone reference clock may signal to downstream clocks that its accuracy is likely to be degraded, and that no other clocks should use its signal as a reference. As described in the preceding section on oscillators, the stability can improve even after the warm-up time has expired. The length of time an oscillator takes to reach its steady state is called the *warm-up time*. When the warm-up time has expired, then the clock may signal the quality based on the technology of its oscillator.

On the other hand, a clock that takes input from a reference source may immediately signal the quality of the reference to downstream slaves, rather than the quality of its own (still warming up) oscillator.

Free-Running

Typically, when equipment powers up, it does not have any reference clock signal or traceability to a reference clock. Free-running is a clock mode where the local oscillator runs without any reference clock as input to guide it. To be in this condition, the local clock either never had a connection to a reference clock or has lost the reference clock for an extended duration. So, the clock runs free without any attempt to control or steer it with an external control circuit (the temperature compensation circuitry of a TCXO is still functioning even in free-running clocks).

Note that the quality of the frequency from a free-running clock (there is no phase alignment) is directly correlated with the quality of the oscillator and how it reacts to changing conditions such as temperature, power supply fluctuations, and aging. A clock with an oscillator in this state will signal a quality level based on the fundamental capabilities of its oscillator. For instance, a free-running clock with a cesium oscillator should signal that it is a better clock than one with a TCXO. In fact, after the warm-up, a standalone cesium clock is only ever in free-running mode (refer to the following section "Clock Types").

Acquiring

The acquiring state is when the slave clock is attempting to align itself with the traceable reference clock. While the clock is acquiring, it may squelch the distribution of its clock signal to downstream slave clocks or signal a clock quality value indicating that it is not yet locked or aligned. Some other clocks, particularly frequency systems, might signal (following some settling time) the quality of the input clock rather than the quality of its own oscillator (like the warm-up case).

Normally, acquiring is the step between free-running state and locked state, but a clock can also drop back to acquiring from out of the locked state.

Locked Mode

After some time acquiring, the clock determines that it is well aligned to the reference and enters locked mode. In this mode, the equipment has traceability to a reference clock signal, and its oscillator is synchronized with that signal. Being locked is the normal state of a slave because aligning to the input from the reference improves the accuracy of the clock.

To stay locked and closely aligned, the control circuitry (see the section "PLLs" in Chapter 5) is constantly adjusting the clock to correct any drift from the reference. When the changes required to stay aligned are too large or too abrupt, the local clock may decide to go back and restart the acquiring process.

Holdover Mode

Consider a scenario where the equipment has been in a locked mode for an extended period. During this time, it has gathered data comparing the behavior of its local oscillator using the input timing signal as a reference. When the equipment loses all connections to a reference clock, it moves into the holdover state. The local clock uses the data previously learned during the locked period to try to maintain the alignment of its oscillator with the reference clock.

Note the difference compared to free-running state: in holdover state, the clock is trying to maintain clock accuracy by applying historical data to its oscillator, whereas with free-running, no attempt is made to train or guide the oscillator. The accuracy of a clock in holdover will generally be much better than one in free-running, at least for a while after the start of holdover.

Clocks generally signal this change in state to downstream clocks so that those clocks can decide whether to continue to be steered by this holdover clock or try to select another, better source that might be traceable to a valid reference.

If the clock in holdover state is measured against a reference clock, the frequency and phase of the clock slowly starts to drift as the local oscillator wanders off its proper frequency. This is because the oscillator in the clock is affected by its physical environment, and so the data the clock is using to steer the oscillator becomes stale and increasingly out of date. As time progresses, the accuracy progressively degrades to a point where any advantage to being in the holdover state is gone.

To guard against this degradation in accuracy being passed to downstream devices, the clock estimates how much accuracy might have degraded since it lost connection to the reference. When it decides that it is no longer accurate enough to remain in holdover, it reverts to free running (and signals that downstream).

Another way that holdover mode terminates is that the equipment reacquires the signal from a traceable reference clock and moves back to locked mode.

ANSI Stratum Levels of Frequency Clocks

Clocks distributing frequency are defined based on their performance, and there are several clock categories used throughout the frequency distribution network, and numerous standards organizations have classified these categories.

One approach, taken by the American National Standard Institute (ANSI), is to classify the performance of clocks into four levels, known as *stratum levels*, with stratum 1 as the highest and stratum 4 the lowest level of performance. Note that this is not the same definition of stratum as used by the Network Time Protocol (NTP). Chapter 2 and Chapter 7 provide more information on NTP.

For the North American market, an ANSI standard, originally called T1.101, "Synchronization Interface Standards for Digital Networks," was first released in 1987. It defined the accuracy and performance requirements for a clock hierarchy in the synchronization network. The latest version of T1.101 is referred to as ATIS-0900101.2013(R2018). Figure 3-11 illustrates the hierarchy of stratum levels from T1.101.

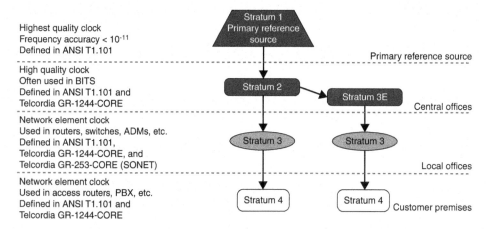

Figure 3-11 *ANSI Frequency Distribution Hierarchy*

The top of the timing pyramid is what T1.101 calls the primary reference source (PRS) or what the ITU-T SDH world calls the PRC—more on the ITU-T approach follows. The job of the PRS is to provide the reference frequency signal to other clocks within a network or part of the network. This is equivalent to a stratum 1 clock, and its output controls other clocks further down the hierarchy, namely stratums 2, 3, and 4. A stratum 1 clock and PRS both have the same accuracy, and the terms are often used interchangeably.

There is one technicality to note here because stratum 1 and PRS are not the same. The difference is that a stratum 1 clock, by definition, is autonomous (meaning it requires no input signal). This means that the normal mode of operation for a stratum 1 clock is free-running rather than locked, as it has no reference signal to acquire or lock to (being autonomous).

However, a PRS can be either autonomous or not. For example, a PRS based on a GNSS receiver is not autonomous (when the GNSS signal disappears, then the quality degrades below stratum 1), and it is not stratum 1 without the GNSS guidance (because it likely does not have a cesium oscillator). But it is *traceable* to both a stratum 1 clock and a source of UTC (in the GNSS system), and it is as accurate as a stratum 1 clock, even though it is not a stratum 1 clock.

On the other hand, a highly accurate clock based on a cesium oscillator that can meet the stratum 1 performance requirement is a combination of a PRS, a stratum 1 clock, and an autonomous clock.

T1.101 defines the stratum levels based on clock performance parameters such as

- **Free-run accuracy:** Maximum long-term deviation from the normal frequency without the use of any external reference (such as the 1×10^{-11} from PRS).

- **Holdover stability:** Maximum change in frequency with respect to time when the clock loses connection with its reference clock and tries to maintain its maximum accuracy (using data learned while being locked). Not applicable for the PRS, because it is always free-running.

- **Pull-in range:** Maximum offset between the reference frequency available to a slave and a specified nominal frequency that will still allow the slave to achieve a lock state. Not relevant to a PRS, because it has no input reference.

- **Hold-in range:** Maximum offset between the reference frequency available to a slave and a specified nominal frequency that will still allow the slave to maintain a lock state as the reference signal varies throughout its allowed range of frequency. Also, not relevant to a PRS, because it has no input reference.

- **Wander:** Long-term variations of the signal (applies to the output of a PRS) where the frequency of the variation is less than 10 Hz.

- **Phase transients:** The characteristic of the clock's time interval error (TIE) during network rearrangement or switch over between reference signals.

For stratum 1 clocks, those based on cesium oscillators are the preferred choice as an autonomous source; however, deploying cesium clocks across the network is not usually done due to the high price as well as the excessive maintenance and operational costs. Consequently, the most common PRS clocks deployed are those based on GNSS receivers (in combination with a good OCXO inside the receiver).

Stratum 2 clocks (G.812 Type II clocks in ITU-T nomenclature) are high-quality clocks that provide accurate time distribution and synchronization in digital networks and are most often found in specialized timing equipment such as BITS systems for ANSI networks and SSU for ITU-T networks. Often, stratum 2 clocks are based on a rubidium oscillator.

Stratum 3E (ITU-T G.812 Type III) and stratum 3 (G.812 Type IV) clocks are generally deployed in better-quality network equipment that is expected to give timing information to devices at the bottom of the hierarchy, with stratum 3E having better performance than stratum 3 (they are required to have better holdover performance and tighter requirements on the filtering of wander).

Stratum 4 and 4E clock systems are deployed on the end device in the synchronization hierarchy. For this level, holdover stability is not defined. Stratum 4 or 4E clock systems are not recommended as a source of timing for any other clock system.

Refer again to Figure 3-11 for a summary of the overall clock types and where they are defined and used.

Clock Types

As well as ANSI (or ATIS), the ITU-T also defines a synchronization hierarchy for timing networks. They are grouped into frequency clocks, which includes the legacy TDM circuits such as SDH as well as the newer ones based on Ethernet and the optical transport network (OTN) system of transport. This second group includes mostly packet clocks that are sometimes frequency only, but increasingly they understand phase and time in addition to frequency.

See Chapter 8 for a lot more detail on the ITU-T recommendations.

Frequency Clocks

The ITU-T recommendations G.811, G.812, and G.813 detail the clock performance requirements in their hierarchy of the frequency network. They define PRCs, synchronization supply units (SSU-A and SSU-B), and synchronous equipment clocks (SEC), respectively.

The G.811 PRC is much like the ANSI PRS, while G.812 covers the higher quality clocks equivalent to BITS for ANSI. At the lower level, G.813 covers the standard network element clocks known as SEC (like stratum 4 in ANSI). There is now a new recommendation, G.811.1, which defines a higher accuracy, enhanced primary reference clock (ePRC), specifying an enhanced level of frequency accuracy of at least 1 part in 10^{12}.

Figure 3-12 captures the synchronization hierarchy of clock quality from different types of synchronization networks, with the ANSI stratum level and SONET clock types included for reference. The ePRC on this diagram would sit above the PRC/PRS layer.

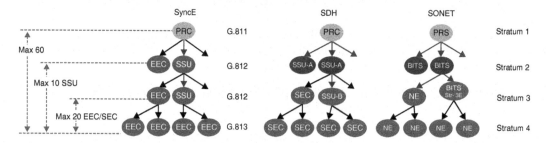

Figure 3-12   *Time Distribution Hierarchies Compared*

Note that the old version of G.812 (November 1988) refers to SSU-Transit (SSU-T) and SSU-Local (SSU-L), which are now referred to as type V and type VI clocks, but were then stratum 2 and stratum 3, respectively.

Table 3-3 provides a quick reference to the defined quality levels required to maintain the accuracy of the different types of clock when deployed in the network. The field of maximum TIE (MTIE) is a measurement of error in the signal which will be covered extensively in Chapter 5.

Table 3-3 *Performance of Clocks in Synchronization Chain*

SI Stratum Level	ITU-T Clock Level	Free-run Accuracy	Holdover Stability	Pull-in/Hold-in Range	Wander Filtering	Phase Transient
—	G.811.1	$\pm 1 \times 10^{-12}$	—	—	—	Not Applicable
1	G.811	$\pm 1 \times 10^{-11}$	—	—	—	Not Applicable
2	G.812 Type II	$\pm 1.6 \times 10^{-8}$	$\pm 1.0 \times 10^{-10}$ / day	$\pm 1.6 \times 10^{-8}$	0.001 Hz	MTIE < 150 ns
-	G.812 Type I	Not defined	$\pm 2.7 \times 10^{-9}$ / day	$\pm 1.0 \times 10^{-8}$	0.003 Hz	MTIE <1 µs
3E	G.812 Type III	$\pm 4.6 \times 10^{-6}$	$\pm 1.2 \times 10^{-8}$ / day	$\pm 4.6 \times 10^{-6}$	0.001 Hz	MTIE <150 ns
3	G.812 Type IV	$\pm 4.6 \times 10^{-6}$	$\pm 3.7 \times 10^{-7}$ / day	$\pm 4.6 \times 10^{-6}$	3.0 Hz	MTIE <1 µs
					0.1 Hz SONET	
—	G.813 Opt 1	$\pm 4.6 \times 10^{-6}$	$\pm 2.0 \times 10^{-6}$ / day	$\pm 4.6 \times 10^{-6}$	1–10 Hz	MTIE <1 µs
SMC	G.813 Opt 2	$\pm 20 \times 10^{-6}$	$\pm 4.6 \times 10^{-6}$ / day	$\pm 20 \times 10^{-6}$	0.1 Hz	MTIE <1 µs
4	Not defined	$\pm 32 \times 10^{-6}$	—	$\pm 32 \times 10^{-6}$	Not defined	Not Defined
4E	Not defined	$\pm 32 \times 10^{-6}$	Not defined	$\pm 32 \times 10^{-6}$	Not defined	MTIE <1 µs

A source of frequency, such as a PRC/PRS, will need to output some form of reference frequency signal for the downstream clocks to use as an input to guide their own oscillators. This is commonly a sine wave. More information on this signal is provided on 2-MHz and 10-MHz frequency signals in the following section "Sources of Frequency."

Frequency, Phase, and Time Clocks

When the topic moves from frequency to phase and time synchronization, the clock that provides the reference signal of frequency, time, and phase synchronization for other clocks in the network or a section of a network is known as the primary reference time clock (PRTC). The addition of the word "time" should give a hint that this source clock is phase and time aware, and not just a highly stable form of frequency.

A PRTC normally consists of some form of GNSS receiver, which may also be mated with a highly accurate PRC/PRS autonomous frequency source. The autonomous source is normally some form of atomic clock and provides an extended accurate holdover should the GNSS signal be interrupted. This is the function of the GNSS receiver, allowing traceability to a source of UTC that can also be used to calibrate the time on the atomic clock.

A standalone, autonomous atomic clock can be a PRTC, but first it needs some form of calibration to align it to UTC, because a source of frequency has no concept of the time. There is little point in having the most accurate and stable clock when it is not set to the correct time. Again, this calibration can come from a GNSS receiver or some reference signal provided as a service (in many cases) from a national physics laboratory.

In addition to frequency, a PRTC needs to be able to output a phase and time signal. More details on that topic are provided later in the chapter in the section "Source of Frequency, Phase, and Time: PRTC" that discusses the 1PPS and time of day (ToD) signals.

The ITU-T has recommendations that define the performance and characteristics of various levels of performance in the PRTC. In line with the introduction of the enhanced version of the PRC, there is now also an enhanced primary reference time clock (ePRTC) offering a more accurate source of phase/time. These clocks are specified in ITU-T G.8272 (PRTC) and G.8272.1 (ePRTC) and can be thought of as the phase/time equivalents of G.811 and G.811.1 for PRC and ePRC clocks. See Chapter 8 for more details on these standards.

Packet Clocks

In the packet-based network, a master clock is the source of time for the connected network elements lower in the synchronization hierarchy, and the packet signal generated by the master clock is provided to downstream slave clocks. These clocks are referred to in some ITU-T documents as packet-based equipment clocks (PEC)—one being the PEC-master (PEC-M) and the other being the PEC-slave (PEC-S). Refer to ITU-T G.8265 for more information.

Earlier chapters mentioned PTP and NTP as methods to distribute timing across the packet network, but there are other options available, as covered in Chapter 9. Mostly, this chapter will focus on PTP as the method to distribute time in a packet network, but be aware that there are other options. Chapter 7 covers PTP in much more detail, and so here, the basic concepts of packet timing will be outlined using PTP as an example.

Within any given part of the PTP network, there is always a single master clock that distributes the time synchronization signals to other clocks in the network. This master clock is called the grandmaster (GM) clock, and it will have a PRTC giving it a stable frequency source and traceability to UTC. The basic rule is that because it is a grandmaster, it cannot take time information from another clock (except, of course, the PRTC reference).

Be aware of the distinction between a PRTC and a GM clock. The PRTC is a source of frequency, phase, and time (aligned to UTC) and the GM is a PTP packet-timing function—it provides timestamped PTP packets. These two different roles can be housed in separate devices; combined into a single device (PRTC+GM); or embedded inside a network element such as a switch or router.

PTP defines a packet interface—called a *port*—that under normal operations performs a principal role based on its state:

- **A port in master state:** A master port on a clock transmits a packet time signal to a slave port on a downstream clock.

- **A port in slave state:** The slave port receives that signal from an upstream master port.

PTP defines any clock with a single port, master or slave, as an ordinary clock (OC). So, the GM in a PTP network is an OC, as it has only a single master port. Typically, the slave at the bottom of the hierarchy is also an ordinary clock, because it can only be a slave clock and has a single slave port.

However, there are clocks with multiple ports that support both functions—being a master clock to downstream clocks as well as a slave to upstream clocks. Such clocks are known as boundary clocks (BC). The PTP BC has the capability to recover the time received on the slave port, use it to discipline its own timing circuits and distribute it to the connected slave network elements. Figure 3-13 illustrates the various ports and clock types.

Note   There is also another PTP clock type called a transparent clock (TC), which will be covered later. Chapter 7 provides comprehensive coverage on PTP clock types, including TC clocks.

Note that the PTP port is a logical entity on a PTP clock that has a number of possible states, the most important two being whether it performs either as a master or a slave. This has little to do with the network interfaces (sometimes also called ports) on a router or other device. The PTP port is a logical entity that participates in a flow of PTP timing messages. With some implementations of PTP, the PTP port can correspond to a physical network interface, but even then, you should treat it as a separate entity entirely.

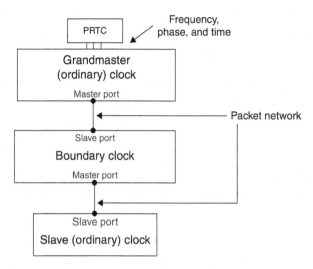

Figure 3-13   *GM, BC, and Slave Clocks with Master and Slave Ports*

Sources of Frequency, Phase, and Time

Now that the components and concepts behind timing distribution have been discussed, it is time to see how those components work together to make a functioning synchronization network. The place to start is at the beginning of the chain, where the components of the time signals are sourced.

The two sources of synchronization information are the PRC/PRS (used as sources for frequency) and the PRTC (used as a source for frequency, phase, and time). The most deployed and cost-effective versions of both device types are based on a GNSS receiver, so this section deals with the satellite systems behind them.

Satellite Sources of Frequency, Phase, and Time

Atomic clocks are very stable and accurate sources of frequency over the long term, which is why they are the basis behind all timing services available from the various GNSS systems. The GNSS systems generally work by sending out specially encoded timing signals from orbiting satellites, as well as orbital information about each of the satellites in the constellation. These signals are radio waves in several specially reserved chunks in the 1.15-GHz to 1.6-GHz frequency range in what is known as the *L band* (they are predominately one-way signals).

With a clear view of the sky, the GNSS receiver can simultaneously receive signals from multiple satellites. Knowing the position of each satellite in the sky and being able to triangulate the signals of at least four satellites based on the time of arrival, the receiver can solve equations to determine position (latitude and longitude), height, and time. The time distributed is based on a predefined timescale, which can vary for each system, but is somehow related to UTC. This timescale is kept aligned by networks of atomic clocks, both on the ground and on the satellites.

So, from the satellite signals, the PRC/PRS or PRTC with an embedded GNSS receiver can constantly determine a value of time that is traceable to a source of UTC (or allows UTC to be determined by some simple conversion). Seeing that a PRC/PRS is normally in a fixed location, further optimizations are possible to gather time from fewer satellites once the receiver position has been fixed.

Like all atomic clocks, the long-term accuracy of the received time is very stable, but over shorter observation intervals, the receiver can see significant variability (due to atmospheric effects on the radio signal reception). For this reason, GNSS receivers being used as a PRC/PRS or PRTC also contain a good quartz oscillator (such as an OCXO) to filter out (smooth) the short-term variability. Sometimes, this combination is referred to as a GDO (GNSS or GPS disciplined OCXO).

For further information on GNSS systems in general, and their use as sources of time for telecommunications networks, a reference source is the ITU-T technical report on GNSS systems, "GSTR-GNSS Considerations on the use of GNSS as a primary time reference in telecommunications." However, the following sections cover the major systems in more than enough detail to gain a good understanding of the characteristics of each.

Because of the popularity and long history of use of GPS in telecommunications, the next section covers GPSs in finer detail as an example of how all these systems work.

GPS

In 1973, the U.S. Department of Defense (DoD) approved the Global Positioning System (GPS), known as NAVSTAR GPS. One of its main goals was to unify several efforts already underway or deployed by various branches of the military. Rockwell International designed and launched the first prototype satellite in 1978, starting the system validation and design qualification process.

Launching of operational satellites started in 1989 and concluded in early 1993, with a total of 24 satellites launched, the original number required for a complete constellation. The medium earth orbit (MEO) satellites are positioned in orbit at about 20,200 Km above the earth. They are arranged into six equally spaced orbital planes in such a way that any GPS receiver on the earth has a clear line of sight to at least four GPS satellites from virtually any point on the earth and at any time. They circle the earth every 12 hours.

The system was declared to have reached full operational capability in 1995, and since then, there have been constant, ongoing improvements to it. For example, recently updated blocks of satellites have introduced additional signals (for civilian use) on new frequencies, which will allow more accuracy and higher resiliency as these satellites replace older models. Having signals in different frequency bands allows the receiver to make better estimates of anomalies in the signal suffered when passing through the upper atmosphere.

The early blocks of satellites had design lives of about 5–7.5 years, but many are lasting two to three times longer than that—there are still eight block IIR satellites operational (February 2021) with the youngest being already 16 years old. They are currently launching the third generation of craft, known as block IIIA, which should be completed by 2023. These latest satellites have design lifetimes ranging up to 15 years. There is already planning underway for the next block, known as IIIF, which should start to launch in the second half of the 2020s.

In June 2011, the U.S. Air Force completed a GPS constellation expansion by adding three more satellites to the orbit by repositioning six satellites. As a result, GPS now effectively operates as a 27-slot constellation that improves the coverage in most parts of the world. In case of any active satellite failure, there are still more satellites reserved to provide redundancy to the system. For that reason, over the past few years, there has been a total of 31 operational satellites in the constellation.

The generations of satellites from oldest to newest are IIA (now all retired), IIR, IIR-M, IIF, and III. Figure 3-14 shows how the satellites are laid out in orbit based on their positions on February 4, 2021.

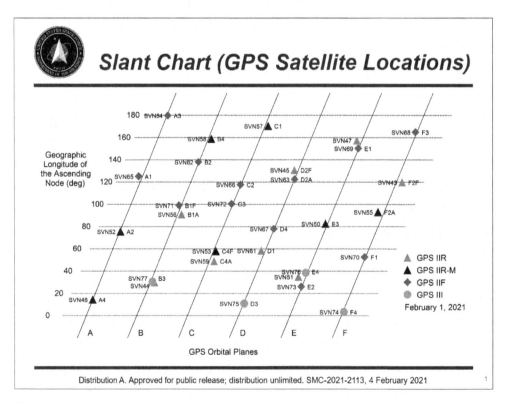

Figure 3-14 *GPS Constellation in Six Orbital Planes*

The GPS system uses its own version of time, known as GPS time, but it also carries key information in its navigation messages to allow the receiver to determine UTC from the received GPS time. The UTC steering the GPS system is based on the UTC(USNO) from the U.S. Naval Observatory (see Chapter 1 for details on UTC and UTC(k) and how GPS can give us time).

Technically, the GPS system (and most others) consists of three segments:

- **Space segment (the satellites):** At the time of writing, there are 31 operational satellites in the GPS constellation, with new Block III satellites currently being integrated into the system.

- **Control segment and monitoring stations:** There are six monitoring stations at U.S. Air Force bases around the world (at Kwajalein Island, Diego Garcia, Ascension Island, Hawaii, and Cape Canaveral, with a master control center at Colorado Springs and a backup control center at Vandenberg). There are also monitoring stations (some run by other agencies) that track the GPS satellites, process navigation signals, make measurements, and collect atmospheric data. Some other stations, with uplink capability, transmit corrections, navigation data, and control information up to the satellites.

- **User segment:** This segment includes the many devices being used for PNT. GPS is free and open for civilian use and has led to the development of hundreds of applications touching every aspect of modern life. New receivers that can take advantage of the new civilian signals are now becoming commonplace, as well as multi-constellation devices that can simultaneously receive signals from other constellations in combination with GPS.

Other GNSS Systems (Galileo, GLONASS, etc.)

As previously discussed in Chapter 1, besides the GPS system, there are another five major GNSS satellite constellations in orbit providing PNT services. This section discusses the basic features of them in turn. These systems are used, either globally or regionally, to provide timing services and are increasingly combined with one of the other constellations (normally GPS at least).

GLONASS

The Soviet Union started launching satellites for the GLObalnaya NAvigazionnaya Sputnikovaya Sistema (GLONASS, or Global Navigation Satellite System) in 1982. Like GPS, it is designed for global coverage and was the second GNSS system deployed after GPS. It is now controlled by the Russian Federation and achieved a full operational constellation in 1995.

Following the collapse of the Soviet Union, GLONASS fell into disrepair owing to a lack of resources, so that by the mid-2000s it was barely functional. With renewed investment in the system, it was slowly rebuilt back to a full constellation and has remained operational since 2011. It is probably the second most popular GNSS system after GPS (although that may change with Galileo).

GLONASS uses 24 MEO satellites in three different orbital planes offering global coverage (and better coverage at the poles than GPS). Most GLONASS receivers also support a combined GPS+GLONASS dual mode for enhanced operation.

The timescale for GLONASS is based on UTC(SU) from Moscow and uses Moscow Standard Time (MSK) of UTC + 3 hours, and like UTC, it also implements leap seconds. Russia has worked over the last few years to massively improve the alignment of UTC(SU) with UTC, so that at the time of writing it is generally less than 3–4 ns.

BeiDou Navigation Satellite System

The Chinese government developed and launched the first generation of BeiDou as a regional system, using geostationary earth orbit (GEO) satellites, in 2000. They declared it operational in 2003 and decommissioned it in 2012. It was basically an experimental system.

In 2006, the Chinese government officially announced the development of the second-generation GNSS called BeiDou-2 (also known as Compass), which ended up with 16 satellites launched. The constellation consisted of five GEO satellites, five inclined geosynchronous satellite orbit (IGSO) satellites, and four MEO satellites This system offered open PNT services with regional coverage to the Asia-Pacific region centered on China's longitude.

In 2015, the Chinese government started working on a third-generation system called BeiDou-3, which has added a constellation of 24 MEO satellites in three orbital planes. This system offers global coverage with a constellation of 35 satellites (in three separate constellations in MEO, IGSO, and GEO orbits) and was declared to be operational in 2020. Most BeiDou receivers support a combined GPS+BeiDou dual mode for enhanced operation, and since BeiDou-2, the system also offers space-based augmentation capabilities (discussed in the upcoming section "Augmentation Systems Enhancing GNSS").

The BeiDou system takes its time, BeiDou time (BDT), from China's National Time Service Center (NTSC), with an epoch of 1 January 2006. BDT is a continuous timescale (no leap seconds) steered to within 50 ns of UTC. BeiDou broadcasts leap second information in the navigation message, allowing receivers to determine UTC. BeiDou also broadcasts offsets between BDT and time from GPS, Galileo, and GLONASS.

Galileo

The European Union, developer of Galileo via the European GNSS Supervisory Agency (GSA), started launching experimental satellites in 2005, followed by operational satellites in 2011. In December 2016, the GSA declared initial operating capability, but the system has not yet reached full operational status. Galileo, like some other systems, also has a strong search and rescue (SAR) capability on board (to receive signals from emergency distress beacons).

Galileo is designed to offer global coverage using up to 24 MEO satellites in 3 different orbital planes with up to 6 spares. As of this writing, there are 22 active operational satellites in the system with 4 more planned to launch in 2020–2021 and 8 more after 2022. As with other systems, there is a significant ground-based control and monitoring segment.

Galileo system time (GST) is a continuous timescale (no leap seconds) maintained by the Galileo Control Segment and synchronized to UTC(k) with a nominal offset below 50 ns. The GST epoch started at 00:00:00 UTC on Sunday, 22 August 1999.

Galileo also broadcasts a signal indicating the difference between GST and UTC as well as the difference between GPS time and GST—a parameter known as the GPS-Galileo time offset (GGTO). This GGTO has a performance goal to be less than 20 ns for 95% of the time. It also broadcasts leap second information in the navigation message, allowing receivers to determine UTC.

The goal is for the recovered time offset from UTC to be less than 30 ns for 95% of the time. At the time of writing, the current performance is that the 95th percentile of the daily average of the disseminated UTC accuracy is less than 15 ns. Like other systems, most Galileo receivers also support combined GPS+Galileo dual constellation mode for enhanced operation.

IRNSS

In 2013, the Indian government started launching the Indian Regional Navigation Satellite System (IRNSS), a regional system centered on India, using three GEO and four IGSO satellites located above the subcontinent. It is now increasingly referred to as "Navigation with Indian Constellation" or NavIC.

IRNSS offers position accuracy of better than 20 m and time accuracy better than 40 ns (using dual frequencies). After the network timing facility became operational in 2018, the goal is to have the timescale traceable to within 20 ns of UTC. The source of UTC for the system is the National Physical Laboratory India (NPLI). The IRNSS System Time epoch is 00:00:00 UT on 22 August 1999, and the system transmits offset information against UTC(NPLI), GPS, and other systems (Galileo and GLONASS).

After some problems with onboard clock failures (also suffered by Galileo), the first failed satellite was eventually replaced in 2018, with five more launches of IGSO satellites planned.

QZSS

The Japanese government started launching the Quasi-Zenith Satellite System (QZSS) in 2010 and placed it into full operation in 2018. QZSS currently offers local coverage with one GEO satellite and three other satellites in a quasi-zenith orbit (QZO) above Japan, Southeast Asia, and Australia. This orbit means that when viewing from southern Japan, the three QZO satellites appear to fly very slow figure-eight patterns, from low in the southern sky to directly overhead (which is exactly where you would want them when using the system between skyscrapers).

QZSS is designed to be compatible with GPS, allowing access to low-cost receivers. The goal of the system is to improve the accuracy of GPS and to address common coverage problems in Japan, such as reception in deep urban canyons. It also augments the GPS reception for single-frequency receivers by allowing atmospheric anomalies to be better modeled for that region rather than globally. In future, probably by 2023, QZSS will operate with seven satellites and will include signals to aid in spoofing detection.

The QZSS system has also assumed the responsibility to transmit the regional Japanese augmentation system on their GEO satellite (see the following section on augmentation).

Multi-constellation

By combining even just the four global systems, GPS, GLONASS, BeiDou, and Galileo, there are over 100 non-geostationary satellites available today, with more being planned. This large number of satellites generally provides the opportunity to deploy multi-constellation and multiband receivers. Users will benefit from the following:

- Ability to compare data across systems to verify signals against each other

- Each system aligned with UTC (and many broadcasting offsets from UTC) allows the receivers to increase the accuracy of their recovered time

- More data points and signal bands to reduce position and timing errors from atmospheric conditions

- Improved coverage by having more satellites distributed all over the sky

- Increasing resistance to spoofing

Already, receivers, even in mobile phones, are increasingly multi-constellation, and this trend will only increase, especially for specialized timing equipment.

Augmentation Systems Enhancing GNSS

Although not widely used in timing applications, there are several other systems, for example space-based augmentation systems (SBAS) that enhance or *augment* existing GNSS systems. This enhancement can help guarantee signal integrity, enhance accuracy, or improve performance to cover special applications, such as guiding aircraft to the runway, docking ships, or precision farming. There are some other local, ground-based augmentation systems, such as differential systems that increase the accuracy for ships navigating around harbors, but they are beyond the scope of this book.

These augmentation systems are generally regional in nature and therefore based on a small number of GEO satellites (that appear fixed in the sky) with numerous monitoring stations on the ground to observe the original GNSS signals and upload correction data. Figure 3-15 illustrates how these systems function using GPS as an example.

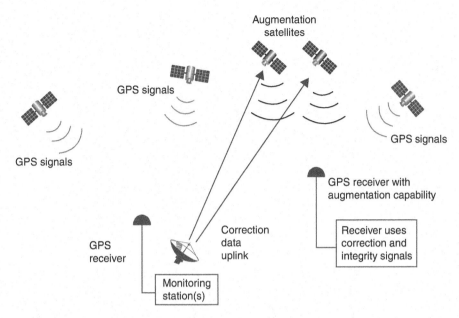

Figure 3-15 *Basic Mechanisms for Space-Based Augmentation Systems*

Almost all these systems currently augment the GPS system and so consist of numerous ground stations widely positioned around the coverage area, which continuously monitor GPS signals received from space. These locations are precisely surveyed, so the monitoring receivers know exactly the solution that they should be calculating from GPS. Any anomalies from the ideal solution can be determined, correction data calculated and transmitted to the augmentation satellites.

Those GEO satellites then broadcast this "correction" signal to the users on the ground (or aircraft trying to get onto the ground) as a "GPS-like" signal. GPS receivers equipped with the additional augmentation capability can use this data to fine-tune the solution they have calculated. This allows the receivers to be much more confident about the signals they are receiving and improves accuracy because ground stations are giving corrections to any position errors caused by, for example, changing atmospheric conditions.

This especially helps aircraft, because most GNSS systems are not as accurate for positioning in the vertical plane (determining the height) as they are in the horizontal plane (latitude and longitude). But to use GPS for aviation, having an accurate value for height is critical, even more so when using it to approach an airport runway.

Finally, these systems give very fast feedback (in the region of a few seconds) to the user if the GPS was somehow broadcasting information that could lead to hazardous positioning errors. This inspires confidence in using GNSS systems for "safety of life" applications. Because of these needs for the aviation industry, most of these SBAS systems have regional or national civil aviation organizations as partners.

The two major space-based augmentation systems are the Wide Area Augmentation System (WAAS) from the U.S. Federal Aviation Administration (FAA) and the European Geostationary Navigation Overlay Service (EGNOS) from the GSA (on behalf of the European Commission). Both systems are heavily involved in enhancing navigation and safety for aircraft operations. These two systems, and others in various stages of deployment, are designed to be interoperable—so that aircraft equipped with one system can use another when in the airspace of another regional system.

The only disadvantage to these GEO augmentation systems is that the satellites, being over the equator, are quite low to the horizon when the receiver is at high latitudes (toward the poles). This makes their use on the ground less than ideal because of obstruction to the line of sight from the satellites (something QZSS has addressed for Japan). However, to assist in those cases, the data is made available in other forms, allowing devices to access the same information from another source, such as the Internet. The signals are mostly fine at the altitudes used by commercial aircraft—anyway, there is only a limited amount of airports in the polar regions.

WAAS is a system of three GEO satellites anchored over the equator above the Pacific Ocean, working with GPS data collected from 38 ground monitoring stations. WAAS is used to provide a GPS-enabled version of an instrument landing system across almost all North America as well as improve navigation across much of South America.

EGNOS is a system of three GEO satellites stationed over the equator (above Africa), and 40 monitoring stations placed in more than 25 countries (not just in Europe).

India's GPS-Aided GEO Augmented Navigation (GAGAN) is a system of three satellites stationed over the Indian subcontinent offering augmentation services for that region. GPS receivers with GAGAN capability use basically the same technology as WAAS.

Japan also has a system based on GEO satellites (most recently MTSAT-2 or Himawari 7) and was known until recently as the MTSAT Satellite Augmentation System (MSAS). However, with the recent decommissioning of MTSAT-2, the MSAS signals have been migrated to the QZSS *Michibiki* geostationary satellite (QZS-3) and so now MSAS stands for Michibiki-Satellite-based Augmentation System.

There are plans from 2023 to add MSAS to a further two QZSS GEO satellites and plans in the future to eventually integrate IGSO QZSS satellites and be able to augment all the major GNSS systems.

Russia has a system called System of Differential Correction and Monitoring (SDCM) that uses three geostationary satellites to provide an augmentation service for GLONASS as well as GPS.

These last three systems, GAGAN, MSAS, and SDCM, provide a similar service to WAAS and EGNOS. There is currently very limited coverage over Africa and South America.

There are several other systems in various stages of development to cover other regions of the planet including the *Korea Augmentation Satellite System* (South Korea), *Agency for Air Navigation Safety in Africa and Madagascar* (ASECNA) for the Indian ocean and Africa, BeiDou SBAS for China, and *Southern Positioning Augmentation Network* for Oceania.

Sources of Frequency

This chapter has gone into considerable detail about the sources of time and frequency and how it is distributed via satellite systems. So, the next step to cover is how time is received and shared at the place it is needed, such as inside the core of an operator's network, or even on top of a radio cell site.

PRC/PRS

As previously described, a source of frequency is the PRC/PRS or stratum 1 clock. This device is either an atomic clock combined with an OCXO or a GNSS receiver combined with an OCXO. Sometimes the two are combined to allow autonomous operation and long holdover performance in case of GNSS failure.

Figure 3-16 shows the typical frequency sources (photos courtesy of Microchip).

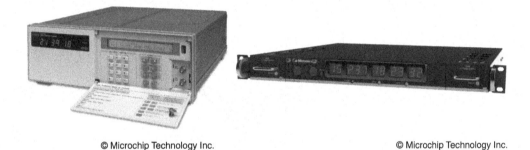

© Microchip Technology Inc.   © Microchip Technology Inc.

Figure 3-16   *Examples of Frequency Sources, Cesium Clock (L) and GPS Receiver (R)*

The atomic clock PRC might have some input ports on it to connect other timing equipment, and the GNSS receiver might have an input to allow it to be guided by an atomic clock (see the following section "10-MHz Frequency").

The GNSS receiver always has an antenna port, which is commonly a subminiature version A (SMA), or a bayonet Neill-Concelman (BNC) connector. It can also be a threaded Neill-Concelman (TNC) connector, which is a threaded form of the BNC. This antenna port commonly supplies a DC voltage up the antenna cable to a pre-amplifier in the base of the antenna (to boost the signal over the length of the antenna cable). Both the TNC/BNC input/output and SMA input connectors on the frequency source are normally female.

Note   A PRC/PRS, although it may not be phase aligned with UTC, may also include the capability to output a 1PPS phase signal. Some devices can recover frequency from this single pulse, or perhaps a network that requires phase synchronization but does not need it aligned with UTC can use it.

The ITU-T has specified performance characteristics for the PRC in G.811 and has recently added the ePRC performance level to that in G.811.1. For more details on these recommendations, see Chapter 8.

Now that the timing distribution network has a source of frequency, how is it used?

10-MHz Frequency

The frequency reference signal that is output from a PRC/PRS is normally a 1-, 2-, 5-, or 10-MHz sine wave, with 10 MHz being the most common. However, 2 MHz is also a popular choice, especially outside North America. Typically, these devices may offer optional modules or product variants that support other types of signal outputs, such as E1/T1.

These reference signals not only are used to frequency synchronize (*syntonize*) networking and telecommunications equipment but are also applied to other applications. As illustrated in Figure 3-17, any device whose measurement accuracy depends on using an oscillator, or counting over time, will normally make provision for a 10-MHz reference signal as input (photos courtesy of Microchip).

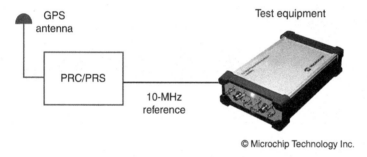

© Microchip Technology Inc.

Figure 3-17   *10-MHz Reference Signal Calibrates a Measurement Instrument*

Some examples include test equipment such as signal generators, spectrum analyzers, frequency counters, oscilloscopes, and so on. This is not surprising, because how can a device accurately measure the frequency of a test signal when its own oscillator is in free-running state?

Of course, these signals are also very important sources to have available when it comes to testing the performance of a timing distribution network, as will be seen in later chapters on testing and deploying timing.

Sine wave reference signals are carried over short distances (less than a few meters) using 50-Ohm connectors and coaxial cable. The connectors for output on the PRC/PRS sources are frequently BNC (female) *jacks*, and larger devices and benchtop equipment will commonly also have BNC (also female), so the cabling is straightforward between them.

The disadvantage to having BNC connectors on (for example) a router or switch is that the BNC connector is very large. Real estate on the front of a switch is normally needed for other purposes, such as network ports or even as holes to allow front-to-back airflow. Many times, engineers will select other, smaller connectors than BNC to save real estate on the faceplate. In that case, some form of dongle or convertor cable will be required between the PRC and other equipment. For example, the 10-MHz input on most Cisco routers is known as a DIN 1.0/2.3 connector that has a push-pull snap on mechanism.

Figure 3-18 shows the route switch processor (RSP) from a Cisco router with two golden DIN 1.0/2.3 connectors for 10 MHz, one for input and one for output (highlighted). The other two DIN 1.0/2.3 connectors are for phase; see the following sections "1PPS" and "Time of Day".

Figure 3-18   *DIN 1.0/2.3 Connectors for 10-MHz Input and Output*

There are recommendations specifying the physical and electrical characteristics of the 10-MHz interface in clause 20 of the ITU-T Recommendation G.703.

2-MHz Frequency and E1, T1 Legacy Circuits

Although the preceding section on 10-MHz frequency mentioned that a PRC/PRS may output a 2-MHz signal and the connector may be BNC, there is another form of this signal. Rather than using a 50-Ohm cable with BNC connectors, an alternative is to use RJ-45 connectors with 120-Ohm E1 cable, which takes up a lot less space on the face-plate. Many routers have a BITS port that will accept this signal using a short run of twisted-pair cable (in Figure 3-18, see the RJ-45 port labeled BITS, second from right). So, the RSP in Figure 3-18 will accept both an input 10-MHz reference signal on the DIN port and/or a 2-MHz signal on the BITS port.

It is possible to buy equipment, called a *balun*, that allows the conversion and impedance matching between the 120-Ohm twisted-pair and the 50-Ohm coaxial versions. Cabling and connector choices are always an issue when working with these signals.

SDH/SONET also provides the capability to multiplex and transport both 2048-kbps based hierarchy (E1, E2, E3, etc.) and 1544-kbps based hierarchy (T1, T2, T3, etc.), and these circuits are also used to carry timing (sometimes without data). Therefore, it is not uncommon for a PRC/PRS to have E1/T1 outputs available or as an option, although it is much more common on SSU/BITS equipment.

SSU and BITS

Chapter 2 already introduced the basic concept of the dedicated timing distribution network deployed solely to synchronize the SDH/SONET network. Figure 3-12 in the earlier section "Frequency Clocks" showed the SSU/BITS network as a clock type with a middle level of performance between the PRC/PRS and the "normal" SEC clocks. These SSU/BITS devices use a PRC/PRS as an input source of frequency or may be combined with a PRC/PRS in the same equipment.

Besides performance, the other place where they differ from a PRC/PRS is that SSU/BITS devices are designed to distribute frequency signals to a very large number of downstream clocks. Frequently, they will have several extension shelves with output ports to distribute their signal. Beyond the 2-MHz option covered in the previous section, these output signals will be normal E1 or T1 signals.

The format and range of connector types for SDH/SONET networks is beyond the scope of this book, but modern systems use 100-Ohm (T1) or 120-Ohm (E1) twisted-pair cable with an RJ-45 style connector (technically, the plug is named RJ-48c).

Obviously, the future for this type of equipment is limited with physical distribution of frequency going to packet systems (PTP), SyncE, or optical systems like OTN; however, many service providers still maintain a large and widespread network of SSU/BITS equipment.

Source of Frequency, Phase, and Time: PRTC

The PRTC differs from the PRC/PRS in that it is also concerned about providing phase and time of day as well as frequency. Like the PRC/PRS, the PRTC provides reference timing signals to other clocks in the network hierarchy. Unlike most PRC/PRS systems, especially those based on autonomous atomic clocks, the PRTC normally provides (and requires) traceability to some source of UTC. This means that the PRTC is always relying on some form of signal that aligns it to UTC.

Primary Reference Time Clock: PRTC

There are two common implementations of a PRTC, either a standalone GNSS receiver or an atomic clock calibrated and aligned to UTC by an external signal. This signal can be from either a local GNSS receiver or an external source, typically an atomic clock at a national time laboratory (which is tracking UTC). See the section "White Rabbit" in Chapter 7 for details on one method to do this.

The time signals output by these devices (beyond the frequency requirements) are expected to be accurate to within ±100 ns of UTC. Up until a few years ago, this was the only level of performance, adopted only because it aligned with the accuracy specified by GPS. With better technology (and improvements to GNSS systems) becoming available, the ITU-T has now defined additional categories for the accuracy of PRTC devices.

Currently, ITU-T defines two levels of PRTCs in G.8272, and new ePRTCs in G.8272.1. The original PRTC specification of ±100 ns was retained but renamed to class A performance (PRTC-A). A new class B (PRTC-B) was added, improving that accuracy to ±40 ns. This improvement is achieved (chiefly) by using better GNSS receivers, typically ones that have the capability of dual-band reception. For example, with GPS, this means a device that receives signals on both the L1 and L2 (or other) bands, but all GNSS systems have an equivalent facility. Note the difference between multiband and multi-constellation.

To minimize environmental factors, such as temperature influence on the clock, the PRTC-B is only intended for locations where it is possible to guarantee satisfactory environment conditions, which means controlled temperature or air-conditioned indoor locations.

Note that there are no long-term holdover requirements in G.8272 for either PRTC-A or PRTC-B. Once the PRTC loses all its input phase and time references, then it enters phase/time holdover, and the clock starts to drift. To attempt to keep the phase/time aligned, the PRTC must rely on the holdover performance of its local oscillator, which may be good but not great.

Many vendors selling dedicated PRTC clocks offer an optional high-quality oscillator (frequently rubidium) to improve the holdover performance. Another option is to connect an optional external input frequency reference obtained from a PRC/PRS. So, many PRTC devices can accept an input 10-MHz signal from a PRC/PRS, which can be used to improve holdover.

The other level of performance available for time sources is the enhanced version, the ePRTC. The level of accuracy required to meet this level of performance is defined as ±30 ns, considerably better than PRTC-B. But unlike the other two cases, the ePRTC has a strict holdover requirement in ITU-T G.8272.1.

To meet the ePRTC levels of performance in a GNSS receiver typically requires the addition of a very good PRC (such as a cesium clock) as a frequency source for the receiver. In this case, when the source of phase/time is lost, the PRC can provide the stability necessary to meet the holdover that the ePRTC is required to meet. In fact, G.8272.1 defines two classes of holdover, with ePRTC becoming ePRTC-A and a new ePRTC-B class of holdover performance (at time of writing, the values are not yet decided). Table 3-4 summarizes the different classes.

Table 3-4   *Classes of PRTC Performance*

Class of Clock	Accuracy	Holdover
PRTC-A	±100 ns	None
PRTC-B	±40 ns	None
ePRTC-A	±30 ns	100 ns / 14 days
ePRTC-B	±30 ns	For further study

Beyond the frequency output (see the preceding section "Sources of Frequency") the PRTC is also required to supply timing signals for phase and time. This is supplied as two additional physical signals in addition to the frequency output, although some implementations multiplex these two signals together.

Other implementations use different pins on a single cable (such as V.11 and RJ-45) to run more than one signal together (commonly 1PPS and ToD). There are ITU-T standards for some of these signals, but there are also many historical and proprietary signals. One example of the former is the Inter-Range Instrumentation Group (IRIG) signal types frequently used in the power industry.

The other option to output these signals is to use PTP to carry the time and phase and optionally the frequency. The frequency, if not carried by the PTP, can be carried with SyncE or as a separate signal (for example 2 MHz or 10 MHz). If so, then these devices must be equipped with an Ethernet packet interface to carry the PTP and optional SyncE. This makes the PRTC simultaneously a grandmaster and PRTC (sometimes written as PRTC+GM).

For the telecoms market, many implementations of PRTC are based on a GNSS receiver, with an embedded GM that uses SyncE for frequency. Their PTP output is compliant with one of the telecom profiles of PTP, and so they are known as Telecom Grandmasters (T-GM) and combined as PRTC+T-GM sources. They are available in PRTC-A or PRTC-B performance levels and come in many form factors, even down to a small form-factor pluggable (SFP). They can be deployed in a central location with a PRC as backup or versions with fewer features (cheaper) can be placed out at the cell site.

Chapter 9 and Chapter 11, "5G Timing Solutions," provide detailed coverage of options and trade-offs to consider when selecting and situating a PRTC to best meet your specific needs.

1PPS

As introduced earlier in the "Clocks" section, 1PPS stands for 1 pulse per second and is an electrical signal used to carry phase. It is most commonly a square wave or, alternatively, a pulse of configurable length (between 100 ns and 500 ms). As the name suggests, the signal has a period of 1 second and the midpoint of the leading edge is defined as the significant instant.

Clause 19 of the ITU-T recommendation G.703 specifies the physical and electrical characteristics of the 1PPS interface. The details are specified for both a 100-Ohm V.11-based signal (with an RJ-45) and a 50-Ohm signal (for coaxial cable of less than 3 m length). There are additional details for the 100-Ohm V.11 version in Annex A of ITU-T G.8271.

The connectors for the 50-Ohm version are not specified, but as with 10 MHz, BNC connectors are typically used on the PRTC side. For similar reasons to the 10-MHz case, the connector on the equipment connected to the PRTC can vary considerably. For example, as was shown in Figure 3-18, most Cisco routers have the DIN 1.0/2.3 connector for 1PPS as well as 10 MHz, but several other connectors are used by other companies and industries.

The 1PPS connector has a very important function as a *measurement* interface, meaning that it allows a quick method of checking the phase offset of a device compared to some source of phase. Figure 3-19 shows how the 1PPS port is used to measure phase alignment.

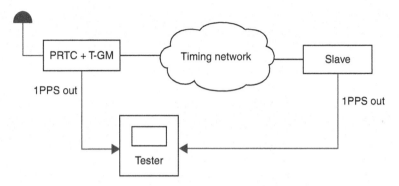

Figure 3-19   *Using 1PPS as a Measurement Port*

Obviously, in this case, the devices must be close to each other because of the distance limitations of the 1PPS signal. Field test personnel use an alternative approach when testing equipment in remote locations. They use a portable tester with its own built-in GNSS receiver, so that the test device can recover the phase from the GNSS radio signal and compare the offset to the pulse on the cable from the 1PPS output port.

This is an important function and one reason why many timing-aware devices have 1PPS output capability as well as the ability to accept 1PPS input. Remember that the 1PPS signal is not used in TDM (SONET/SDH) networks or in SyncE, because they are both only concerned with frequency.

Time of Day

The ToD signal is made up of a string of characters that encodes the time in a manner that allows the receiving device to reconstruct UTC. So, while 10 MHz delivers frequency and 1PPS indicates the instant when the second commences, the ToD says what is the actual second that has just started.

This signal often uses ITU-T V.11 (RS-422) or something similar like RS-232, transmitting ASCII, binary coded decimal (BCD) strings, or timestamp information over short-distance cabling. But the main problem with delivering ToD is agreeing upon the format and content of the messages carrying the date and time information. Transferring the time of day between machines is a long-running need, and so there are many historical versions of this signal, including IRIG and those from the National Marine Electronics Association (NMEA).

There are also agreed-upon formats available for the ToD message depending on the industry. Many companies and organizations have their own format, for reasons going back many decades. For example, NMEA sentences are used to transfer time (plus position and a lot of other information) between navigation systems (such as Long-Range Navigation [LORAN] or GPS receivers) and other equipment on the bridge of a ship. For telecommunications and packet-switched networks, one frequently encounters UBX (defined by the uBlox company), various NMEA sentences, NTP driver format, and even a Cisco-developed format.

Obviously, there are considerable interoperability issues with having so many electrical and physical versions of this signal, without even considering message format. For this reason, the ITU-T has developed a recommendation to base ToD on a V.11 signal with defined message formats. The details of this format for ToD, as well as the 1PPS signal, are outlined in Annex A of G.8271.

The ITU-T specifies that the ToD signal should not start before the 1PPS it represents, should take no longer than 0.5 seconds to transmit, and be sent once per second. They define three message types, but only one, the Time Event Message, is used to transfer time. The time is sent as a PTP/TAI timestamp which is a 48-bit unsigned integer representing the number of seconds since the PTP epoch (1 January 1970). The message also contains enough additional information (for example, leap second offset) to be able to reconstruct UTC.

The ToD signal is not used in TDM (SONET/SDH) networks nor in SyncE, since they are both only concerned with frequency.

Timing Distribution Network

Until now, this chapter has covered all the components needed to build a timing distribution network, so now it is time to see how the timing signal can be transmitted from its source to where it is needed. This chapter has shown some physical techniques using short cable runs to distribute time signals locally to a node. But of course, it is possible to use digital technologies to distribute frequency and phase/time over a network. For simplicity and cost reasons, it is also an advantage to carry the timing information using the same network as that used to carry data.

Previous generations of networks were mostly synchronous in nature, for example SDH/SONET (but there are more), and so they required frequency synchronization. Usually, they derive frequency either from the carrier signal itself, or from some time code or other information modulated onto the carrier. But to carry accurate phase and time over the wide area was more difficult.

It becomes even more complex because digital networks can be based on a wide array of media: coaxial cable; twisted copper pairs; optical fiber; radio signals; and so on. And these different methods use different technologies to encode and transport signals. And so, engineers needed to design frequency and phase/time solutions for many different types of transport.

But to support the evolution of modern applications, it became increasingly necessary to distribute phase and time as well as frequency on packet-based networks. The time of day (of medium accuracy) was already carried using a packet mechanism like NTP, but phase timing, especially accurate phase timing, is much more challenging.

Transport of Time and Sync

Distributing timing over various transmission media can be accomplished in several ways. There are mainly three categories of solutions available for time synchronization distribution:

- **Physical layer:** Frequency distribution in SDH/SONET and SyncE, as well as digital subscriber line (DSL), passive optical network (PON), DOCSIS cable, and so on

- **Packet-based (Layer 2 and Layer 3) distribution:** NTP, PTP, White Rabbit, adaptive clock recovery (ACR), and so on

- **Radio or satellite navigation systems:** GNSS systems, but also other radio systems such as LORAN-C, IRIDIUM, WWVB, DCF77 and many others

The GNSS and radio systems have been treated extensively above as a source of time, but this section outlines the different approaches and the tradeoff with each of them.

Physical Layer

In this solution approach, nodes in the network hierarchy carries the synchronization signals over in layer 1, as part of the communications method itself. It has been shown that legacy TDM and SDH/SONET networks carry their frequency reference at the physical layer. It uses a hierarchy of clocks to distribute network synchronization from the PRC/PRS down to the network element clocks.

As shown previously in Figure 3-13, the highest accuracy clock, the PRC/PRS, is deployed at the top of the hierarchy. The next level of hierarchy is the SSU/BITS layer that supports better holdover and stability. The third level is SEC or SONET minimum clock (SMC). And this three-layer hierarchy will deliver a frequency that is traceable to the PRC/PRS, so long as the path is not broken.

In this hierarchy, the PRC and SSU/BITS are standalone timing elements, while SEC/SMC devices are implemented as a component inside the network elements.

SyncE has been designed along the same principles as SDH, where the Ethernet physical layer provides traceability to a PRC/PRS frequency reference. Because the first version of Ethernet was originally designed for asynchronous data traffic; it was initially not suitable to carry synchronous signals. In fact, the original 10 Mbps Ethernet cannot provide a synchronization signal over physical layer because it periodically stops transmission.

These mechanisms are only used to carry frequency. There are systems that use physical signals to carry phase (you could say that 1PPS is one) although they have problems over distances longer than a few meters.

There is a lot more detail on physical frequency distribution in Chapter 6.

Radio Distribution Systems

Quite early in the twentieth century, techniques to distribute time over radio signals were developed as engineers discovered that low frequency (LF) radio signals (30–300 kHz) are ideal for broadcasting time signals. Today, many countries use LF and high frequency (HF) radio signals to offer time and frequency synchronization at a regional level.

The original purpose for these clocks was to use them as a frequency reference, although this function has been increasingly replaced by GNSS-based systems. Historically, these signals were used as a reference to synchronize the frequency of radio transmitters. The radio transmitters belonging to broadcasters needed to be calibrated to stop them drifting off frequency.

There were also such things as radio-disciplined oscillators, much as GNSS-disciplined oscillators are used today. Besides frequency, the signals were also used as a reference for time intervals, but in many cases, the ability to broadcast actual time of day signals was added later (starting in the 1960s). Of course, even with these new functions, they continue to broadcast on a very strictly controlled frequency to allow their signals to be used as a frequency reference.

As one example of many, in the United States, the *National Institute of Standards and Technology* (NIST) operates a LF radio station (WWVB) from Fort Collins, Colorado at a frequency of 60 kHz. It broadcasts a time code that contains the year, day of the year, hour, minute, second, and values that indicate the status of the daylight-saving time, leap years, and leap seconds. The other systems around the world work similarly, another example being the DCF77 transmission on 77.5 kHz, covering much of Europe from Frankfurt, Germany and the MSF transmission on 60 kHz covering the United Kingdom.

It was known that radio-based time signals are delayed when they travel the path from a terrestrial transmitter to a terrestrial receiver, and the accuracy of the received signal can be no better than the knowledge of the path delay. As seen with GNSS systems, these path delays are difficult to measure due to continuously changing ionospheric conditions.

Despite all this, the recovered time accuracy for a clock with good reception could be around 5–10 μs. Of course, this does not take account of propagation delay, which would need to be corrected to gain high accuracy. In practical use, instrumented and calibrated receivers can readily achieve in the low milliseconds of accuracy for phase/time. For frequency, one example is that a receiver tuned to WWVB, with one day of averaging, could achieve a frequency uncertainty of less than 1 part in 10^{-11} across most of the United States.

Various consumer electronic products like wristwatches and wall clocks can receive and decode the time signals, as can public radio-controlled clocks, such as those at railway stations. There are also many other uses such as synchronizing all manner of industrial devices.

However, as has been shown, time and frequency are now distributed globally using GNSS systems. At almost any location on the planet, GNSS and other radio receivers provide a source of frequency and time, in a topology sometimes referred to as a distributed PRTC. Figure 3-20 illustrates how this works in practice.

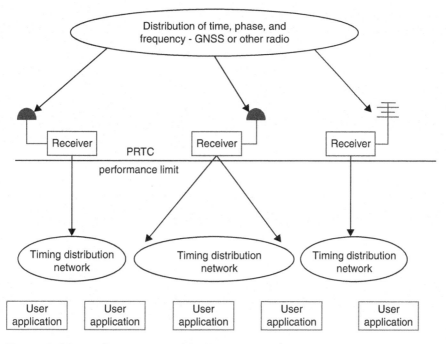

Figure 3-20 *Radio Waves Used for Timing Distribution*

One of the main advantages of the distributed-PRTC approach is that the reference time is available globally, and it is a flat distribution method without any hierarchy and hence it provides a simple and loop-free design.

The main disadvantages of this approach are that it requires infrastructure everywhere and has the usual issues inherent with any radio reception. Issues specific to GNSS have already been covered—it requires an antenna with a wide-angle view of the sky as well as protection against lightning and weather; it is sensitive to signal jamming and can suffer signal degradation problems due to poorly routed antenna cables.

Resilience

One radio-based navigation system that started during World War II is the LORAN system, which evolved with several generations of signals, finally ending up as LORAN-C. Over the last 10–20 years, LORAN has been undergoing decommissioning, and many of the remaining transmitters (at least outside the far east FERNS timing chain) were shut down at the end of 2015.

However, during the time that its use was declining, there were proposals to upgrade this system into what is referred to as enhanced LORAN (eLORAN). eLORAN would use a similar approach to the LORAN-C system, with very high-power LF (90–110 kHz) transmitters broadcasting signals for PNT services. This would allow the users' receiver to have access to a signal millions of times stronger than GPS and available in demanding locations, such as inside buildings or even underground. Testing was carried out in the United States, and some pilot stations even became operational in the United Kingdom.

The idea was to use eLORAN as a backup to space-based GNSS systems in case of any outages caused by events such as jamming or space weather. It was planned for eLORAN to broadcast signals that would allow receivers to recover UTC time with an accuracy of 50 ns, as well as meet the stratum 1 frequency standard.

Such a system has the following very definite potential advantages for such a role:

- The high power (hundreds of kilowatts) would make it almost impossible to jam.

- The long wavelength would enable it to propagate very well around the earth and easily penetrate buildings, making an outside antenna unnecessary.

- Its terrestrial base would (hopefully) make it more resistant to space-based weather events.

Alas, the decommissioning of the French and Norwegian stations made the continuing pilot in the United Kingdom untenable, and so the most promising initiative to adopt it widely was abandoned.

Despite that, some countries, such as South Korea, who feel especially vulnerable to GNSS interruptions, have continued to develop a PNT alternative using eLORAN. During 2019, the UrsaNav company, working for the Korea Research Institute of Ships and Oceans Engineering, installed a temporary eLORAN test transmission system.

There also are plans by the Korea Ministry of Oceans and Fisheries to convert their two existing LORAN stations to eLORAN. They also intend to deploy differential eLORAN stations to improve accuracy down to about 5 m within a local reception area (within 50–60 km of the differential station).

There are also reports that the Chinese government (and others) continue to work on either deploying new transmitter stations or converting existing ones to eLORAN, although there is not much information publicly available.

Elsewhere, the timing and telecommunications community, especially in the United States, has lobbied politicians to take an interest in legislating some backup alternative to the U.S. GPS system. After several false starts and legislative attempts, in late 2018, President Trump signed the *National Timing Resilience and Security Act of 2018*, which seeks to address the problem.

The president also issued Executive Order 13905 titled *Strengthening National Resilience Through Responsible Use of Positioning, Navigation, and Timing Services* to increase efforts to provide backups to GPS. Plans to build a backup to space-based systems seem to be starting slowly, but momentum is building in the U.S. government and elsewhere to address the issue.

Packet-Based Distribution

Ever since the development of packet-based transport, engineers have devised methods to carry time and synchronize the clocks running on different nodes in the network. One of the early Internet protocols developed by the *Internet Engineering Task Force* (IETF) was RFC 868, "Time Protocol," which allows a user to query time on a node (via the rtime(3) system call on Linux systems). The main packet-based protocols in use today are NTP and PTP. See Chapter 7 for a comparison of these methods.

NTP was first proposed as RFC 958 in the mid-1980s, and it is now up to its fourth generation. For telecommunications, NTP is mainly used to carry time of day, although there are applications using it to carry frequency for mobile. The Simple Network Time Protocol (SNTP) is a lighter version subset of the standard NTP, used where the performance of the full NTP implementation is not needed. NTPv4 and SNTP are defined in RFC 5905, and together they are probably the most popular and widely deployed time transport used worldwide (mainly for setting system time on devices connected to data networks such as the Internet).

In order to carry frequency using packets, one method that has been widely deployed is based on one of the PTP telecom profiles specifically designed for this role. See the section "G.8265.1—Telecom Profile for Frequency Distribution" in Chapter 7 for details of this approach. Other methods, such as ACR, exist to carry frequency over packets, and they are covered in detail in Chapter 9.

However, when it comes to carrying time and phase, the best-in-class tool available is a packet-based distribution method based on the PTP protocol. PTP and all its profiles are covered extensively in Chapter 7, so refer there for all the details. Basically, when there is not a radio solution for distributing time, the key alternative is a two-way time transfer protocol on a packet network, using PTP.

Transport and Signaling of Quality Levels

No matter whether the clock is distributed over the physical media or with packet-based methods, an additional mechanism is required to exchange clock quality information and traceability of synchronization signals. Depending on the method that is used in the timing distribution network, there are different methods to signal the clock traceability and quality.

As seen previously, SSM is used in SDH to exchange the quality level of the frequency clock between the network elements—carried with several bits located in the frame header. SyncE, on the other hand, uses the Ethernet Synchronization Message Channel (ESMC), which is a packet-based mechanism, to exchange clock quality information. PTP also deploys messages, the Announce message, to exchange information about clock traceability and quality levels.

See the preceding section "Clock Quality Traceability" and Table 3-1 for information on traceability through a network. Chapter 6 covers SSM and ESMC in much more detail, so it is enough just to summarize them here with one example. Figure 3-21 indicates how ESMC clock quality is transferred with SyncE (ESMC is specified in ITU-T G.8264).

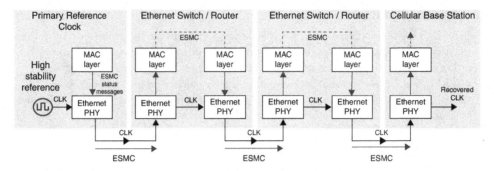

Figure 3-21   *ESMC Distribution in a SyncE-Enabled Network*

Consumer of Time and Sync (the End Application)

Of course, the whole goal behind building a timing distribution network is to allow some application that needs the synchronization service to function correctly. It is vitally important, before deploying such a solution, that the requirements of the end application are clearly understood. This cannot be overstated, as it is not unusual to see engineers build a timing distribution network that is not what was needed.

A key component, then, is to understand what the timing requirements are from the end application. The following list should be some of the first questions asked before starting a project:

■ What forms of synchronization are required: frequency, phase, or time?

■ What is the timing budget, meaning what level of performance and accuracy is needed at the end application?

■ What is the requirement for holdover performance—how long must the end application continue to function correctly when the synchronization is lost?

■ What are the existing timing methods, and how can they be reused or repurposed?

■ What is the requirement for resiliency and failover protection?

■ What form of transport is available, and can it carry time signals accurately?

■ How will the end application "consume" the synchronization? Does it need some form of physical signal and, if so, what is it? Will it take packets, and are those packets PTP? What version or profile of PTP is required?

One axiom that the authors like to share when working with customers on timing networks is, "This is a timing problem; you need to design a timing solution." Meaning, it is not just a problem of transporting data packets from A to B, but one of transporting time.

Summary

This chapter dealt with numerous synchronization concepts. First, you learned about the different network types, then the clock types and the operational state of the clocks. Next, you learned about sources of time, including various levels of oscillators and the many different GNSS systems.

The chapter continued with a look at the evolution of various time distribution techniques, from TDM and physical methods to packet-based techniques and how they are related. Next, you learned about the timing distribution network and how information on the quality and traceability of the clock is maintained during transport.

Chapter 5 goes into timing error, and clock noise performance, and Chapter 12, "Operating and Verifying Timing Solutions," covers the measurable characteristics of clocks and the testing and measurement of clock parameters. But the next step is into Chapter 4, "Standards Development Organizations," that takes a small diversion and examines the major organizations that are setting the standards for both synchronization and mobile communications.

References in This Chapter

Alliance for Telecommunications Industry Solutions (ATIS). "GPS Vulnerability." *ATIS-0900005 Technical Report*, 2017. https://access.atis.org/apps/group_public/download.php/36304/ATIS-0900005.pdf

Alliance for Telecommunications Industry Solutions (ATIS). "Synchronization Interface Standard." *ATIS-0900101:2013(R2018)*, Supersedes ANSI T1.101, 2013. https://www.techstreet.com/atis/standards/atis-0900101-2013-r2018?product_id=1873262

American National Standards Institute (ANSI). "Synchronization Interface Standards for Digital Networks." *ANSI T1.101-1999*, Superseded by ATIS-0900101.2013(R2018).

Amphenol RF. "1.0/2.3 Connector Series." *Amphenol RF*, 2020. https://www.amphenolrf.com/connectors/1-0-2-3.html

BeiDou. "BeiDou Navigation Satellite System." *BeiDou*. http://en.beidou.gov.cn/

BeiDou. "Development of the BeiDou Navigation Satellite System (Version 4.0)." *China Satellite Navigation Office*, 2019. http://en.beidou.gov.cn/SYSTEMS/Officialdocument/

European Commission. "The History of Galileo." *European Commission*. https://ec.europa.eu/growth/sectors/space/galileo/history_en

European Global Navigation Satellite Systems Agency (GSA). "What Is EGNOS?" GSA, 2020. https://www.gsa.europa.eu/egnos/what-egnos

European Union. "European GNSS (Galileo) Open Service Signal-In-Space Interface Control Document (OS SIS ICD)." *European Global Navigation Satellite Systems Agency (GSA)*, Issue 1.3, 2016. https://www.gsc-europa.eu/sites/default/files/sites/all/files/Galileo-OS-SIS-ICD.pdf

European Union. "Galileo – Open Service – Service Definition Document (Galileo OS SDD)." *European Global Navigation Satellite Systems Agency (GSA)*, Issue 1.1, 2019. European Union, https://www.gsc-europa.eu/sites/default/files/sites/all/files/Galileo-OS-SDD_v1.1.pdf

Global Navigation Satellite System (GLONASS). "Information and Analysis Center for Positioning, Navigation and Timing." *GLONASS*. https://www.glonass-iac.ru/en/

Government of Japan Cabinet Office. "QZSS, "Quasi-Zenith Satellite System (QZSS)." *Cabinet Office National Space Policy Secretariat*. https://qzss.go.jp/en/

Indian Space Research Organization (ISRO). "Indian Regional Navigation Satellite System – Signal in Space Interface Control Document (ICD) for Standard Positioning Service." *ISRO*, Version 1.1, 2017. https://www.isro.gov.in/sites/default/files/irnss_sps_icd_version1.1-2017.pdf

Institute of Electrical and Electronics Engineers (IEEE). "IEEE Standard for Ethernet" *IEEE Std 802.3-2018*, 2018. https://ieeexplore.ieee.org/document/8457469

International Bureau of Weights and Measures (BIPM). "Metrological Traceability." *BIPM*. https://www.bipm.org/en/bipm-services/calibrations/traceability.html

International Telecommunication Union (ITU). *Satellite Time and Frequency Transfer and Dissemination* [ITU-R Handbook]. *ITU*, Edition 2010. https://www.itu.int/pub/R-HDB-55

International Telecommunication Union Telecommunication Standardization Sector (ITU-T):

"G.703: Physical/electrical characteristics of hierarchical digital interfaces." *ITU-T Recommendation*, 2016. https://www.itu.int/rec/T-REC-G.703/en

"G.781: Synchronization layer functions for frequency synchronization based on the physical layer." *ITU-T Recommendation*, 2020. https://www.itu.int/rec/T-REC-G.781/en

"G.803, Architecture of transport networks based on the synchronous digital hierarchy (SDH)." *ITU-T Recommendation*, 2000. https://www.itu.int/rec/T-REC-G.803/en

"G.810, Definitions and terminology for synchronization networks." *ITU-T Recommendation*, 1996. https://www.itu.int/rec/T-REC-G.810/en

"G.8260, Definitions and terminology for synchronization in packet networks." *ITU-T Recommendation*, 2020. https://handle.itu.int/11.1002/1000/14206

"ITU-T GSTR-GNSS: Considerations on the use of GNSS as a primary time reference in telecommunications" *ITU-T Technical Report*, 2020-02. http://handle.itu.int/11.1002/pub/815052de-en

Jespersen, J. and J. Fitz-Randolph. "From Sundial to Atomic Clocks, Understanding Time and Frequency." *National Institute of Standards and Technology (NIST)*, ISBN 978-0-16-050010-7, 1999. https://tf.nist.gov/general/pdf/1796.pdf

Joint Committee for Guides in Metrology (JCGM). "International Vocabulary of Metrology – Basic and General Concepts and Associated Terms (VIM). *JCGM*, 3rd Edition, 2012. https://www.bipm.org/utils/common/documents/jcgm/JCGM_200_2012.pdf

Langley, R. "Innovation: GLONASS — past, present and future." *GPS World*, 2017. https://www.gpsworld.com/innovation-glonass-past-present-and-future/

Lombardi, M. and G. Nelson. "WWVB: A Half Century of Delivering Accurate Frequency and Time by Radio." *Journal of Research of the National Institute of Standards and Technology (NIST)*, Volume 119, 2014. http://dx.doi.org/10.6028/jres.119.004

National Coordination Office for Space-Based Positioning, Navigation, and Timing. "GPS: The Global Positioning System." *GPS*. https://www.gps.gov

NIST Physical Measurement Laboratory, Time and Frequency Division. "Common View GPS Time Transfer." *National Institute of Standards and Technology (NIST)*, 2016. https://www.nist.gov/pml/time-and-frequency-division/time-services/common-view-gps-time-transfer

NIST Physical Measurement Laboratory, Time and Frequency Division. "Time and Frequency from A to Z, H." *National Institute of Standards and Technology (NIST)*, 2016. https://www.nist.gov/pml/time-and-frequency-division/popular-links/time-frequency-z/time-and-frequency-z-h

SDCM ICD. "Interface Control Document, Radio signals and digital data structure of GLONASS Wide Area Augmentation System, System of Differential Correction and Monitoring." *Russian Space Systems*, Ed. 1, 2012. http://www.sdcm.ru/smglo/ICD_SDCM_1dot0_Eng.pdf

Telcordia Technologies. "Clocks for the Synchronized Network: Common Generic Criteria." *GR-1244-CORE*, Issue 4, 2009. https://telecom-info.njdepot.ericsson.net/site-cgi/ido/docs.cgi?ID=SEARCH&DOCUMENT=GR-1244&

Telcordia Technologies. "Synchronous Optical Network (SONET) Transport Systems: Common Generic Criteria" *GR-253-CORE*, Issue 5, 2009. https://telecom-info.njdepot.ericsson.net/site-cgi/ido/docs.cgi?ID=SEARCH&DOCUMENT=GR-253&

U.S. Coast Guard Navigation Center. "GPS Constellation Status" and "Slant Chart (Satellite Locations)." https://www.navcen.uscg.gov/?Do=constellationStatus

U.S. Department of Defense. "Global Positioning System Standard Positioning Service Performance Standard." U.S. Department of Defense, 5th Edition, 2020. https://www.navcen.uscg.gov/pdf/gps/geninfo/2020SPSPerformanceStandardFINAL.pdf

U.S. Federal Aviation Administration (FAA). "Satellite Navigation – Wide Area Augmentation System (WAAS)." *FAA*, 2020. https://www.faa.gov/about/office_org/headquarters_offices/ato/service_units/techops/navservices/gnss/waas/

Chapter 3 Acronyms Key

The following table expands the key acronyms used in this chapter.

Term	Value
1PPS	1 pulse per second
ACR	adaptive clock recovery
ANSI	American National Standards Institute
ATIS	Alliance for Telecommunications Industry Solutions
ASCII	American Standard Code for Information Interchange
ASECNA	Agency for Air Navigation Safety in Africa and Madagascar
BC	boundary clock
BCD	binary coded decimal
BDT	BeiDou time
BITS	building integrated timing supply (SONET)
BNC	bayonet Neill-Concelman (connector)
DIN	Deutsches Institut für Normung (German Institute for Standardization)
DOCSIS	Data-Over-Cable Service Interface Specification
DoD	Department of Defense (USA)
DSL	digital subscriber line (a broadband technology)
eESMC	enhanced Ethernet synchronization messaging channel
EGNOS	European Geostationary Navigation Overlay Service
eLORAN	enhanced Long Range Navigation
ePRC	enhanced primary reference clock
ePRTC	enhanced primary reference time clock
ePRTC-A	enhanced primary reference time clock—class A
ePRTC-B	enhanced primary reference time clock—class B
ESMC	Ethernet synchronization messaging channel
eSyncE	enhanced synchronous Ethernet
FAA	Federal Aviation Administration (USA)
FERNS	Far East Radio Navigation Service
GAGAN	GPS-Aided GEO Augmented Navigation
GDO	GNSS or GPS disciplined OCXO

Term	Value
GEO	geostationary earth orbit
GGTO	GPS-Galileo time offset
GLONASS	GLObalnaya NAvigazionnaya Sputnikovaya Sistema (Global Navigation Satellite System)
GM	grandmaster
GNSS	global navigation satellite system
GPS	Global Positioning System
GSA	Global Navigation Satellite Systems Supervisory Agency (Europe)
GST	Galileo System Time
HF	high frequency
IEEE	Institute of Electrical and Electronics Engineers
IETF	Internet Engineering Task Force
IGSO	inclined geosynchronous satellite orbit
IoT	Internet of things
INRSS	Indian Regional Navigation Satellite System
IRIG	Inter-Range Instrumentation Group
ITU	International Telecommunication Union
ITU-T	ITU Telecommunication Standardization Sector
LF	low frequency
LORAN	Long Range Navigation
maser	microwave amplification by stimulated emission of radiation
MEO	medium earth orbit
MSAS	Michibiki Satellite-based Augmentation System
MSAS	MTSAT Satellite-based Augmentation System (historical)
MSK	Moscow Standard Time
MTIE	maximum time interval error
MTSAT	multi-functional transport satellite
NavIC	Navigation with Indian Constellation
NIST	National Institute of Standards and Technology (USA)
NMEA	National Marine Electronics Association
NPLI	National Physical Laboratory India
NTP	Network Time Protocol

Term	Value
NTSC	National Time Service Center (China)
OC	ordinary clock
OCXO	oven-controlled crystal oscillator
OTN	optical transport network (optical transport technology)
PDV	packet delay variation
PEC	packet-based equipment clock
PEC-M	packet-based equipment clock—master
PEC-S	packet-based equipment clock—slave
PMA	Physical Medium Attachment
PNT	positioning, navigation and timing
PON	passive optical networking (optical broadband)
PRC	primary reference clock (SDH)
PRS	primary reference source (SONET)
PRTC	primary reference time clock
PRTC-A	primary reference time clock—class A
PRTC-B	primary reference time clock—class B
PTP	precision time protocol
QL	quality level
QZO	quasi-zenith orbit
QZSS	Quasi-Zenith Satellite System
RJ-45, RJ-48c	registered jack 45, 48c
RSP	route switch processor
SAR	search and rescue (service)
SBAS	space-based augmentation system
SDCM	System of Differential Correction and Monitoring
SDH	synchronous digital hierarchy (optical transport technology)
SEC	synchronous equipment clock
SEC	SDH equipment clock
SFN	single frequency network
SFP	small form-factor pluggable (transceiver)
SI	International System of Units
SMA	SubMiniature version A (connector)

Term	Value
SMC	SONET minimum clock
SNTP	Simple Network Time Protocol
SONET	synchronous optical network (optical transport technology)
SSM	synchronization status message
SSU	synchronization supply unit (SDH)
SSU-A	primary level synchronization supply unit (SDH)
SSU-B	secondary level synchronization supply unit (SDH)
SSU-L	synchronization supply unit-local (SDH)
SSU-T	synchronization supply unit-transit (SDH)
SyncE	synchronous Ethernet—a set of ITU-T standards
TAI	Temps Atomique International
TC	transparent clock
TCXO	temperature-compensated crystal oscillator
TDM	time-division multiplexing
T-GM	Telecom Grandmaster
TIE	time interval error
TNC	threaded Neill-Concelman (connector)
ToD	time of day
USNO	United States Naval Observatory
UBX	u-Blox
UTC	Coordinated Universal Time
UTC(k)	Coordinated Universal Time at laboratory k
WAAS	Wide Area Augmentation System

Further Reading

Refer to the following recommended sources for further information about the topics covered in this chapter.

GPS: The Global Positioning System: https://www.gps.gov/systems/

U.S. Coast Guard Navigation Center Constellation: https://www.navcen.uscg.gov/?Do=constellationStatus

European GNSS Service Centre (for the Galileo GNSS system): https://www.gsc-europa.eu/

Chapter 4

Standards Development Organizations

This chapter provides brief introductions to various standards development organizations (SDO) involved in telecommunications and synchronization. It includes an overview of their organizational structures and the processes they use to define standards. For each SDO, the chapter also outlines its contribution to time synchronization specifications and standards.

This chapter also introduces some of the telecom groups and industry alliances that contribute to the open interfaces and disaggregation architecture supporting the 5th generation (5G) radio access network (RAN). Many of the details on the standards themselves are covered in the following chapters.

This chapter is focused on those SDOs whose area of interest and the contributions they make involve either time synchronization or mobile architectures, or both. Understanding the position these SDOs take, and their role in defining standards, should help you to understand the purpose of each standard as well as the customer requirements or architectural problem that the standard is meant to address.

Innovators, organizations, and industries benefit from having a common definition that simplifies transactions, allows interoperability, and enables people to work toward common goals. SDOs, also known as standards setting organizations (SSO), define and maintain standards to meet industry needs. The terms SDO and SSO are widely used interchangeably.

SDOs are organizations that focus on developing, maintaining, amending, and publishing technical standards using open and transparent processes. Some SDOs are large, formally recognized organizations with a global membership consisting of national representatives (public or private)—although companies and individuals may also participate. Others are based on a community of professionals with or without a formal membership. Still others are organizations that support or certify standards developed elsewhere and may even author none themselves.

There are standards bodies at the international, regional, and national levels. For example, the International Telecommunication Union (ITU) is an international SDO, the European Telecommunications Standards Institute (ETSI) is a regional SDO, and the American National Standards Institute (ANSI) is more like a national SDO. There are more than 200 SDOs developing standards for the information and communication technology (ICT) sector alone. Of course, in search of mutual goals, many of these standards bodies frequently collaborate with each other.

A *standard* can be defined as a set of technical specifications that specifies common design requirements for a product, process, service, or system. Some standards might be backed by international treaty law, some by government legislation, and others might have no legal force at all. Some organizations, such as the ITU, totally avoid the concept of compulsion and refer to their specification documents as recommendations.

This chapter covers some of these organizations and their contributions toward defining standards in the fields of both telecommunications and time synchronization.

International Telecommunication Union

The ITU, as the United Nations specialized agency for ICT, develops and publishes international telecommunications recommendations (standards). The ITU has representation from 193 member states and membership of over 900 companies, universities, and international and regional organizations.

As the ITU says about itself:

> *Established over 150 years ago in 1865, ITU is the intergovernmental body responsible for coordinating the shared global use of the radio spectrum, promoting international cooperation in assigning satellite orbits, improving communication infrastructure in the developing world, and establishing the worldwide standards that foster seamless interconnection of a vast range of communications systems. From broadband networks to cutting-edge wireless technologies, aeronautical and maritime navigation, radio astronomy, oceanographic and satellite-based earth monitoring as well as converging fixed-mobile phone, Internet and broadcasting technologies, ITU is committed to connecting the world.*

The ITU organizes conferences to provide opportunities to learn more about specific topics, and workshops to exchange ideas and innovations. The outcomes of these workshops become inputs to what are called study groups (SG). The standardization work is carried out by the technical SGs, in which delegates from the ITU membership develop and approve specifications, or what the ITU calls *recommendations.*

The ITU focuses on three main sectors: the Radiocommunication Sector (ITU-R); the Telecommunication Standardization Sector (ITU-T); and the Telecommunication Development Sector (ITU-D).

ITU-R

The ITU Radiocommunication Sector ensures the rational, efficient, and economical use of radio spectrum and satellite orbits by facilitating international collaboration. Its activities include the following:

- Conducts conferences and seminars to adopt radio regulations and regional agreements for efficient use of radio frequency spectrum

- Coordinates activities to eliminate radio interference between radio stations of different countries

- Maintains the Master International Frequency Register (MIFR)

- Carries out studies and collates recommendations on radiocommunication matters

- Approves recommendations, technical characteristics or specifications, and operational procedures developed by ITU-R study groups

Note Use of the terms "master" and "slave" is ONLY in association with the official terminology used in industry specifications and standards, and in no way diminishes Pearson's commitment to promoting diversity, equity, and inclusion, and challenging, countering, and/or combating bias and stereotyping in the global population of the learners we serve.

The ITU-R organizes world radiocommunication conferences (WRC) to review and revise the Radio Regulations—the international treaty governing regional or national radio frequency spectrum and satellite orbits. It also organizes regional radiocommunication conferences (RRC) to establish regional agreements and plans for radio-based services.

Based on the decisions taken in the WRCs, the ITU-R study groups develop global standards, reports, and guidelines on radiocommunication matters. The ITU-R study groups mainly focus on

- Efficient management and use of spectrum by terrestrial services

- Efficient management and use of orbital resources by space services

- Characteristics and performance specifications of radio systems

- Radio station operations

- Safety aspects of radiocommunications

There are six study groups that work on *questions* assigned to these specialist areas:

- **SG 1**: Spectrum management (www.itu.int/ITU-R/go/rsg1)

- **SG 3**: Radiowave propagation (www.itu.int/ITU-R/go/rsg3)

- **SG 4**: Satellite services (www.itu.int/ITU-R/go/rsg4)

- **SG 5:** Terrestrial services (www.itu.int/ITU-R/go/rsg5)

- **SG 6:** Broadcasting service (www.itu.int/ITU-R/go/rsg6)

- **SG 7:** Science services (www.itu.int/ITU-R/go/rsg7)

The ITU-R has been instrumental in terms of providing and defining the basic specifications for International Mobile Telecommunications: IMT-2000 (3G radio technology), IMT-Advanced (4G radio technology), and IMT-2020 (5G radio technology). ITU-R also publishes specifications for digital television and sound, including digital audio broadcasting (DAB), high definition television (HDTV), ultra-high definition television (UHDTV), and the high dynamic range (HDR) enhancement.

ITU-T

The ITU Telecommunication Standardization Sector is responsible for defining international recommendations (standards) for telecommunication network operations and interworking. There are over 4000 recommendations in force defining services, technical specifications, network architectures and security requirements for many network systems, including broadband, digital subscriber line (DSL), passive optical network (PON) systems, optical transport networks (OTN), next-generation networks, as well as Internet protocol (IP).

The framework for the work of the ITU-T is decided by the following two organizations:

- **World Telecommunication Standardization Assembly (WTSA):** Provides overall direction and defines policies, structure, and the responsibilities of study groups. WTSA is also responsible for appointing chairs and vice-chairs and approving any programs during their appointment.

- **Telecommunication Standardization Advisory Group (TSAG):** Reviews program priorities, financial matters, and strategies for the sector. It coordinates between study groups and provides guidelines and organizational working procedures for them to follow. The TSAG also provides advice and necessary assistance to the director of the Telecommunication Standardization Bureau (TSB). The TSB acts as a secretariat to the ITU-T Sector and provides support to coordinate and manage the study groups and maintains the publication records for recommendations.

The study groups are the heart of the ITU-T. The SGs are responsible for doing the actual work of developing standards in different areas of interest. Each of the study groups is further divided into numerous *questions* (up to 20 per study group) that tackle specific areas of technology. Some of the study groups in this sector are as follows:

- **SG2:** Operational aspects of service provision and telecommunications management

  SG2 is the home for defining E.164, the numbering standard that provides structure and functionality of telephone numbers. SG2 also worked on electronic numbering (ENUM), an Internet Engineering Task Force (IETF) protocol for entering E.164 numbers into the Internet Domain Name System (DNS).

- **SG3:** Tariff and accounting principles including economic and policy issues related to telecommunications

  SG3 is responsible for studying international telecommunication policy, economic issues, tariffing, and accounting matters, and defining regulatory models and frameworks.

- **SG5:** Environmental and circular economy

  SG5 is responsible for studying the effect of ICT on climate change and publishing guidelines for using ICT in an ecologically responsible manner.

- **SG9:** Broadband cable and TV

  SG9 develops standards on voice, video, and data IP applications over cable TV networks. This includes cable modems, interactive cable television services, high-speed data services, IP-based TV (IPTV), and video-on-demand.

- **SG11:** Signaling requirements, protocols, and test specifications

  SG11 was responsible for defining signaling standards for national and international telephone calls as well as data calls. It is also responsible for conformance and interoperability testing. Some of the work of SG11, such as intelligent networks, was adopted by the 3rd Generation Partnership Project (3GPP) for circuit-switched networks. Some of the admission control procedures and protocols defined by SG11 are based on IETF standards: protocols like Diameter, Simple Network Management Protocol (SNMP), and Common Open Policy Service (COPS).

- **SG12:** Performance, QoS, and QoE

  SG12 develops standards that focuses on the operational aspects of performance, quality of service (QoS), and quality of experience (QoE) for the ICT industry. This group identifies the impact of different parameters like jitter, packet loss, and latency on various multimedia services.

- **SG13:** Future networks, with focus on IMT-2020, cloud computing, and trusted network infrastructure

  SG13 has led the standardization work for next-generation networks (NGN) and now increasingly focuses on future networks. NGN networks are based on packets rather than circuits, and SG13 looks at facets beyond that. The group also looks at network aspects of mobile communications, including IMT-2000, IMT-Advanced, and IMT-2020.

  SG13 develops standards enabling multi-cloud management and monitoring of services that spread across multiple service provider domains. This group is now active in the network aspects of the Internet of things (IoT) and especially aspects of cloud computing supporting IoT.

- **SG15:** Networks, technologies, and infrastructures for transport, access, and home

  SG15 is one of the largest and most active study groups in the ITU-T. For that reason, the work of SG15 is distributed into three working parties (WP) and each WP contains between six and seven questions that belong to SG15. The three WPs consist of

  - **WP1:** Focuses on the transport aspects of optical and metallic access systems, home networks, and smart grid networks. Within this framework, WP1 studies performance, field deployment, and installation aspects of fiber-optic cables. WP1 is also interested in aspects of network resiliency and recovery during disaster situations.

  - **WP2:** Focuses on physical interfaces, transmission characteristics, maintenance, and operation of OTNs. This covers features for network equipment, such as routers, switches, repeaters, multiplexers, amplifiers, transceivers, and so forth.

    WP2 covers maintenance and operations of transport networks, including network protection and restoration, resource optimization, scalability, and network agility with Software-Defined Networks (SDN).

    The network transport media and technologies covered by WP2 are optical fiber cables; dense and coarse wavelength division multiplexing (DWDM and CWDM) optical systems; OTN, Ethernet, and other packet-based data services.

  - **WP3:** Focuses on the logical layers of the transport network and their characteristics. This program addresses management and control of transport systems covering Operations, Administration, and Maintenance (OAM), data center interconnect, and home networking services. WP3 also focuses on synchronization for frequency and precise time and interworking of access technologies and their characteristics to support 5G backhaul and fronthaul networks. Most of the recommendations cited by this book come from Question 13, which is part of WP3 (itself a part of SG15). There is more specific information on Q13 (as it is referred to) in Chapter 8, "ITU-T Timing Recommendations."

- **SG16:** Multimedia coding, systems, and applications

  SG16 is responsible for defining all aspects of multimedia standardization, including terminal architecture, protocols, security, mobility, interworking, and QoS. SG16 covers various digital services, including telepresence, IPTV, audio and video coding, signal processing, visual surveillance, accessibility to communications for disabled people, e-health, intelligent transport systems, and so on.

- **SG17:** Security

  SG17 coordinates the security-related requirements across all ITU-T SGs. It works very closely with other SDOs and deals with a broad range of standardization issues in areas like cybersecurity, security management, security architectures, and identity management. A focus area of this SG addresses the security architecture requirements for IoT, smart grid, smartphones, SDN, cloud networking, web services, analytics, social networking, mobile financial systems, IPTV, and healthcare teleservices.

There is much more information in Chapter 8 on the processes the ITU-T uses to develop recommendations and specific details on those in SG15 recommendations related to timing and synchronization.

ITU-D

The ITU Telecommunication Development Sector examines telecommunication questions as they relate to countries that are underserved by telecommunications services. The ITU-D SGs conduct surveys and provide opportunities for member states and sector members (associates and academia) to exchange views, present ideas, and share experiences. The ITU-D is responsible for sharing final reports, guidelines, and recommendations based on the input received from study group members.

There are two study groups under ITU-D:

- **SG1**: Enabling environment for development of telecommunications/ITCs

  SG1 focuses on the national telecommunications/ICT policy, regulation, and technical and strategic aspects of telecommunications to drive benefits for developing countries. These policies and strategies also cover infrastructure requirements to support broadband services, cloud computing, Network Functions Virtualization (NFV), consumer protection, and future telecommunication services that contribute to sustainable growth. SG1 covers the following:

  - Economic policies and methods of determining costs of services

  - Policies and strategies to provide telecommunication services to rural and remote areas, providing access to those with disabilities, and other people with special needs

  - Policies on adopting, implementing, and migrating digital broadcasting services

- **SG2**: ICT services and applications for the promotion of sustainable development

  Beside the applications of ICT, SG2 also focuses on the ICT services and applications and building more confidence and security into the use of ICT. This includes

  - The use of ICT applications to monitor and mitigate the impact of climate change particularly on developing countries. This also covers preparing for and managing the impact and relief from natural disasters.

  - Studies the impact of electromagnetic fields on humans and develops procedures for the safe disposal of electronic waste. Also, securing telecoms and ICT devices against forgery, counterfeiting or theft.

  - Implement conformance and interoperability test methodology for devices and equipment.

International Mobile Telecommunications

The ITU-R defines the radio frequency (RF) spectrum for various services, including mobile services. Within the ITU-R, working party 5D (WP5D) has the responsibility for the overall radio system. WP5D has been working with various governments and industry participants to develop an International Mobile Telecommunications system, known as IMT. IMT provides a global platform to define the requirements and specifications to build the next generations of mobile services.

The initial set of IMT standards was approved by ITU in 2000 and was known as IMT-2000 and was used as the basis for the third generation (3G) mobile network. In January 2012, the ITU defined a detailed specification for the fourth generation (4G) mobile networks as part of IMT-Advanced (see the ITU-R M.2012-4 entry in the "References in This Chapter" section at the end of the chapter).

The document ITU-R M.2083 (also referenced at end of this chapter) describes a detailed framework for the development of IMT for 2020 and beyond. The scope of IMT-2020 is focused upon the fifth generation (5G) mobile network. It not only covers the enhancement to the traditional mobile broadband system but extends it to new application use cases.

As shown in Figure 4-1, IMT-2020 defines three usage scenarios: enhanced mobile broadband (eMBB), massive machine-type communications (mMTC), and ultra-reliable and low-latency communications (uRLLC).

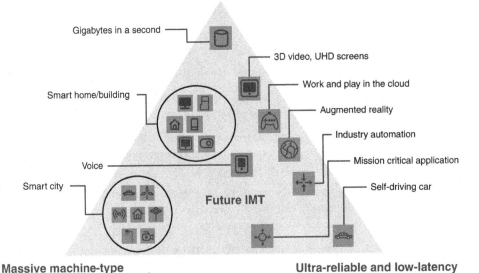

Figure 4-1   *IMT-2020 Usage Scenarios*

The ITU-R defined a set of capabilities needed for an IMT-2020-compatible technology to support the 5G use cases and scenarios. This is an important distinction: the IMT develops the use cases and capabilities, whereas other organizations propose and standardize the technology to achieve those goals. In total, they defined 13 capabilities, out of which 8 are highlighted as *key capabilities*. These eight capabilities are illustrated in the two spider web diagrams shown in Figure 4-2 and Figure 4-3.

Figure 4-2 shows the key capabilities together with the indicative target numbers as the foundational guidelines to develop detailed specifications for 5G radio access technology. It compares the targets with those from the previous generation of mobile technology (outlined in the IMT-Advanced recommendation).

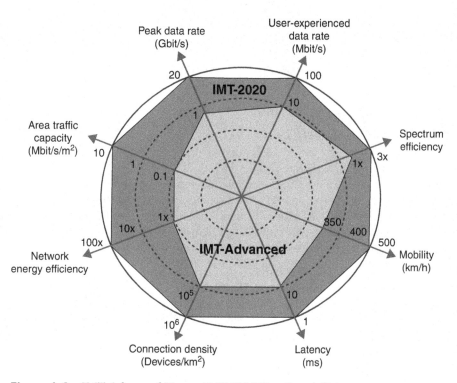

Figure 4-2 *IMT-Advanced Versus IMT-2020 Key Capabilities*

Figure 4-3 illustrates the correlation between the use cases and the key requirements.

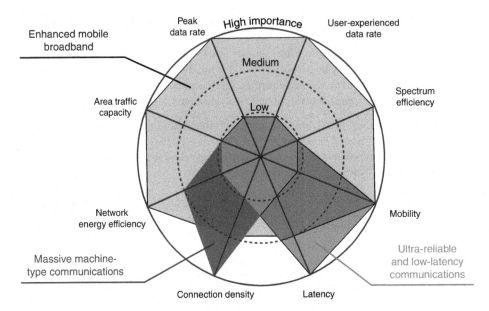

Figure 4-3 *Key Requirements to Usage Scenarios*

The process defined by the ITU-R to finalize the IMT-2020 vision has been a multistep journey involving leadership by WP5D over a period of more than eight years. In the middle of 2020, the ITU-R determined the final candidate technology submissions to be accepted as qualified IMT-2020 technologies. As of the time of writing, the final IMT-2020 specifications (known as IMT-2020.SPECS) are going through final agreement before forwarding to member states for adoption and approval. Please see the link at the end of the chapter for further reading on IMT-2020.SPECS.

The reason this process is important is that IMT sets the requirements that 3GPP aims to meet with its specification of the mobile technology. Only when the 3GPP specification release meets the IMT requirement is it qualified to be called 3G, 4G, or 5G. For example, the IMT-Advanced requirements were met by 3GPP in Release 10, which has become known as LTE-Advanced. This IMT document becomes the goal sheet that 3GPP will endeavor to implement, so the next section covers 3GPP and its standardization processes.

3rd Generation Partnership Project

The 3rd Generation Partnership Project (3GPP) was formed in December 1998 with an original goal to produce technical specifications and technical reports for a 3G mobile system based on an evolution of the core network and radio access technologies from the Global System for Mobile communications (GSM). The scope was subsequently amended to include maintenance and development of the GSM system including evolution in the radio access, such as General Packet Radio Service (GPRS) and Enhanced Data rates for GSM Evolution (EDGE).

3GPP develops technical specifications, and these specifications are converted into standards by seven separate SSOs for telecommunications, as shown in Figure 4-4. These organizations are ETSI (Europe and rest of the world), ATIS (United States), ARIB and TTC (Japan), CCSA (China), TSDSI (India), and TTA (South Korea).

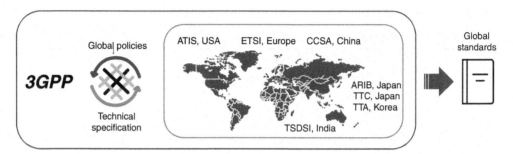

Figure 4-4 *3GPP and Regional SSOs*

The 3GPP project encompasses the end-to-end specification of the cellular system including radio access technologies, the cellular system, and the service framework. To address these technology areas, 3GPP has three technical specification groups (TSG):

- **Radio Access Network (TSG RAN):** Responsible for the radio parts of the network (mobile user equipment and base stations) for both frequency division duplex (FDD) and time division duplex (TDD) radio types. TSG RAN specifies not only performance and RF characteristics but also testing for conformance. It is made up of six working groups, including WGs dedicated to radio layer 1 (WG1), radio layers 2 and 3 (WG2), and operations and maintenance (WG3).

- **Core Network and Terminals (TSG CT):** Responsible for specifying the core network components of 3GPP systems, including the signaling interfaces that control factors such as mobility, policy, call connection control, and so forth. TSG CT also specifies terminal (user equipment) characteristics, QoS, and subscriber identity module (SIM) technologies.

- **Service and System Aspects (TSG SA):** Responsible for the definition, evolution, and maintenance of the overall system architecture, including the assignment of functions between various subsystems. It is made up of six working groups, including one for security.

3GPP technologies have evolved through every generation of the mobile systems. After defining 3G, 3GPP delivered Long-Term Evolution (LTE), LTE-Advanced, LTE-Advanced Pro, and the current body of work on 5G. 3GPP has become the single technology body that defines technical specifications for global mobile. 3GPP defines the time and frequency synchronization requirements for radio access networks to address various mobility use cases, but it does not define the time and frequency implementation methods. The 3GPP specifications are structured as *releases*, whereby each release includes features added to the previous release. Table 4-1 outlines the 3GPP releases and the associated mobile technology for each of them. Note that this table is very much simplified so as not to introduce excessive amounts of new terminology and acronyms. If you are not familiar with these technologies, Chapter 10, "Mobile Timing Requirements," discusses them in some detail.

Table 4-1   *3GPP Releases*

Release	Milestone
R99	3G specifications taken from a consolidation of the underlying GSM specifications combined with the development of a new RAN
R4	3G (UMTS) with all IP core network
R5	High Speed Downlink Packet Access (HSDPA)
R6	High Speed Uplink Packet Access (HSUPA)
R7	HSPA+ or evolved HSPA, sometimes marketed as 3.5G
R8	Long Term Evolution (LTE)
R9	LTE broadcast (enhanced Multimedia Broadcast Multicast System, known as eMBMS), Voice over LTE (VoLTE), LTE positioning
R10	LTE-Advanced, Carrier Aggregation (CA), and advanced multiple input and multiple output (MIMO) antenna technology
R11	Coordinated Multipoint (CoMP)
R12	FDD-TDD Carrier Aggregation (CA)
R13	LTE-Advanced Pro, IoT technologies, LTE in unlicensed bands
R14	Enhanced Licensed Assisted Access (LAA) allowing LTE to use 5-GHz unlicensed spectrum (also used by Wi-Fi) for uplink and downlink
R15	5G Phase 1, introducing the first IMT-2020 enhancements within existing spectrum; also includes aspects of machine-type communications (MTC) and IoT
R16	5G Phase 2, completing the initial 5G for IMT-2020 submission
R17	Under development over the 2021–22 period (schedule may be further affected by COVID) and includes enhancements to 5G (discussed in following text)
R18	Studies just beginning as of time of writing (early 2021)

3GPP uses a system of parallel releases to provide a stable platform for implementation while at the same time allowing the addition of new features. A release not only contains the newly implemented features but builds upon the previous releases to ensure backward compatibility. For example, LTE was first introduced in Release 8, and it is still evolving today with Release 14 and Release 15. Meanwhile in parallel, 3GPP introduced the 5G new radio (NR) in Release 15.

3GPP started the study of 5G NR in September 2015 with the goal and vision to address IMT-2020 requirements. 3GPP technical report (TR) 38.913 describes scenarios, key performance requirements, and requirements for architecture, migration, supplemental services, operations, and testing. This study completed in March 2017 with 3GPP TR 38.912 from Release 14 that captures a set of features and capabilities for the new radio access technology along with a study of their feasibility.

3GPP began to develop the 5G specification in two phases. Release 15 marks the first phase of 5G specifications, whereas Release 16 covers the second phase. Even work from Release 15 on NR access got split into three phases addressing different network operator demands. This is because some operators wanted to deploy 5G radio technology in combination with their existing LTE networks—what became known as non-standalone (NSA) mode:

- **Early Release 15 drop:** Focused on the NSA NR architecture that allows operators to add an NR base station (called the gNB) to existing LTE base stations (eNB) in a system architecture still based on the 4G mobile core (known as the Evolved Packet Core [EPC]).

- **Main Release 15 freeze:** Focused on a standalone NR (SA NR) architecture in which NR gNBs connect to the 5G Core network (5GCN) without any LTE involvement. This would be more suitable to new network deployments that did not need to coexist with any existing LTE network infrastructure.

- **Late Release 15 drop:** Included support for an architecture where LTE base stations can be part of an SA NR network with two different deployment options: 1) control plane is assisted via the NR base station; 2) control plane is assisted via LTE base station.

Please refer to 3GPP TR 38.801 (see the "References" section) if you are interested in the details of the architecture options.

The 5G NR specifications were introduced in 3GPP Release 15. However, satisfying the full set of IMT-2020 requirements is addressed only with the completion of Release 16, and further 5G system enhancements are to be included as part of Release 17. They include items such as NR MIMO, enhancements to positioning, RAN/network slicing, and support for drones and satellites. Refer to https://www.3gpp.org/release-17 for the latest details of the work items that make up Release 17.

As with 2G, 3G, and 4G LTE in the past, 5G will continue to evolve in the future to address the industry and customer demands, and 3GPP will lead that effort.

Institute of Electrical and Electronics Engineers

The Institute of Electrical and Electronics Engineers (IEEE) was founded in 1884 by a small group of individuals in the electrical profession to support colleagues in their bourgeoning field and to aid them in their efforts to apply innovation for the betterment of humanity. Today the IEEE is the world's largest professional technical organization and now consists of many societies (39) that each focus on specialized technology areas. Out of these societies, the IEEE Computer Society is the world's leading organization of computing and information technology professionals.

The IEEE Standards Association (IEEE SA) is led by the Standards Association Standards Board (SASB), which is responsible for coordinating and encouraging the development of standards. This includes approving the launch of projects and ensuring that they maintain consensus, due process, openness, and balance. The SASB is also responsible for giving the final approval for standards.

Besides telecommunications, the IEEE SA contributes to a wide range of additional fields such as consumer technology, transport, electronics, information technology, healthcare and medical, robotics, and many more. The standards development process includes a formal balloting process on the draft before appraisal by a review committee of the SASB and final approval by the SASB itself.

One of the most notable successes of the IEEE SA is the IEEE 802 family of standards. As of today, the IEEE 802 family of standards consists of more than 70 published standards with at least 50 more under development. IEEE 802 covers all things Ethernet, as well as local-area networks (LAN), wireless local-area networks (WLAN), metropolitan-area networks (MAN), and more. One section of the 802.1 group is the Time-Sensitive Networking (TSN) Task Group, which among other things, developed a precision time protocol (PTP) profile, IEEE 802.1AS. More information on TSN is provided in the upcoming section, "IEEE TSN," and Chapter 7, "Precision Time Protocol," includes in-depth coverage of PTP profiles.

One IEEE society, the Instrumentation and Measurement Society, formed the Precise Networked Clock Synchronization (PNCS) Working Group, also known as P1588, to work on precise time distribution in Ethernet-based networks. The following section briefly discusses the contribution of these IEEE groups to time synchronization standards.

IEEE PTP

In 2002, IEEE introduced the 1588 standard that defined a precision clock synchronization protocol to synchronize distributed clocks using a packet network. IEEE 1588, widely known as PTP, was designed to improve timing accuracy in local systems compared to other common protocols such as Network Time Protocol (NTP).

The first approved version of IEEE 1588, published in November 2002, was titled "Standard for a Precision Clock Synchronization Protocol for Networked Measurement and Control Systems." However, many people refer to the original 1588-2002 as IEEE 1588v1.

IEEE 1588-2002 was later revised with the IEEE 1588-2008 standard, also referred to as IEEE 1588v2 or PTP version 2 (PTPv2). As of writing, the IEEE has added another version, so there are three generations of the 1588 standard titled, "Standard for a Precision Clock Synchronization Protocol for Networked Measurement and Control Systems":

■ **IEEE 1588-2002:** The original edition

- **IEEE 1588-2008:** The second version improves accuracy, precision, and robustness but is not backward compatible with the original 2002 version. It is also better adapted to wide-area deployments and introduces extensibility through the concept of PTP *profiles*.

- **IEEE 1588-2019:** Five years later, in June 2013, a project was authorized to revise the IEEE-2008 standard to improve security of PTP and also provide an implementation for applications needing sub-nanosecond time accuracy. In November 2019, the IEEE published IEEE 1588-2019, informally known as PTPv2.1, which was designed to be fully backward compatible with IEEE 1588-2008.

Currently, the P1588 Working Group is working on six additional projects, approved by the IEEE SA, to amend IEEE Std 1588-2019. These projects are described in what the IEEE calls project authorization requests (PARs) and are known by the titles P1588a through P1588g. These projects are expected to complete through the end of 2022. Consult https://sagroups.ieee.org/1588/active-projects/ for details on these amendments.

This standards-based method for synchronizing clocks is cost effective, supports heterogenous networks, and provides nanosecond-level synchronization precision for clock distribution. Nowadays, PTP is applied to a large array of different application areas including test and measurement, industrial automation, power industry, telecommunications, aerospace and automobile, audio-video distribution, and so on.

Note that none of the three releases of the IEEE 1588 standards, PTPv1, PTPv2, or PTPv2.1, are part of the Ethernet specifications. Chapter 7 discusses PTP and all the major profiles in detail.

IEEE TSN

The Time-Sensitive Networking (TSN) Task Group (TG) (part of IEEE 802.1 Working Group) is responsible for a subset of standards under the umbrella of the IEEE 802 LAN/MAN Standards Committee. IEEE 802.1 TSN focuses on enhancing IEEE Ethernet standards to address cross-industry requirements for applications that are time-sensitive. The TSN Task Group (TG) has evolved from the Audio Video Bridging (AVB) TG and so some of its standards still contain elements of that history.

The principal charter of the TSN Task Group is to develop support for deterministic services to be transported over IEEE 802 networks. *Deterministic* means it has a guaranteed, predictable level of performance with accurate timing, bounded (low) latency, low Packet Delay Variation (PDV), and low packet loss. These deterministic characteristics are very important to multiple industries, including aerospace, automotive, manufacturing, transportation, utilities, 5G mobile fronthaul networks, and so forth.

TSN is a collective name for a set of standards. It is more like a toolbox for the network designer to pick what is needed within the network for their targeted application. The key standards for TSN includes the following:

- **IEEE Std 802.1Q-2018:** Bridges and Bridged Networks (specifies Dot1q VLAN tags on IEEE 802.3 Ethernet networks)

- **IEEE Std 802.1AB-2016:** Station and Media Access Control Connectivity Discovery (specifies the Link Layer Discovery Protocol [LLDP])

- **IEEE Std 802.1AS-2020:** Timing and Synchronization for Time-Sensitive Applications (specifies a PTP profile)

- **IEEE Std 802.1AX-2020:** Link Aggregation Groups (known as LAG groups)

- **IEEE Std 802.1CB-2017:** Frame Replication and Elimination for Reliability

Many times, TSN develops amendments to one of the base 802.1 standards that eventually get rolled into updates to the base standards—as is the case with the 802.1Q VLAN specification. Subsequent versions of the larger standard incorporate many standalone documents that were originally published separately.

Because there are so many technologies and options available to apply to a specific problem area, the TSN TG has introduced the concept of TSN profiles (not to be confused with PTP profiles). TSN profiles can be categorized based on the industry where they apply (a *P* in front of the standard identifier means it is an ongoing project):

- **IEEE 802.1BA-2011:** TSN for Audio Video Bridging Systems

- **IEEE 802.1CM-2018:** TSN for Fronthaul

- **IEC/IEEE 60802:** TSN Profile for Industrial Automation

- **IEEE P802.1DG:** TSN Profile for Automotive In-Vehicle Ethernet Communications

- **IEEE P802-1DF:** TSN Profile for Service Provider Networks

- **IEEE P802.1DP:** TSN for Aerospace Onboard Ethernet Communications

The idea of a TSN profile is to limit to number and scope of options and features to ease deployment and improve interoperability and compliance (same goal as PTP Profiles). By making intelligent decisions about how to build a network for a specific role, the TSN is encouraging industries to adopt their solution.

Note that in terms used by these standards, a *bridge* is basically a switch (as opposed to an *end station* that is not involved in forwarding). The TSN profiles for other use cases, such as vehicles and industrial automation as well as service provider networks, are in various stages of development.

For the topic of this book, the 802.1CM-2018 standard defining the TSN profile for RAN fronthaul is the most relevant standard. IEEE 802.1CM mandates use of the ITU-T G.8275.1 PTP profile as part of a fronthaul network. There is a follow-on project, P802.1CMde, developing enhancements to this standard to support new developments in RAN fronthaul.

TSN structured the standards to address four main pillars of network architecture—time synchronization, bounded low latency, ultra-reliability, and resource management. Table 4-2 lists the associated standard or ongoing projects for each of these groups. (Again, references beginning with a *P* before the identifier are ongoing projects.)

Table 4-2 *TSN Standards, Projects, and the Associated Category*

Category	TSN Standard/Project
Time synchronization	Timing and Synchronization (802.1AS-2020); includes a profile of IEEE 1588
	Hot Standby Mechanism for 802.1AS (P802.1ASdm)
	YANG Data Model for 802.1AS (P802.1ASdn)
Bounded low latency	Credit Based Shaper (802.1Qav)
	Frame Preemption (802.3br-2016 and 802.1Qbu-2016)
	Scheduled Traffic (802.1Qbv-2015)
	Cyclic Queuing and Forwarding (802.1Qch-2017)
	Asynchronous Traffic Shaping (802.1Qcr-2020)
	Quality of Service Provisions (P802.1DC)
Ultra-reliability	Frame Replication and Elimination (802.1CB-2017)
	Path Control and Reservation (802.1Qca-2015)
	Per-Stream Filtering and Policing (802.1Qci-2017)
	Reliability for Time and Sync (802.1AS-2020)
Dedicated resources, models, and application programming interfaces	Stream Reservation Protocol (SRP) (802.1Qat-2010)
	SRP Enhancements and Performance Improvements (802.1Qcc-2018)
	Basic YANG Data Model (802.1Qcp-2018)
	Link-local Registration Protocol (P802.1CS)
	Resource Allocation Protocol (P802.1Qdd)
	TSN Configuration Enhancements (P802.1Qdj)
	LLDPv2 for Multiframe Data Units (P802.1ABdh)
	YANG for Connectivity Fault Management (CFM) (802.1Qcx-2020)
	YANG for LLDP (P802.1ABcu)
	YANG for Qbv, Qbu, and Qci (P802.1Qcw)
	YANG and MIB for FRER (P802.1CBcv)
	Extended Stream Identification (P802.1CBdb)
	YANG Data Model for 802.1AS (P802.1ASdn)

You have previously seen that some applications require a common understanding of time across interconnected network devices to function properly. Similarly, a network designed to deliver deterministic behavior may also require time alignment to behave correctly. This situation would normally be addressed by applying a PTP-based solution.

The problem with IEEE 1588 is that it describes numerous loosely defined parameters and deployment options. One could describe 1588 as only outlining what is possible for you to do, rather than what you should do. This lack of strict definition can impact hardware and software choices and increases the chances of incompatibility in deployment.

For these reasons, TSN specified a profile that uses applicable PTP technology in the context of IEEE Std 802.1Q over full-duplex Ethernet links (some other transport types are also included). The TSN TG published the IEEE 802.1AS-2011 profile with the title of "Timing and Synchronization for Time-Sensitive Applications in Bridged Local Area Networks." This specification facilitates the implementation of accurate time synchronization based on 1588/PTP and is commonly referred as *generalized PTP* or gPTP. The latest version of the profile is IEEE 802.1AS-2020.

IEEE 802.1AS specifies a PTP profile using layer 2 to transport synchronized time over physical or virtual bridged LANs. It also allows synchronization over other media types such as Wi-Fi (IEEE 802.11). Chapter 7 discusses IEEE 802.1AS in more detail along with the other important PTP profiles.

The 802.1AS profile is one of the several key 802.1 standards that were designed for networked low-latency applications, most specifically audio-video (remember that the TSN group started out as AVB). As mentioned previously, they are brought together in the TSN profile (not a PTP profile) of IEEE Std 802.1BA-2011. So, a TSN profile includes a PTP profile.

Of course, this book is dedicated mainly to mobile applications, which (at least so far) have made almost no use of the IEEE 802.1AS PTP profile—therefore, it is of limited further interest. But it is worth looking at several TSN technologies that can be used in one of the TSN profiles or even as a standalone technology (see the TSN TG website, https://1.ieee802.org/tsn/ for much more details on these standards).

TSN improves the real-time performance of time-sensitive network applications by adding two techniques: frame scheduling (IEEE 802.1Qbv) and frame preemption (IEEE 802.1Qbu). *Frame scheduling* is a technique that periodically allocates transmission time to different classes of traffic. The scheduler allocates each class of traffic a time slot during which it is allowed exclusive use of the Ethernet network, ensuring that the priority traffic can be granted guaranteed and uninterrupted transmission. This makes the path through the network much more predictable and deterministic.

Frame preemption defines a mechanism to allow time-critical messages to interrupt ongoing transmissions that are not time critical. Frame preemption is linked with another standard, IEEE 802.3br, "Specification and Management Parameters for Interspersing Express Traffic," which allows splitting longer frames into smaller fragments when sharing a single physical link. This feature allows the interface to temporarily interrupt the transmission of a long frame and send out a time-critical *express* frame before resuming transmission of the long frame. The interrupted frame is reassembled at the other end of the link (so both ends of the link need to support this feature).

Of course, this technique helps to reduce latency, especially where short, highly time-critical traffic is mixed with traffic using very long (jumbo) frames. As traffic interface speeds increase, the positive gain from frame preemption and express interleaving tends to reduce because even jumbo frames transmit quickly at higher speeds (consider 1 GE versus 100 GE).

In the 5G RAN, meeting deterministic jitter and latency requirements of the packet-based fronthaul network is quite a challenge. For now, just know that fronthaul is the name given to the transport network that delivers traffic over the *last mile* to the cell sites at the very edge of the mobile network. Fronthaul is covered in much more detail in several of the following chapters. For now, just understand that 5G has strict requirements for devices to support low latency and low PDV when used to build RAN fronthaul networks.

The TSN profile standard, IEEE 802.1CM, was defined to provide recommendations and specifications of TSN and timing techniques to use in fronthaul deployments based on Ethernet bridges. As discussed in the preceding section, another factor is that 5G Releases 15 and 16 define time-sensitive use cases for mobile, which further adds to the push for deterministic, low-latency transport networks. Please refer to Chapter 10 for more details on the 5G fronthaul network and timing and latency requirements for it.

In summary, TSN builds foundation technologies to address real-time and deterministic service behavior over Ethernet bridged networks. TSN is continuously evolving to meet an increasing interest in time-sensitive networking from an expanding number of unrelated industries.

IEEE and IEC

The International Electrotechnical Commission (IEC) is a global organization, founded in 1906, that develops international standards and operates conformity assessment systems in the fields of electrical, electronics, and related technologies. In the early 19th century, the IEC had a very important function in unifying the electrical systems of the globe and even contributed to the present system of measurement by introducing units concerned with electricity and magnetism (including the Ohm, Volt, and Ampere).

The IEEE and IEC have an agreement in place to increase their cooperation in developing international standards. The IEC/IEEE dual logo agreement encourages the joint development of standards by optimizing the resources and pooling the expertise to shorten the time to develop standards and publish them.

Related to time synchronization, the following three specifications are quite prominent: IEC 62439-3, IEC/IEEE 61850, and IEC/IEEE 60802.

■ **IEC 62439-3**: Part of the IEC 62439 series "Industrial Communication Networks—High availability automation networks," specifies the Parallel Redundancy Protocol (PRP) and High-availability Seamless Redundancy (HSR) protocol that provide seamless recovery in case of single failure of either a bridge in the network or a link between bridges.

To further improve resilience in the network, the (redundancy) principles behind PRP and HSR are also extended to network clocks. Annex C of the IEC 62439-3 (2016) standard specifies two profiles of a precision clock for industrial automation: one using the layer 3, end-to-end (L3E2E) delay measurement and the other using the layer 2, peer-to-peer (L2P2P) delay measurement. See Chapter 7 for more information.

- **IEC/IEEE 61850**: A substation automation standard that defines a communication protocol for intelligent electronic devices (IED) deployed in power grid substations. For correct functioning (and other benefits), devices within the substation are required to be time synchronized to a time server (such as a GPS receiver or PTP grandmaster).

  The IEEE/IEC also defined a pair of PTP profiles known as the Power Utility automation Profile (PUP) and the PTP Industry Profile (PIP). Subsequently, they developed IEEE C37.238-2017, known as the *Power Profile*, to distribute time within power stations, electrical substations, and across the distribution grid. Again, see Chapter 7 for much more detail on all these standards.

- **IEC/IEEE 60802**: A joint project of IEC and IEEE 802 to define TSN profiles for industrial automation. Like the other TSN profiles, this standard selects features, options, configurations, defaults, protocols, and procedures of bridges, end stations, and LANs to build industrial automation networks. IEC/IEEE 60802 provides the foundational guidelines to design converged multi-vendor TSN network infrastructure for industrial automation.

  60802 uses TSN techniques to improve upon standard (IEEE 802.1 and 802.3) Ethernet networks by providing guaranteed data transport with low PDV, bounded low latency, zero packet loss for critical traffic, and high availability.

  The current draft of 60802 defines two types of devices: Class A and Class B. Class A devices support a wide range of TSN functions including Scheduled Traffic, Frame Preemption, and numerous others. Some features are optional for Class B devices, which allows a more relaxed implementation than Class A.

Chapter 7 provides further coverage of the PTP profiles to support the power industry.

European Telecommunications Standards Institute

The European Telecommunications Standards Institute (ETSI) is a European standards organization that deals with telecommunications, broadcasting, and other electronic communications networks and services. ETSI is a much more recent organization, having only been created at the behest of the European Commission in 1988. Although its special role is that it is one of only three official European standards organizations, its impact is increasingly global. It now consists of more than 900 member organizations that are drawn from more than 60 countries.

ETSI has numerous Technical Committees that address standardization work within specific areas of technology. There are also other parts of the organization that work more with market segments or industry groupings rather than technologies. Still others work with their major partners, which are currently 3GPP and oneM2M.

Note   For details on the standards setting process in ETSI and a comparison with the ITU-T, see the ETSI reference *Understanding ICT Standardization: Principles and Practice* at the end of the chapter.

ETSI is one of the seven members (organizational partners) that make up 3GPP. 3GPP offers an established development environment allowing the production of standards, technical reports, and specifications about the technology underpinning mobile communications. Once these documents are approved by 3GPP, each organizational partner adopts these standards as their own and publishes them using identical text. If you ever browse the 3GPP or ETSI websites for technical reports or standards, you should notice this commonality. If asked by the European Commission, these publications can be adopted as a European standard (EuroNorm).

The other partner of ETSI, oneM2M, was created to consolidate standardization of machine-to-machine (M2M) and IoT technologies. As with 3GPP, oneM2M is made up of regional standards-setting organizations including ETSI. The purpose and goal of oneM2M is to develop technical specifications for a common service layer that can be embedded into the hardware and software of devices. This would enable devices in the field to more easily communicate with M2M application servers equipped with the same capabilities. Like the arrangement with 3GPP, the partner organizations republish the oneM2M specifications and reports.

ETSI also provides testing and interoperability assistance to 3GPP and oneM2M technical committees. The ETSI Center for Testing and Interoperability (CTI) supports testing interoperability, conforming protocol specifications, and validating standards defined by 3GPP and oneM2M.

ETSI also hosted the formation of an industry specification group (ISG) to contribute to standardizing specifications for NFV. The ETSI ISG NFV community, originally started with seven operators, has contributed over 100 specifications and reports to the topic. These specifications are gathered and published as releases roughly every two years. The latest, NFV Release 4, was formally approved and published in late 2020.

Note   See https://www.etsi.org/technologies/nfv for more details on the ETSI contribution toward NFV.

Internet Engineering Task Force

The Internet Engineering Task Force (IETF) is a standards body, under the legal umbrella of the Internet Society (ISOC), that focuses on the development of protocols for IP-based (Internet) networks. The IETF is a totally open community with no formal membership and is primarily a volunteer organization. The IETF is responsible for the development and quality of the Internet standards, whereas ISOC facilitates the legal and organizational issues for the IETF. The IETF organization structures and functions are outlined in numerous RFCs, such as RFC 8712.

The standardization process of the IETF is often described as "rough consensus and running code." Engineers (globally) assemble into working groups dedicated to a particular area and exchange their ideas and wisdom to arrive at a rough consensus on a specification. The goal then is to implement the standards through running code—an approach specifically designed to allow rapid development of interoperable protocols. The standardization process is outlined in RFC 2026 (and subsequent errata and updates).

Internet specifications (and many other related documents) are formally published as one of a series of documents known as *request for comments* (RFC). The RFC document tradition, going back to the original RFP 1, has a history stretching back over 50 years. The RFC editor is directly responsible for the publication of the RFC, under the general direction of the Internet Architecture Board (IAB).

To get adopted, a specification undergoes a period of development and cycles of review and updates by the Internet community. This is not an easy process, because even the status of "Draft Standard" requires at least two completely independent implementations of the standard that have been demonstrated to interoperate. Eventually the revised document is adopted as a standard and is published, but only when it has gained substantial operational maturity and a significant number of successful implementations.

The IETF divides its work into various areas, and each focus area has its own set of working groups:

- **Applications and Real-time Area (art):** Develops application protocols and architectures for "real-time" applications—voice, video, instant messaging, and so forth, applications that may be more tolerant to delay—HTTP, email, FTP, and so forth; and applications that have a wide range of use across real-time and nonreal-time deployments—Uniform Resource Identifier (URI) schemes, Multipurpose Internet Mail Extensions (MIME) types, authentication mechanisms, codecs, and so forth.

- **General Area (gen):** Focuses on updating and maintaining the IETF standards development process.

- **Internet Area (int):** Includes the IP layer (both IPv4 and IPv6), co-existence between IP versions, issues associated with IPv4 address space limitations, DNS, Virtual Private Networks (VPN), and pseudo-wires and Multiprotocol Label Switching (MPLS)-related issues, mobility, and so on.

- **Operations and Management Area (ops):** Includes network management, Authentication, Authorization, and Accounting (AAA), and various operational issues with the DNS, IPv6, security, and routing. The IETF further divides ops into two separate functions:

  - Network Management covers Internet management, protocols related to AAA (such as NETCONF, SNMP, RADIUS, and Diameter), and data modeling languages such as YANG.

  - Operations is largely responsible for incorporating operator feedback and input regarding the IETF work.

- **Routing Area (rtg):** Is responsible for ensuring continuous operation of the Internet routing system and developing new protocols, extensions, and bug fixes in a timely manner.

- **Security Area (sec):** Focuses on security protocols that provide security services like integrity, authentication, non-repudiation, confidentiality, and access control.

- **Transport Area (tsv):** Covers a range of technical topics related to data transport on the Internet. It supports mechanisms that detect, react to, and manage congestion on the Internet, for protocols such as Transmission Control Protocol (TCP), Forward Error Correction (FEC), quality of service, congestion control and management, and so forth.

The IETF collaborates and coordinates closely with the IEEE, most especially the 802 part of the organization. With this arrangement, the IETF gains improved access to IEEE 802 expertise in the very widely deployed LAN technologies, while IEEE 802 gains better access to IETF expertise on IP encapsulation, routing, and transport. See RFC 7241 for details on the relationship between the two organizations.

Well before the IEEE defined PTP/1588, the IETF developed NTP through its RFC process. The IETF started specifying NTP starting in the mid-1980s (RFC 958), and its current version, version 4 (RFC 5905), is a very mature technology. Study on NTPv5 has been initiated with the goal of better accuracy, and the work is currently in RFC draft stage.

RFC 5905 and RFC 4330 also define a subset of NTP called the Simple Network Time Protocol (SNTP) used to support networks and applications that do not require the complete range of NTP features and performance. Another notable contribution of the IETF to the area of timing is RFC 3339, which defines a date and time format for use in IP based on ISO 8601.

There is a working group within the IETF called Timing over IP Connection and Transfer of Clock (TICTOC), which is responsible for aspects of timing for the Internet technologies. The TICTOC WG is the driving force behind about four or five published (but currently still draft) RFCs related to PTP and timing. Most importantly, TICTOC currently defines a PTP profile for corporate networks known as the *Enterprise PTP Profile*. An example of likely adopters is financial trading companies, where more precise time accuracy is needed than available from current implementations of NTP.

The contributions of TICTOC also cover the areas of defining an SNMP MIB and YANG data model for PTP as well as RFCs on PTP security and multipath synchronization. Draft RFC 8575 defines a YANG model to configure PTP clocks as specified in IEEE 1588 (for the default profiles). Please refer to Chapter 12, "Operating and Verifying Timing Solutions," for more details on PTP monitoring, management, and assurance. Chapter 7 offers further information on the Enterprise PTP profile, and there are additional references at the end of the chapter.

Another relevant part of the IETF is in the area of deterministic networking with the Deterministic Networking (DETNET) Working Group. This working group focuses on mechanisms to define data paths that are bounded in latency, packet loss, PDV (jitter), and reliability over layer 2 and layer 3 networks. The DETNET WG collaborates with IEEE TSN to define TSN for next-generation technologies (such as 5G) and industrial automation. Based on the work of IEEE TSN, which is responsible for layer 2 operation, the DETNET WG acts to define a common architecture for both layer 2 and layer 3.

Of course, beyond the specialized areas outlined here, the IETF is the overarching guiding hand behind much of the Internet-, IP-, and MPLS-based transport networks that connect everything together. These technologies are key components of transport networks, not only to carry timing signals but also to build networks that connect the mobile base station to the rest of the Internet. So, the next section deals with the organizations that define standards for the radio access network.

Radio Access Network

5G networks need to address a wide range of use cases and therefore must be based on a flexible architecture that can adapt to multiple deployment scenarios. Traditionally, the radio access network (RAN) consisted of base stations, radios, and antennas. This equipment normally consisted of dedicated hardware or proprietary appliances from a radio vendor. There are significant changes now occurring to this model in the RAN, of which the two most important are virtualization and centralization.

One change is that some functions in the base station (for example, radio baseband processing) can be virtualized and deployed on the commercial off-the-shelf (COTS) hardware. This is moving the RAN away from the hardware appliance toward a (increasingly open) software model. RAN software virtualization also allows operators and vendors to build new feature sets to address various industry use cases, including IoT, industrial automation, and so forth.

The second evolution is to move some of the radio processing to a more centralized site instead of it all being hosted out at the cell site. This has advantages for allowing low-latency coordination between neighboring cell sites and reduces the footprint of equipment and infrastructure needed at remote cell sites. The radio components remaining at the cell site are connected to the centralized components via a network—what is now called the *fronthaul* network.

Because of latency requirements in radio processing, this fronthaul network is not widely spread; typically it is restricted to a radius of up to 20 km. The centralized sites tend to be more like a mini-datacenter and are increasingly operated using cloud-type techniques. To address any ultra-low-latency use cases (autonomous vehicles, for example), application infrastructure can be deployed closer to the end user in these mini-datacenters. This architecture provides the flexibility to deploy RAN software along with packet core functions and application software within a single cloud infrastructure.

One of the consequences of the move to virtualization and the change to a 5G architecture is the increased contribution by software and the reduction in proprietary hardware. This has generated a strong desire by the operators to move to more open systems in the RAN. These operators and many industry vendors came together and formed various alliances to define 5G RAN functions using open protocols as well as specify COTS platform features. The following section covers the most important organizations behind these evolutions in the RAN.

Common Public Radio Interface

Common Public Radio Interface (CPRI) is a cooperation between five companies: Ericsson, Huawei, NEC, Nokia (previously Siemens), and Nortel. Nortel left the organization in December 2009 after contributing to the early releases of CPRI. The CPRI specification defines an interface that connects radio antennas to the 3G/4G base station. In their terminology, CPRI is a specification for the interface that connects the Radio Equipment Controller (REC) to the Radio Equipment (RE). The latest version of the CPRI specification (7.0) was published in October 2015.

Note   The terminology between different SDOs and even each generation of mobile is a cause of severe confusion for those ramping up in this field. At this point of reading, the book tends toward simplicity, avoiding some terms until later in the book. But do not be surprised to see completely different terms from different organizations and to see that terminology changes for each new generation of mobile specifications.

Although the layer 1 transport is standardized, the CPRI specification reserves data fields for the transmission of vendor-specific data, allowing manufacturers to customize their deployment of CPRI. Due to this vendor-specific customization, the industry regards implementations of CPRI interfaces as proprietary.

5G introduced new features, increased performance, and a need for more flexible deployment scenarios, which requires an improved specification for CPRI. In August 2017, CPRI published the first version of the enhanced CPRI specification, known as eCPRI, with several advantages:

- Tenfold reduction in the fronthaul interface bandwidth
- Statistical multiplexing benefits

- Support of packet-based transport such as Ethernet

- Support for real-time user applications using sophisticated coordination algorithms to guarantee radio performance

- Provision to add new functionality by software updates

In May 2019, CPRI released eCPRI version 2.0, a specification to support legacy CPRI (version 7.0) over Ethernet, allowing for CPRI and eCPRI interworking. Chapter 10 covers CPRI and eCPRI in some detail.

The eCPRI specifications became a foundation for many alliances wishing to define open packet-based fronthaul interfaces for 5G deployments. The move toward open specifications makes it easier for operators to mix and match vendor equipment and accelerate innovation in their 5G RAN networks.

xRAN and O-RAN Alliance

The xRAN Forum was formed in 2016 with the goal of developing, standardizing, and promoting a software-based extensible RAN (xRAN). The xRAN Forum gained tremendous industry momentum with leadership from operators AT&T, Deutsche Telecom, KDDI, NTT DOCOMO, SK Telecom, Telstra, and Verizon. At the same time, the xRAN Forum has added contributing members from suppliers, innovators, and research leaders including Intel, Texas Instruments, Mavenir, Cisco, Altiostar, Fujitsu, NEC, leading universities, and so forth.

Researchers from this grouping focused on three areas:

- Decoupling the RAN control plane from the user plane

- Building a modular base station software stack that uses COTS hardware

- Publishing open northbound and southbound interfaces

On April 12, 2018, the xRAN Forum released its first specification, "xRAN Fronthaul Specification version 1.0," dedicated to virtualizing the fronthaul portion of the network. The specification is intended to drive interoperability between the base station or baseband units and the remote radio heads, even if they are from different vendors.

In February 2018, the xRAN Forum announced its intent to merge with the C-RAN Alliance to form a worldwide, carrier-led effort toward openness in the RAN of next-generation mobile systems. The new alliance is called the O-RAN Alliance. The xRAN open fronthaul interface specifications are now under Working Group 4 of the O-RAN alliance.

The O-RAN Alliance, originally founded by AT&T, China Mobile, Deutsche Telekom, NTT DOCOMO, and Orange, held its first board meeting at MWC Shanghai in June 2018. Seven new members were also approved, including Bharti Airtel, China Telecom, KT, Singtel, SK Telecom, Telefonica, and Telstra. The articles constituting the O-RAN Alliance are cosigned by delegates from 12 operators as the official foundation of the operator-driven initiative.

The O-RAN architecture (see Figure 4-5) defines standardized interfaces to enable an open, interoperable ecosystem that complements the standards promoted by 3GPP and other SDOs.

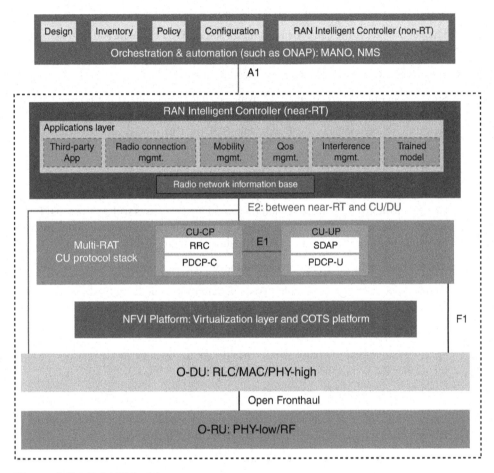

Figure 4-5 *O-RAN Architecture*

The O-RAN Alliance defines nine technical working groups (WG), all of them under the supervision of the Technical Steering Committee. The technical WGs have specific focus areas and are open to all members and contributors. The workgroups are as follows:

- **WG1**: Use Cases and Overall Architecture Workgroup

  Identifies tasks to be completed within the scope of the architecture and use cases, assigns task group leads, and drives completion of the task in cooperation with other O-RAN work groups.

- **WG2**: The Non-RealTime (RT) RAN Intelligent Controller (RIC) and A1 Interface Workgroup

Focuses on radio resource management, higher-layer procedure optimization, policy optimization in RAN, and providing artificial intelligence/machine learning (AI/ML) models to near-RT RIC.

- **WG3:** The Near-real-time RIC and E2 Interface Workgroup

  Enables near-real-time control and optimization of RAN elements and resources via data collection and actions over E2 interface.

- **WG4:** The Open Fronthaul Interfaces Workgroup

  Specifies an open fronthaul interface, which enables true multivendor interoperability in the RAN. Provides guidelines for time synchronization requirements on the various RAN components: the Radio Unit (RU), the Distributed Unit (DU), and the Control Unit (CU).

- **WG5:** The Open F1/W1/E1/X2/Xn Interface Workgroup

  Provides fully operable multivendor specifications (compliant with 3GPP specification) for the F1/W1/E1/X2/Xn interfaces and in some cases will propose 3GPP specification enhancements. These interfaces are a collection of virtual connections between components in the RAN (specified originally by 3GPP).

- **WG6:** The Cloudification and Orchestration Workgroup

  Defines specification to decouple RAN software from underlying hardware to leverage the benefits of being able to use commodity hardware platforms.

- **WG7:** The White-box Hardware Workgroup

  Specifies and releases a complete reference design for open hardware platforms for the various RAN components, including Picocells.

- **WG8:** Stack Reference Design Workgroup

  Develops the software architecture, design, and release plan for the O-RAN open DU and open CU as per 3GPP specifications for the NR protocol stack.

- **WG9:** Open X-haul Transport Work Group

  Focuses on the 5G transport design and provides a reference architecture blueprint for network deployment.

There is also another organization to mention: The O-RAN Software Community (SC) is a collaboration between the O-RAN Alliance and Linux Foundation with the mission to support the creation of software for the RAN.

The O-RAN Alliance is making great strides and swiftly moving to reengineer the RAN for the mobile industry. From a timing point of view, the most interesting is probably WG4, because that WG is concerned with the timing requirements for the fronthaul network. One of WG4's documents, "O-RAN Fronthaul Control, User and Synchronization Plane Specification," is recommended reading (version 5.0 as of writing, referenced at the end of the chapter).

TIP OpenRAN

The Telecom Infra Project (TIP) was cofounded in 2016 by Facebook, Intel, Nokia, Deutsche Telekom, and SK Telecom. Its mission is to accelerate the development and deployment of open, disaggregated, and standards-based solutions to deliver high-quality connectivity. Its more than 1000 members include a large number of Internet companies, telecom operators, technology vendors, research organizations, and system integrators.

TIP has two main categories of project groups:

- **Product Project Groups:** Have three main projects that effectively make up an end-to-end network: Access, Transport, and Core and Services. These groups are focused on innovating and building the right products.

- **Solution Group:** Aims to codify open, disaggregated, interoperable network elements into a broad range of end-to-end solutions.

As shown in Figure 4-6, each main project is further divided into project and solution groups more focused on a particular product area or solution type.

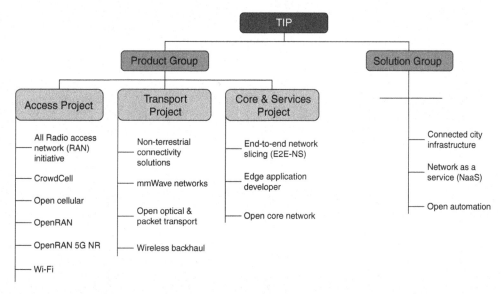

Figure 4-6   *Telecom Infra Project Organizational Structure*

As part of the Access project, TIP initiated an OpenRAN project group with several subcomponents for each part of the RAN. The main goal of this initiative is to develop a comprehensive RAN solution based on COTS, general-purpose computing platforms. These RAN components would be interconnected using open interfaces based on open, vendor-neutral software.

The objective is to have RAN solutions that benefit from the flexibility and pace of innovation associated with software-driven developments on fully programmable platforms.

This goal should sound very similar to that of another organization, and so it is no surprise that in February 2020, the OpenRAN project from TIP and O-RAN Alliance announced a liaison agreement.

Compared to TIP, O-RAN is more focused on developing and driving the adoption of standards so that equipment can work together. On the other hand, TIP is more focused on implementation, field testing, and deployment, so TIP encourages interoperability events such as PlugFests. This approach aims to improve the ability of software and hardware equipment from different vendors to work together.

One additional difference is that TIP is more interested in all generations of mobile technology, while O-RAN is much more focused on 4G and especially 5G.

For more information on TIP and OpenRAN, see the references at the end of the chapter.

MEF Forum

Founded in 2001, the MEF Forum (MEF) focused on Carrier Ethernet (CE) services and technology covering scalability, reliability, quality of service, and service management attributes. Later, MEF extended its scope and included optical, CE, IP, and Software-Defined Wide-Area Network (SD-WAN) services, as well as orchestration of the service life cycle. MEF is composed of 200+ member companies, including service providers, carriers, network equipment and software vendors, and other networking companies from across the globe.

Compared to networking SDOs like the IETF and IEEE 802.1, MEF dedicates its efforts to aligning the standards with corporate objectives and defining network architectures, deployment scenarios, and test suites. The biggest single aspect that differentiates MEF is that it defines networks from the point of view of delivering transport as a service. MEF defines service interfaces to enable easy interconnection between two main entities:

- **Subscriber:** The organization that purchases CE services
- **Service provider (SP):** The organization that provides CE services

Carrier is a term that is used to describe large telecommunications service providers. *Carrier Ethernet* (CE) describes extensions to base Ethernet standards that allow SPs to provide and market transport services to their customers using Ethernet technology. A good example of such a feature would be Q-in-Q (originally specified in IEEE 802.1ad, now incorporated in IEEE 802.1Q), which allows double VLAN tags and allows much better separation of individual customer traffic in the SP network.

MEF has published over 70 technical specifications and implementation agreements that have contributed to technology interoperability, operational efficiency, and innovation. MEF standards are public and can be freely downloaded without charge. MEF also has a certification program that provides conformity testing against the MEF specifications.

New MEF standards are numbered sequentially as they are published—for example, MEF 1, MEF 2, MEF 3, and so on. Complete revision of an existing standard retains the original number, with a dot and numeral indicating the revision number. For example, MEF 22 is replaced by MEF 22.1, and MEF 22.1 is replaced by MEF 22.2. Any amendment to an existing standard uses a third level of numbering; for example, MEF 22.2.1 is an amendment to MEF 22.2.

For the mobile and synchronization markets, MEF produced MEF 22.1, "Mobile Backhaul Implementation Agreement – Phase 2," which provides a definition of Ethernet services for use in mobile backhaul networks. There are subsequent updates:

- MEF 22.1 includes support for physical layer synchronous Ethernet (SyncE) and packet-based PTPv2 to transport frequency synchronization across Carrier Ethernet networks to support mobile backhaul.

- MEF 22.2 incorporates service requirements for small cells and heterogeneous radio network deployments covering tight radio synchronization over a wide range of mobile network deployments.

- MEF 22.3 incorporates phase/time synchronization as service deployment over the SP network. The latest amendment MEF 22.3.1 covers the requirements for Ethernet services that can be used as transport for 5G mobile networks and discusses fronthaul and network slicing applicability.

Society of Motion Picture and Television Engineers and Audio Engineering Society

There are two main standards bodies that contribute to the audio-video media environment:

- **Society of Motion Picture and Television Engineers (SMPTE):** Founded in 1916, SMPTE is a global association of engineers, technologists, and executives working in the media and entertainment industry. SMPTE has more than 800 standards and engineering guidelines for broadcasting, filmmaking, digital cinema, audio recording, information technology, and medical imaging.

  When pieces of audio, video, or musical equipment are required to work together, some means of synchronization is required to ensure they play in time with each other. SMPTE defines a time code, or SMPTE code, which is an electronic signal that is used to identify a precise location in digital systems and on time-based media like audio or video tape.

  Every frame of video or audio has a time reference that is recorded while capturing it. Time code on a video frame is like a page number in a book, and each page in a book is like a video frame in a video file. Time codes are added to film, video, or audio materials and used to accurately blend or synchronize the audio (recorded by the audio recorder), video (recorded by the video recorder), music instruments, or theatrical production work.

SMPTE 2059 defines how to synchronize video equipment over an IP network and uses a 1588v2 profile for synchronization. SMPTE 2059 is published in two parts:

- SMPTE 2059-1 defines signal generation based on time information distributed by PTP.

- SMPTE 2059-2 includes techniques for media synchronization using IEEE 1588v2 (as a PTP profile).

- **Audio Engineering Society (AES):** Founded in 1948, the AES is the only professional society devoted exclusively to audio technology. The AES is a global organization of audio engineers, creative artists, scientists, and students that promotes and contributes to research in audio technologies.

   AES67 is a standard for audio over IP and audio over Ethernet interoperability. This profile maintains synchronization of high-performance audio streams at the receiver when transmitted over low-latency IP multicast or unicast transport. AES67 Annex A specifies the use of the IEEE 1588-2008 default profile for synchronization.

In May 2016, the AES published AES-R16-2016, a report explaining interoperability between AES67 and SMPTE 2059-2 profiles. See the AES-R16-2016 reference in the upcoming list and refer to Chapter 7 for more details on these profiles and updates to them.

Summary

Along with SDOs, there are many alliances and nonprofit organizations that contribute to technology awareness, business value creation, and promotion of standards. Next Generation Mobile Networks Alliance (NGMN) and GSM Association (GSMA) are also prominent alliances that represent the interests of mobile network operators globally.

SDOs benefits industry in different ways:

- Prevent overlap and fragmentation of work.

- Combine efforts to capture requirements, define specifications, and establish architectures, which allows the operator or user to deploy multivendor solutions that are known to interoperate with a defined quality. This leads to improved diversity in suppliers and increased cost effectiveness.

- Provide a legal and business framework that ensures fair practices in licensing and publication rights and establishes clear processes to resolve conflicts of interest or issues around intellectual property rights.

SDOs exist to drive continuous innovation, feature developments, and standardization across industry deployments, and their contribution to connecting the world is incalculable.

Now that you have an overview of the organizations that define the solutions for the timing and synchronization problems that the industry faces, it is time to look at the key components of the timing solution and the metrics used to quantify their performance, as described in Chapter 5, "Clocks, Time Error, and Noise."

References in This Chapter

3GPP

"Evolved Universal Terrestrial Radio Access (E-UTRA); Base Station (BS) radio transmission and reception." *3GPP*, 36.104, Release 16, (16.5.0), 2020. https://www.3gpp.org/DynaReport/36104.htm

"Study on New Radio (NR) access technology." *3GPP* 38.912, Release 15 (15.0.0), 2018. https://portal.3gpp.org/desktopmodules/Specifications/SpecificationDetails.aspx?specificationId=3059

"Study on new radio access technology: Radio access architecture and interfaces." *3GPP*, 38.801, Release 14 (14.0.0), 2017. https://portal.3gpp.org/desktopmodules/Specifications/SpecificationDetails.aspx?specificationId=3056

"Study on scenarios and requirements for next generation access technologies." *3GPP*, 38.913, Release 14 (14.3.0), 2017. https://portal.3gpp.org/desktopmodules/Specifications/SpecificationDetails.aspx?specificationId=2996

Abdelkafi, N. et al. *Understanding ICT Standardization: Principles and Practice.* ETSI, 2018. https://www.etsi.org/images/files/Education/Understanding_ICT_Standardization_LoResWeb_20190524.pdf

Audio Engineering Society (AES)

"AES standard for audio applications of networks – High-performance streaming audio-over-IP interoperability." *AES*, AES67-2018, 2018. http://www.aes.org/publications/standards/

"PTP parameters for AES67 and SMPTE ST 2059-2 interoperability." *AES*, AES-R16-2016, 2016. http://www.aes.org/publications/standards/

Internet Engineering Task Force (IETF)

Arnold, D. and H. Gerstung. "Enterprise Profile for the Precision Time Protocol With Mixed Multicast and Unicast Messages." *IETF*, Internet-Draft draft-ietf-tictoc-ptp-enterprise-profile. https://tools.ietf.org/id/draft-ietf-tictoc-ptp-enterprise-profile-18.txt (current draft at time of writing)

Bradner, S. "The Internet Standards Process – Revision 3." *IETF*, RFC 2026, 1996. https://tools.ietf.org/html/rfc2026

Camarillo, G. and J. Livingood. "The IETF-ISOC Relationship." *IETF*, RFC 8712, 2020. https://tools.ietf.org/html/rfc8712

Dawkins, S, P. Thaler, D. Romascanu, and B. Aboba. "The IEEE 802/IETF Relationship." *IETF*, RFC 7241, 2014. https://tools.ietf.org/html/rfc7241

Jiang, Y., J. Xu, and R. Cummings. "YANG Data Model for the Precision Time Protocol (PTP)." *IETF*, RFC 8575, https://tools.ietf.org/html/rfc8575

Klyne, G. and C. Newman. "Date and Time on the Internet: Timestamps." *IETF*, RFC 3339, 2002. https://tools.ietf.org/html/rfc3339

Mills, D. "Simple Network Time Protocol (SNTP) Version 4 for IPv4, IPv6 and OSI." *IETF*, RFC 4330, 2006 (obsoleted by RFC 5905). https://tools.ietf.org/html/rfc4330

Mills, D., J. Martin, J. Burbank, and W. Kasch. "Network Time Protocol Version 4: Protocol and Algorithms Specification." *IETF*, RFC 5905, 2010. https://tools.ietf.org/html/rfc5905

Common Public Radio Interface (CPRI)

"Common Public Radio Interface: Requirements for the eCPRI Transport Network." *CPRI*, eCPRI Transport Network V1.2, 2018. http://www.cpri.info/downloads/Requirements_for_the_eCPRI_Transport_Network_V1_2_2018_06_25.pdf

"Common Public Radio Interface: eCPRI Interface Specification." *CPRI*, eCPRI Specification V1.2, 2018. http://www.cpri.info/downloads/eCPRI_v_1_2_2018_06_25.pdf

"Common Public Radio Interface: eCPRI Interface Specification." *CPRI*, eCPRI Specification V2.0, 2019. http://www.cpri.info/downloads/eCPRI_v_2.0_2019_05_10c.pdf

"Common Public Radio Interface (CPRI): Interface Specification." *CPRI*, CPRI Specification V7.0, 2015. http://www.cpri.info/downloads/CPRI_v_7_0_2015-10-09.pdf

IEEE Standards Association

"IEC/IEEE International Standard – Communication networks and systems for power utility automation – Part 9-3: Precision time protocol profile for power utility automation." *IEC/IEEE Std 61850-9-3:2016*, 2016. https://standards.ieee.org/standard/61850-9-3-2016.html

"IEEE Standard for a Precision Synchronization Protocol for Networked Measurement and Control Systems." *IEEE Std 1588:2002*, 2002. https://standards.ieee.org/standard/1588-2002.html

"IEEE Standard for Precision Synchronization Protocol for Networked Measurement and Control Systems." *IEEE Std 1588:2008*, 2008. https://standards.ieee.org/standard/1588-2008.html

"IEEE Standard for Precision Synchronization Protocol for Networked Measurement and Control Systems" *IEEE Std 1588:2019*, 2019. https://standards.ieee.org/standard/1588-2019.html

"IEEE Standard for Local and Metropolitan Area Networks – Timing and Synchronization for Time-Sensitive Applications." *IEEE Std 802.1AS-2020*, 2020 https://standards.ieee.org/standard/802_1AS-2020.html

"IEEE Standard for Local and Metropolitan Area Networks – Timing and Synchronization for Time-Sensitive Applications in Bridged Local Area Networks." *IEEE Std 802.1AS-2011*, 2011. https://standards.ieee.org/standard/802_1AS-2011.html

"IEEE Standard for Local and Metropolitan Area Networks – Time-Sensitive Networking for Fronthaul." *IEEE Std 802.1CM-2018*, 2018. https://standards.ieee.org/standard/802_1CM-2018.html

International Electrotechnical Commission (IEC). "Industrial communication networks – High availability automation networks – Part 3: Parallel Redundancy Protocol (PRP) and High-availability Seamless Redundancy (HSR)." *IEC Std 62439-3:2016*, Ed. 3, 2016. https://webstore.iec.ch/publication/24447

International Organization for Standardization (ISO). "Date and time – Representations for information interchange – Part 1: Basic rules." *ISO*, 8601-1:2019, 2019. https://www.iso.org/standard/40874.html

International Telecommunication Union Radiocommunication Sector (ITU-R)

"Detailed schedule for finalization of the first release of new Recommendation ITU-R M.[IMT-2020.SPECS] 'Detailed specifications of the terrestrial radio interfaces of International Mobile Telecommunications-2020 (IMT-2020)'." *ITU-R*, Contribution 21, Revision 1, 2020. https://www.itu.int/md/R15-IMT.2020-C-0021/en

"M.2012-4: Detailed specifications of the terrestrial radio interfaces of International Mobile Telecommunications Advanced (IMT-Advanced)." *ITU-R Recommendation*, M.2012-4, 2019. https://www.itu.int/rec/R-REC-M.2012

"M.2083-0: IMT Vision – Framework and overall objectives of the future development of IMT for 2020 and beyond." *ITU-R Recommendation*, M.2083-0, 2015. https://www.itu.int/rec/R-REC-M.2083

International Telecommunication Union Telecommunication Standardization Sector (ITU-T). "Q13/15 – Network synchronization and time distribution performance." *ITU-T Study Groups*, Study Period 2017–2020. https://www.itu.int/en/ITU-T/studygroups/2017-2020/15/Pages/q13.aspx

MEF Forum

"Amendment to MEF 22.3: Transport Services for Mobile Networks." *MEF Amendment*, 22.3.1, 2020. https://www.mef.net/wp-content/uploads/2020/04/MEF-22-3-1.pdf

"Transport Services for Mobile Networks." *MEF Implementation Agreement*, 22.3, 2018. https://www.mef.net/wp-content/uploads/2018/01/MEF-22-3.pdf

"O-RAN Fronthaul Control, User and Synchronization Plane Specification 5.0." *O-RAN*, O-RAN.WG4.CUS.0-v05.00, 2020. https://www.o-ran.org/s/O-RAN.WG4.CUS.0-v05.00.pdf

SMPTE. "SMPTE Profile for Use of IEEE-1588 Precision Time Protocol in Professional Broadcast Applications." SMPTE ST 2059-2:2015, 2015. https://www.smpte.org/

Chapter 4 Acronyms Key

The following table expands the key acronyms used in this chapter.

Term	Value
3G	3rd generation (mobile telecommunications system)
3GPP	3rd Generation Partnership Project (SDO)
4G	4th generation (mobile telecommunications system)
5G	5th generation (mobile telecommunications system)
5GCN	5G Core network
AAA	Authentication, Authorization, and Accounting
AES	Audio Engineering Society
AI	artificial intelligence
ANSI	American National Standards Institute
ARIB	Association of Radio Industries and Businesses
ATIS	Alliance for Telecommunications Industry Solutions
AVB	Audio Video Bridging
CA	Carrier Aggregation
CCSA	China Communications Standards Association
CE	Carrier Ethernet
CFM	Connectivity Fault Management
codec	coder-decoder
CoMP	Coordinated Multipoint
COPS	Common Open Policy Service
COTS	commercial off-the-shelf
CPRI	Common Public Radio Interface
CT	Core Network and Terminals
CTI	ETSI Center for Testing and Interoperability
CU	Central Unit (3GPP)/Control Unit (O-RAN) (5G radio access network)
CWDM	Course Wavelength Division Multiplexing (optical technology)
DAB	digital audio broadcasting (digital radio)
DETNET	Deterministic Networking (Working Group)
DNS	Domain Name System

Term	Value
DSL	digital subscriber line (a broadband technology)
DU	Distributed Unit (5G radio access network)
DWDM	Dense Wavelength Division Multiplexing (optical technology)
eCPRI	enhanced Common Public Radio Interface
EDGE	Enhanced Data rate for GSM Evolution
eLAA	enhanced Licensed Assisted Access
eMBB	enhanced mobile broadband
eMBMS	enhanced Mobile Broadcast Multicast System
eNB	evolved Node B (E-UTRAN)
ENUM	electronic numbering
EPC	Evolved Packet Core
ETSI	European Telecommunications Standards Institute
FDD	frequency division duplex
FEC	Forward Error Correction
FRER	Frame Replication and Elimination for Reliability
FTP	File Transfer Protocol
GE	gigabit Ethernet
gNB	next (5th) generation node B
GPS	Global Positioning System
GPRS	General Packet Radio Service
gPTP	generalized precision time protocol (IEEE 802.1AS)
GSM	Global System for Mobile communications
GSMA	GSM Association
HDR	high dynamic range
HDTV	high definition television
HSDPA	High Speed Downlink Packet Access
HSPA	High Speed Packet Access
HSR	High-availability Seamless Redundancy
HSUPA	High Speed Uplink Packet Access
HTTP	Hypertext Transfer Protocol
IAB	Internet Architecture Board

Term	Value
ICT	information and communication technology
IEC	International Electrotechnical Commission
IED	intelligent electronic devices
IEEE	Institute of Electrical and Electronics Engineers
IEEE SA	IEEE Standards Association
IETF	Internet Engineering Task Force
IMT	International Mobile Telecommunication
IoT	Internet of things
IP	Internet protocol
IPTV	Internet protocol television
IPv4, IPv6	Internet protocol version 4, Internet protocol version 6
ISG	industry specification group
ISO	International Organization for Standardization
ISOC	Internet Society
ITU	International Telecommunication Union
ITU-D	ITU Telecommunication Development Sector
ITU-R	ITU Radiocommunication Sector
ITU-T	ITU Telecommunication Standardization Sector
L2P2P	layer 2 peer-to-peer
L3E2E	layer 3 end-to-end
LAA	Licensed Assisted Access
LAN	local-area network
LLDP	Link Layer Discovery Protocol
LTE	Long Term Evolution (mobile communications standard)
M2M	machine-to-machine
MAN	metropolitan-area network
MEF	MEF Forum (formerly Metro Ethernet Forum)
MIB	Management Information Base
MIFR	Master International Frequency Register
MIME	Multipurpose Internet Mail Extensions
MIMO	multiple input multiple output

Term	Value
ML	machine learning
mMTC	massive machine-type communications
MPLS	Multiprotocol Label Switching
MTC	machine-type communications
NETCONF	Network Configuration
NFV	Network Function Virtualization
NGMN	Next Generation Mobile Networks Alliance
NGN	next generation network
NR	New Radio
NSA	non-standalone
NTP	Network Time Protocol
O-RAN	Open Radio Access Network
OAM	Operations, Administration, and Maintenance
OTN	optical transport network (optical transport technology)
PAR	project authorization request
PDV	packet delay variation
PIP	PTP Industry Profile
PNCS	Precise Networked Clock Synchronization
PON	passive optical networking (optical broadband)
PRP	Parallel Redundancy Protocol
PTP	precision time protocol
PTPv1	precision time protocol according to 1588-2002
PTPv2	precision time protocol according to 1588-2008
PTPv2.1	precision time protocol according to 1588-2019
PUP	Power Utility Automation Profile
QoE	quality of experience
QoS	quality of service
RADIUS	Remote Authentication Dial In User Service
RAN	radio access network
RE	Radio Equipment (CPRI)
REC	Radio Equipment Control (CPRI)

Term	Value
RF	radio frequency/function
RFC	Request for Comments (IETF document)
RIC	Radio Intelligent Controller
RRC	regional radiocommunication conference
RT	RealTime
RU	Radio Unit (5G radio access network)
SA	standalone
SA	Services and System Aspects
SASB	Standards Association Standards Board
SC	Software Community (O-RAN)
SD-WAN	Software-Defined Wide Area Network
SDN	Software-Defined Network
SDO	standards development organization
SG	study group
SIM	subscriber identity module
SMPTE	Society of Motion Picture and Television Engineers
SNMP	Simple Network Management Protocol
SNTP	Simple Network Time Protocol
SP	service provider (for telecommunications services)
SRP	Stream Reservation Protocol
SSO	standards setting organization
SyncE	synchronous Ethernet—a set of ITU-T standards
TCP	Transmission Control Protocol
TDD	time division duplex
TG	Task Group
TICTOC	Timing over IP Connection and Transfer of Clock (Working Group)
TIP	Telecom Infra Project
TR	technical report
TSAG	Telecommunication Standardization Advisory Group
TSB	Telecommunication Standardization Bureau
TSDSI	Telecommunications Standards Development Society, India

Term	Value
TSG	technical specification group
TSN	time-sensitive networking
TTA	Telecommunications Technology Association (South Korea)
TTC	Telecommunication Technology Committee (Japan)
TV	television
UHDTV	ultra-high definition television
UMTS	Universal Mobile Telecommunications System
URI	Uniform Resource Identifier
uRLLC	ultra-reliable and low-latency communications
VLAN	virtual local-area network
VoLTE	Voice over Long-Term Evolution (LTE)
VPN	Virtual Private Network
WG	working group
WLAN	wireless local-area network
WP	working party
WP5D	working party 5D
WRC	world radiocommunication conference
WTSA	World Telecommunication Standard Assembly
xRAN	extensible Radio Access Network (Forum)
YANG	Yet Another Next Generation (data modelling language)

Further Reading

Refer to the following recommended sources for further information about the topics covered in this chapter.

3GPP: https://www.3gpp.org/

3GPP Release 15: https://www.3gpp.org/release-15

3GPP Release 16: https://www.3gpp.org/release-16

3GPP Release 17: https://www.3gpp.org/release-17

CPRI: http://www.cpri.info/

ETSI: https://www.etsi.org/

ETSI NFV: https://www.etsi.org/technologies/nfv

IEEE: https://www.ieee.org/

IEEE 802: https://www.ieee802.org/

IEEE TSN Task Group: https://1.ieee802.org/tsn

IEEE P1588 Working Group, Active Projects: https://sagroups.ieee.org/1588/active-projects/

IETF: https://www.ietf.org/

IMT-2020: https://www.itu.int/en/ITU-R/study-groups/rsg5/rwp5d/imt-2020/Pages/default.aspx

ITU: https://www.itu.int

MEF: https://www.mef.net/

oneM2M: https://www.onem2m.org/

O-RAN: https://www.o-ran.org/

O-RAN Software Community: https://o-ran-sc.org/

O-RAN specifications: https://www.o-ran.org/specifications

TICTOC: https://tools.ietf.org/wg/tictoc/

TIP: https://telecominfraproject.com/

TIP OpenRAN, Telecom Infra Project OpenRAN: https://telecominfraproject.com/openran/

Chapter 5

Clocks, Time Error, and Noise

In this chapter, you will learn the following:

- **Clocks:** Covers clocks, clock signals, and the key components of clocks.
- **Time error:** Explains what time error is, the different types of metrics to quantify time error, and how these metrics are useful in defining clock accuracy and stability.
- **Holdover performance:** Explains *holdover* for clocks, applicable whenever a synchronization reference is temporarily lost, and why the holdover capability of a clock becomes critical to ensure optimal network and application functioning.
- **Transient response:** Examines what happens when a slave clock changes its input reference.
- **Measuring time error:** Describes how to determine and quantify the key metrics of time error.

In earlier chapters, you read that any clock can only be near perfect (none are perfect)—there are always inherent errors in clocks. In this chapter, you will discover the various components of a clock and understand how these components contribute to removing or introducing certain errors. Because these errors can adversely impact the consumer of the synchronization services (the end application), it is important to track and quantify them. Once measured, these errors are compared against some defined performance parameters to determine how much (if any) impact they have on the end application. This chapter also explains the different metrics used to quantify time error and how to measure them.

Clocks

In everyday usage, the term *clock* refers to a device that maintains and displays the time of day and perhaps the date. In the world of electronics, however, *clock* refers to a microchip that generates a *clock signal*, which is used to regulate the timing and speed of the components on a circuit board. This clock signal is a waveform that is generated either by a clock generator or the clock itself—the most common form of clock signal in electronics is a square wave.

This type of clock is able to generate clock signals of different frequencies and phases as may be required by separate components within an electronic circuit or device. The following are some examples showing the functions of a clock:

- Most sophisticated electronic devices require a clock signal for proper operation. These devices require that the clock signal delivered to them adheres to a core set of specifications.

- All electronics devices on an electronic circuit board communicate with each other to accomplish certain tasks. Every device might require clock signals with a different specification; providing the needed signals allows these devices to interoperate with each other.

In both cases, a clock device on the circuit board provides such signals.

Note   Use of the terms "master" and "slave" is ONLY in association with the official terminology used in industry specifications and standards, and in no way diminishes Pearson's commitment to promoting diversity, equity, and inclusion, and challenging, countering and/or combating bias and stereotyping in the global population of the learners we serve.

When discussing network synchronization or designing a timing distribution network, the timing signals need to travel much further than a circuit board. In this case, nodes must transfer clock signal information across the network. To achieve this, the engineer designates a clock as either a master clock or a slave clock. The master clock is the source for the clock signals, and a slave clock then synchronizes or aligns its clock signals to that of the master.

A clock signal relates to a (hardware) clock subsystem that generates a clocking signal, but often engineers refer to it simply using the term *clock*. You might hear the statement, "the clock on node A is not synchronized to a reference clock," whereas the real meaning of clock in this sentence is that the *clock signals* are not synchronized. So, clock signal and clock are technically different terms with different meanings, but because the common usage has made one refer to the other, this chapter will also use the term *clock* to refer to a clock signal.

Oscillators

An electronic *oscillator* is a device that "oscillates" when an electric current is applied to it, causing the device to generate a periodic and continuous waveform. This waveform can be of different wave shapes and frequencies, but for most purposes, the clock signals utilized are sine waves or square waves. Thus, oscillators are a simple form of clock signal generation device.

There are a few different types of oscillators (as described in Chapter 3, "Synchronization and Timing Concepts"), but in modern electronics, *crystal oscillators* (referred to as XO) are the most common. The crystal oscillator is made up of a special type of crystal, which is *piezoelectric* in nature, which means that when electric current is applied to this crystal, it oscillates and emits a signal at a very specific frequency. The frequency could vary from a few tens of kHz to hundreds of MHz depending on the physical properties of the crystal.

Quartz is one example of a piezoelectric crystal and is commonly used in many consumer devices, such as wristwatches, wall clocks, and computers. Similar devices are also used in networking devices such as switches, routers, radio base stations, and so on. Figure 5-1 shows a typical crystal commonly utilized in such a device.

Figure 5-1 *16-MHz Crystal Oscillator*

The quartz that is being used in crystal oscillators is a naturally occurring element, although manufacturers grow their own for purity. The natural frequency of the clock signal generated by a crystal depends on the shape or physical properties (sometimes referred to as the *cut*) of the crystal.

On the other hand, the stability of the output signal is also heavily influenced by many environmental factors, such as temperature, humidity, pressure, vibration, and magnetic and electric fields. Engineers refer to this as the *sensitivity* of the oscillator to environmental factors. For a given oscillator, the sensitivity to one factor is often dependent on the sensitivity to another factor, as well as the age of the crystal or device itself.

As a real-life example, if your wristwatch is using a 32,768-Hz quartz crystal oscillator, the accuracy of the wristwatch in different environmental conditions will vary. The same behavior also applies to other electronic equipment, including transport devices and routers in the network infrastructure. This means that when electronic devices are used to synchronize devices to a common frequency, phase, or time, the environmental conditions adversely impact the stability of synchronization.

Note   The accuracy for wristwatches is usually measured in seconds or minutes, while for other tasks involving network transport, the accuracy requirement is frequently small fractions of a second—microseconds (millionths of a second) or even nanoseconds (billionths of a second).

There have been many innovations to improve the stability of crystal oscillators deployed in unstable environmental conditions. One common approach in modern designs is for the hardware designer to design a circuit to vary the voltage being applied to the oscillator to adjust its frequency in small amounts. This class of crystal oscillator is known as *voltage-controlled crystal oscillators* (VCXO).

Of the many environmental factors that affect the stability and accuracy of a crystal oscillator, the major one is temperature. To provide better oscillator stability of crystal against temperature variations, two additional types of oscillators have emerged in the market:

■ *Temperature-compensated crystal oscillators* (TCXO) are crystal oscillators designed to provide improved frequency stability despite wide variations in temperature. TCXOs have a temperature compensation circuit together with the crystal, which measures the ambient temperature and compensates for any change by altering the voltage applied to the crystal. By aligning the voltage to values within the possible temperature range, the compensation circuit stabilizes the output clock frequency at different temperatures.

■ *Oven-controlled crystal oscillators* (OCXO) are crystal oscillators where the crystal itself is placed in an *oven* that attempts to maintain a specific temperature inside the crystal housing, independent of the temperature changes occurring outside. This reduces the temperature variation on the oscillator and thereby increases the stability of the frequency. As you can imagine, oscillators with additional heating components end up bulkier and costlier than TCXOs.

The basic approach with the TCXO is to compensate for measured changes in temperature by applying appropriate changes in voltage, whereas for the OCXO, the temperature is controlled (by being elevated above the expected operating temperature range). Figure 5-2 shows a typical OCXO.

An oscillator is the core component of a clock, which alone can significantly impact the quality of the clock. An approximate comparison of stability between these different oscillators suggests that the stability of an OCXO might be 10 to 100 times higher than a TCXO class device. Table 3-2 in Chapter 3 outlines the characteristics of the common types of oscillator.

The stability of an oscillator type also gets reflected in the cost. As a very rough estimate, cesium-based oscillators cost about $50,000 and rubidium-based oscillators around $100, whereas an OCXO costs around $30 and a TCXO would be less than $10.

Figure 5-2 *Typical OCXO*

PLLs

A phase-locked loop (PLL) is an electronic device or circuit that generates an output clock signal that is phase-aligned as well as frequency-aligned to an input clock signal. As shown in Figure 5-3, in its simplest form, a PLL circuit consists of three basic elements, as described in the list that follows.

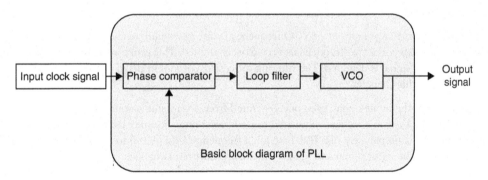

Figure 5-3 *PLL Building Blocks*

- **Voltage-controlled oscillator (VCO):** A special type of oscillator that changes frequency with changes in an input voltage (in this case from the loop filter). The frequency of the VCO with a nominal control signal applied is called the *free-running frequency*, indicated by the symbol f_0.

- **Phase comparator:** Compares the phase of two signals (input clock and local oscillator) and generates a voltage according to the phase difference detected between the two signals. This output voltage is fed into the loop filter.

■ **PLL loop filter:** Primary function is to detect and filter out undesired phase changes passed on by the phase comparator in the form of voltage. This filtered voltage is then applied to the VCO to adjust the frequency. It is important to note that if the voltage is not filtered appropriately, it will result in a signal that exactly follows the input clock, inheriting all the variations or errors of the input clock reference. Thus, the properties of the loop filter directly affect the stability and performance of a PLL and the quality of the output signal.

When a PLL is initially turned on, the VCO with a nominal control signal applied will provide its free-running frequency (f_0). When fed an input signal, the phase comparator measures the phase difference compared to the VCO signal. Based on the size of the phase difference between the two signals, the phase comparator generates a correcting voltage and feeds it to the loop filter.

The loop filter removes (or filters out) the noise and passes the filtered voltage to the VCO. With the new voltage applied, the VCO output frequency begins to change. Assuming the input signal and VCO frequency are not the same, the phase comparator sees this as a phase shift, and the output of the loop filter will be an increasing or decreasing voltage depending on which signal has higher frequency.

This voltage adjustment causes the VCO to continuously change its frequency, reducing the difference between VCO and input frequency. Eventually the size of changes in the output voltage of the loop filter are also reduced, resulting in ever smaller changes to the VCO frequency—at some point achieving a "locked" state.

Any further change in input or VCO frequency is also tracked by a change in loop filter output, keeping the two frequencies very closely aligned. This process continues as long as an input signal is detected. The filtering process is covered in detail in the following section in this chapter.

Recall that the input signal, even one generated from a very stable source (such as an atomic clock), would have accumulated noise (errors) on its journey over the clocking chain. The main purpose of a PLL is to align frequency (and phase) to the long-term average of the input signal and ignore (filter out) short-term changes.

Now a couple of questions arise:

■ Will the loop filter react and vary the voltage fed to the VCO for every phase variation seen on the input signal?

■ When should the PLL declare itself in locked state or, conversely, if already in locked state, under what conditions could the PLL declare that it has lost its lock with the input reference?

The first question raised is answered in the next section, but the second question needs a discussion on PLL states and the regions of operation of a PLL. The PLL is said to be in the *transient* state when the output is not locked to the input reference and it is in the process of locking to the input signal. Alternatively, the *steady* state is when the PLL is locked with the input reference. As explained earlier, even during steady-state operations, the VCO will keep adjusting the frequency to match the input frequency based on the differential voltage being fed from the loop filter.

These states are governed by three different frequency regions, known as hold-in range, pull-in range, and pull-out range. Figure 5-4 illustrates these frequency ranges.

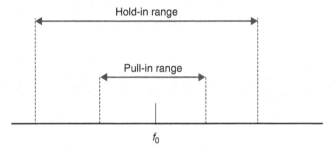

Figure 5-4   *Frequency Ranges of PLL Operations*

These frequency ranges are always quoted relative to the free-running frequency of the VCO (f_0) and are specified in parts per million (ppm) or parts per billion (ppb). Because these are ranges, there needs to be both minimum and maximum values. If not explicitly specified, the range is interpreted as including everything between the positive and negative values of the specified number. For example, if the range is specified as 4.6 ppm (or 4.6×10^{-6}), the range is assumed to cover a range between −4.6 ppm and +4.6 ppm around f_0.

As per ITU-T G.810, the definitions of these frequency regions are as follows:

- **Hold-in range:** "The largest offset between a slave clock's input signal and the nominal (natural) frequency of its VCO, within which the slave clock maintains lock as the frequency varies arbitrarily slowly over the frequency range."

  This is the range of difference between the nominal frequency and input frequency for which the PLL can steadily maintain phase tracking while in a steady (or locked) state. If the frequency of the input reference is slowly reduced or increased within the range, the PLL can still track it. The edge of the hold-in range is the point at which the PLL will lose the locked state.

- **Pull-in range:** "The largest offset between a slave clock's input signal and the nominal (natural) frequency of its VCO, within which the slave clock will achieve locked mode."

  This is the range of difference between the nominal frequency and the input frequency for which the PLL will always become locked throughout an acquisition (or tracking) process. Note that during this acquisition process there might be more than one cycle slip, but the PLL will always lock to the input signal. This is the range of frequencies within which a PLL can transition from the transient state to steady (locked) state.

■ **Pull-out range:** "The offset between a slave clock's input signal and the nominal (natural) frequency of its VCO, within which the slave clock stays in the locked mode and outside of which the slave clock cannot maintain locked mode, irrespective of the rate of the frequency change."

This can be seen as the range of the frequency step, which if applied to a steady-state PLL, the PLL still remains in the steady (or locked) state. The PLL declares itself not locked if the input frequency step is outside of this range.

Taking an example from ITU-T G.812, both the pull-in and hold-in ranges for Type III clock type is defined as 4.6×10^{-6}, which is the same as ±4.6 ppm. Table 3-3 in Chapter 3 provides a quick reference of these frequency ranges for different types of clock nodes.

When the PLL is tracking an input reference, it is the loop filter that is enforcing the limits of these frequency ranges. The loop filter is usually a *low-pass filter*, and as the name suggests, it allows only low-frequency (slow) variations to pass through. That means it removes high-frequency variation and *noise* in the reference signal. Conversely, a *high-pass filter* allows high-frequency variations to pass and removes the low-frequency changes.

While these filters are discussed in the next section in detail, it is important to note that "low-frequency variations" does not refer to the frequency of the clock signal, but the rate with which the frequency or phase of the clock signal varies. Low rate (less frequent) changes are a gradual *wander* in the signal, whereas high rate changes are a very short-term *jitter* in the signal. You will read more about jitter and wander later in this chapter.

Because PLLs synchronize local clock signals to an external or reference clock signal, these devices have become one of the most commonly used electronic circuits on any communication device. Out of several types of PLL devices, one of the main types of PLL used today is the digital PLL (DPLL), which is used to synchronize digital signals.

It is worthwhile noting that, just like any other electronic circuits and devices, PLLs have also been evolving. Designers of modern communications equipment are incorporating the latest PLL devices into them, circuits that now contain multiple PLLs (analogue or digital).

To reduce real-estate requirements on circuit boards, newer-generation devices can operate with lower-frequency oscillators that can replace expensive high-frequency oscillators. They also output signals with ultra-low levels of jitter that is required for the tight jitter specifications required by some equipment designs (remembering that jitter is high-frequency noise).

Low-Pass and High-Pass Filters

A *low-pass filter* (LPF) is a filter that filters out (removes) signals that are higher than a fixed frequency, which means an LPF passes only signals that are lower than a certain frequency—hence the name low-pass filter. For the same reasons, sometimes LPFs are also called *high-cut filters* because they *cut off* signals higher than some fixed frequency. Figure 5-5 illustrates LPF filtering where signals with lower frequency than a cut-off frequency are not *attenuated* (diminished). The *pass band* is the range of frequencies that are not attenuated, and the *stop band* is the range of frequencies that are attenuated.

Meanwhile, the range up to the cut-off frequency becomes the *clock bandwidth*, which also matches the width of a filter's pass band. For example, as shown in Figure 5-5, in the case of an LPF, the clock bandwidth is the range of frequencies that constitute the pass band. The clock bandwidth is typically a configurable parameter based on the capabilities of a PLL.

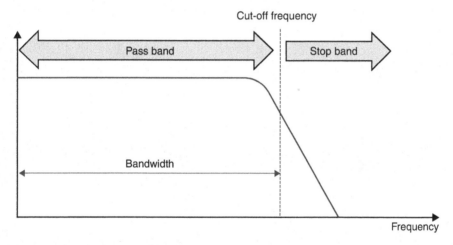

Figure 5-5   *Low-Pass Filter*

Similarly, a *high-pass filter* (HPF) is a device that filters out signals that are lower than some fixed frequency, which means that an HPF passes only signals that are higher than a certain frequency. And again, HPFs are also sometimes called *low-cut filters* because they *cut off* lower than a fixed frequency signal. Figure 5-6 depicts an HPF, showing that the pass band and stop band are a mirror image of the LPF case.

It is interesting to note that combining both filters (LPF and HPF) on a circuit, you could design a system to allow only a certain range of frequencies and filter out the rest. Such a combination of LPF and HPF behaves as shown in Figure 5-7 and is called a *band-pass filter*. The *band-pass* name comes from the fact that it allows a certain band of frequencies (from lower cut-off to higher cut-off) to pass and attenuates the rest of the spectrum.

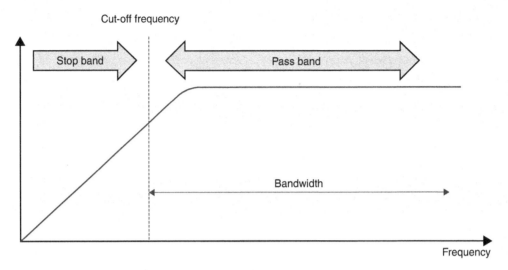

Figure 5-6   *High-Pass Filter*

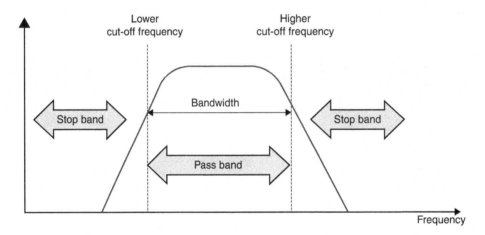

Figure 5-7   *Band-Pass Filter*

These filters are of the utmost importance when trying to filter errors out from a clock signal. To understand this process in more detail, this chapter next defines these errors, primarily jitter and wander, more technically.

Jitter and Wander

Jitter and wander are phase variations in the clock signal timing instants, as illustrated in Figure 5-8. Also, refer to the introduction to jitter and wander in Chapter 3. This variance and error, commonly called *clock signal noise*, can be caused by many factors, one being the quality of the clock components. Another factor is the noise accumulating from one node to the next when distributing timing signals through a chain of clocks.

Low rate (less frequent, slower) changes are a gradual *wander* in the signal, whereas high rate changes are a very short-term *jitter* in the signal. The ITU-T specifies (in G.810) that 10 Hz is the dividing line between jitter and wander (and has been a convention in the telecom industry for some time). And so, phase variations occurring at rate higher than 10 Hz are described as jitter, and the variations lower than 10 Hz are described as wander.

Jitter

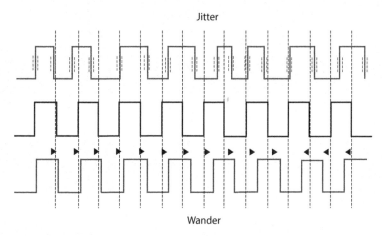

Wander

Figure 5-8 *Jitter Versus Ideal Versus Wander*

When all the phase variations of a clock signal as compared to its reference signal are measured and plotted in a graph, for jitter the resultant frequency (or rate) of the phase variations is higher than 10 Hz (ten variations per second). Figure 5-9 shows one such example of jitter, where the y-axis shows the phase variations (in ns), and the x-axis shows the time of variation (in seconds) itself. As depicted in Figure 5-9, the rate of phase variations recorded is much higher than 10 Hz, and such phase variations are classified as jitter.

For wander, in a similar approach, a graph plotting the phase variations will show the frequency (or rate) of less than 10 Hz. It is important to note that the rate of phase variations for wander could go down to mHz or μHz (rate down to once in several minutes or hours). For this reason, it is always recommended to run wander measurement tests for long periods of time (hours or days).

As you read in Chapter 3, the jitter (and wander to some extent) can be filtered out, and to do that, filters are used. If one configures an LPF with 10 Hz as the cut-off frequency, it will eliminate phase variation with a rate of variation higher than 10 Hz. Because jitter is defined as phase variation above 10 Hz, any LPF configured this way filters out jitter.

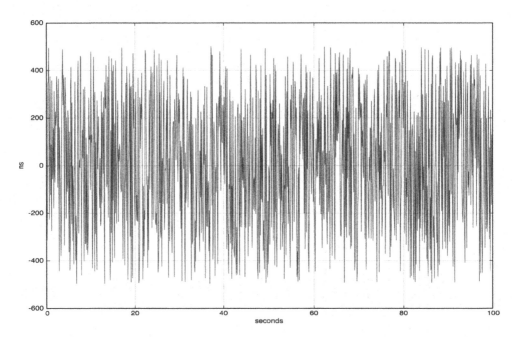

Figure 5-9   *Plot of Phase Deviations Showing Jitter in a Clock Signal*

Similarly, an HPF configured with a 10-Hz cut-off will filter out wander, because the HPF will filter out variations with a rate of 10 Hz or lower.

Phase variations from a real-life test are captured in Figure 5-10, Figure 5-11, and Figure 5-12, which show, respectively: 1) phase deviations of a clock signal with jitter and wander present with no filter; 2) with an HPF filter applied; and 3) with an LPF filter applied. In these figures, the y-axis shows phase variations (in ns), and the x-axis shows the time elapsed since the measurement was started (in minutes).

The graph shown in Figure 5-10 captures all the phase variations of a clock signal (low and high rate phase variations plotted in a single graph), and so it is not easy to visualize the jitter and the wander of a clock signal.

In order to clearly visualize (and analyze) the jitter and wander, filters are applied to the phase variation measurements. In Figure 5-11, you can see that after the HPF is applied (which filters out the wander), the remaining noise is jitter (frequency of phase variations higher than 10 Hz).

Signal with jitter and wander present

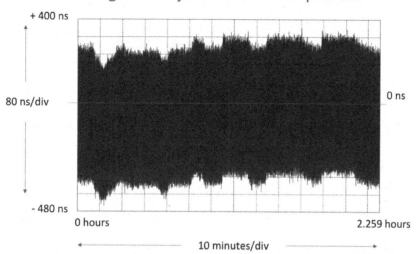

Figure 5-10   *Plot of Phase Deviations of a Clock Signal Without Any Filters*

Jitter: Filter out low frequency components with HPF

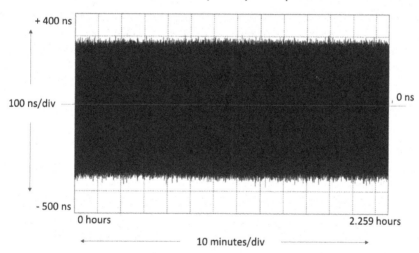

Figure 5-11   *Plot of Phase Deviations of a Clock Signal After HPF Applied*

Similarly, in Figure 5-12, after the LPF is applied (which filters out the jitter), the remaining noise is wander.

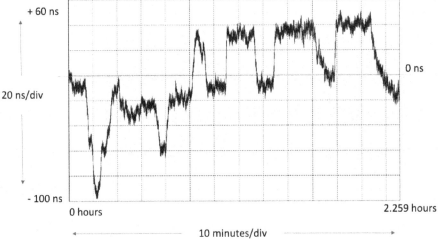

Figure 5-12   *Plot of Phase Deviations of a Clock Signal After LPF Applied*

After reading the preceding section about clock bandwidth and the PLL loop filter, there are two questions that could arise. First, why not keep the LPF cut-off frequency very low to filter the jitter and also limit the wander? Recall that jitter is filtered with the LPF. And if the LPF cut-off frequency is kept low, it could also filter some range of wander. Of course, this means that the pass band for the LPF becomes very small. Secondly, why not do the same on every clock node in the chain?

To understand the answer to the second question, you first need to appreciate the following aspects of PLLs:

- Not all PLLs can keep LPF cut-off frequency very low. Wander is classified as low-rate phase variations, which can reach extremely small values—10 Hz down to microhertz. So, there will always be some phase noise (in this case wander) within the LPF clock bandwidth and so will always be tracked by the PLL.

- A PLL combines two signals: 1) the input reference signal and 2) the clock signal from the local oscillator (VCO) to output a synchronized signal. When the PLL loop filter (LPF) blocks the signal from the input reference, the output signal is constructed using the local oscillator.

So the process of a PLL filtering the noise from the input signal is substituting noise from the local oscillator. Taken to a theoretical corner case, if the clock bandwidth of an LPF is made zero, all that the PLL will output is the signal from the local oscillator—which defeats the purpose of having a reference signal.

So, if using a very low cut-off frequency for the LPF, the PLL needs to be matched to a good-quality local oscillator, so that the noise added by the local oscillator is reduced. For the hardware designer, this has obvious cost ramifications—to get better noise filtering, you need to spend more money on the oscillator.

■ The time taken for a PLL to move from transient state to the steady (or locked) state depends on the clock bandwidth (as well as the input frequency and quality of the local oscillator). The narrower the clock bandwidth for the LPF, the longer time it takes for the PLL to move to the steady state.

   For example, a synchronization supply unit (SSU) type I clock or building integrated timing supply (BITS) stratum 2 clock with an LPF configured for bandwidth of 3 mHz will take several minutes to lock to an input signal. However, telecom networks are very widely distributed and can consist of long chains of clock nodes. If all the clocks in the chain had low bandwidth, the complete chain could take many hours to settle to a steady state. Similarly, it might take several hours for the last node of the chain to settle down after any disruption to the distribution of clock.

It is for these reasons that a clock node with better filtering capabilities should have a good-quality oscillator and should be placed in a chain of clock nodes at selected locations. These factors also explain why SSU/BITS clock nodes (which have stratum 2–quality oscillators and better PLL capabilities) are recommended only after a certain number of SDH equipment clock (SEC) nodes. The section "Synchronization Network Chain" in Chapter 6, "Physical Frequency Synchronization," covers this limit and recommendations by ITU-T in greater detail.

To ensure interoperability between devices and to minimize the signal degradation due to jitter and wander accumulation across the network, the ITU-T recommendations (such as G.8261 and G.8262) specify jitter and wander performance limits for networks and clocks. The normal network elements (NE) and synchronous Ethernet equipment clocks (EEC) are usually allocated the most relaxed limits.

For example, ITU-T G.8262 specifies the maximum amount of peak-to-peak output jitter (within a defined bandwidth) permitted from an EEC. This is to ensure that the amount of jitter never exceeds the specified input tolerance level for subsequent EECs. Chapter 8, "ITU-T Timing Recommendations," covers the ITU-T recommendations in greater detail.

Frequency Error

While jitter and wander are both metrics to measure phase errors, the frequency error (or accuracy) also needs to be measured.

The *frequency error* (also referred to as *frequency accuracy*) is the degree to which the frequency of a clock can deviate from a nominal (or reference) frequency. The metric to measure this degree is called the *fractional frequency deviation* (FFD) or sometimes just the *frequency offset*. This offset is also referred to as the *fractional frequency offset* (FFO).

The basic definition is given in ITU-T G.810 by the following equation:

$$y(t) = \frac{v(t) - v_{nom}}{v_{nom}}$$

where:

- $y(t)$ is the FFD at time t
- $v(t)$ is the frequency being measured and
- v_{nom} is the nominal (or reference) frequency

FFD is often expressed in parts per million (ppm) or parts per billion (ppb). For example, the free-running frequency accuracy of a synchronous Ethernet (SyncE) clock is < 4.6 ppm, while the carrier frequency accuracy required in cellular mobile radio is < 50 ppb.

Taking an example calculation, if the nominal frequency of an oscillator is 20 MHz (which is 20,000,000 Hz and represents v_{nom} as per the previous ITU-T G.810 formula) and the measured frequency is 20,000,092 Hz, then the FFD for this case (in ppm) will be

$$FFD = \frac{(20,000,092 - 20,000,000)}{(20,000,000)} \times 1 \times 10^6 \; ppm = +4.6 \; ppm$$

Using the same formula, a measured frequency of 19,999,08 Hz and nominal frequency of 20,000,000 Hz will give the FFD as −4.6 ppm.

FFD can also be expressed in units of time. For example, an FFD of +4.6 ppm means that the measured signal is accumulating 4.6 µs of time for each second, whereas an FFD of −4.6 ppm means that the measured signal is losing 4.6 µs of time for each second. After 10 s of real time, the signal would be off by 46 µs compared to the reference signal. This is known as the *frequency drift* and the relation is as follows:

$$FFD \; (in \; ppm) = \frac{X \, \mu s}{sec}$$

ITU-T clock specifications such as ITU-T G.812, define the holdover specification (also called *long-term transient response*) using frequency accuracy. This sets the limits of the drift of a clock away from the frequency to which it was previously locked. Also, other ITU-T clock specifications, such as ITU-T G.813, define the frequency accuracy of the free-running clock. This defines the required accuracy of the frequency of a clock at start-up, before it has been locked to any other reference.

Time Error

So far, this chapter has outlined the mechanisms that help to synchronize a local clock to a given external reference clock signal. However, even when a local clock is determined to be synchronized in both phase and frequency to a reference clock, there are always errors that can be seen in the final synchronized clock signal.

These errors can be attributed to various factors such as the quality and configuration of input circuit being used as a reference; any environmental factors that influence the synchronization process; and the quality of oscillator providing the initial clock.

To quantify the error in any synchronized clock, various parameters have been defined by the standard development organizations (SDO). To understand the errors and the various metrics to quantify these errors, refer to Figure 5-13, which illustrates the output of a clock synchronized to an external reference clock. Because the measurement is taken from the synchronized clock that is receiving a reference signal, this synchronized clock will be referred to as the *measured clock*.

It is also important to note that the error being referred to here is an error in significant instants or phase of a clock. This is called *phase error*, which is a measure of the difference between the significant instants of a measured signal and a reference signal. Phase error is usually expressed in degrees, radians, or units of time; however, in telecommunication networks, phase error is typically represented as a unit of time (such as ns).

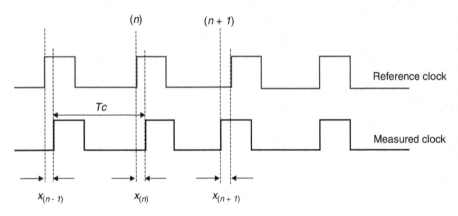

Figure 5-13 *Comparing Clock Signals*

The two main aspects that you need to keep in mind for understanding clock measurements are as follows:

■ Just as the measured clock is synchronized to a reference clock, the measured clock is always compared to a reference clock (ideally the same one). Therefore, the measurement represents the relative difference between the two signals. Much like checking the accuracy of a wristwatch, measurement of a clock has no meaning unless it is compared to a reference.

■ The error in a clock varies over time, so it is important that clock errors are measured over a sufficiently long time period to thoroughly understand the quality of a clock. To illustrate, most quartz watches gain/lose some time (seconds) daily. For such a device, a measurement done every few minutes or every few hours might not be a good indication of the quality. For such watches, a good time period might be one month, over which it could wander by up to (say) 10 seconds.

For illustration purposes, Figure 5-13 shows only four clock periods or cycle times, T_c, of a clock. These clock periods are shown as $(n-1)$, (n), and $(n+1)$. Notice that the measured clock is not perfectly aligned to the reference clock, and that difference is called *clock error* or *time error* (TE). The error is marked as $x_{(n-1)}$ showing the error for instance $(n-1)$ of the clock period, $x_{(n)}$ for period (n), and so on. In the figure, the TEs at each successive instance of the clock period are marked as $x_{(n-1)}$, $x_{(n)}$, and $x_{(n+1)}$. The TE is simply the time of the measured clock minus the time of the reference clock.

In Figure 5-13, the measured clock signal at instance $(n-1)$ and (n) lags (arrives later than) the reference clock signal and is therefore by convention a *negative time error*. And for the interval $(n+1)$, the clock signal is leading the reference and is referred to as a *positive time error*. Of course, the time error measured at varying time periods varies and can be either negative or positive.

TE measurements in the time domain are normally specified in seconds, or some fraction of, such as nanoseconds or microseconds. As mentioned previously, there could be positive or negative errors, and so it is imperative to indicate this via a positive or negative sign. For example, a time error for a certain interval might be written as +40 ns, whereas another interval might be −10 ns.

An engineer can measure the TE value (against the reference) at every edge (or significant instant) of the clock over an extended observation interval. If these values are plotted in a graph against time on the horizontal axis, it might look something like the graph shown in Figure 5-14. Each additional measurement (+ or −) is added to the graph at the right. A value of zero means that for that measurement, the offset between the signals was measured as zero.

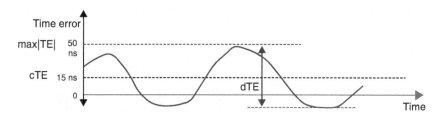

Figure 5-14   *Graph Showing cTE, dTE, and max|TE|*

This figure also includes several statistical derivations of TE, the main one being max|TE|, which is defined as the maximum absolute value of the time error observed over the course of the measurement. The following sections explain these measures and concepts. Refer to ITU-T G.810 and G.8260 to get more details about the mathematical models of timing signals and TE.

Maximum Absolute Time Error

The *maximum absolute time error*, written as max|TE|, is the maximum TE value that the measured clock has produced during a test measurement. This is a single value, expressed in units of seconds or fractions thereof, and represents the TE value that lies furthest away from the reference. Note that although the TE itself can be positive or negative, the max|TE| is taken as an absolute value; hence, the signed measured value is always written as a positive number. In Figure 5-14, the max|TE| is measured as 50 ns.

Time Interval Error

The *time interval error* (TIE) is the measure of the change in the TE over an observation interval. Recall that the TE itself is the phase error of the measured clock as compared to the reference clock. So, the change in the TE, or the TIE, is the difference between the TE values of a specific observation interval.

The observation interval, also referred to as τ (tau), is a time interval that is generally fixed before the start of a measurement test. The TE is measured and recorded at the beginning of the test and after each observation interval time has passed; and the difference between the corresponding time error values gives the TIE for that observation interval.

As an example, assuming an observation interval consists of k cycle times of a clock, the calculations for TE and TIE for the clock signals shown in Figure 5-15 are as follows (the measurement process starts at clock period n or $(n)^{th}$ T_c):

- TE at $(n)^{th}$ T_c = difference between measured clock and reference = $x_{(n)}$

- TE at $(n + 1)^{th}$ T_c = difference between measured clock and reference = $x_{(n + 1)}$

- And similarly, TE at $(n + k)^{th}$ $T_c = x_{(n + k)}$ and so on

- TIE for first observation interval (k cycles between n and $n + k$) = $x_{(n + k)} - x_{(n)}$

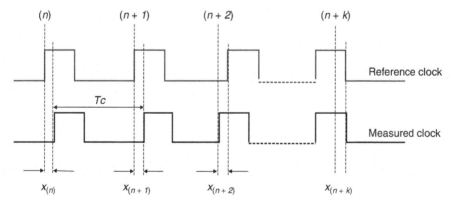

Figure 5-15 *Time Error for* k *Cycle Periods as an Observation Interval*

To characterize the timing performance of a clock, the timing test device measures (observes) and records the TIE as the test proceeds (as the observation interval gets longer and longer). So, if the clock period (T_c) is 1 s, then the first observation interval after the start of the test will be at $t = 1$ s, then the second will be at $t = 2$ s, and so on (assuming the observation interval steps are the same as the clock period).

This way, TIE measures the total error that a clock has accumulated as compared to a reference clock since the beginning of the test. Also, because TIE is calculated from TE, TIE is also measured using units of seconds. For any measurement, by convention the value of TIE is defined to be zero at the start of the measurement.

In comparison, the TE is the instantaneous measure between the two clocks; there is no interval being observed. So, while the TE is the recorded error between the two clocks at any one instance, the TIE is the error accumulated over the interval (length) of an observation. Another way to think of it is that the TE is the relative time error between two clocks at a point (instant) of time, whereas TIE is the relative time error between two clocks accumulated between two points of time (which is an interval).

Taking a simple example where the steps in the observation interval for a TIE measurement is the same as the period of a TE measurement, the TIE will exactly follow the contour of the TE. One case might be where the TE is being measured every second, and the TIE observation interval is increasing by one second at every observation. Although the curve will be the same, there is an offset, because the TIE value was defined to be at 0 at the start of the test.

Figure 5-16 shows this example case, where the plotted TIE graph looks the same as a plot of time error with a constant offset because the TIE value starts at 0. The list that follows explains this figure in more detail.

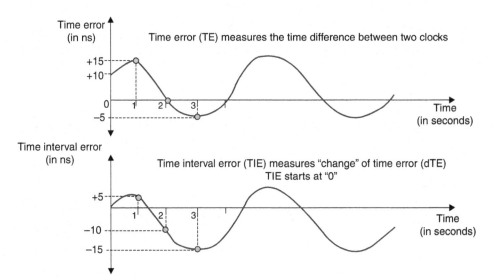

Figure 5-16 *Graph Showing TE and TIE Measurements and Plots*

- As shown in Figure 5-16, the TE is +10 ns at the start of a measurement (TE_0), at $t = 0$. Therefore, TE_0 is +10 ns (measured), and the TIE is zero (by convention).

- Assume that the observation interval is 1 second. After the first observation interval, from Figure 5-16, TE_1 is 15 ns. The TIE calculated for this interval will be $(TE_1 - TE_0) = 5$ ns.

- After the second observation interval, assume TE_2 is 0 ns. The TIE calculated now will be $(TE_2 - TE_0) = -10$ ns. Note that this calculation is for the second observation interval and the interval for calculations really became 2 seconds.

- The observation interval keeps increasing until the end of the entire measurement test.

There are other clock metrics (discussed later in this chapter), such as MTIE and TDEV, that are calculated and derived from the TIE measurements. TIE, which is primarily a measurement of phase accuracy, is *the* fundamental measurement of the clock metrics.

Some conclusions that you can draw based on plotted TIE on a graph are as follows:

- An ever-increasing trend in the TIE graph suggests that the measured clock has a frequency offset (compared to the reference clock). You can infer this because the frequency offset will be seen in each time error value and hence will get reflected in TIE calculations.

- If TIE graph shows a large value at the start of the measurement interval and starts converging slowly toward zero, it might suggest that the clock or PLL is not yet locked to the reference clock or is slow in locking or responding to the reference input.

Note that the measurement of the change in TE at each interval really becomes measurement of a short-term variation, or jitter, in the clock signals, or what is sometimes called *timing jitter*. And as TIE is a measurement of the change in phase, it becomes a perfect measurement for capturing jitter.

Now it is time to revisit the concepts of *accuracy* and *stability* from Chapter 1, "Introduction to Synchronization and Timing." These are two terms that have a very precise meaning in the context of timing.

For synchronization, accuracy measures how close one clock is to the reference clock. And this measurement is related to the maximum absolute TE (or max|TE|). So, a clock that closely follows the reference clock with a very small offset is an accurate clock.

On the other hand, stability refers to the change and the speed of change in the clock during a given observation interval, while saying nothing about how close it is to a reference clock.

As shown in Figure 5-17, an accurate clock might not be a very stable clock, or a very stable clock might not be accurate, and so on. As the end goal is to have the most stable as well as the most accurate clock possible, a clock is measured on both aspects. The metrics that quantify both these aspects become the *timing characteristics* of a clock.

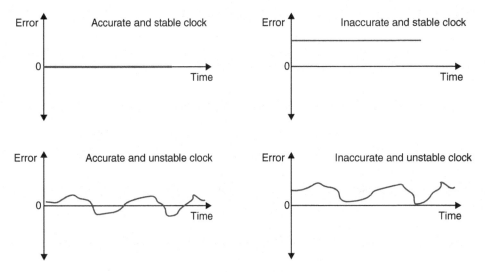

Figure 5-17   *Considering Clock Accuracy and Stability*

As you should know by now, the timing characteristics of a clock depend on several factors, including the performance of the in-built oscillator. So, it is possible to use that to differentiate between different classes of performance that a clock exhibits. Consequently, the ITU-T has classified clocks by the difference in expected timing characteristics—based on the oscillator quality. For example, a primary reference clock or primary reference source (PRC/PRS), based on a cesium atomic reference, is expected to provide better accuracy and stability when compared to a clock based on a stratum 3E OCXO oscillator. Chapter 3 thoroughly discussed these different classes of oscillator, so now it is time to define some metrics to categorize these qualities more formally.

Constant Versus Dynamic Time Error

Constant time error (cTE) is the mean of the TE values that have been measured. The TE mean is calculated by averaging measurements over either some fixed time period (say 1000 seconds) or the whole measurement period. When calculated over all the TE measurements, cTE represents an average offset from the reference clock as a single value. Figure 5-14 showed this previously, where the cTE is shown as a line on the graph at +15 ns. Because it measures an average difference from the reference clock, cTE is a good measure of the accuracy of a clock.

Dynamic time error (dTE) is the variation of TE over a certain time interval (you may remember, the variation of TE is also measured by TIE). Additionally, the variation of TE over a longer time period is known as wander, so the dTE is effectively a representation of wander. Figure 5-14 showed this previously, where the dTE represents the difference between the minimum and maximum TE during the measurement. Another way to think about it is that dTE is a measure of the stability of the clock.

These two metrics are very commonly used to define timing performance, so they are important concepts to understand. Normally, dTE is further statistically analyzed using MTIE and TDEV; and the next section details the derivations of those two metrics.

Maximum Time Interval Error

As you read in the "Time Interval Error" section, TIE measures the change in time error over an observation interval, and the TIE is measured as the difference between the TE at the start and end of the observation interval. However, during this observation interval, there would have been other TE values that were not considered in the TIE calculation.

Taking the example as shown in Figure 5-18, there were k time error values (k also being the observation interval). These time error values were denoted as $x_{(n)}$, $x_{(n+1)}$, $x_{(n+2)}$,..., $x_{(n+k)}$. The TIE calculation only considered $x_{(n)}$ and $x_{(n+k)}$ because the observation interval was k cycles starting at cycle n. It could be that these two values do not capture the maximum or minimum time error values. To determine the maximum variation during this time, one needs to find the maximum TE ($x_{(max)}$) and minimum TE ($x_{(min)}$) during the same observation interval.

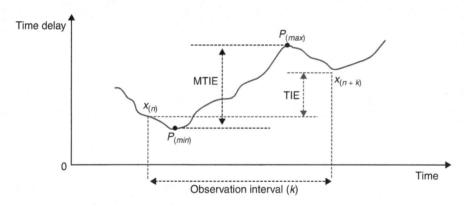

Figure 5-18   *TIE and MTIE Example*

This maximum variation of the TIE within an observation interval is known as the *maximum time interval error* (MTIE). The MTIE is defined in ITU-T G.810 as "the maximum peak-to-peak delay variation of a given timing signal" for a given observation interval. As this essentially represents the variation in the clock, it is also referred to as the *maximum peak-to-peak clock variation*.

The peak-to-peak delay variation captures the two inverse peaks (low peak and high peak) during an observation interval. This is calculated as the difference between the maximum TE (at one peak) and minimum TE (another peak) for a certain observation interval. Figure 5-18 shows an example of TIE and of MTIE for the same observation interval, illustrating the maximum peak-to-peak delay variation.

The example in Figure 5-18 clearly shows that during an observation interval k, while the TIE calculation is done based on the interval between $x_{(n)}$ and $x_{(n+k)}$, there are peaks during this interval that are the basis for the MTIE calculations. These peaks are denoted by $P_{(max)}$ and $P_{(min)}$ in the figure.

Note also that as the MTIE is the *maximum* value of delay variation, it is recorded as a real maximum value; not only for one observation interval but for all observation intervals. This means that if a higher value of MTIE is measured for subsequent observation intervals, it is recorded as a new maximum or else the MTIE value remains the same. This in turn means that the MTIE values never decrease over longer and longer observation intervals.

For example, if the MTIE observed over all the 5-second periods of a measurement run is 40 ns, then this value will not go to less than 40 ns when measured for any subsequent (longer) periods. For example, any MTIE that is calculated using a 10-s period will be the same or more than any maximum value found for the 5-s periods. This applies similarly for larger values.

Therefore, MTIE can only stay the same or increase if another higher value is found during the measurement using a longer observation interval; hence, the MTIE graph will never decline as it moves to the right (with increasing tau).

The MTIE graph remaining flat means that the maximum peak-to-peak delay variation remains constant; or in other words, no new maximum delay variation was recorded. On the other hand, if the graph increases over time, it suggests that the test equipment is recording ever-higher maximum peak-to-peak delay variations. Also, if the graph resembles a line increasing linearly, it shows that the measured clock is not locked to the reference clock at all, and continually wandering off without correction. MTIE is often used to define a limit on the maximum phase deviation of a clock signal.

A simplified algorithm to record and plot MTIE is explained as follows:

1. Determine all the TE values for every 1-second interval (the values may have been sampled at a higher rate, such as 1/30 of 1 second).

2. Find the maximum peak-to-peak delay variation within each observed 1-s observation interval (MTIE values for 1-s intervals).

3. Record the highest value at the 1-s horizontal axis position in the MTIE graph.

4. Determine all the TE values for all 2-s intervals.

5. Find the maximum peak-to-peak delay variation within each observed 2-s interval (MTIE values for 2-s intervals).

6. Record the highest value against the 2-s horizontal axis position in the MTIE graph.

7. Repeat for other time intervals (4 s, 8 s, 16 s, etc.).

Figure 5-19 shows the MTIE limit for a primary reference time clock (PRTC), as specified in ITU-T G.8272.

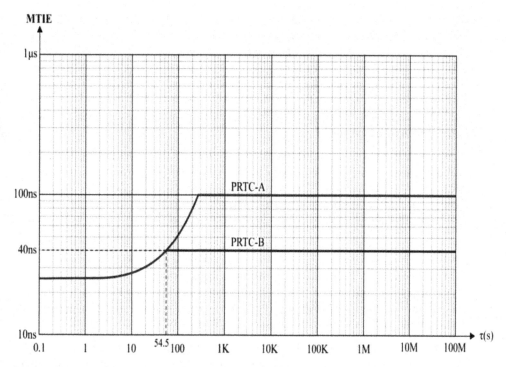

Figure 5-19   *MTIE Limits for Two Classes of PRTC Clock (Figure 1 from ITU-T 8272)*

The flat graph required by the recommendation for the MTIE values suggests that after a certain time period, there should not be any new maximum delay variation recorded for a PRTC clock. This maximum MTIE value, as recommended in ITU-T G.8272, is 100 ns for a PRTC class A (PRTC-A) clock and 40 ns for a PRTC class B (PRTC-B) clock. This line on the graph represents the limits below which the calculated values must remain and is referred to as a *mask*.

The standards development organizations have specified many MTIE masks—for example, the ITU-T has written them into recommendations G.812, G.813, G.823, G.824, and G.8261 for frequency performance of different clock and network types.

To summarize, MTIE values record the peak-to-peak phase variations or fluctuations, and these peak-to-peak phase variations can point to frequency shift in the clock signal. Thus, MTIE is very useful in identifying a shift in frequency.

Time Deviation

Whereas MTIE shows the largest phase swings for various observation intervals, *time deviation* (TDEV) provides information about phase stability of a clock signal. TDEV is a metric to measure and characterize the degree of phase variations present in the clock signal, primarily calculated from TIE measurements.

Unlike MTIE, which records the difference between the high and low peaks of phase variations, TDEV primarily focuses on how frequent and how stable (or unstable) such phase variations are occurring over a given time—note the importance of "over a given time." This is important because both "how frequent" and "how stable" parameters would change by changing the time duration of the measurements. For example, a significant error occurring every second of a test can be considered a frequently occurring error. But if that was the only error that occurred for the next hour, the error cannot be considered to be a frequently occurring error.

Consider a mathematical equation that divides the total phase variations (TE values) by the total test time; you will realize that if there are less phase variations over a long time, this equation produces a low value. However, the result from the equation increases if there are more phase variations during the same test time. This sort of equation is used for TDEV calculations to determine the quality of the clock.

The low frequency phase variations could appear to be occurring randomly. Only if such events are occurring less often over a longer time can these be marked as outliers. And this results in gaining confidence in the long-term quality of the clock. Think of tossing five coins simultaneously, and with the first toss, all five coins come down heads! Without tossing those coins a lot more times, you cannot be sure if that was an outlier (a random event) or if something strange is wrong with the coins.

The TDEV of a clock defines the limit of any *randomness* (or outliers) in the behavior of a clock. The degree of this randomness is a measure of the low frequency phase variation (wander), which can be statistically calculated from TIE measurements and plotted on graphs, using a TDEV calculation.

According to ITU-T G.810, TDEV is defined as "a measure of the expected time variation of a signal as a function of integration time." So, TDEV is particularly useful in revealing the presence of several noise processes in clocks over a given time interval; and this calculated measurement is compared to the limit (specified as a mask in the recommendations) for each clock type.

Usually, instability of a system is expressed in terms of some statistical representation of its performance. Mathematically, *variance* or *standard deviation* are usual metrics that would provide an adequate view to express such instability. However, for clocks, it has been shown that the usual variance or standard deviation calculation has a problem—the calculation does not *converge*.

The idea of convergence can be explained with a generic example. A dataset composed of a large collection of values with a Gaussian distribution will have a specific mean and variance. The mean shows the average value, and the variance shows how closely the values cluster around the mean—much like the stability of a clock. Now, if one were to calculate these values using a small subset or sample of these values, then the level of confidence in those metrics would be low—much like the toss of five coins. And as these parameters are calculated using larger and larger samples of this example dataset, the confidence in these metrics increases. The increase in confidence here means that variance moves toward a true value for the given dataset—this is referred to as *convergence of variance*. However, for clocks and oscillators, no matter how large the sample of variations is, apparently there is no evidence of this variance value converging. To deal with this deficiency, one of the alternative approaches is to use TDEV.

Rather than understanding the exact formula for calculating TDEV, it might be better to see the TDEV measure as a root mean square (RMS) type of metric to capture clock noise. Mathematically, RMS is calculated by first taking a mean (average) of square values of the sample data and then taking a square root of it. The key aspect of this calculation is that it averages out extremes or outliers in the sample (much more than the normal mean). Interestingly, the extremes average out more and more as the sample data increases.

Thus, TDEV becomes a highly averaged value. And if the noise in the clock is occurring rarely, the degree (or magnitude) of this rarely occurring error starts diminishing over longer periods of measurement. An example might be something occurring rarely or seldom, such as errors introduced by environmental changes over the course of the day (diurnal rhythms), such as the afternoon sun striking the exterior metal panel of the clock.

So, to gain confidence in the TDEV values (primarily to discard the rare noise occurrence in clock), the tests are usually quite long, at least several times longer than the usual MTIE observation intervals. For example, ITU-T G.811 calls for running TDEV tests for 12 times the maximum observation interval.

Compared to TDEV, you might notice that MTIE is perfectly suited to catch outliers in samples of TIE measurements. MTIE is a detector of peaks, but TDEV is a measure of wander over varying observation time intervals and is well suited to characterization of wander noise. It can do that because it is very good at averaging out short-term and rare variations.

So, it makes sense that TDEV is used alongside MTIE as metrics to characterize the quality of clock. As is the case for MTIE, ITU-T recommendations also specify TDEV masks for the same purpose. These masks show the maximum permissible values for both TDEV and MTIE for different clock types and in different conditions.

Noise

In the timing world, *noise* is an unwanted disturbance in a clock signal—the signal is still there, and it is good, but it is not as perfect as one would like. Noise arises over both the short term, say for a single sampling period, as well as over much longer timeframes—up to the lifetime of the clock. Because there are several factors that can introduce noise, the amount and type of noise can vary greatly between different systems.

The metrics that are discussed in this chapter try to quantify the noise that a clock can generate at its output. For example, max|TE| represents the maximum amount of noise that can be generated by a clock. Similarly, cTE and dTE also characterize different aspects of the noise of a clock.

Noise is classified as either jitter or wander. As seen in the preceding section "Jitter and Wander," by convention, jitter is the short-term variation, and wander is the long-term variation of the measured clock compared to the reference clock. The ITU-T G.810 standard defines jitter as phase variation with a rate of change greater than or equal to 10 Hz, whereas it defines wander as phase variation with a rate of change less than 10 Hz. In other words, slower-moving, low-frequency jitter (less than 10 Hz) is called wander.

One analogy is that of a driver behind the wheel of a car driving along a straight road across a featureless plain. To keep within the lane, the driver typically "corrects" the steering wheel with very short-term variations in the steering input and force (unconsciously, drivers do these corrections many times per second). This is to correct the "jitter" of the car moving around within the lane.

But now imagine that the road is gently turning to the right and the driver is beginning to impact a little more force on the steering wheel in that direction to affect the turn. This is an analogy for wander. The jitter is the short-term variance trying to take you out of your lane, but the long-term direction being steered is wander.

The preceding sections explained that MTIE and TDEV are metrics that can be used to quantify the timing characteristics of clocks. Therefore, it is no surprise that the timing characteristics of these clocks based on the noise performance (specifically noise generation) have been specified by ITU-T in terms of MTIE and TDEV masks. These definitions are spread across numerous ITU-T specifications, such as G.811, G.812, G.813, and G.8262. Chapter 8 covers these specifications in more detail.

Because these metrics are so important, any timing test equipment used to measure the quality of clocks needs to be able to generate MTIE and TDEV graphs as output. The results on these graphs are then compared to one of the masks from the relevant ITU-T recommendation. However, it is imperative that the engineer uses the correct mask to compare the measured results against.

For example, if the engineer is measuring SyncE output from a router, the MTIE data needs to be compared to the SyncE mask from G.8262 and not from some other standard like G.812 (which covers SSU/BITS clocks).

Figure 5-20 represents the various types of categories of noise performance and the point at which they are relevant (and could be measured).

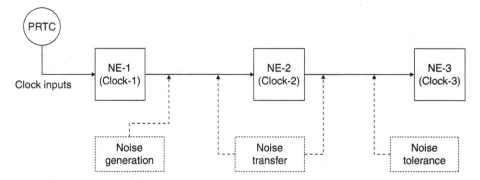

Figure 5-20 *Noise Generation, Transfer, and Tolerance*

The sections that follow cover each of these categories in more detail.

Noise Generation

Figure 5-21 illustrates the noise generation at the output of Network Element-1 (NE-1) and hence the measurement point for noise generation. *Noise generation* is the amount of jitter and wander that is added to a perfect reference signal at the output of the clock. Therefore, it is clearly a performance characteristic of the quality or fidelity with which a clock can receive and transmit a time signal. Figure 5-21 depicts some amount of noise that gets added and is observed at output, when an ideal (perfect) reference signal is fed to the clock.

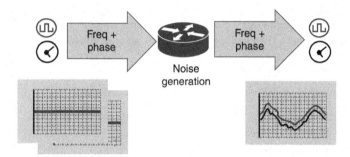

Figure 5-21   *Noise Generation*

One analogy that might help is that of a photocopier. If you take a document and run it though a photocopier, it will generate a good (or even excellent) copy of the document. The loss of quality in that process is an analogy for noise generation. But now, take a photocopy of that copy of the document, and a photocopy of that copy. Repeat that process, and after four to five generations of copies, the quality will have degraded substantially. The lower the noise generated in each cycle of the copy process, the more generations of copies can be made with acceptable quality.

For transport of time across a network, lower noise generation in each clock translates into a better clock signal at the end of the chain, or the ability to have longer chains. That is why noise generation is a very important characteristic for clocks.

Noise Tolerance

A slave clock can lock to the input signal from a reference master clock, and yet every clock (even reference clocks) generates some additional noise on its output. As described in the previous section, there are metrics that quantify the noise that is generated by a clock. But how much noise can a slave clock receive (tolerate) on its input and still be able to maintain its output signal within the prescribed performance limits?

This is known as the *noise tolerance* of a clock. Figure 5-20 shows the noise tolerance at the input of Network Element-3 (NE-3). It is simply a measure of how bad an input signal can become before the clock can no longer use it as a reference.

Going back to the photocopier analogy, the noise tolerance is the limit at which the photocopier can only just read the input document well enough to produce a readable copy. Any further degradation and it cannot produce a satisfactory output.

Like noise generation, noise tolerance is also specified as MTIE and TDEV masks for each class of clock performance. Up until the maximum permissible values of noise specified in ITU-T recommendations is applied at the input of a clock, the output of the clock should continue to operate within the expected performance limits. These masks are specified in the same ITU-T specifications that cover noise generation, such as G.812, G.813, G.823, G.824, and G.8262.

What happens when the noise at the input of a clock exceeds the permissible limits—what should be the action of a clock? The clock could do one or more of the following:

- Report an alarm, warning that it has lost the lock to the input clock

- Switch to another reference clock if there are backup references available

- Go into holdover (operating without an external reference input) and raise an alarm warning of the change of status and quality

Noise Transfer

To synchronize network elements to a common timing source, there can be at least two different approaches—external clock distribution and line timing clock distribution.

An external clock distribution or frequency distribution network (refer to the section "Legacy Synchronization Networks" in Chapter 2, "Usage of Synchronization and Timing") becomes challenging because the network elements are normally geographically dispersed. So, building a dedicated synchronization network to each node could become very costly, and hence this method is often not preferred.

The other approach is to cascade the network elements and use the line clocking (again refer to the section "Legacy Synchronization Networks" in Chapter 2), where the clock is passed from one network element to the next element, as in a daisy chain. Figure 5-20 shows the clock distribution model using the line clocking approach, where NE-1 is passing the clock to NE-2 and then subsequently to NE-3.

From the previous section on noise generation, you know that every clock will generate additional noise on the output of the clock compared to the input. When networks are designed as a daisy chain clock distribution, the noise generated at one clock is being passed as input to the next clock synchronized to it. This way, noise not only cascades down to all the clocks below it in the chain, but the noise gets amplified during transfer. To try to lessen that, the accumulated noise needs to be somehow reduced or filtered at the output of the clock (on each network element).

This propagation of noise from input to output of a node is called *noise transfer* and can be defined as the noise that is seen at the output of the clock due to the noise that is fed to the input of the clock. The filtering capabilities of the clock determines the amount of noise being transferred through a clock from the input to the output. Obviously, less noise transfer is desirable. Figure 5-20 shows the noise that NE-2 transferred to NE-3 based on noise it was subjected to from NE-1.

Again, as with the ITU-T clock specifications for noise generation and tolerance, MTIE and TDEV masks are used to specify the allowed amount of noise transfer. Remember that each clock node has a PLL with filtering capabilities (LPF and HPF) that can filter out noise based on the clock bandwidth (refer to the earlier section "Low-Pass and High-Pass Filters").

The main metric used to describe how noise is passed from input to output is clock bandwidth, which was explained in the section "Low-Pass and High-Pass Filters." For example, the noise transfer of an Option 1 EEC is described in clause 10.1 of ITU-T G.8262 as follows: "The minimum bandwidth requirement for an EEC is 1 Hz. The maximum bandwidth requirement for an EEC is 10 Hz."

Holdover Performance

Given an accurate reference clock as an input signal, one can achieve a well-synchronized clock. This is the fundamental principle of timing synchronization whereby a master clock drives the slave clock, always assuming that the master clock itself is stable and accurate.

Occasionally, there will be cases where the master clock becomes unavailable. During such periods, the local clock is no longer disciplined by a master (or reference) clock. Most synchronization-based applications need to know the amount of frequency drift that can occur during such events.

The time during which a clock does not have a reference clock to synchronize to is called clock *holdover*. In this state, the clock behaves like a flywheel that keeps spinning at a constant speed even when it is not being actively driven, so this is sometimes also referred to as flywheel mode. And the measure of the speed with which a slave clock drifts away from the reference or master clock is called its *holdover performance*. The clock's ability to maintain the same frequency over an interval of time without a reference frequency being available is called *frequency holdover*. Similarly, time or *phase holdover* is the ability to maintain phase accuracy over an interval of time without the external phase reference.

This synchronization across network elements is achieved either via a physical signal (such as SyncE) or via precision time protocol (PTP) packets carrying timing synchronization information. There is even the ability to combine both approaches in a hybrid clocking architecture, whereby frequency is carried with SyncE and phase synchronization with PTP packets. Chapter 7, "Precision Time Protocol," delves into this a little more.

A physical link failure (or fault) will break the frequency distribution via SyncE or interrupt the PTP packet path between slave and master. During such events, the slave clock will start slowly "drifting" away from the reference or master clock. This frequency and phase drift, besides depending on many external factors, also depends on components used for the local clock and the characteristics of the PLL circuitry.

Obviously, it is desirable for the drift from the reference clock during holdover to be small, although "small" is not a helpful measure and so it is defined in more quantitative terms. The measurement of holdover performance is done when either one, both, or all clock sources (frequency and phase) are removed, and the output of the slave clock is measured against a reference signal from the master clock.

The ITU-T recommendations specify MTIE and TDEV masks to compare the accepted holdover performance for each clock type (the masks are not as strict as the masks used when the clocks are locked to an external reference clock).

Two major factors determine holdover performance. These are primarily the internal components of a clock or PLL and secondly the external factors that could impact the clock output signals. Among the components, the oscillator (chiefly its stability) plays the most important role in providing better holdover performance from a clock. For example, a clock based on a rubidium (stratum 2) oscillator or an OCXO (stratum 3E) oscillator will provide better holdover characteristics than a stratum 4 oscillator.

Similarly, smaller variability in environmental conditions (especially temperature) during the holdover period will result in better holdover characteristics. For example, rubidium oscillators are very stable, but they are susceptible to being less so when exposed to temperature changes. Figure 5-22 provides a very rough comparison of holdover performance of clocks with different classes of oscillator. Note that Figure 5-22 is a graphical representation to show the magnitude of deviation with respect to time; either positive or negative deviations are possible for different oscillators.

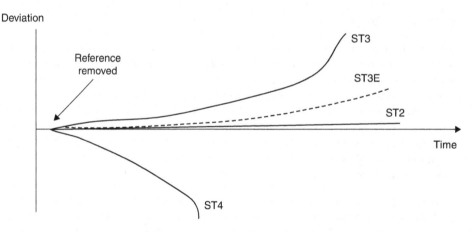

Figure 5-22   *Relative Holdover Performance for Different Classes of Oscillator*

Hardware designers are constantly innovating to improve holdover performance. One such innovation is to observe and record the variations in the local oscillator while the reference signal is available, sometimes also referred to as *history*. When the reference signal is lost, this historical information is used to improve the preservation of the accuracy of the output signal. When using such techniques, the quality of this holdover data plays an important role in the holdover performance.

There are cases where multiple paths of synchronization are available, and not all are lost simultaneously. For the hybrid synchronization case, where frequency synchronization is achieved via physical network (SyncE) and phase synchronization is achieved via PTP packets, it may be that only one of the synchronization transport paths breaks (such as, only the PTP packet path or just the frequency synchronization path). For such cases, one of the synchronization modes could move into holdover.

For example, it could be that the PTP path to the master is lost (for example, due to logical path failure to the PTP master) but frequency synchronization (SyncE) is still available. In this case, the frequency stays locked to the reference and just the phase/time moves into holdover.

Why is holdover performance important, and how long should good holdover performance be required?

As an example, let us examine the case of a 4G LTE-Advanced cell site, which may have a phase alignment requirement to be always within ±1.5 μs of its master. This cell site is equipped with a GPS receiver and external antenna to synchronize its clock to this level of phase accuracy. Suppose this GPS antenna fails for some reason (say, a lightning strike). In that case, the clock in this cell site equipment will immediately move into holdover. The operation of this cell site can only continue so long as its clock phase alignment of that cell site radio stays within ±1.5 μs of accuracy.

Should this GPS antenna failure happen during the night, it might take some time for operations staff to discover and locate the failure, and then dispatch a technician to the correct location. If it was a long weekend, it might take quite some time for the technician to reach the location and fix the issue (assuming the technician knows what the problem is and has spares for the GPS antenna readily available).

Of course, it would be very convenient if, during this whole duration, the clock was able to stay within a phase alignment of ±1.5 μs and provide uninterrupted service to that area. If the maximum time to detect and correct this fault is (say) 24 hours, then for uninterrupted service, the clock on this cell site requires a holdover period of 24 hours for ±1.5 μs. In fact, the generally accepted use case at the ITU-T is that a clock should need holdover performance for up to 72 hours (considering a long weekend).

Note that this requirement (24 hours for ±1.5 μs) is quite a strict holdover specification and just about all currently deployed systems (cell site radios and routers) are not able to achieve this level of holdover. Being able to do it for 72 hours is very difficult and requires a clock with a very high-quality oscillator, such as one based on rubidium.

Adding to the difficulty, the weather conditions at a cell site (up some mountaintop) might be extremely challenging with only minimal control of environmental conditions for the equipment. Temperature is a very significant factor contributing to the stability of an oscillator, and that stability feeds directly into holdover performance. For this reason, the ITU-T G.8273.2 specification for a PTP boundary and slave clock defines the required holdover performance in absence of PTP packets for both cases—constant temperature and variable temperature.

Transient Response

Holdover is the state that a clock transitions to whenever the input reference clock is lost. It has been shown that providing good, long-duration holdover to a failure of the reference signal is neither easy nor inexpensive. However, there could be multiple reference clocks simultaneously available to a slave clock, offering the ability to provide backup signals in the case of failure. Consequently, network designers prefer a design where multiple references are available to the network elements—providing redundancy and backup for timing synchronization.

The question arises, how should the clock behave when an input reference clock is lost and the clock decides to select the second-best clock source as the input reference? The behavior during this event is characterized as the *transient response* of a clock.

Multiple reference inputs to a clock could be traceable (via different paths) to the same source of time (such as a PRTC) or back to different sources. The clock quality information (such as Ethernet synchronization messaging channel (ESMC) and PTP packets) sent through the network shows the quality of the source of time that these separate references offer. The software managing the slave clocks then decides which backup reference clock to use based on the quality information. As expected, a high-quality reference is preferred over a lower-quality reference clock.

For cases where there are multiple inputs, the network designer might assign the input references a priority, so that the slave clock can decide in what order the input is preferred. So, when the highest priority (say priority 1) reference clock fails, the slave must decide whether there is another reference available to provide an input signal. The slave selects one of the remaining sources available out of those signaling the highest quality level. If more than one source signal is available at that quality level, the slave selects the available reference with the highest priority (say, priority 2). The slave will then start to acquire this new input signal and continue to provide an output signal but now aligned to the new reference.

This switchover, referred to as *reference clock change*, is known as a *transient event* and the change in timing characteristics of a clock during such an event is referred to as *transient response*. This is a very different event from a clock moving into the holdover state, because the change in reference clock happens very quickly, resulting in the slave clock being in the holdover state for only a very brief time.

These transients are thus divided in the ITU-T recommendations into two categories: long-term transients and short-term transients. Switching between two reference inputs is an example of a short-term transient. *Long-term transient* is another name for holdover.

Transient response specifies the noise that a slave clock generates during such events. Slave clocks are unlikely to be transparent during such events and could either transfer or generate additional noise, which is measured as the transient response. Again, for each clock type, there are MTIE and TDEV masks in the ITU-T standards to specify the accepted transient response. For example, you can find MTIE and TDEV masks for transient response for clocks supporting SDH (E1) and SONET (T1) in clause 11 of ITU-T G.812.

Measuring Time Error

This chapter explores several definitions and limits of time error, and numerous metrics to quantify it, as well as examining the impact of time error on applications. In summary, time error, essentially the difference between a slave clock and a reference or master clock, is primarily measured using four metrics: cTE, dTE, max|TE|, and TIE. From those basic measurements, engineers have statistically derived MTIE and TDEV from TIE, in combination with the application of different filters.

ITU-T recommendations specify the allowed limits for these different metrics of time error. The metrics cTE and max|TE| are defined in terms of time (such as nanoseconds). dTE, being the change of the time error function over time, is characterized using TIE and is then compared to a mask expressed in MTIE and TDEV, which defines the permissible limits for that type of clock.

Some ITU-T recommendations define allowed limits for a single standalone clock (a single network element), and others specify clock performance at the final output of a timing chain in an end-to-end network. You should not mix these two types of specifications, because limits that apply to the standalone case do not apply to the network case. Similarly, masks used to define the limits on a standalone clock do not apply to the network. This is especially the case when discussing the testing of timing performance, as in Chapter 12, "Operating and Verifying Timing Solutions."

For each type and class of standalone clock, all four metrics, max|TE|, cTE, dTE, and TIE, are specified by these standards, while for the network case, max|TE| and dTE are specified. The cTE is not listed for network limits because the limits of max|TE| for the network case automatically include cTE, which might be added by any one clock in the chain of clocks.

The ITU-T specifies the various noise and/or time error constraints for almost all the clock types. For each one of the numerous clock specifications, the ITU-T recommendations define all five aspects for timing characterization of a clock:

- Noise generation

- Noise tolerance

- Noise transfer

- Transient response

- Holdover performance

Chapter 8 will cover these specifications in more detail.

Topology

As shown in Figure 5-23, the setup required for time error measurement testing includes

- Timing testing equipment that can synthesize *ideal* reference clock signals.

- Clock under test that would normally be embedded in a network element.

■ Clock capture and measurement equipment. Most often this is the same equipment that generates the reference clock so that it can do an easy comparison to the output signal returned from the clock under test.

Note that if the tester is a different piece of equipment from the reference input clock, the same clock signal needs to be also passed to the tester to allow it to compare the measured clock to the reference clock. You cannot use two separate signals, one as reference and one in the tester.

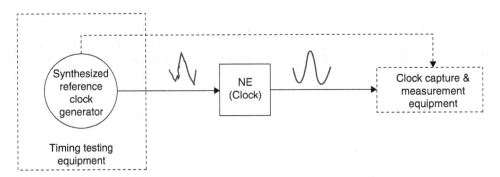

Figure 5-23   *Time Error Measurement Setup*

The testing equipment that is generating the reference clock is referred to as a *synthesized reference clock* in Figure 5-23 because of the following:

■ For noise transfer tests, the reference clock needs to be synthesized such that jitter and wander is introduced (under software control). The tester generates the correct amount of jitter and wander, according to ITU-T recommendations, for the types of clock under test. The clock under test is required to filter out the introduced input noise and produce an output signal that is under the permissible limits from the standards. This limit is defined as a set of MTIE and TDEV masks.

■ For noise tolerance tests, the reference clock needs to emulate the input noise that the clock under test could experience in real-world deployments. And for this purpose, the testing equipment simulates a range of predefined noise or time errors in the reference signal to the clock under test. To test the maximum noise that can be tolerated by the clock as input, the engineer tests that the clock does not generate any alarms; does not switch to another reference; or does not go into holdover mode.

■ For noise generation tests, the reference clock that is fed to the clock under test should be ideal, such as one sourced from a PRTC. So, for this measurement, there is no need for artificially synthesizing the reference clock, unlike the noise transfer and noise tolerance case, where the clock is synthesized. The testing equipment then simply compares the reference input signal to the noisy signal from the clock under test.

References in This Chapter

Open Circuit. "16Mhz Crystal Oscillator HC-49s." Open Circuit. Picture. https://opencircuit.shop/Product/16MHz-Crystal-Oscillator-HC-49S

The International Telecommunication Union Telecommunication Standardization Sector (ITU-T).

"G.781: Synchronization layer functions for frequency synchronization based on the physical layer." *ITU-T Recommendation*, 2020. http://handle.itu.int/11.1002/1000/14240

"G.810: Definitions and terminology for synchronization networks." *ITU-T Recommendation*, 1996. http://handle.itu.int/11.1002/1000/3713

"G.811: Timing characteristics of primary reference clocks." *ITU-T Recommendation*, 1997. http://handle.itu.int/11.1002/1000/4197

"G.812: Timing requirements of slave clocks suitable for use as node clocks in synchronization networks." *ITU-T Recommendation*, 2004. http://handle.itu.int/11.1002/1000/7335

"G.813: Timing characteristics of SDH equipment slave clocks (SEC)." *ITU-T Recommendation*, 2003. http://handle.itu.int/11.1002/1000/6268

"G.8261: Timing and synchronization aspects in packet networks." *ITU-T Recommendation*, Amendment 1, 2020. http://handle.itu.int/11.1002/1000/14207

"G.8262: Timing characteristics of a synchronous equipment slave clock." *ITU-T Recommendation*, Amendment 1, 2020. http://handle.itu.int/11.1002/1000/14208

"G.8262.1: Timing characteristics of an enhanced synchronous equipment slave clock." *ITU-T Recommendation*, Amendment 1, 2019. http://handle.itu.int/11.1002/1000/14011

Chapter 5 Acronyms Key

The following table expands the key acronyms used in this chapter.

Term	Value
µs	microsecond or 1×10^{-6} second
4G	4th generation (mobile telecommunications system)
BITS	building integrated timing supply (SONET)
cTE	constant time error
DPLL	digital phase-locked loop
dTE	dynamic time error
E1	2-Mbps (2048-kbps) signal (SDH)
EEC	Ethernet equipment clock

Term	Value
ESMC	Ethernet synchronization messaging channel
FFD	fractional frequency deviation
FFO	fractional frequency offset
GPS	Global Positioning System
HPF	high-pass filter
ITU	International Telecommunication Union
ITU-T	ITU Telecommunication Standardization Sector
LPF	low-pass filter
LTE	Long Term Evolution (mobile communications standard)
max|TE|	maximum absolute time error
MTIE	maximum time interval error
NE	network element
OCXO	oven-controlled crystal oscillator
PLL	phase-locked loop
ppb	parts per billion
ppm	parts per million
PRC	primary reference clock (SDH)
PRS	primary reference source (SONET)
PRTC	primary reference time clock
PRTC-A	primary reference time clock—class A
PRTC-B	primary reference time clock—class B
PTP	precision time protocol
RMS	root mean square
SDH	synchronous digital hierarchy (optical transport technology)
SDO	standards development organization
SEC	SDH equipment clock
SONET	synchronous optical network (optical transport technology)
SSU	synchronization supply unit (SDH)
SyncE	synchronous Ethernet—a set of ITU-T standards
T1	1.544-Mbps signal (SONET)
TCXO	temperature-compensated crystal oscillator

Term	Value
TDEV	time deviation
TE	time error
TIE	time interval error
VCO	voltage-controlled oscillator
VCXO	voltage-controlled crystal oscillator
XO	crystal oscillator

Further Reading

Refer to the following recommended sources for further information about the topics covered in this chapter.

ANSI: https://www.ansi.org

ETSI: https://www.etsi.org

ITU: https://www.itu.int

Physical Frequency Synchronization

In this chapter, you will learn the following:

- **Evolution of frequency synchronization:** Covers how frequency synchronization strategies evolved over time and what are the different approaches to achieve frequency synchronization.

- **BITS and SSU:** Describes these pieces of timing equipment and their role in achieving network-wide frequency synchronization.

- **Clocking hierarchy:** Provides an overview of a clocking hierarchy within a frequency synchronization network.

- **Synchronous Ethernet:** Covers how Ethernet evolved the ability to distribute frequency over Ethernet links.

- **Enhanced synchronous Ethernet:** Describes the need for higher-performance synchronization and how synchronous Ethernet is being enhanced to achieve it.

- **Clock traceability:** Covers techniques to transport the quality of a clock to all network nodes and how nodes use this information to select the best timing source available.

- **Synchronization network chain:** Describes the principles to aid in deploying and managing network nodes to build a frequency distribution network.

- **Clock selection process:** Provides an overview of the methods a clock uses to select the best source of reference timing signal.

- **Timing loops:** Covers what timing loops are and how timing loops can be avoided in networks.

- **Standardization:** Provides an overview of the standards development organizations that are applicable to frequency synchronization.

In the previous chapters, you have read that synchronization mechanisms are essential in both mobile and fixed communications. Synchronization is one of the fundamental technologies that telecommunication operators use to build and deliver network services, and they have introduced it into many of their networks. Chapter 3, "Synchronization and Timing Concepts," introduced the different types of synchronization, classified as frequency, time, and phase synchronization. You also learned that the state in which the clock frequencies of different systems match is called *frequency synchronization*, and the state in which timings between clocks agree is called *phase synchronization*. This chapter focuses primarily on how frequency synchronization is achieved across clocks that are connected to each other in a network. The central focus will be on those aspects specific to frequency synchronization.

Evolution of Frequency Synchronization

Frequency synchronization had an immense impact on the evolution of traditional telecommunication, especially in enabling the transition of transmission and switching from analog to digital. Traditional public switched telephone networks (PSTN) started the digitization process with isolated digital transmission links between analog switching machines or radio transmission systems. The fact that digital technology was being applied was transparent to the interfaces, so the internal clock rate in one system was not required to be synchronized to the internal clock rate of another system.

Even networking systems that combined higher bit-rate multiplexing with lower-speed tributaries did not require synchronization across systems. It was the introduction of circuit-switched data networks and the Integrated Services Digital Network (ISDN) that first needed more stringent synchronization. With this development, the evolution of network synchronization began.

A few different network synchronization strategies evolved, but for frequency synchronization, the one that became most prevalent was for the clock distribution of frequency signals via a separate external network (refer to the section "Legacy Synchronization Networks" in Chapter 2, "Usage of Synchronization and Timing"). This led all the major network service providers to build national frequency synchronization networks to distribute a common timing reference to each node of their telecommunication networks.

At the same time, various standard bodies, such as the International Telecommunication Union Telecommunication Standardization Sector (ITU-T), the European Telecommunications Standards Institute (ETSI), and the American National Standards Institute (ANSI), started developing new standards specifying more stringent and complex requirements for jitter and wander in synchronization networks.

As a result, operators settled on two basic approaches to achieve frequency synchronization:

- Using the physical layer, which involves distributing frequency electrically or optically over the network interface. The synchronous digital hierarchy (SDH), synchronous optical networking (SONET), and synchronous Ethernet (SyncE) are examples of technologies using the physical layer from node to node. SDH/SONET and SyncE provide predictable carrier-class frequency distribution but require system/equipment changes in case any nodes in the network are not capable of supporting compatible physical interfaces. This is usually not an issue with legacy SDH/SONET network elements, but it becomes a challenge for Ethernet-based systems if they are not capable of supporting SyncE.

■ Frequency can also be distributed using packet-based networks. Over the recent past, engineers have developed several protocols, such as structure-agnostic TDM over packet (SAToP), circuit emulation over packet switched network (CESoPSN), circuit emulation over packet (CEP), Network Time Protocol (NTP), and precision time protocol (PTP), to distribute frequency via packets. At first glance, the packet method of distributing frequency looks flexible and simple; however, carrying frequency information over nondeterministic packet-based networks requires careful engineering and can be more difficult than using physical methods.

This chapter focuses mainly on frequency distribution using physical interfaces. Frequency distribution over packet is covered in Chapter 9, "PTP Deployment Considerations."

BITS and SSU

For traditional PSTN, frequency synchronization is based on the concept of the *building or office clock*, where an entire building or office is supplied with frequency to all the installed equipment, including digital switching exchanges, digital cross-connects, and terminal equipment. Such a clock for the complete building (referred to as a central office or CO) is known as the building timing integrated supply (BITS) in the ANSI standards and synchronization supply units (SSU) or standalone synchronization equipment (SASE) in the ITU-T and ETSI standards. With recent advances, the ITU-T has incorporated both ANSI and ETSI considerations, so the rest of this chapter uses the BITS and SSU terms interchangeably.

Figure 6-1 shows the distribution of a BITS or SSU timing circuit into a building through either a GPS receiver or a signal from an atomic clock, such as a primary reference clock (PRC). Without a centralized source such as SSU/BITS, you can potentially have many different timing sources within the same building because of the number of synchronous links that might exist throughout a building.

The SSU/BITS system operates by serving as a clock synchronization receiver that extracts a stable timing signal from a reference clock and then distributes it internally within the office or building to all equipment requiring a traceable frequency reference. Thus, the BITS node effectively becomes the master timing supply for all deployed equipment requiring frequency synchronization within a network location. It can also provide external timing links to structures within the same campus, such as adjacent buildings.

Most of these connections are deployed as T1 or E1 circuits, and many vendors carry these signals by using either coaxial or unshielded twisted-pair (UTP) cabling. Typically, the BITS system has multiple timing reference inputs, which act as redundant timing sources in case the primary timing source is lost. During such failures, the BITS system continues to supply timing to help achieve the required equipment uptime.

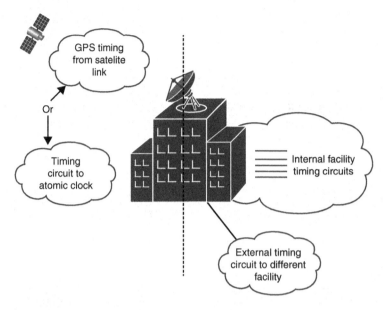

Figure 6-1 *BITS or SSU Timing Distribution*

SSU/BITS systems extract the timing information from an incoming signal, which could be a T1, E1, 2-MHz, or 64-kbps signal from a local or remote dedicated timing source. This timing signal then becomes the master timing reference for all downstream network elements. Figure 6-2 shows a typical BITS unit and a few potential connections using T1 as distribution to different equipment in the building.

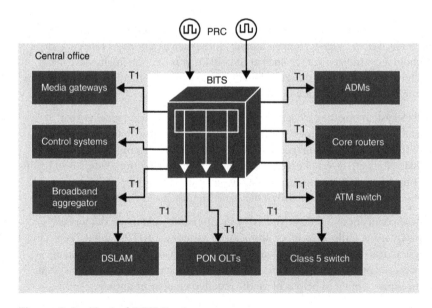

Figure 6-2 *Typical BITS Equipment*

Figure 6-2 shows connections that carry only timing information to different network elements. This is usually referred to as *external timing*, where the connections carry only timing information and no user data. So, when a piece of digital equipment is timed externally, it means the equipment supports a dedicated interface that is designed to receive a timing signal from the timing distribution network. Figure 6-3 shows how these devices are connected to a master or centralized clock, which could be either a BITS or SSU system.

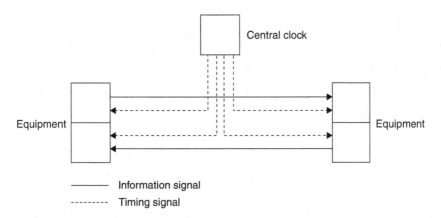

Figure 6-3 *Central Clock Distributing Timing to Equipment (Figure 6-2 from ITU-T G.703)*

This network equipment can then use the timing signal to both regulate the speed of data communications and carry the timing information multiplexed with user/customer data to downstream devices. This mode where user traffic is mixed with a timing signal is referred to as *line timing*, and because these nodes also transport data, they are considered part of the transport network. Such devices are referred to as SDH or synchronous equipment clocks (SEC) or network elements (NE).

Later in this chapter and in the section "Definitions, Architecture, and Requirements" in Chapter 8, "ITU-T Timing Recommendations," you will see that over time, the definition of SEC has evolved to include other types of clocks as well. Note that this equipment is different from the SSU/BITS systems because they are part of the timing network and not part of the transport network.

Figure 6-4 shows an example network with connections for external timing and line timing. The solid arrows between the elements represent the transfer of timing information, and in external timing mode, only timing signals pass. In the line timing mode, both timing and data pass over the interface. The data transfer between SECs is represented as dashed lines.

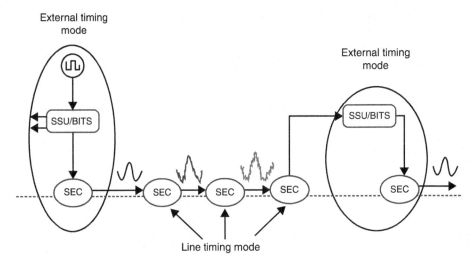

Figure 6-4 *Partial Network with External Timing and Line Timing*

Figure 6-4 also shows that the clock signal starts degrading while being carried over a network of SEC nodes. Each SEC node on the path recovers and regenerates the clock using its own oscillator and phase-locked loop (PLL). During this process, the clock starts accumulating jitter and wander at a rate based on the quality of oscillators being used in SEC clocks. Refer to Chapter 5, "Clocks, Time Error, and Noise," for more information on jitter and wander of clocks.

On other hand, the SSU/BITS equipment either sources its clock from a stratum 1–traceable reference or contains a better oscillator, such as a stratum 2 rubidium. By using higher quality timing sources and/or better oscillators, SSU/BITS nodes can filter out the impairments or noise introduced into the timing signal. Figure 6-4 shows such a case where an SSU/BITS node in the middle of a SEC chain is *cleaning up* the jitter that has accumulated in the clock distribution network across the stratum 3 nodes. This process is sometimes referred as *clock cleanup* in the timing industry.

Note   Use of the terms "master" and "slave" is ONLY in association with the official terminology used in industry specifications and standards, and in no way diminishes Pearson's commitment to promoting diversity, equity, and inclusion, and challenging, countering, and/or combating bias and stereotyping in the global population of the learners we serve.

The SEC clock, like any clock slaved to another, is defined to have three basic modes of operation: locked, free-run, and holdover. As described in Chapter 3, in locked mode, the output of the clock is considered *synchronized* with the input clock, such that the output frequency is tracking (or aligned) to the input reference frequency. This state of operation is the *locked* condition for a slave clock.

If the input reference is lost, the slave clock can enter a *holdover* state. The holdover state is achieved by using stored data, acquired during locked mode, to control the frequency and phase of the output clock. In this mode, the output clock is a "reproduction" of what was the input reference. A transition back to the locked mode would occur when the input reference becomes available again.

The final mode of operation is called *free running*. The output clock in this mode of operation is based on the oscillator of the slave clock and is not tracking the input reference signal or using stored data. This mode is the typical mode of operation when the slave clock is first powered on. It is also achieved when there is no stored data available in which to transition to the holdover state.

Refer to the section "Clock Modes" in Chapter 3 for a detailed description of the different modes of operation of a clock and the different possible transitions between the modes.

Clocking Hierarchy

The basic task of network synchronization is to distribute the reference signal of the most accurate clock available to all network elements that require synchronization. Different methods (such as external timing and line timing) that distribute the reference signal in the network mostly use a single strategy, known as master-slave synchronization.

This method uses master clock and slave clock nodes whereby a slave clock always receives timing information from a master clock. And this strategy is then replicated throughout the network. Note that this means that a device slaved to an upstream master clock can also take the master clock role to other downstream equipment. The rule of this master and slave clock relationship is that a slave clock must only be "slaved" to a clock of higher or equal stability.

This becomes a hierarchical model, which is sometimes referred to as *synchronization hierarchy*, where clocks are divided into different levels or quality that define the accuracy and stability for the timing (again, refer to Chapter 3 for a detailed description). Figure 6-5 shows a hierarchy of clocks belonging to different stratum levels mixed with transport elements (SECs/NEs).

The stratum level 1 primary reference clock (PRC) or primary reference source (PRS) is hosting the reference clock for the whole network. SSU/BITS systems are typically stratum 2 systems and the SECs/NEs typically have stratum 3E or stratum 3 clocks (or lower).

Note that BITS and SSU could also have stratum 3E clocks, which makes them like SECs/NEs in terms of accuracy and stability. Thus, the performance specifications for such systems are tied to the clock type these systems belong to. For example, a BITS system with stratum 2 is required to comply with performance specifications defined by ITU-T for stratum 2 systems, which are Type II clocks as specified in ITU-T G.812, whereas a BITS system with stratum 3E should comply with Type III clocks as specified in ITU-T G.812.

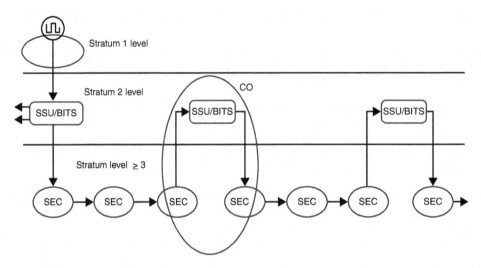

Figure 6-5   *Clock Hierarchy with Different Stratum Levels*

Figure 6-6 shows the synchronization hierarchy for SDH and SONET nodes along with the ITU-T specifications that cover the performance requirements for clock types. Note that G.781 refers to SSU-A as Type I or V clocks and to SSU-B as Type VI clocks. The performance specifications for these clock types are defined in G.812.

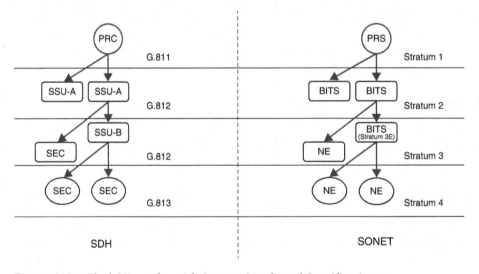

Figure 6-6   *Clock Hierarchy with Stratum Levels and Specifications*

Synchronous Ethernet

For a long time, even though Ethernet became a dominant technology for data transmission, frequency distribution over Ethernet never emerged. Ethernet itself was originally intended for transmission of asynchronous data, and either did not need to synchronize the components in an Ethernet circuit or needed the synchronization only on a link-by-link basis.

As you read in Chapter 3, to achieve link-by-link synchronization, Ethernet selects one end of the link to be the master and the other end becomes the slave. The master end uses its local free-running oscillator to transmit data to the slave; and the slave *turns around* the frequency of the incoming signal and uses that same frequency to clock its transmission back toward the master. However, these Ethernet nodes do not need to pass timing information from one node to another, which is required to achieve timing distribution across the whole network.

Due to its simplicity, low cost, and bandwidth flexibility, Ethernet soon became the default choice of physical medium for data transmission and the service providers started migration from SDH/SONET or T1/E1 to Ethernet.

While the data transmission (and TDM to packet) was easily migrated, a need appeared for the timing network to also migrate to Ethernet and packet technology. Therefore, Ethernet needed to support frequency distribution in the same or a similar way to SDH/SONET networks. So, the gaining popularity of Ethernet and demand for the benefits of synchronization in networks led to the development of Ethernet to carry frequency.

The technology that was proposed to carry frequency using the packet network is *synchronous Ethernet* (SyncE). SyncE uses the physical interface to pass timing from node to node—meaning that it uses the frequency of the pulses of light (or electrical waves in copper), much like SDH/SONET or E1/T1. Because the frequency is carried over a physical medium, SyncE can provide at least equivalent reliability and accurate frequency distribution as does SDH/SONET or T1/E1.

SyncE uses line coding of the Ethernet interface to transfer timing. The timing information on the slave clock is extracted from the *preamble* of the frame (refer to the section "Asynchronous Networks" in Chapter 3), which is used to transfer packets over optical or electrical interfaces. Thus, SyncE does not use packets (the packet layers of Ethernet) to distribute a frequency signal; hence, there is no impact on the quality of SyncE when experiencing high rates of packet traffic.

Note that the legacy 10-Mbps Ethernet is not capable of a synchronization signal transfer over the physical layer because the clock edges (or pulses) that are used to extract timing information are not sent continuously, and these signals are stopped during times when there is no data to transmit (called *idle periods*). And if there is no data during these periods, there is no timing information carried across physical links.

Within the node as well, SyncE operates in a similar method to SDH/SONET, whereby the slave node extracts timing information from the physical interface and routes it to its own phase-locked loop (PLL). This PLL then locks to this recovered frequency and drives all the other Ethernet or non-Ethernet interfaces with the same frequency, allowing the timing information to be passed to other downstream nodes.

Obviously, enabling SyncE on a platform requires hardware that supports these functions as well as the signal traces (wires) to carry the frequency signals between the timing sub-system and the Ethernet ports. Besides performing normal data transmission, equipment supporting SyncE also has the ability to transfer frequency, and such devices are referred to as *Ethernet equipment clocks* (EEC).

Figure 6-7 shows some EEC nodes that are timed externally from the SSU/BITS and the line timing mode using SyncE between EEC nodes.

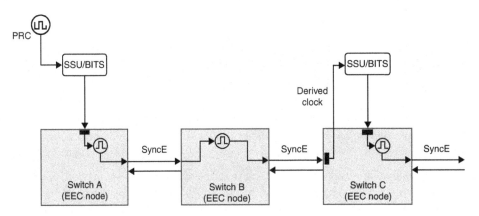

Figure 6-7   *SSU/BITS Distributing Timing to EEC Nodes*

Note the similarity of EEC nodes to SEC nodes or NEs. Like the SEC node, the EEC node can take an input reference from PRC or SSU/BITS equipment. They can also pro-vide line timing to downstream clocks as SEC nodes can. In fact, the SDH clock hierarchy shown in Figure 6-6 also applies to EEC nodes (substituting EEC instead of SEC).

Further details on SyncE and the EEC clock can be found in ITU-T G.8262.
See Chapter 8 for more details on the ITU-T recommendations.

The evolving 5G wireless architecture and some of the newer features and services that it makes possible will require higher performance of timing and frequency synchronization. To define and meet the much tighter synchronization requirements of 5G in fronthaul networks, the ITU-T is defining a new set of *enhanced clocks*. These newer clocks have around an order of magnitude better performance than the clocks defined for pre-5G net-works. The frequency synchronization aspects of these "enhanced clocks" are covered by the new *enhanced synchronous Ethernet*.

Enhanced Synchronous Ethernet

Over the past several years, the ITU-T defined (in G.8262.1) the *enhanced* Ethernet equipment clock (eEEC) that supports an improved version of SyncE called *enhanced* synchronous Ethernet (eSyncE). This eEEC is the newer, more accurate alternative to the older EEC clock that was defined in ITU-T G.8262.

As explained in Chapter 5, different ITU-T specifications define the performance requirements for each type of clock. These performance requirements are primarily for clock accuracy, noise transfer, holdover performance, noise tolerance, and noise generation. For the eEEC, the ITU defines performance specifications that are much stricter than those for the EEC to ensure that the new clock delivers better performance.

The ITU-T G8262.1 recommendation is relatively recent (January 2019), and the authors expect that it might take some time for the equipment to fully comply with the performance requirements specified in this recommendation.

Note that you must take care when deploying frequency synchronization over Ethernet using a mixture of clock nodes with different performance levels. Imagine a network where, in a chain of nodes, only a few pieces of equipment comply with eEEC (G8262.1 clock type) performance requirements, and the rest of the equipment complies with EEC (G.8262 clock type). For such cases, even though several nodes operate at the higher performance level, the end-to-end performance of the network might not provide the higher performance required by some 5G services.

For more details on deploying networks of mixed clock quality, see ITU-T G.8261 Appendix XIV, "Interoperability guidelines for interworking between synchronous equipment clocks and enhanced synchronous equipment clocks."

However, with the improvement in hardware over recent years, it is likely that most modern equipment that is compliant with SyncE approaches the performance levels of eSyncE. Remember that SyncE is aligned with SDH/SONET levels of quality, and those standards go back decades. Despite that, you should expect that the highest performance can only be achieved in networks where eEEC or better clocks are implemented throughout.

Clock Traceability

As discussed previously in the section "Clocking Hierarchy," there exists a hierarchical relationship between stratum levels and a master-slave relationship is used to distribute frequency. Remember that the basic rule is that a slave should lock to a clock of higher or equal stability. This does require a method to communicate the stability level of a master clock to a slave clock along with the frequency signal itself.

Because the ability to receive a stable timing signal has become a critical need, a good timing distribution network is designed such that there are multiple frequency sources (or master clocks) available to the slave clock. This design is used to provide a backup timing signal during clock outages as well during physical link failures. In such cases, a decision is required at the slave clock to pick the best clock signal from those available to it. This decision to pick the best available clock at the slave node is also sometimes referred to as the *clock selection algorithm*.

Figure 6-8 shows a simple case of a network with SEC nodes, and Figure 6-9 shows a possible condition where during a fault, the clock available at different SECs will be traceable to different stratum levels.

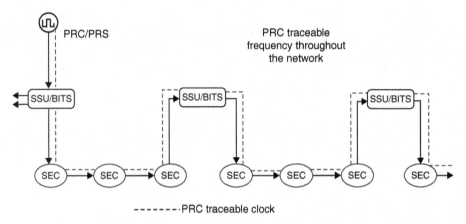

Figure 6-8   *SEC Nodes Timing Distribution with SSU/BITS*

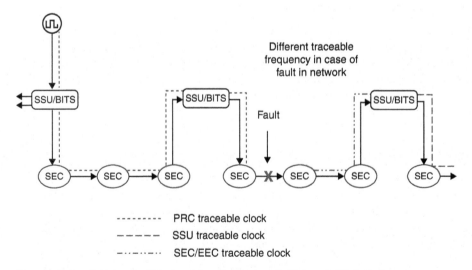

Figure 6-9   *SEC Nodes Timing Distribution with Network Link Failure*

To know the clock stability level at any piece of equipment, it is necessary that every clock be able to trace its received reference signal back to its source. For example, as shown in Figure 6-8, the reference clock that all the NEs are receiving during the no-fault condition is traceable to the PRC/PRS. However, as shown in Figure 6-9, after a physical interface fault, a few NEs in the network will receive a SEC traceable clock signal while others will receive SSU/BITS.

Synchronization Status Message

To achieve traceability, the physical interfaces, in addition to carrying timing information, also need to carry a clock *quality level* (QL) indicator from the master to the slave node. This is achieved via the *synchronization status message* (SSM) signals for the SDH/SONET and T1/E1 physical interfaces. SDH/SONET frames have an overhead frame, which is not considered as part of regular data or protocol data unit (PDU), that carries extra information from one end to the other. This overhead frame is used to carry SSM information.

These signals (sometimes called "messages," though in networking terminology these are not really messages) are defined in ANSI T1.105 and ITU-T G.781. They carry the quality level of the clock signal throughout the network to deliver end-to-end clock traceability. The exact mechanism for transporting of SSM is as follows:

- For optical lines, there are two S bytes in the overhead frame for both SDH and SONET, and the SSM is transported in the S1 byte of this overhead frame.

- If the SSM is carried over an electrical line:

    - For SDH, the SSM is transported in the Sa bits of the E1.

    - For SONET, the SSM is transported out-of-band in an extended super frame (ESF) data link, in addition to transmitting the in-band signal in the payload.

It is important to note that ITU-T specifications classify SDH and packet networks by means of Options I, II, and III processing. *Option I* applies to SDH networks optimized for the 2048-kbps hierarchy. *Option II* applies to SDH networks optimized for the 1544-kbps hierarchy that includes the rates 1544 kbps, 6312 kbps, and 44,736 kbps. *Option III* applies to SDH networks optimized for the 1544-kbps hierarchy that includes the rates 1544 kbps, 6312 kbps, 33,064 kbps, 44,736 kbps, and 97,728 kbps.

For each of these different options, ITU-T G.781 also specifies the following clock source quality levels:

- **Option I synchronization networking:** The clock source quality levels that are defined under Option-I are QL-ePRTC, QL-PRTC, QL-ePRC, QL-PRC, QL-SSU-A, QL-SSU-B, QL-eSEC, QL-SEC, and QL-DNU. Here, QL-SEC represents the same quality level as QL-EEC1 as defined by ITU-T G.8264.

- **Option II synchronization networking:** Clock source quality levels of Option II were defined initially as seven quality levels. This set of seven quality levels is also referred to as *First Generation*, or *GEN1*. These seven levels were expanded to nine levels, and this extended level set is referred to as *Second Generation*, or *GEN2*. Later the GEN2 quality levels were further enhanced to include enhanced PRTC (ePRTC) and PRTC as well. The GEN1 clock quality set is a subset of GEN2 because GEN1 quality levels do not define QL-ST3E and QL-TNC as separate quality levels, and QL-PROV was identified as QL-RES. Table 6-1 lists the different clock quality levels, in order from highest to lowest levels of clock quality.

Table 6-1   *Clock Quality Levels for Option II Synchronization*

Quality Level	Description	Order
QL-ePRTC	Enhanced PRTC Traceable	Highest
QL-PRTC	PRTC Traceable	|
QL-PRS	PRS Traceable	|
QL-STU	Synchronized (and) Traceability Unknown	|
QL-ST2	Stratum 2 Traceable	|
QL-TNC	Traceable to Transit Node Clock	|
QL-ST3E	Stratum 3E Traceable	|
QL-ST3	Stratum 3 Traceable	|
QL-SMC	SONET Minimum Clock Traceable	|
QL-ST4	Stratum 4 Traceable (freerun)	|
QL-PROV	Provisionable by the Network Operator	|
QL-DUS	Do not Use for Synchronization	Lowest

Source: ITU-T G.781, Table 2

- **Option III synchronization networking:** The only clock source quality levels that are defined in the synchronization process of an Option III network are QL-ePRTC, QL-PRTC, QL-UNK, QL-eSEC, and QL-SEC, corresponding to four levels of synchronization quality. QL-UNK is used to indicate an unknown clock source.

The exact encoding of SSM depends on the physical medium in the network being used and configured. For example, encoding of an SSM in T1 is different from encoding in E1, and so on. But irrespective of the encoding scheme, the only information that an SSM carries is the QL information from the master clock to the slave clock. This QL information is basically equivalent to the stratum level of the master clock to which the quality level on that physical link can be traced. Table 6-2 shows the SSM codes in an Option II SDH synchronization network.

Table 6-2 *Clock Quality Levels and SSM Coding in Option II SDH Synchronization Network*

Quality Level	SSM Usage	GEN2 SSM		GEN1 SSM	
		SSM coding [MSB..LSB] in STM-N signal (BINARY)	SSM coding [MSB..LSB] in 1544-kbps signal with ESF (HEX)	SSM coding [MSB..LSB] in STM-N signal (BINARY)	SSM coding [MSB..LSB] in 1544-kbps signal with ESF (HEX)
QL-PRS	Enabled	0001	04FF	0001	04FF
QL-STU	Enabled	0000	08FF	0000	08FF
QL-ST2	Enabled	0111	0CFF	0111	0CFF
QL-TNC	Enabled	0100	78FF	1010	10FF
QL-ST3E	Enabled	1101	7CFF	1010	10FF
QL-ST3	Enabled	1010	10FF	1010	10FF
QL-SIC	Enabled	1100	22FF	1100	22FF
QL-ST4	Enabled	—	28FF	—	28FF
QL-PROV	Enabled	1110	40FF	1110	40FF
QL-DUS	Enabled	1111	30FF	1111	30FF
—	Disabled	1111	08FF	1111	08FF

Source: ITU-T G.781, Table 6

Note that there are shaded fields in Table 6-2 for the first generation (GEN1) SSM codes. As explained previously, GEN1 SSM codes are a subset of the GEN2 SSM codes, and the shaded fields are not differentiated in the GEN1 case, but they are split out in the GEN2 case. For example, SSM coding for both QL-TNC and QL-ST3E are the same as QL-ST3 for GEN1 SSM encoding, whereas these are separated with different codes when using GEN2 SSM encoding.

There is one point that the network engineer should be aware of when deploying a QL distribution network. Because the codes for different SSM values vary between Option I, II, and III circuits, it is imperative that the router connecting to these circuits is configured with the correct option code under frequency synchronization. This is especially so in North American markets, as Cisco routers assume Option I is the default encoding unless changed via configuration (because it should be Option II in North America).

Ethernet Synchronization Messaging Channel

For SyncE, the clock traceability is achieved via the transmission of special packets, in a method called the Ethernet synchronization messaging channel (ESMC). An ESMC is a logical communication channel running over the Ethernet transport layer. It transmits

QL information, which is the quality level traceable from the transmitting synchronous Ethernet equipment clock (EEC), by using ESMC PDUs. The ESMC is defined by ITU-T G.8264 and is covered in more detail in Chapter 8.

Even though an ESMC is sometimes called "SSM for Ethernet," there are some differences between the SSM and ESMC mechanisms. Although SSM messages are carried in a fixed location within an SDH/SONET frame, the same information is carried via ESMC PDUs in Ethernet because it does not have a fixed frame. The QL values are carried using a special Ethernet method known as a *slow protocol*, with its own Ethertype (88–09), similar to other overhead functions for Ethernet, such as OAM, Link Aggregation Control Protocol (LACP), and so on. Specifically, the ITU-T defines ESMC using one of the slow protocol mechanisms called an organization specific slow protocol (OSSP), originally specified in IEEE 802.3ay. Note that the definition of OSSP is now included in Annex 57A and 57B of IEEE 802.3-2018.

The ESMC packet is composed of the standard Ethernet header for an OSSP frame; the Organizationally Unique Identifier (OUI) representing the ITU-T; an ESMC-specific header; a flag field; and a type, length, value (TLV) structure. ESMC packets, like SSM, carry a QL identifier that identifies the timing quality of the synchronization trail. QL values in the QL-TLV structures are the same QL values defined for the SDH/SONET SSM messages. Table 6-3 shows the ESMC PDU format.

Table 6-3   *ESMC PDU Format*

Octet Number	Size	Field
1–6	6 octets	Destination address = 01–80–C2–00–00–02
7–12	6 octets	Source address
13–14	2 octets	Slow protocol Ethertype = 0x88–09
15	1 octet	Slow protocol subtype = 0x0A
16–18	3 octets	ITU-OUI = 0x00–19–A7
19–20	2 octets	ITU-T subtype = 0x00–01 = ESMC
21	4 bits	Version = 0x1
	1 bit	Event flag
	3 bits	Reserved
22–24	3 octets	Reserved
25–1514	36–1490 octets	Data and padding
Last 4	4 octets	Frame check sequence (FCS)

Source: ITU-T G.8264, Table 11-3

The *ITU-OUI* and *ITU-T subtype* are important fields in the OSSP PDU for QL values because these fields differentiate the ESMC PDU from the other OSSP PDUs. The ITU-T subtype (octets number 19–20 in the ESMC PDU) is defined by G.8264 as a fixed value of 0x00–01 for ESMC frames.

There are two types of ESMC PDUs that are generated by a clock. The ESMC *information* PDUs carrying QL information are generated once per second. On the other hand, an ESMC *event* PDU (which has the *event flag* shown in Table 6-3 set to 0x1) is generated upon events that involve a change in QL (such as physical link failure or detection of a change in QL from an incoming ESMC PDU). This helps to ensure the latest QL information is passed to the peer EEC as quickly as possible following a change in clock status.

The QL information itself is carried in a QL TLV structure in an ESMC PDU. Table 6-4 shows the format of the QL TLV and how the SSM is encoded within the QL TLV as a 4-bit field.

Table 6-4   *QL TLV Format*

Octet Number	Size/bits	Field
1	8 bits	Type: 0x01
2–3	16 bits	Length: 00–04
4	bits 7:4	0x0 (unused)
	bits 3:0	SSM code

Source: ITU-T G.8264, Table 11-4

As specified in Annex 57A and 57B of IEEE 802.3-2018 (specifying slow protocols) and restated by ITU-T G.8264 (describing ESMC), in no case should more than ten ESMC PDUs (information and/or event) be generated in any 1-second period.

ESMC PDUs are based on slow protocols that are applied to single physical links. However, the case of multiple parallel links, such as Link Aggregation Group (LAG) between two NEs, brings up an interesting problem (LAG is specified in IEEE 802.1AX-2020—formerly IEEE 802.3ad). Each physical link of a LAG can exchange its own independent ESMC PDUs. So, when using a LAG, the QL value in ESMC PDUs sent from one NE to another on one physical link can get returned by the other NE on a different physical link that belongs to the same LAG.

This condition has a potential for creating a critical condition for timing networks, referred to as a *timing loop* (see the section "Timing Loops" later in this chapter), where an NE is synchronized to the delayed version of its own clock that has returned to it from another connected NE.

For cases where multiple links share the same synchronization source, the concept of a *bundle of ports* is introduced in ITU-T G.781. And the recommendation is that an NE is required to generate QL-DNU (do not use) or QL-DUS (do not use for synchronization) on all the ports of the bundle, when any one of them has been selected as the reference source. This use of DNU/DUS on all ports of a bundle makes sure that timing loops are avoided for links made up of a bundle of ports.

Enhanced ESMC

Although ESMC has worked well to carry quality levels in Ethernet packet networks, now there are new, more accurate, enhanced clocks being specified by the ITU-T. The previously existing ESMC specification was not able to describe the quality of these new clocks, so the ITU-T has defined new *enhanced* ESMC (eESMC) values to describe the new clock types. These clock types include enhanced versions of both the PRC/PRS and the PRTC. This change was made to reflect the new SSM values that the ITU-T introduced into G.781 for the same reason.

In addition, ITU G.8262.1 specifies an enhanced Ethernet equipment clock (eEEC), which also needs a new ESMC value to describe it. Amendment 1 of ITU-T G.8264 (2018-03) defines these new eESMC values for these enhanced clock types. The new SSM codes for eESMC are referred to as enhanced SSM (eSSM) codes, and the value is carried in an *enhanced QL TLV*. Table 6-5 describes the formats of the fields in the enhanced QL TLV from the amended G.8264.

Table 6-5   *Enhanced QL TLV Format*

Octet Number	Size/bits	Field
1	8 bits	Type: 0x02
2–3	16 bits	Length: 00–14
4	8 bits	Enhanced SSM code
5–12	64 bits	SyncE clockIdentity of the originator of the extended QL TLV
13	8 bits	Flag
14	8 bits	Number of cascaded eEECs from nearest SSU/PRC/ePRC
15	8 bits	Number of cascaded EECs from nearest SSU/PRC/ePRC
16–20	40 bits	Reserved

Source: ITU-T G.8264, Table 11-5

Note that the normal QL TLV (refer to Table 6-4) is always sent as the first TLV in all the ESMC PDUs (for both EECs and eEECs), and any clock supporting eESMC includes the enhanced QL TLV. This is done to allow interoperability of ESMC and eESMC within the same network. While eEECs are always required to support eESMC, there could be EECs in the same synchronization chain that either support eESMC or do not; mandating the first TLV as a normal QL TLV makes sure that the EECs that do not support eESMC can coexist in the same synchronization chain.

The main difference between the enhanced SSM codes compared to normal SSM codes is that the enhanced version is extended to 8 bits of value as compared to 4 bits of SSM code value as specified in ITU-T G.781 and ITU-T G.8264. Table 6-6 shows the eSSM codes for the enhanced clock types.

Table 6-6 *Enhanced SSM Codes for SyncE*

Clock	Quality Level	SSM Code
EEC1	QL-EEC1	0xFF
EEC2	QL-EEC2	0xFF
As per [ITU-T G.781]/ [ITU-T G.8264]	QL as per [ITU-T G.781]/ [ITU-T G.8264]	0xFF
PRTC	QL-PRTC	0x20
ePRTC	QL-ePRTC	0x21
eEEC	QL-eEEC	0x22
ePRC	QL-ePRC	0x23

Source: ITU-T G.8264, Table 11-6

Synchronization Network Chain

There are many network deployment models for timing distribution that can exist, and one such deployment is shown in Figure 6-10. The figure shows a sample network with a chain of nodes carrying timing information; however, there can be different network models as well.

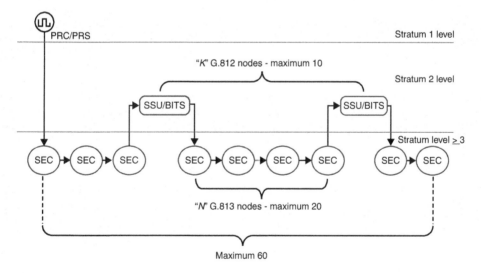

Figure 6-10 *Synchronization Reference Network Chain*

The following points explain the different elements of the network chain and cover the variations that can be seen in those chains:

■ The network consists of several quality levels related to the height of the clock within the synchronization hierarchy and its position along the chain.

- The highest level contains the PRC/PRS of the network, and there could be additional PRC/PRS sources for backup purposes. Note that there are alternative network models that allow multiple PRC/PRSs to be active at the same time.

- On level 1, the PRC/PRS provides distribution of the synchronization reference signal to one or more SSU/BITS clocks, which forms the next synchronization level. ITU-T G.811 specifies the performance requirements for these reference clocks.

- Each of these SSU/BITS clocks supplies timing to a subnetwork in the second synchronization level. The role of the SSU/BITS for the subnetwork is like the role of the PRC/PRS for the whole network. If the SSU/BITS clocks lose reference signals coming from the PRC/PRS, the SSU/BITS takes over as the reference source for the subnetwork. Hence this role requires high-quality clocks with good holdover performance (their characteristics are specified in ITU-T G.812).

- The EECs/SECs or NEs become the leaf level of the synchronization network chain, receiving clocking information from SSU/BITS. Note that the SECs can be further chained together in a network to pass timing information from one node to another, achieving network-wide frequency synchronization. ITU-T G.813 (and G.8261) specifies the performance requirements for these clocks.

Just as networks are planned and designed to optimize various networking data protocols, the networks also need to be designed (and engineered) specifically for synchronization purposes. And like the deployment of data networks, the engineering of the synchronization aspects of a network should be done prior to the equipment being commissioned. The result of this engineering process is often described in a synchronization plan or timing plan.

The timing plan should specify a synchronization network design that reliably distributes timing using a combination of clock types and failover mechanisms. And the plan should contain the physical and logical topology of the equipment, carefully considering the different quality levels of the clock equipment connected to each other. There are several very important principles that this plan should include:

- The synchronization network forms a pyramid hierarchy with the master clock at the top of the tree and supplying timing information to other nodes using a master-slave methodology. Note that this hierarchy is a logical topology, which can span different physical topologies (such as rings).

- No parts of the synchronization network should operate isolated from the master clock. During cases of clock reference failures (due to any faults in the transport or the network), the clock nodes should be able to rearrange themselves such that there exists a new master clock on the top of the newly formed tree. This new master clock may receive a reference clock signal from an alternate source or supply a good quality holdover (such as an SSU/BITS clock with stratum 2 performance).

- A clock should never lock to a reference of lower quality. If a case arises where only a clock of lower quality is available, even during physical link failures, this clock should revert to a holdover mode of operation.

- Loops in the timing signal must not be present nor be allowed to form following failures. More on this in the upcoming section "Timing Loops."

- The lengths of the branches of the synchronization tree are kept as short as possible. This branch length refers to the synchronization links between SSU/BITS and EEC/SEC nodes connected to each other in the network chain. The longer the path through a specific synchronization chain becomes, the more susceptible it will be to impairments and wander accumulation.

While some of these points are very specific, the last point provides a general guideline. Each clock node adds its own *noise* (different types of errors) while recovering the reference clock signal and before passing that signal to the next clock. So, any errors that arise accumulate over the chain—the longer the chain, the more error. Refer to Chapter 5 for descriptions of parameters defining clock performance and error.

Besides the general guideline of keeping the network chain as short as possible, ITU-T G.803 also defines the maximum number of nodes that can exist in such network chain. These limits on the number of nodes in the chain are also shown in Figure 6-10 for each clock type.

The main points of the synchronization reference network chain (from Figure 6-10) are as follows:

- The node clocks are interconnected via N network elements, each being a clock compliant to ITU-T G.813. The maximum value of N is determined by the quality of timing required by the last network element in the chain. This also means that the quality of timing synchronization deteriorates for each hop as the number of synchronization links increase.

- The longest chain should not exceed K SSU/BITS type clocks, those compliant to ITU-T G.812. As shown in Figure 6-10, there could be N number of G.812 clock nodes between $(K - 1)^{th}$ and K^{th} SSU/BITS clock nodes. Note that only one type of clock is shown because the difference in holdover performance between SSU-T and SSU-L (the two clock types defined under ITU-T G.812) is no longer relevant in SDH network synchronization.

- The worst-case synchronization reference chain values allowed are $K = 10$ and $N = 20$. And the total number of SDH network element clocks in one synchronization network chain is limited to a maximum of 60 elements.

Clock Selection Process

A good timing distribution network is designed such that more than one frequency source (or master clock) is available to the slave clock, providing one or more backup timing signals during clock outages, which can occur for a variety of reasons. When more than one frequency source is available, the slave clock needs to make a decision to pick the best clock from those available to it. This decision is a result of a well-defined process, referred as the *clock selection process*.

Two different modes of selection process are possible: QL-enabled mode and QL-disabled mode. As the names suggest, *QL-enabled mode* is relevant in deployments where SSM messages are exchanged and interpreted by the NEs. It means that there is QL information available to assist in making a clock selection decision. With *QL-disabled mode*, either QL information is not available or the engineer chooses not to use it. The process itself does not change much between these two modes, except for the fact that SSM messages carrying QL information are not considered for QL-disabled mode. ITU-T G.781 Annex A captures the detailed description of this process.

It is important to note that QL-enabled and QL-disabled modes suggested in G.781 are applicable only to SDH/SONET deployments—no similar approach exists for SyncE—a clock generating a SyncE signal but without an ESMC stream is not be considered as a reference source.

ITU-T G.8264 describes only synchronous and non synchronous modes of operation for SyncE. Synchronous mode is always accompanied with ESMC messages, and the receiver can recover the frequency for synchronization purposes, whereas in non synchronous mode, the Ethernet port is not considered as a candidate reference for the clock selection process, and the frequency is not used for clock synchronization (but it can be used to drive the frequency of that single link).

The clock selection process gets triggered by occurrence of one or more of the following events:

- Change in received QL (as signaled by SSM or ESMC)

- Signal failure detected on the currently selected source

- Configuration changes by operator; for example, addition or deletion of assigned frequency source(s)

In general terms, once a physical port is available as a frequency source, the selection process considers the following data to select the best clock:

- **Quality of a clock:** The clock with the highest QL value is always a winner, as the clock is traceable to the highest-quality source. Note that a change in QL can occur anytime during normal operations. The slave clock acts on a QL change event only if a better QL than the current selection is discovered.

- **Availability of a clock signal:** A frequency source must be always available for it to stay as the selected source. In case of physical link failures, the frequency source is marked as signal fail (SF), and the QL for this source is marked as QL-FAILED. If the frequency source moves to QL-FAILED state, the clock starts the process of selecting the next best source.

- **Synchronization source priority:** Operators can allocate priority values to the assigned frequency sources to design a preferred network synchronization flow. Prioritization of the sources helps in the cases where the same QL value is received from two (or more) different frequency sources but the operator has a preference. Of course, setting equal priorities indicates that neither is favored over the other.

- **External configuration commands:** ITU-T G.781 recommends multiple configuration commands that operators can use to influence the clock selection process. The primary use of these commands is for maintenance purposes (such as maintenance of a physical port). These directives can be categorized as commands to do the following:

  - Detach one or more physical ports from the clock selection process. These are referred to as *lockout* commands. If a port is configured for lockout, the clock selection process does not consider the port for any frequency selection decisions. G.781 recommends both a *set* and *clear* operation to enable and disable the lockout for physical ports.

  - Switch to a different frequency source from the one currently chosen by the clock selection process. G.781 recommends that vendors implement a *forced switch* command for an interface that can override the currently selected synchronization source and force selection of another (assuming the port is question is not in locked-out state). It should support both a *set* and *clear* operation.

  - They also suggest a *manual switch* command for a port that would influence the clock selection process to switch to the port in question. This command is accepted only if the configured port is enabled, not locked out, not in signal fail condition, and has a better QL than DNU (in QL-enabled mode); otherwise it is rejected. It should also support both a *set* and *clear* operation.

In summary, the clock selection process selects the frequency source with the highest QL, provided it is not experiencing a signal fail condition. If multiple sources have the same QL, the source with the highest priority is selected. In the case of a tie in priority, the currently selected source is retained.

Timing Loops

All NEs in a synchronous network must be synchronized to the same clock frequency to prevent data loss. However, as noted in previous sections, there could be failures of the physical links that are carrying the timing signal. During such events, the nodes that are impacted directly by the physical link failure run a clock selection algorithm to choose the next best available clock. This process is known as a *clock rearrangement*.

During the clock rearrangement process, it is possible for the node to lose synchronization to the master reference clock (such as PRC/PRS) and to become part of a *timing loop*. In this state, synchronization is looped back through the network so that all nodes affected by the timing loop get synchronized to the delayed version of their own clock. This causes the network clock frequency inside the timing loop to drift outside of the frequency range acceptable to the synchronous network. This is a major failure condition that can have adverse effects on the data traffic throughput and stability of the network.

Figure 6-11 shows a typical configuration of four NE nodes synchronized within a ring topology. Note that unlike what is shown in Figure 6-11, most real network topologies can have many more NE nodes, spanning across wide geographical areas. The figure only illustrates a simple example whereby all NE nodes are EEC nodes, whereas real networks can have many different types of nodes and physical links carrying the timing signal.

Switch A is synchronized to an onsite SSU/BITS, which itself is synchronized to a PRC traceable clock. Switch A then makes the traceable frequency signal available on both of its output interfaces. The other switches in the ring recover the timing from their adjacent node, so that the timing signal propagates around the ring. However, for redundancy, each node can accept synchronization from either of its neighbors (assuming the port is capable of it).

The operators can influence the direction of flow by configuring the preferred source as having the highest priority (see the preceding section on the clock selection process). So, Switch C can extract timing and synchronize itself to either Switch B or Switch D. In this case, the operator configures it in a clockwise direction—with Switch B synchronized to Switch A, Switch C to Switch B, and Switch D to Switch C. This capability of being able to synchronize to either input (changing the direction of flow) does increase the reliability of the network but at the same time also raises the possibility of timing loops.

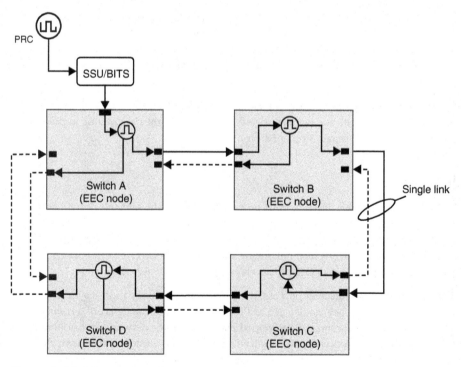

Figure 6-11   *Timing Distribution in a Ring Topology*

These points expand on the details in Figure 6-11:

- Each link connecting the switches is shown as two unidirectional arrows; this is to show the timing transfer from one switch to the other in both directions. The direction of the arrows shows the transfer of timing information from one switch to another.

- All links carry timing information in the example. The solid arrows indicate the direction of a link that is selected as the input timing reference to a switch. The dashed arrows indicate the direction of a link that is not selected as the input timing reference but will be available for selection during a clock rearrangement.

  For example, Switch A is transferring the frequency toward Switch B and Switch D; the solid arrow from Switch A to Switch B shows that Switch B selected this link as its input timing reference; whereas the dashed arrow from Switch A to Switch D shows that Switch D has not selected this link as input timing reference.

- This example shows Switch D selects Switch C as its preferred source of frequency even though Switch A might be "closer" to the PRC/PRS. This is because the network engineer configures the preferred priority in the order of possible frequency sources. Frequency traceability schemes such as SSM do not have a mechanism to determine the distance, or hop count, between a clock node and the PRC/PRS (looking at Table 6-5 shows that this changes with the eESMC mechanism).

In case of synchronization failures, such as that caused by physical link failure, the impacted nodes go through the clock rearrangement process to select a new clock source. If such a clock rearrangement is automatic, this process could pick up clock references in such a fashion that it creates timing loops. One such possibility is shown in Figure 6-12, where the physical link between Switch A and Switch B suffers a fault, and although Switch B picks up another clock reference, it happens to be coming from Switch C. However, because Switch C was itself locked to Switch B, this situation creates a timing loop.

One of the ways to prevent timing loops is to enable the SSM or ESMC mechanism on the links. As outlined earlier in this chapter, the SSM and ESMC both carry QL information, which denotes clock quality using values such as PRC/PRS, stratum level, SSU, and so on. This can be used by the nodes during a clock rearrangement to determine the best possible input frequency source.

One special value specified by the SSM or ESMC is *do not use* (DNU) for Option I, or *do not use for synchronization* (DUS) for Option II. This SSM value is sent upstream when a node selects an input frequency as a timing reference signal. It indicates that the downstream receiving node is locked onto that transmitted signal.

In the example shown in Figure 6-13, Switch B sees that its link to Switch A is traceable to a PRC, so it selects it as an input frequency reference. When it does this, Switch B sends DNU back to Switch A to indicate that Switch A should not use this signal as a candidate for clock input. This mechanism warns the transmitting node not to use the clocking information, as it originated from itself, thereby preventing a local timing loop. Figure 6-13 shows the same network topology as shown previously in Figure 6-11, along with the SSM values that are carried in each direction (clockwise and counterclockwise). This example uses the SSM/QL values for an Option I network.

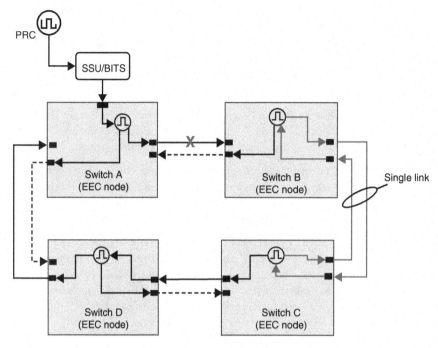

Figure 6-12   *Timing Distribution in Ring Topology with Link Failure*

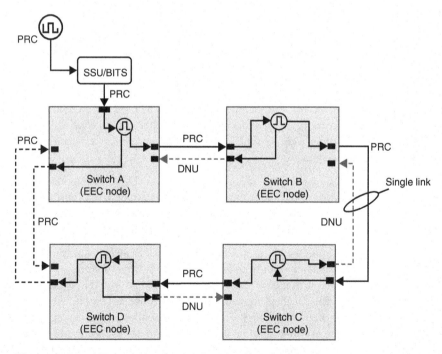

Figure 6-13   *Timing Distribution in Ring Topology with SSM*

Figure 6-14 shows the clock rearrangement in the case of a physical link failure, and the way the affected nodes go through the clock rearrangement process to select a new clock source. As the figure illustrates, the clock rearrangement settles on a topology in accordance with the QL values being received. The clock selection algorithm uses the QL values to select the correct reference clock. In addition to timing loop prevention, with this clock selection algorithm the best possible rearrangement is achieved, considering the timing information flow in the network.

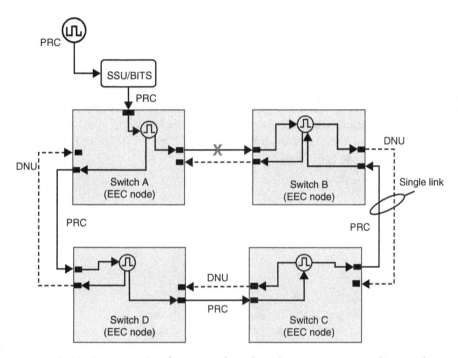

Figure 6-14   *Timing Distribution with Link Failure in a Ring Topology with SSM*

Figure 6-13 also shows another case whereby a timing loop can arise. Under normal conditions, Switch A receives the PRC from the SSU/BITS, but Switch A is also receiving the PRC from Switch D, which is just a copy of its own clock signal that has gone around the ring. There is no mechanism stopping Switch A from preferring that link to Switch D over the link to the SSU/BITS device (there may be a tie-breaker mechanism that will decide which one).

The operator must set a local priority on the correct link so that Switch A will select the SSU/BITS source of frequency above all others, even if they are all signaling PRC/PRS. If not, it is possible that a timing loop can develop. In addition, when designing a frequency distribution network, the network engineer must also ensure that timing signals can never flow upward in the clock distribution hierarchy.

But what happens when the input from the higher SSU/BITS layer is lost? As Switch A starts the clock selection algorithm, it will see that Switch D is sending it a PRC/PRS signal, so it will select that as the best clock available. But the signal from Switch D is only traceable back to itself, and there is no PRC/PRS involved in this frequency signal. Now, once again, a timing loop has developed.

If this scenario should ever occur, the clocks will all indicate that they have a frequency input that is traceable to a PRC/PRS source, but in fact, they do not because the timing signal is going around in circles, so the frequency will start to wander. An operator might not be able to determine this state exists until they notice strange things happening in their network.

To avoid this, an operator must configure Switch A to tell it that an input from Switch D is not to be selected. In this case, another source must be found, or the clocks must go into holdover. This is an analogous situation to other data protocols that need to have *a break in the ring* to avoid situations with loops.

Figure 6-15 shows the case where the input to Switch A from the SSU/BITS layer is lost, and Switch A was configured not to allow the input from Switch D. Because no better source is available to it, Switch A goes into holdover, which means SEC or EEC quality, and it sends that out as an SSM value.

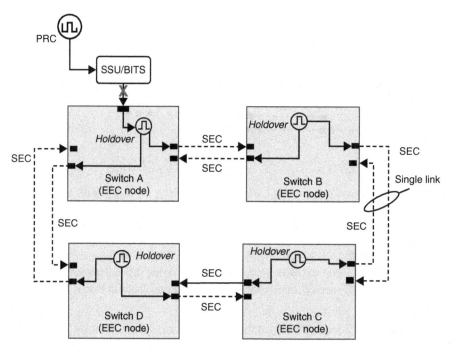

Figure 6-15   *Timing Distribution with Link Failure from Timing Reference Node*

It is also interesting to note in Figure 6-15 that because all the clocks are of SEC quality, all the remaining clocks also go into holdover, so none of the clocks are aligned with each other; they are all relying on the holdover performance of their own oscillator (using saved data).

Standardization

This chapter has referenced numerous recommendations from several standards development organizations, such as ITU-T, IEEE, and ANSI. (Chapter 4, "Standards Development Organizations," outlines the major SDOs and their relevance to both the fields of mobile and synchronization.)

Chapter 8 covers the functioning of one of the major SDOs in this space, the ITU-T, in much more detail. It also outlines the processes behind the development of standards, and the concise summary of specific standards that are applicable to synchronization and timing. Chapter 8 also provides a detailed outline of all the applicable standards for frequency synchronization, covered in separate sections of that chapter.

Summary

At a broad level, the mechanism to achieve frequency synchronization can be categorized as frequency distribution using the physical layer and packets. The primary focus of this chapter was frequency distribution using the physical layer. The chapter explained the SSU/BITS devices and their role in achieving and distributing frequency synchronization.

Based on the accuracy and stability that a clock can provide, different clock levels are defined within a timing network hierarchy. These levels are segregated primarily based on different stratum levels, such as PRC/PRS belongs to stratum 1, SSU/BITS usually falls into stratum 2, and SECs/NEs to stratum 3 level. This chapter explained the hierarchical relationship between stratum levels, referred to as clocking or synchronization hierarchy.

For proper frequency synchronization, a clock should always lock to another clock with a higher or at least equal stratum level. This requires a method to communicate the stability level of a master clock to a slave clock along with the frequency signal itself and is achieved via a clock quality (QL) indicator. The chapter covered the QL mechanism in detail, and how it is carried as SSM, ESMC, or enhanced ESMC from the master to the slave nodes. This forms the basis of clock traceability, with which every clock can trace its received reference signal from the source.

A timing network chain is a connected network of nodes transporting timing information. The latter part of this chapter explained the different elements of this network chain and the variations that can be seen in these chains. Even though it is recommended to keep the network chain as short as possible, ITU-T G.803 defines the maximum number of nodes that can exist in such chains and explains the restrictions based on different clock types belonging to different stratum levels.

As discussed in this chapter, there can be deployments where NEs are connected into different topologies and are often configured into ring topologies. If not designed properly for frequency distribution, such deployments can suffer from a degraded synchronization signal, where frequency is looped back through the network so that all nodes get synchronized to the delayed version of their own clock—a timing loop. The chapter discussed an example of NEs connected in a ring topology and explained the formation of timing loops, which could cause critical issues for timing networks.

The goal of this chapter was to explain the technology, deployments, and restrictions of frequency distribution via the physical layer, which is one of the main aspects of timing networks. Chapter 7, "Precision Time Protocol," focuses in detail on the protocols and exact mechanism of distributing timing information using packets.

References in This Chapter

Alliance for Telecommunications Industry Solutions (ATIS). "Synchronization Interface Standard." *ATIS-0900101:2013(R2018)*, Supersedes ANSI T1.101, 2013. https://www.techstreet.com/atis/standards/atis-0900101-2013-r2018?product_id=1873262

American National Standards Institute (ANSI). "Synchronization Interface Standards for Digital Networks." *ANSI T1.101-1999*, Superseded by ATIS-0900101.2013(R2018). https://www.techstreet.com/atis/standards/atis-0900101-2013-r2018?product_id=1873262

IEEE Standards Association. "IEEE Standard for Ethernet." *IEEE Std. 802.3:2018*, 2018. https://standards.ieee.org/standard/802_3-2018.html

Internet Engineering Task Force (IETF)

Bryant, S. and P. Pate. "Pseudo Wire Emulation Edge-to-Edge (PWE3) Architecture." *IETF*, RFC 3985, 2005. https://tools.ietf.org/html/rfc3985

Malis, A., P. Pate, R. Cohen, and D. Zelig. "Synchronous Optical Network/Synchronous Digital Hierarchy (SONET/SDH) Circuit Emulation over Packet (CEP)." *IETF*, RFC 4842, https://tools.ietf.org/html/rfc4842

Stein, Y., R. Shashoua, R. Insler, and M. Anavi. "Time Division Multiplexing over IP (TDMoIP)." *IETF*, RFC 5087, 2007. https://tools.ietf.org/html/rfc5087

Vainshtein, A. and Y. Stein. "Structure-Agnostic Time Division Multiplexing (TDM) over Packet (SAToP)." *IETF*, RFC 4553, 2006. https://tools.ietf.org/html/rfc4553

Vainshtein, A., I. Sasson, E. Metz, T. Frost, and P. Pate. "Structure-Aware Time Division Multiplexed (TDM) Circuit Emulation Service over Packet Switched Network (CESoPSN)." *IETF*, RFC 5086, 2007. https://tools.ietf.org/html/rfc5086

ITU Telecommunication Standardization Sector (ITU-T)

"G.781: Synchronization layer functions for frequency synchronization based on the physical layer." *ITU-T Recommendation*, 2020. http://handle.itu.int/11.1002/1000/14240

"G.803: Architecture of transport networks based on the synchronous digital hierarchy (SDH)." *ITU-T Recommendation*, 2000. http://handle.itu.int/11.1002/1000/4955

"G.811: Timing characteristics of primary reference clocks." *ITU-T Recommendation*, 1997. http://handle.itu.int/11.1002/1000/4197

"G.812: Timing requirements of slave clocks suitable for use as node clocks in synchronization networks." *ITU-T Recommendation*, 2004. http://handle.itu.int/11.1002/1000/7335

"G.813: Timing characteristics of SDH equipment slave clocks (SEC)." *ITU-T Recommendation*, 2003. http://handle.itu.int/11.1002/1000/6268

"G.8261: Timing and synchronization aspects in packet networks." *ITU-T Recommendation*, Amendment 1, 2020. http://handle.itu.int/11.1002/1000/14207

"G.8262: Timing characteristics of a synchronous equipment slave clock." *ITU-T Recommendation*, Amendment 1, 2020. http://handle.itu.int/11.1002/1000/14208

"G.8262.1: Timing characteristics of an enhanced synchronous equipment slave clock." *ITU-T Recommendation*, Amendment 1, 2019. http://handle.itu.int/11.1002/1000/14011

"G.8264: Distribution of timing information through packet networks." *ITU-T Recommendation*, Amendment 1, 2018. http://handle.itu.int/11.1002/1000/13547

Chapter 6 Acronyms Key

The following table expands the key acronyms used in this chapter.

Term	Value
5G	5th generation (mobile telecommunications system)
ANSI	American National Standards Institute
BITS	building integrated timing supply (SONET)
CESoPSN	circuit emulation over packet switched network
CEP	circuit emulation over packet
CO	central office
DNU	do not use (quality level, Option I)
DUS	do not use for synchronization (quality level, Option II)
E1	2-Mbps (2048 kbps) signal (SDH)
EEC	Ethernet equipment clock
eEEC	enhanced Ethernet equipment clock
eESMC	enhanced Ethernet synchronization messaging channel
ESF	extended super frame
ESMC	Ethernet synchronization messaging channel
eSSM	enhanced synchronization status message
eSyncE	enhanced synchronous Ethernet
ETSI	European Telecommunications Standards Institute
FCS	frame check sequence
GEN1	Generation 1 (Option II quality levels)

Term	Value
GEN2	Generation 2 (Option II quality levels)
IEEE	Institute of Electrical and Electronics Engineers
ISDN	Integrated Services Digital Network
ITU-T	ITU Telecommunication Standardization Sector
kbps	kilobits per second
LACP	Link Aggregation Control Protocol
LAG	Link Aggregation Group
NE	network element
NTP	Network Time Protocol
OAM	Operations, Administration, and Maintenance
OSSP	organization specific slow protocol
OUI	Organizationally Unique Identifier
PDU	protocol data unit
PLL	phase-locked loop
PRC	primary reference clock (SDH)
PRS	primary reference source (SONET)
PSTN	public switched telephone network
PTP	precision time protocol
QL	quality level
SASE	standalone synchronization equipment
SAToP	structure-agnostic TDM over packet
SDH	synchronous digital hierarchy (optical transport technology)
SDO	standards development organization
SEC	SDH equipment clock
SEC	synchronous equipment clock
SF	signal fail
SONET	synchronous optical network (optical transport technology)
SSM	synchronization status message
SSU	synchronization supply unit (SDH)
SSU-A	primary level synchronization supply unit (SDH)
SSU-B	secondary level synchronization supply unit (SDH)

Term	Value
ST2	stratum 2 (ANSI T1.101)
ST3E	stratum 3E (ANSI T1.101)
SyncE	synchronous Ethernet—a set of ITU-T standards
T1	1.544-Mbps signal (SONET)
TDM	time-division multiplexing
TLV	type, length, value
UTP	unshielded twisted-pair

Further Reading

Refer to the following recommended sources for further information about the topics covered in this chapter.

ANSI: http://www.ansi.org

ETSI: http://www.etsi.org

ITU: https://www.itu.int

Chapter 7

Precision Time Protocol

In this chapter, you will learn the following:

- **History and overview of PTP:** Explains how PTP originated and provides an overview of the imperatives that led to its development.

- **PTP Versus NTP:** Compares these commonly used time protocols.

- **IEEE 1588-2008 (PTPv2):** Examines the many features of the standard and the principal mechanisms of its operation.

- **PTP clocks:** Examines the types of clocks defined in IEEE 1588 and the characteristics of each clock type.

- **Profiles:** Provides details of the three telecom profiles and the other major industry profiles.

- **PTP security:** Describes how security has been handled in PTP and explores future developments on the topic.

- **IEEE 1588-2019 (PTPv2.1):** Introduces the newest edition of the PTP standard and explains how it differs from the previous versions.

Up until this point, you have explored the basic concepts required to understand the sources of time and the distribution of synchronization using physical systems and methods (such as cables and wires). Now the focus moves onto another popular method to carry timing information, one based upon packet transport. This chapter covers the details of one commonly used packet distribution method, namely the precision time protocol (PTP). Chapter 9, "PTP Deployment Considerations," will cover other methods of transporting synchronization using packets. Network engineers implementing high-precision timing solutions increasingly deploy PTP to replace the dedicated time distribution circuits used in older technologies.

History and Overview of PTP

In 2002, the Institute of Electrical and Electronics Engineers (IEEE) defined the first version of a packet-based protocol, known as PTP, to accurately carry time. The IEEE designed PTP chiefly for industrial applications over short-distance communications—such as a local-area network (LAN). The protocol proposed in this standard specifically addressed the following needs of measurement and control systems:

- Spatially localized (meaning situated close to each other)
- Microsecond to sub-microsecond accuracy and precision
- Administration-free operation
- Accessible for both high-end devices and low-cost, low-end devices

The standard was assigned the number 1588, and most people refer to the first version of the standard as IEEE 1588-2002, or more colloquially, PTPv1, named after the version of the protocol.

The IEEE updated the PTP standard in 2008 to version 2 of the protocol (PTPv2). Members of numerous industries, including the industrial, utility, and telecommunications sectors, who use PTP to provide high-accuracy time between machines or radio base stations, supported the revision. The update added features such as the ability to support non-LAN implementations and to allow for extensibility through profiles. There were also changes made to improve the practical accuracy of PTP time transfer.

The drawback of the 2008 update was the limited backward compatibility with the previous version, which meant translation was required to ensure meaningful coexistence of PTPv1- and PTPv2-capable equipment in the same network. For example, one significant difference is that PTPv1 did not use separate information notifications (called *Announce* messages, introduced in PTPv2) to communicate valuable data about the source and quality of the upstream clock.

Nowadays, interoperability is no longer an issue for most applications, as PTPv1 is now rarely used. In fact, a new version of PTP, IEEE 1588-2019, was published in 2020. It is also referred to as PTPv2.1, at least in part to emphasize the very strong desire by the IEEE committee to ensure backward compatibility with PTPv2. More details on PTPv2.1 are provided toward the end of this chapter.

To summarize, at the time of writing, there are three separate releases of the IEEE 1588 standard, from 2002, 2008, and 2019. Although people refer to them colloquially as PTPv1, PTPv2, and PTPv2.1, technically the version of the protocol is a separate issue from the edition of the standard.

Note that the PTP specifications use the term *message* rather than *packet*, although many other timing specifications refer to packet-based timing. For compatibility with the existing editions of the IEEE 1588 standards, this chapter follows the convention of using the term *message* when referring to packets carrying PTP information.

PTP Versus NTP

According to the IEEE 1588-2002 specification, PTP was designed to provide microsecond-level time synchronization between devices that were located closely to each other. For example, although PTPv1 supported User Datagram Protocol (UDP)/ Internet protocol (IP) transport, it was required to use layer 3 (L3) multicast with messages not to be forwarded through a router (meaning the Time to Live [TTL] was set to zero). This made PTPv1 unsuitable for a WAN deployment but fine around the factory floor or across a laboratory.

The Network Time Protocol (NTP), on the other hand, was ideal for synchronization of widely distributed computing systems and was almost always routed. The use case for NTP is to allow all the routers in a network, every laptop on a college campus, or the home router on your broadband connection to align their time of day. One example where this is important might be to align the syslog timestamps of many machines, which is a necessary precondition to effectively find and trace faults across a network. For current usage, the most prevalent deployments of NTP are used to deliver time synchronization down to the millisecond level.

In contrast to the IEEE defining IEEE 1588, it was the Internet Engineering Task Force (IETF) standards development organization (SDO) who developed NTP through its Request for Comment (RFC) process. NTP had been specified starting in the mid-1980s (RFC 958) and is now at version 4 (RFC 5905)—so it is a very mature technology. See the following link for further details on the IETF process for creating Internet standards: https://www.ietf.org/standards/process/.

There is also a somewhat simplified version of NTP known as *Simple Network Time Protocol* (SNTP), which has less-demanding requirements for accuracy and reliability than NTP. SNTP is fine for simply updating the clock on remote machines, but its simplified approach leaves out features that might be desired. For example, it does not have continuous system clock adjustment methods, normally used to prevent steps and jumps in the clock, but for everyday usage (such as your home laptop) it is perfectly adequate. SNTPv4 is specified in RFC 4330, although that functionality has now been folded into the single NTPv4 specification in RFC 5905, thereby obsoleting RFC 4330.

NTP was typically implemented only in software, which limited the potential accuracy it could achieve, whereas PTP was normally built around hardware support, including specialized design for accurate timestamping (see Chapter 5, "Clocks, Time Error, and Noise"). The problem is that if timestamping is software based, it is likely to be frequently interrupted and delayed by other functions in the operating system, which affects the accuracy.

Furthermore, if the (software) timestamping is done by the main CPU, a lot can happen to delay a time-sensitive packet between being timestamped and finally being transmitted at the egress port. On the other hand, if the Ethernet PHY can timestamp the packet as it is being transmitted at egress, there is very little that can delay that packet (at least until it gets to the next router). Of course, the packet transport or some optical component can also add some unpredictable delay—more on that later.

In spite of that, NTP is accurate enough for many purposes, such as aligning the time-stamps in log files, and general sub-second time agreement. NTP (and SNTP) is widely used to synchronize perhaps billions of devices around the world. In fact, the whole Internet is time synchronized by NTP/SNTP. NTP can be used for more specialized applications; for example, some companies use special versions of NTP to frequency synchronize mobile radio equipment. However, to get more accurate time, something else was needed, and PTP is what the industry has settled upon.

There are some differences between NTP and PTP in the values used to carry time—their timescales. The most basic difference is that the NTP epoch starts on 1 January 1900, while the PTP (and POSIX/Unix) epoch starts on 1 January 1970. As they both count seconds since their respective epochs, there is a constant difference of 70 years and 17 leap days (2,208,988,800 seconds) between NTP and POSIX/Unix, representing the number of Gregorian seconds between those two dates. The offset to PTP is affected by leap seconds, which is discussed later in the chapter in the section "Timestamps and Timescales."

Note also that PTP uses a 48-bit timestamp for seconds, whereas NTP uses several timestamp formats, including one that uses only a 32-bit value for seconds. Therefore, the largest representation for seconds using this NTP timestamp is 2^{32} seconds or 136 years, meaning that there will be a rollover event for NTP during the day of 7 February 2036. The IETF has developed mechanisms to deal with this; see section 6 of RFC 5905 for more details.

IEEE 1588-2008 (PTPv2)

The IEEE 1588-2008 standard defines a protocol for precise synchronization of clocks that communicate with each other over a packet network. At its simplest, the clocks in the network form a master-slave relationship with one another, during which a master gives timing information to a slave. All clocks in the PTP network ultimately derive their time from the ultimate clock source known as the *grandmaster clock* (GM clock), which sources its time from a reference clock, such as a satellite receiver.

Note   Use of the terms "master" and "slave" is ONLY in association with the official terminology used in industry specifications and standards, and in no way diminishes Pearson's commitment to promoting diversity, equity, and inclusion, and challenging, countering and/or combating bias and stereotyping in the global population of the learners we serve.

There is no single transport type or encapsulation method defined by PTPv2; rather, it is a toolbox of possible ways to solve the timing synchronization problem. So, the messages being exchanged can be multicast or unicast, layer 2 or layer 3, Ethernet or UDP/IP, or even something else entirely. The goal is to allow heterogeneous systems to get their time from a grandmaster clock, no matter what their implementation, operating system, time accuracy, or cost. Even relatively simple implementations should be able to recover the clock in the microsecond range without a lot of administrative intervention or burdensome management.

One important feature of PTPv2 is its extensibility, a capability that allows interested organizations to select only the protocol features they want and design an industry-specific sub-version—known as a *profile*. These profiles allow implementations of PTP that only support the features needed for a specific use case, which eases interoperability and simplifies deployment.

Finally, although PTPv2 claims to be capable of delivering microsecond to sub-microsecond accuracy, the achievable accuracy is obviously dependent on the quality of the implementation. In fact, it is scalable to higher accuracies, as some implementations can deliver synchronization accuracies of less than 1 ns—one example being *White Rabbit*—but even standard PTP can deliver accuracy in the range of tens of nanoseconds. For information on White Rabbit and other profiles, see the section "Profiles" later in this chapter.

As mentioned in the preceding overview, the IEEE has published a newer version of PTP, known as IEEE 1588-2019 (PTPv2.1). So, over the next several years, we will see SDOs start to adopt features from the new version and develop updates to their own profiles incorporating new features from this updated version of PTP.

Later in this chapter, we cover the definitions and use of specific profiles of PTP defined by the ITU-T, called *telecom profiles*. But first, we address the concepts and common features from the IEEE 1588-2008 standard before addressing the more specialized versions. Note that some features of IEEE 1588-2008 are not used in the telecom profiles, so we will not cover those specific features in depth.

General Overview

PTP is a two-way time protocol (like NTP) and functions by sending timestamps between the source of time (a *master*) and a receiver of time (a *slave*). The method involves sending timestamps from the master to the slave, so that the slave can *recover the clock*, meaning (1) frequency syntonize its oscillator, (2) align its clock phase, and (3) recognize and track the correct time of day.

As the slave receives the time information, it has an exact measurement of what the time was on the master, representing the time that the message was timestamped. What it does not know is the amount of time that elapsed from when the message was timestamped until it was received, meaning how long it took for the message to travel the intervening path (the *transit time*).

To figure out that transit time, the slave receives timestamps from the master in order to calculate the *round-trip time* (much like the two-way delay measurement [2DM] introduced in Chapter 2, "Usage of Synchronization and Timing"). Dividing the round-trip time by two gives an estimate of the one-way time, what PTP calls the *mean path delay*. But for this calculation to be accurate, the forward transit time (master to slave) must be exactly equal to the reverse transit time (slave to master). The difference between the path transit time in either direction is known as *delay asymmetry*. There will always be some delay asymmetry present, even if in some circumstances it is almost negligible.

The two-way transit time can be quite accurately determined, but the accuracy of the one-way value relies on the correctness of our assumption. This assumption and the consequences of delay asymmetry are the most important thing to understand about packet-based timing.

As a convention, IEEE 1588-2008 (7.4.2) defines delay asymmetry as positive when the master→slave transit time is longer than the slave→master transit time. Thus:

transit (master→slave) = mean path delay + delay asymmetry

transit (slave→master) = mean path delay − delay asymmetry

However, without an independent measurement of the delay asymmetry using some external tool or reference clock, PTP cannot detect it. Therefore, the mean path delay is used as an estimate of the one-way (forward) transit time, with the precondition that the network engineer must design a network that does not introduce excessive asymmetry.

Any asymmetry will lead to direct time error on the slave clock because asymmetry undermines the assumption about how long the timestamp has been "in transit" between the master and the slave. If it is excessive, this time error will consume all the phase alignment budget that is allowed for the application. However, if you can measure the asymmetry, then PTP could be configured to correct for it and eliminate that error from the calculations.

Now that the slave has an accurate timestamp from the master, and an estimate of the delay in a packet getting from the master to the slave, the slave can calculate the present time. This process occurs repeatedly (depending on the profile) but in the range of between once every 8 seconds to 128 times per second.

Overview of PTP Clocks

To read about and comprehend the core components of PTP, you need to understand the features and functions of PTP clocks; however, to discuss PTP clocks in much detail requires significant knowledge of other concepts. For this reason, this section provides a simplified overview of PTP clocks and the subject is revisited in much more detail in the section "PTP Clocks" later in the chapter.

In IEEE 1588-2008, there are three basic types of PTP clocks (it gets more nuanced later):

- The ordinary clock (OC)

- The boundary clock (BC)

- The transparent clock (TC)

Clocks contain *PTP ports*, which are simply communications channels to other clocks. In most cases, a PTP port is either a master port (giving out time) or a slave port (recovering it).

IEEE 1588 defines ordinary clocks as those that have only one PTP port, and under normal conditions, that port can be placed into either master or slave state via algorithmic control. The clock can also be restricted to be a slave-only clock, in which case it can never enter master state. As previously mentioned, the ultimate master clock in the timing network is called the grandmaster clock.

In some PTP profiles, the port state is fixed, and therefore, there are two types of ordinary clock: an ordinary master with one master port and an ordinary slave clock with a single slave port. In this case, master clocks cannot take time from another clock, and slave clocks cannot give time to another clock.

Note that in much documentation and discussion on this topic, a PTP port in master state is referred to as a *PTP master port* and, similarly, a PTP port in slave state is referred to as a *PTP slave port.* Just realize that based on the profile or configuration, the port may be limited to adopt only one of several states or might be capable of changing state under PTP control.

So, the simplest PTP network has two ordinary clocks, a master clock and a slave clock with one port each, and a PTP message exchange between them. Normally, you would also want the master to have a primary reference clock (PRC) or primary reference time clock (PRTC) as a reference source of time for its timestamps (see Chapter 3, "Synchronization and Timing Concepts").

Boundary clocks contain more than one PTP port, one of which must become a slave port (to get time from an upstream master port) and one or more PTP ports that then become master ports, feeding time downstream to slave ports on other clocks. Of course, this means that BCs can be "stacked" in a chain, with one feeding into the next.

Transparent clocks are clocks that help maintain the accuracy of the timing signal through the network. There are several types of TC available in PTP, which are covered in more detail later in the chapter in the section "Transparent Clocks." For this general overview, suffice it to say that TCs are "transparent" in that they allow the important PTP time messages to pass through them. Figure 7-1 illustrates the interaction of the various clock types.

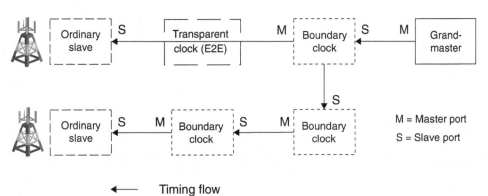

Figure 7-1 *Overview of PTP Clocks*

PTP Clock Domains

The most basic concept in IEEE 1588 is that of the PTP domain, which at its simplest is a network of clocks where one grandmaster is delivering its time over the hierarchy of clocks belonging to that domain. There can be multiple domains in a network, and a PTP clock can belong to multiple domains (in theory), although a PTP port can belong to only one single domain. The operation of PTP in one domain is independent from the operation of PTP in other domains in the same network—meaning they do not talk to one another.

The value for a domain can vary between 0–255, but values of 128–255 are reserved (although 254 is used for the C37.238 Power Profile) and 0 is normally known as the default domain. The legal values for domain number are typically specified (and limited) by the PTP profile, and most profiles take a range of domain numbers from between the values of 4–127.

Generally, a limited number of domains are deployed throughout the network, at least in the case of the telecom profiles, which normally include only one or maybe two domains. The use cases of some industries (such as MiFID-II compliance) can deploy several independent domains and allow them all to flow through the network. The slave devices can recover the clock not only from several PTP domains but also NTP and other time transports. They use the diversity of sources to "elect" the time sources most accurately agreeing with each other about the value of the current time—mostly represented by Coordinated Universal Time (UTC).

Message Rates

Most message rates for PTP fit within the range of one message every 64 seconds to one every 128th of a second. PTP expresses these message rates as a logarithmic value (to base 2) of the time *interval* between messages. This gives a value range for the log of inter-message period of −7 to +6 as outlined in Table 7-1.

Table 7-1   *Log of Inter-Message Interval Rates*

Message Interval	Log of Inter-Message Interval	Message Interval	Log of Inter-Message Interval
1/128 second	−7	1 second	0
1/64 second	−6	2 seconds	1
1/32 second	−5	4 seconds	2
1/16 second	−4	8 seconds	3
1/8 second	−3	16 seconds	4
1/4 second	−2	32 seconds	5
1/2 second	−1	64 seconds	6

Definitions of message rate in the IEEE 1588-2008 standard are always expressed in these terms, not as a *rate*, but as the logarithm (base 2) of the *interval* between messages. The field names used to represent message intervals are also expressed in terms of logarithms—so that in the standard, the message rate of an Announce message has the name *logAnnounceInterval*.

This convention carries over into the operation of a PTP node, so that on some implementations, an operator can configure the message rates on the clock only as log intervals rather than message rates.

Message Types and Flows

PTP consists of two major types of messages. One type where the time is recorded at transmission, the message transported without delay, and then a timestamp recorded on reception (meaning the messages are time sensitive while in flight). These are known as *event messages.*

Other messages that are not time sensitive are known as *general messages* and can carry data such as control, signaling, timestamps, and management information. Depending on the message type, both event and general messages may or may not carry accurate timestamps within them. Table 7-2 outlines the messages, which categories they fall into, and where they are covered in the IEEE 1588-2008 specification

Table 7-2   *PTP Message Types*

Event Messages	General Messages
Sync (13.6)	Announce (13.5)
Delay_Req (13.6)	Follow_Up (13.7)
Pdelay_Req (13.9)	Delay_Resp (13.8)
Pdelay_Resp (13.10)	Pdelay_Resp_Follow_Up (13.11)
	Signaling (13.12)
	Management (13.13 and 15)

Some of the general messages (such as Follow_Up) may carry timestamp information but only as more of a reporting function—they do not need to traverse the network in the minimal time possible, because their in-flight time is not being measured. On the other hand, the event messages are sensitive to delay while in transit because the duration of their in-flight time affects the time equations on the slave.

The message types beginning with "Pdelay" are specifically for the peer-to-peer (P2P) delay mechanism (IEEE 1588-2008 section 11.4), and Delay_Req, and Delay_Resp messages are used for the end-to-end (E2E) delay response mechanism (1588-2008 section 11.3). These two approaches are covered in more detail in the text that follows. Sync and Follow_Up are used in both methods.

Because this book is focused mainly on the telecom profiles, and the telecom profiles use the E2E delay response mechanism, more attention will be paid to those message types. Therefore, it is these five messages that are the most significant (there are also signaling messages for negotiation that is covered later):

- Announce

- Sync

- Follow_Up

- Delay_Req

- Delay_Resp

The Announce message is the only outlier here because it is used to announce the quality and accuracy of the master clock offering the PTP service to the slaves. It is the equivalent of a billboard advertisement that might say something like: "I am a master clock, and am phase aligned to an upstream grandmaster (that is three hops away) with my time sourced from a PRTC connected to a GPS receiver with 100 ns accuracy."

They are sent periodically from the master to the slave (the possible message rates are configurable, but the range of allowed rates may be limited by a profile) and allow downstream clocks to help select the most accurate master. The Announce messages are shown being regularly transmitted from master to slave in Figure 7-2. More details are provided in the section "The Announce Message" later in the chapter.

Meanwhile, Figure 7-2 shows how the other four messages work together to transport time.

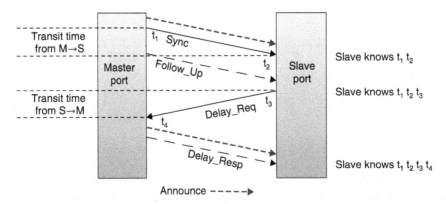

Figure 7-2 *Message Flow Between PTP Master and Slave Ports*

Before looking at the mathematics, it is better to discuss the mechanism and the concepts. The first thing to happen is that the master informs the slave of the current time. To do so, the master transmits a Sync message and notes the time, t_1, that the message is sent. The master can either place that t_1 timestamp inside the Sync message itself or send it a very short time later in a Follow_Up message. There might be good implementation reasons for either approach, but maximizing the accuracy of the timestamp is the main factor (some devices might find it difficult to accurately timestamp a message while trying to transmit it).

When the Sync message arrives at the slave, its time of arrival (according to the slave clock) is noted as t_2, but of course the message took some unknown transit time to get there (what we will call transit-time master→slave, or t_{ms}). How the Sync message is handled in-flight directly affects the accuracy of the time that the slave can recover, so for accurate timing, the Sync message needs to be handled with utmost priority. Needless delay "ages" the timestamp with the consequence that the timestamp data quickly becomes useless.

Remember that the protocol allows that the Sync message does not need to carry t_1, allowing it to arrive a little later in the Follow_Up. The Follow_Up message is not a time-sensitive message, but more of a reporting function, "the master is sending the t_1 that reflects the departure time of the last Sync message."

So, at that point, the slave knows the time on the master when the Sync message was sent, and the slave believes it knows what time it is when the message arrives. If the transit time from the master to the slave was 0 ns, then the slave could simply set its clock to be the same as the t_1, and it would be perfectly aligned. But of course, t_{ms} is greater than zero, so the slave must try to figure out how long the Sync message was in flight so that it knows how "old" the t_1 timestamp was when it was received, and it can correct or compensate for it.

To do this, the slave sends another time-sensitive Delay_Req message (with a transmit timestamp of t_3) back to the master. That Delay_Req message also has a non-zero transit time, which is known as t_{sm}, and again, must not be unnecessarily delayed in order to recover the most accurate time. The master records the time of arrival of the Delay_Req as t_4 and reports that value back to the slave in a Delay_Resp message. As with Follow_Up, the Delay_Resp is more of a reporting message, "this is the last t_4 from the master."

It is the duty of the slave to periodically generate the Delay_Req message, and the master is required to answer with a Delay_Resp. With this protocol flow, the slave is in possession of all the t_1, t_2, t_3, and t_4 timestamps that allows it to determine the time it should be on the slave when the Sync message was received.

This transaction is repeated frequently, with the exact value determined by a combination of the use-case needs or the limits of the profile. For some profiles, it is fixed at 16 messages per second, but for some others, it can range from a high of 64–128 per second down to once every 8 seconds or so.

The (Simple) Mathematics

Now that the slave has the t_1, t_2, t_3, and t_4 timestamps, it has enough information to recover the time. Remember that t_1 and t_4 are measured using time on the master, while t_2 and t_3 are measured using the time according to the slave.

There are several terms defined in 1588 that are used on PTP implementations, so it is helpful to introduce them here as they may appear in the output from PTP commands. The main one is known as *offsetFromMaster* (offset from master) and is defined as

offset from master = <time on the slave clock> – <time on the master clock>

Note that IEEE 1588-2008 defines the offset to be slave time – master time; should you compute it the other way around, you would see the same value but with the sign swapped. However, sticking to this convention, a negative value means the slave clock is slow, and positive means it is fast.

Looking back at the mathematics a little more, you can conclude the following from Figure 7-2:

$t_2 = t_1 + t_{ms} + $ offset from master

$t_4 = t_3 + t_{sm} - $ offset from master

Note   Because the offset from master is positive if the slave time is fast, we need to add that offset to obtain t_2 because the slave clock will be timestamping with a value more than would be expected if the offset was zero. Similarly, for t_4 the value of the timestamp on the master clock will be less than expected as the slave clock (and thus t_3) is ahead of the master. Therefore, we must subtract the offset to determine t_4 from t_3 values.

What we have been calling transit time is known in IEEE 1588 as *one-way propagation delay* and is calculated as a value called *meanPathDelay*, where *mean* has the meaning of average—because it is an average of the forward and reverse transit times. Now, applying the assumption that the transit time is the same in both directions (meaning that $t_{ms} = t_{sm} = $ one-way delay = *meanPathDelay*):

$t_2 = t_1 + $ meanPathDelay + offset from master

$t_4 = t_3 + $ meanPathDelay – offset from master

If we combine these two equations and solve for *meanPathDelay*, the offset from master cancels out and we get

meanPathDelay = $(t_2 - t_1 - $ offset from master $ + t_4 - t_3 + $ offset from master$) / 2$

meanPathDelay = $(t_2 - t_1 + t_4 - t_3) / 2$

If we combine these two equations and solve for the offset from master, the meanPathDelay cancels out and we get

offset from master = $(t_2 - t_1 - $ meanPathDelay $ + t_3 - t_4 + $ meanPathDelay$) / 2$

offset from master = $(t_2 - t_1 + t_3 - t_4) / 2$

For a better understanding, see Figure 7-3 as a practical example.

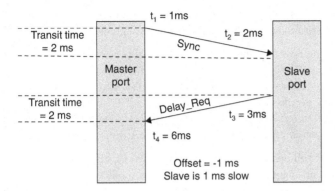

Figure 7-3 *Example Timestamps Between Master and Slave*

The assumption from Figure 7-3 is that the slave clock is 1 ms slow and that the forward and reverse transit times are also 2 ms in each direction. So, you can start with t_1 as 1 ms, t_2 as 2 ms, t_3 as 3 ms, and t_4 as 6 ms (t_2 and t_3 are 1 ms less than they should be because of the offset). Filling in the values for the equation gives

offset from master = $(t_2 - t_1 + t_3 - t_4) / 2 = (2 - 1 + 3 - 6) / 2 = -1$

Therefore, the equation shows that the slave clock does have a –1 ms offset from the master (meaning it is slow). Note that it does not matter how long the transit time is or how far the slave clock is wrong; the equation still gives an accurate solution to the value of the offset from the master.

Should there be asymmetry in the forward and reverse paths, and that asymmetry is known and static (unchanging), then there are mechanisms to include that data in the equations to increase the accuracy of the recovered clock. This is covered in more detail in the upcoming section "Asymmetry Correction."

Finally, there exists another mechanism to improve the recovered clock accuracy by using the *correctionField* (correction field), which we address in the section "The Correction Field" and when talking about a transparent clock. To avoid overcomplicating the equations at this stage, those corrections are not included here. Suffice to say that there is a field available in the header of messages that allows PTP clocks to signal corrections arising from known errors in the timestamps or delays in the network (it is not used in Announce, signaling, and management messages).

Asymmetry and Message Delay

The amount of time that the Sync and Delay_Req messages take to cross a packet path is not important; it could be microseconds, milliseconds, or even minutes. This means that PTP still delivers time accurately, even over long-distance cable runs. Of course, a very long delay means that the slave clock will take longer to lock because of the time that the messages are in transit, but theoretically it will still work. What is important is that the length of time that the packet is in flight is symmetrical in both directions. The golden rule is that half the asymmetry shows up as time error in the slave alignment.

If the transit time is 10 ms in one direction and 15 ms in another, this would lead to a (15 − 10) / 2 = 2.5-ms phase error. Similarly, accurate recovery of time with PTP does not automatically degrade with distance, although things like *chromatic dispersion* can introduce small asymmetry over long-distance fiber runs. If the PTP messages arrive without undue delay and are symmetrical in transit time, there is no theoretical limit to the distance one can recover accurate time using PTP.

So, what is "undue delay"? This will be covered in more detail when we discuss packet delay variation (PDV) in following chapters. For now, a practical example will suffice.

Imagine that a Sync message is timestamped with the current master time and takes 100 µs to arrive at the slave. Assume that this is the absolute minimal time possible to transit that path, counting distance through fiber, and best-case switching time through devices. When the Sync message arrives, this is very valuable time data, since it has been stamped by a master with an accurate time source, and the message has arrived in the minimum time possible. The data is very "fresh."

A clock getting a fair number of these messages (and the Delay_Req in the other direction) would quickly figure out that the t_{ms} was 100 µs (from 200 µs two-way delay) and that the timestamp was only 100 µs old. So, add 100 µs to the t_1 timestamp, and that is what the slave clock time should be.

Now imagine that the next Sync message (1/16 second later) is timestamped but is held up by some event and arrives 400 µs after timestamping—rather than 100 µs. That data is useless because it will show the slave time as 300 µs off compared to the previous message. The data that contains outlier values in messages will therefore be ignored and discarded. If too many messages arrive like this, PTP will struggle to recover accurate time.

So, the transport of time with PTP requires a small but steady stream of these "lucky" messages to allow a slave to recover time accurately. But the actual transit time does not matter; PTP can work over any practical distance if either the transit is symmetrical or any asymmetry is controlled or compensated for.

Asymmetry Correction

You have seen the effect that asymmetry has on the recovered time accuracy on the downstream slave. In many cases, some proportion of any observed asymmetry can be static (meaning that it never changes, even following a node or link restart) and the rest of it will be dynamic. Dynamic asymmetry is mostly caused by changes in network conditions, even where no local reason for it is apparent. An example of static asymmetry could be cable length mismatch between fiber pairs, whereas dynamic asymmetry could arise from traffic conditions causing message queueing in a non-PTP interface upstream of the clock.

In the case of total asymmetry, it is possible to measure it with specialized testing equipment that carries onboard a source of time (for example, a global navigation satellite system [GNSS] receiver). The network designer may also be able to estimate the amount of asymmetry because of their understanding of the likely sources of it. For example, a long run of bidirectional (BiDi) fiber will have different forward and reverse path delay because of the different speed of the forward and reverse lambdas in the fiber. Those values can be known or estimated to a considerable degree of accuracy.

If the operator can measure or calculate the asymmetry, it is possible to configure this expected asymmetry in the PTP port characteristics so that the time-recovery algorithm can permanently correct for this error. The only difficulty can be understanding what proportion of the asymmetry is static or constant and what is dynamic. Generally, carrying PTP over many nodes (using IP) that do not process PTP can introduce an excess of dynamic asymmetry. Dynamic asymmetry, by its very nature, cannot be corrected for, although it can be filtered to reduce its impact.

So, the calculation used to determine the offset from master and the one-way delay can include calculations that include this extra correction factor. Returning to Figure 7-3, if the transit time from master to slave was known to be 4 ms instead of 2 ms, then an extension to the equations could easily accommodate that knowledge and compensate for the error and allow accurate time to be recovered. The IEEE 1588-2008 standard does this with the use of the *delayAsymmetry* attribute.

The attribute delayAsymmetry is defined as follows (you can see that the value is defined to be positive when the master-to-slave propagation time is longer than the slave-to-master propagation time):

t_{ms} = <meanPathDelay> + delayAsymmetry

t_{sm} = <meanPathDelay> − delayAsymmetry

This delayAsymmetry is added to the correctionField (covered in the next section) before the calculations of meanPathDelay and offsetFromMaster are done.

It is the authors' opinion that outlining these expanded equations to include compensation is not required to understand how PTP time recovery works or how to deploy it. To include these extra calculations just makes the mathematics more complex and is an impediment to learning. If you are interested in an example of asymmetry compensation, see the following section "Asymmetry Correction" and the section "Correction and Calibration" in Chapter 9.

For now, it is enough to recognize that asymmetry compensation can be done if the operator understands the likely error and that PTP implementations include the ability to configure and compensate for this error

See clauses 11.6 and 7.4.2 of the IEEE 1588-2008 standard for more details.

The Correction Field

The earlier section on the mathematics of PTP mentioned the use of the correctionField as a mechanism to improve the accuracy of the recovered clock. The correctionField is a 64-bit field in the common PTP header. Any nanosecond value is multiplied by 2^{16} before being represented in the correctionField, which really means the lower 16 bits are sub-nanoseconds and the other 48 are nanoseconds.

You could think of it as two subcomponents, nanoseconds and sub-nanoseconds, and the PTP dissector on Wireshark splits them and displays the correctionField split into two values. Wireshark labels the upper 48 bits as *correction.ns* and the lower 16 as *correction.subns*.

One example of the use of this (combined) field is to allow a TC to report a delay (the residence time measured by the TC) in the Sync message to the downstream slave, so that the slave can remove that delay from the calculation of meanPathDelay. For a two-step clock, the adjustment is made to the Follow_Up message rather than the Sync message.

Beyond this example, the correctionField has several functions in a PTP network:

- **Reporting residence time:** The preceding paragraph gave an example of the use of the field to report residence time of Sync messages in TCs. The same approach can be used not only on Sync messages, but also on Follow_Up, Delay_Req, Delay_Resp, Pdelay_Req, Pdelay_Resp, and Pdelay_Resp_Follow_Up messages.

- **Asymmetry correction:** Any configured delayAsymmetry is combined with the correctionField before the calculations of meanPathDelay and offsetFromMaster are done. In the E2E delay mechanism, this is done on the transmitting port for a Delay_Req or the receiving port for the Sync (or Follow_Up) message. In a case of a non-transparent clock, this is the slave port.

- **To carry sub-nanoseconds:** The PTP timestamp fields are integer fields for seconds (since the epoch) and nanoseconds. There is no field to carry sub-nanosecond values. The correctionField, having 16 bits to represent fractions of a nanosecond, is used to do this.

The field is used differently in the case of E2E transparent clocks (E2E TC) from the peer-to-peer transparent clock (P2P TC) delay mechanism, which is covered in detail in the sections "Peer-to-Peer Versus End-to-End Delay Mechanisms" and "Transparent Clocks" later in the chapter.

Please refer to sections 11.5 and 11.6 of the IEEE 1588-2008 standard for more details on residence time correction and asymmetry correction.

PTP Ports and Port Types

PTP has the concept of a "port," which is an interface that connects to the network for the transfer of time; the port is the window from the PTP clock to the network. Any one PTP port will support only a single version of the PTP protocol and use a single transport protocol; however, a physical network interface might support multiple PTP ports. Another important point is that a PTP port can belong to only a single domain and, therefore, the communications path between that port and another port can belong to only one domain.

PTP ports (per 1588-2008 section 7.5) have several characteristics, the most important of which is the *portIdentity*, made up from

- *clockIdentity* (64-bit number representing the identity of the clock)
- *portNumber* (16-bit value counting upward starting from 1)

There is also a *versionNumber* that is 2 for the 2008 version of the standard.

There is no need to spend too much time on these fields; they are only mentioned because the port identity and port number can be visible in the command-line interface (CLI) outputs, and both the clock identity and port number can sometimes become a tie-breaking factor in selecting the best master under the best master clock algorithm (BMCA), which is discussed in depth later in the chapter.

The PTP port will be in a port state that is governed by the PTP node state machine (and the BMCA), which is outlined in IEEE 1588-2008, section 9.2. Table 7-3 outlines the possible port states (the state always starts off in initializing).

Table 7-3   *PTP Port States*

PTP State	Meaning
Initializing	The port is initializing its state and preparing the communication path, and no traffic is flowing. This is the state where the port begins.
Faulty	The port is in a fault state and ignored, although it can send management messages.
Disabled	Any traffic on this port is ignored except for management messages.
Listening	The port is listening for an Announce message, although it can send peer-delay, signaling, and management messages.
Pre_master	The port is behaving as a master but not sending timing messages, although it can send peer-delay, signaling, and management messages.
Master	The port is sending timing information and Announce messages to downstream slaves.
Passive	The port is not sending timing messages, although it can send peer-delay, signaling, and management messages.
Uncalibrated	A master port has been detected and selected, and the local port is preparing to synchronize to that port.
Slave	The port is being synchronized by a selected master port.

Under normal operations, the deployment engineer would most likely see the states of master, passive, and slave, as the other states are mostly transient (unless something goes wrong).

When the PTP node is initialized, then the PTP ports are all in the initializing state as the timing system prepares itself, fills the data structures, and prepares the communications paths. Then, the ports should go into listening state, whereby they listen for Announce messages to determine what master clock they can synchronize with. Then, as the master is selected, the highest-priority port will go through uncalibrated before entering slave state. As the PTP node becomes phase aligned to the remote master, the other ports may become master (to a downstream slave) or passive (neither a master to another clock nor synchronized to a master).

Transport and Encapsulation

The IEEE 1588-2008 standard defines numerous methods of transporting and encapsulating PTP messages, which are outlined in the Annexes to the standard shown in Table 7-4. Note that the equivalent sequence of Annexes in IEEE 1588-2019 have been reduced by one, so that they start at Annex C.

Table 7-4   *PTP Transport and Encapsulation*

Transport and Encapsulation	Reference in IEEE 1588-2008
UDP over IPv4	Annex D
UDP over IPv6	Annex E
IEEE 802.3/Ethernet	Annex F
DeviceNET (from www.odva.org)	Annex G
ControlNET	Annex H
IEC 61158 Type 10 (PROFINET)	Annex I

At layer 3, UDP/IP can carry PTP messages as either unicast or multicast or even some mixture of the two depending on the profile and the use case; generally, these encapsulations are informally referred to as PTPoIP (PTP over Internet protocol). Event messages use UDP port number 319 and general messages use 320, which makes it quite easy to discern between the two in an application tool like Wireshark.

Although PTP profiles for other industries commonly use L3 multicast, the telecom profiles that use IP as a transport (G.8265.1 and G.8275.2) are unicast only and mandate support for IPv4, whereas IPv6 is optional to implement.

One-Step Versus Two-Step Clocks

The preceding section showed the master port sending the Sync message plus the t_1 timestamp and the optional Follow_Up message. If the designer of the PTP node feels that the accuracy of the timestamp cannot be assured when it is being sent inside the Sync message, then the implementation allows the timestamp to be included in the Follow_Up sent a short time later. To indicate that this has been done, PTP sets a flag in the Sync message (*twoStepFlag*) to indicate that the correct timestamp is coming in the later Follow_Up message.

So, the options are send the Sync with the t_1 timestamp or send the Sync without the t_1 and generate the Follow_Up including it. A PTP clock that sends just the Sync message is known as a *one-step clock*, whereas one that uses a subsequent Follow_Up is known as a *two-step clock*.

Whichever approach is implemented is completely at the discretion of the master port. The downstream slave is *required* to accept either approach without problem. The authors have run across implementations (especially some radio devices) that cannot accept PTP messages from a two-step master, which is clearly a faulty implementation of PTP.

Some implementations allow the operator to configure either a one-step or two-step clock. However, the accuracy of the timestamp (and consequently the accuracy of the time recovered by the downstream slave) will be better with the method that suits the hardware design and implementation. The PTP node will perform better with one or the other option, so the network engineer needs to understand what clock method is better suited to their hardware and use that. So, do not treat this value as a configuration item to simply choose based on some other external factor.

One aspect that will not be covered in detail is that there are one-step and two-step behavior differences that also apply to transparent clocks. There are a lot of examples in Annex C of IEEE 1588-2008 showing the interaction between each combination of one-step and two-step clock for master, transparent, and slave. It is enough to understand that there are such things as one-step and two-step TCs, and you know where to go to find more information should you ever need to know more about them.

Peer-to-Peer Versus End-to-End Delay Mechanisms

Two mechanisms are used in PTP to measure the propagation delay between PTP ports, and ordinary and boundary clocks can be implemented using either mechanism. In general, a network using one of the methods cannot coexist with ports using the other one. The methods are as follows:

- **Delay request-response mechanism:** Determines the mean path delay between ports and is more focused on the E2E delay between clocks. It uses the messages Sync, Delay_Req, Delay_Resp, and the optional Follow_Up. This method is specified as a default profile in Annex J.3 of IEEE 1588-2008, and is the same mechanism used in the telecom profiles.

- **Peer-to-peer delay mechanism:** Uses the link delay between ports. It is specified as a default profile in Annex J.4 of IEEE 1588-2008 and uses an additional set of Pdelay messages:

  - Pdelay_Req

  - Pdelay_Resp

  - Pdelay_Resp_Follow_Up (if required)

This latter mechanism is used in several industry profiles, with one example being the Power Profile (IEEE C37.238:2017) using the layer 2 (Ethernet) transport. But again, just note that no telecom profiles use this mechanism.

The P2P delay mechanism is covered in a little more detail when addressing TCs because there are special TCs that support and enable the P2P method. For now, accept that a P2P TC is one that provides accurate transit time information for PTP event messages and processes the Pdelay messages to implement the P2P delay mechanism.

One-Way Versus Two-Way PTP

As stated earlier in this chapter, PTP is a two-way time protocol, and to recover frequency, phase, and time, it needs to be two-way—chiefly to estimate the one-way delay time (meanPathDelay) between the master and slave. However, if PTP is being used only to recover the frequency from the master, then the slave does not need to know the one-way delay time and therefore does not need to send messages to the master. By recovering frequency only, the clock is concerned that the rate of passage of time on the slave is the same as the master, and not worried about the offset in phase from the master clock. Therefore, for frequency-only synchronization, PTP can be a one-way protocol.

How is this possible? Well, in frequency synchronization, the task of the slave is to ensure that the oscillator in the slave is running at very close to its nominal frequency. The way to do this is to get timestamps from the master over an extended period and see if the local clock is gaining or losing time compared to the master. If it is gaining time compared to the master, then the oscillator is running too quickly, and if it is losing time, it is running too slowly. The slave doesn't really need to know the exact time and phase, just whether time is passing at the correct rate.

One could imagine doing the same with a wristwatch and using a similar method to determine whether the watch gains or loses time. The wristwatch wearer would check the time against a relevant source (say, at the precise start of the evening TV news) and then do the same thing one day later. By seeing how much the watch had gained or lost, one could ascertain how fast/slow the oscillator was going.

Obviously, doing frequency synchronization by this method is somewhat slow, since the slave must compare time with a reference clock over an extended period to detect small gains and losses in the local clock. This means that PTP can be somewhat slow to recover frequency and accurately track it (alignment between a slave and master might take 30–40 mins in some profiles).

Because the slave only needs the timestamps from the master, only the Sync messages (and optional Follow_Up for two-step clocks) need to pass in the master→slave direction and, therefore, PTP frequency-only synchronization becomes a one-way protocol. The IEEE 1588-2008 recommendation covers this feature in Chapter 12, where it discusses frequency-only synchronization (meaning syntonization) versus synchronization.

For real-world deployments, there are better ways, such as using a physical signal as a frequency reference, which allows almost instantaneous frequency alignment. This approach commonly uses synchronous Ethernet (SyncE), although it is possible to use a frequency signal from a dedicated timing node (SSU/BITS) or from a GNSS source (for example, a 10 MHz signal). See Chapter 6, "Physical Frequency Synchronization," for more details on this topic.

Simultaneously using SyncE to synchronize frequency and PTP to synchronize phase and time is the best of both worlds, and the case of PTP clocks using both methods together is known as *hybrid mode*.

Timestamps and Timescales

Chapter 1, "Introduction to Synchronization and Timing" introduced the internationally accepted systems of time, namely Temps Atomique International (TAI) and Coordinated Universal Time (UTC). The topic now is to examine how PTP uses these forms of time to implement its timestamping mechanisms.

The PTP timescale uses an epoch of 1 January 1970 00:00:00 TAI, which means that PTP started counting at zero seconds from then. If you initialize a PTP grandmaster with no reference time input, it might send out timestamps counting from zero, and the slaves will think it is January 1970.

Note   Using this date and time means that PTP shares almost the same epoch as POSIX time. However, the POSIX timescale is different from the PTP timescale because POSIX is based on UTC ("sort of" recognizing leap seconds), yet PTP is based on TAI (and so it is monotonic, with no leap seconds). This means that the POSIX epoch started 8 seconds after PTP because there were 8 leap seconds current at that time.

Subtracting the then-current leap second value from a PTP timestamp and applying the POSIX time algorithm results in UTC. If you are writing some code or using a website to calculate time and notice a discrepancy anywhere between 8 (1970) and 37 seconds (2020), you will know where to look for an explanation. See IEEE 1588-2008 section 7.2 and Annex B.2 for more information.

The date/time of 1 January 1970 00:00:00 TAI was also almost exactly 31 December 1969 23:59:52 UTC. There was then an approximately 8-second difference between UTC and TAI, which increased to 10 leap seconds by the time the current leap second system was introduced on 1 January 1972.

Therefore, the PTP timescale is always ahead of UTC by 10 seconds plus the number of leap seconds since 1972 (27 as of 2021), meaning 37 total seconds up until 31 December 2021 at least (IERS has announced that there is no leap second scheduled before this date). Note that as outlined in Chapter 3, the epoch used by GPS (like other GNSS systems) is different from both TAI and UTC.

The PTP timestamp consists of two parts:

- *secondsField*: Seconds since the Epoch (6 bytes or a 48-bit unsigned integer)
- *nanosecondsField*: Number of nanoseconds following the second in the first part (4 bytes or a 32-bit unsigned integer)

Expected values for the seconds timestamp, based on the PTP epoch, should be around 1,610,000,000 (Jan 2021) to 1,700,000,000 (Nov 2023), and these numbers increase by approximately 2,600,000 per month or 31,100,000 per year.

The nanosecond field can range from zero to 999,999,999 even though it is a 32-bit integer.

PTP can run one of two timescales:

- **PTP timescale:** The epoch is the PTP epoch.

- **ARB timescale:** Arbitrary = the epoch is specific to the implementation.

The ARB timescale can be basically whatever the operator or engineer desires it to be. Under normal operations, it is, like PTP, a continuous timescale. However, under normal deployments in the telecoms world, you should not see ARB, so seeing it means that something is wrongly configured or there is an operational issue.

The Announce Message

The Announce message is sent out regularly to announce the capabilities of the PTP port, and depending on the profile, it is transmitted at a rate between eight every second to once every eight seconds. There is also a timeout mechanism whereby the operator configures the slave port with a timeout value that fixes the maximum number of messages allowed to be "missed." For proper operation, it is recommended that the Announce timeout should be uniform across a domain.

For example, if the configured rate of Announce is four per second, and the timeout is four messages, then the slave would declare a timeout after not receiving any Announce messages for one second. According to IEEE 1588-2008, the allowable timeout range is from 2–255 messages but can be further constrained by a PTP profile. IEEE 1588 suggests that the minimum value should be 3. See IEEE 1588-2008 section 9.2.6.11 on the Announce receipt timeout expiry for more details.

The Announce message contains the following fields (see IEEE 1588-2008 section 13.5):

- **Common PTP header:** Includes the domain number, clock identity, and port identity. There are also flag fields in the header that are used in combination with the Announce message. This includes information about upcoming leap events, whether the GM is traceable to sources of time and frequency, and whether the UTC offset value is valid.

- **Origin timestamp:** A timestamp from the originating node within ±1 second. This is not like a timestamp in an event message, as it is not really time sensitive, although the format is the same. Therefore, the timestamp consists of a 6-byte (48-bit) seconds field, with a 4-byte (32-bit) nanoseconds field.

- **Current UTC offset:** Total number of leap seconds currently in effect (37 as of 2021). In combination with an accurate timestamp based on TAI, this allows the PTP clock to accurately calculate an estimate for UTC.

- **Grandmaster priority1:** Value in range 0–255 used as a priority order for GM clock selection, whereby lower values are higher priority. The allowable range can vary (or be fixed at a specific value) when specified by a PTP profile.

- **Grandmaster clock quality:** Made up of three separate values that indicate the inherent accuracy of the grandmaster; the traceability of the grandmaster to a time source; and the accuracy of that time (total of 32 bits). See the following list for a detailed breakdown.

- **Grandmaster priority2:** Value in range 0–255, used as a tiebreaker in GM clock selection when priority1 values agree. As with priority1, lower values are higher priority, and the allowable values can be varied by a profile.

- **Grandmaster identity:** A 64-bit address *clockIdentity* that is either an IEEE extended unique identifier (EUI-64) individually assigned number or some constructed value (see IEEE 1588-2008 section 7.5.2.2). This should be a unique number that can act as the final tiebreaker for the BMCA.

- **Steps removed:** *stepsRemoved* is a count of the number of BC steps from the grandmaster. Every time the PTP flow is passed through a BC, the steps removed count is increased by one.

- **Time source:** *timeSource* is a value that indicates the GM's source of time, such as GPS, NTP, ATOMIC_CLOCK, or INTERNAL_OSCILLATOR (see IEEE 1588-2008 section 7.6.2.6 for all possible values of 16-bit value). GPS refers to any of the GNSS systems rather than having to be specifically GPS.

The grandmaster clock accuracy field is made up of three values:

- **Grandmaster clock class:** The *clockClass* value in the range of 0–255 (8 bits) indicates the traceability to a time source, meaning how well the master is synchronized to a source of time. As an example, in many profiles the value 6 indicates traceability to a PRTC, but the allowed values can vary depending on the profile.

- **Grandmaster clock accuracy:** The *clockAccuracy* metric is based on the quality of the source of time on the grandmaster in the *timeSource* attribute below (8 bits). For example, traceability to GPS would result in a *clockAccuracy* value of 0x21 = *time is accurate to within 100 ns.* Values are enumerated in Table 6 of IEEE 1588-2008.

- **Grandmaster offset scaled log variance:** The *offsetScaledLogVariance* is an estimate of the stability of the GM (16 bits). This is an estimate of the variations of the clock and thus the precision of the timestamps (see section 7.6.3 of IEEE 1588-2008 for a math-heavy treatment). For example, the value of 0x4E5D indicates a GM connected and locked to a PRTC, but the value can vary between 0x0000 and 0xFFFF.

Perhaps the most important field in the Announce message is *clockClass*. In some PTP profiles, as well as IEEE 1588 PTP default profiles, a clock class of 6 indicates a clock that is synchronized to a PRTC. If the clock loses the connection to the PRTC and goes into holdover, the clockClass value will change to a higher number indicating the loss to traceability to a source of time. More details on clock classes are included in the following section "Telecom Profiles."

Note that the Announce message is *not* forwarded by a boundary clock. It is used to update the datasets (see the upcoming section "PTP Datasets") on the BC and new Announce messages are created to advertise service to downstream slave ports. This means some values can be updated (one example is stepsRemoved that will be increased by +1 to indicate another additional PTP hop back to the grandmaster).

The next section examines how the Announce message is used.

Best Master Clock Algorithm

The PTP specification includes the *Best Master Clock Algorithm* (BMCA), which is designed to build a valid master-slave hierarchy across the network. Although the IEEE 1588 standard defines a BMCA mechanism known as the *default* BMCA (see section 9.3), any PTP profile can also specify a different *alternate* BMCA (aBMCA). Because the algorithm runs on all the clocks in the PTP domain, the BMCA also ensures that the best master clock in that domain becomes the grandmaster.

The BMCA runs on every PTP port of an ordinary and boundary clock and complements the PTP clock's state machine to ensure that each clock selects the best available master as its source of time. The goal is to select one and only one master, based on the information available to it. The BMCA can use information passed in the Announce message to decide the state of each of the ports in the clock (there is no negotiation between clocks to determine the master).

Also, the BMCA adapts continually to changing conditions to ensure that a new master-slave hierarchy will quickly emerge following a network or clock outage. The speed of this reconvergence is dependent on the network topology as well as the rate at which Announce messages are sent (the Announce interval).

The BMCA will gather this data and evaluate it to decide what state each of the PTP ports will be in. For most clocks, this means selecting a single PTP port as the slave port to communicate with the best available master. With a BC containing more than one PTP port, the other non-slave ports could then enter the master state to provide time to downstream slave ports. The state table for the BMCA compares many parameters in order to determine the best master clock and which port to select as the slave port for the clock. As an example, Figure 7-4 simplifies it to two parameters, namely *clockClass* and *stepsRemoved*, to demonstrate the concept.

Should the BMCA fail to select the best clock based on the values of priority1, clockClass, clockAccuracy, and offsetScaledLogVariance, the priority2 attribute allows 256 additional priorities to be used before the BMCA goes to a tiebreaker. The tiebreaker is based on the clockIdentity field.

The *clockClass* (from the Announce message) signals the traceability of clock quality. The value of 6 indicates that the clock has a traceable path back to a PRTC source of time, meaning the grandmaster at the top of the figure. In Figure 7-4, they all have this traceability.

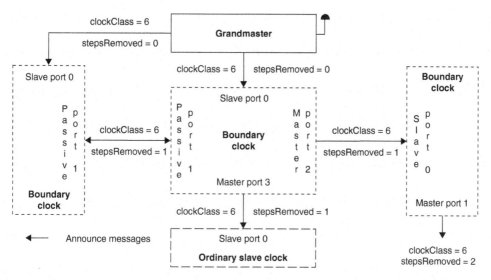

Figure 7-4 *BMCA Running on a Multi-Port BC*

The *stepsRemoved* indicates how many different PTP boundary clocks have been crossed in the way from the grandmaster to the current clock. Therefore, the grandmaster sends out an Announce message with the value set to zero, while the left and center boundary clocks will send it out with the value of one and the right boundary clock with value of two.

Given that the clock class in this network is the same (they are all traceable to a PRTC) for every Announce message, BMCA will select the port receiving the Announce message with the smallest value for the steps removed, and the master and slave ports will align themselves (master, slave, and passive) as shown in the figure.

PTP Datasets

Both ordinary and boundary clocks maintain two types of datasets, known as clock datasets and port datasets. In IEEE 1588-2008 terminology, the fields in a dataset are known as *members*. The clock dataset names and the corresponding sections of 1588-2008 that cover them are as follows:

■ **Default dataset or** *defaultDS*: Attributes describing the entire clock and common to every port (section 8.2.1).

■ **Current dataset or** *currentDS*: Attributes related to the current state of the clock resulting from the mechanisms of synchronization, such as the calculated mean path delay (section 8.2.2).

■ **Parent dataset or** *parentDS*: Attributes describing the parent (the clock to which the clock synchronizes) and the grandmaster clock at the root of the converged master-slave hierarchy (section 8.2.3). The values for members of this dataset (such as clockClass) are set by the grandmaster based on its source of time and transmitted to the current PTP clock with Announce messages via any boundary clocks. The intervening boundary clocks may update some values (such as stepsRemoved) when they create their Announce messages to downstream slave ports.

■ **Time properties dataset** or *timePropertiesDS*: Attributes of the timescale including information about the UTC offset and details of upcoming leap second events (section 8.2.4).

■ **Foreign master dataset** or *foreignMasterDS*: Data describing a foreign master from the grandmaster fields of a received Announce message (an example is the clock quality fields of the grandmaster).

■ **Port dataset** or *portDS*: A separate dataset for every port that contains the attributes of that port, including the PTP state (section 8.2.5).

An example of a single member of the default dataset, namely the clock class of the clock, is *defaultDS.clockQuality.clockClass*. Members of the datasets can have static values (fixed by the profile, the protocol, or some inherent properties). They can also be dynamic, as a result of the operation of the PTP clock mechanisms or the PTP protocol, or they can be configurable, meaning selected by the operator. Obviously, the members of the current dataset are always dynamic since the values update based on the operation of the PTP clock.

The Parent and Time Properties datasets are also dynamic as they are based on data from the upstream master and ultimately the attributes of the grandmaster. These datasets are filled in for the current clock with Announce messages from the upstream master. Table 7-5 shows an example of a member or two from each dataset.

Table 7-5   *Example Dataset Members*

Member Name	Type	Value	How Set?
defaultDS.domainNumber	Configurable	0–255	Operator configuration
defaultDS.clockQuality.clockClass	Dynamic	0–255	Announce from current master
defaultDS.twoStepFlag	Static	T/F	Property of clock implementation
currentDS.stepsRemoved	Dynamic	0–255	Announce from current master
parentDS.grandmasterPriority1	Dynamic	0–255	Announce from current master
portDS.portIdentity	Static	Hex	Clock ID and port number
portDS.portState	Dynamic	Enum	Clock processing and BMCA
portDS.logAnnounceInterval	Configurable	–3 to 4	Operator configuration/profile
timePropertiesDS.currentUtcOffset	Dynamic	~37	Announce from current master
timePropertiesDS.leap61	Dynamic	T/F	Announce from current master

Transparent clocks can also maintain transparent clock datasets: one Default dataset (section 8.3.2) for the clock and a Port dataset for each port (section 8.3.3). These two datasets are known as *transparentClockDefaultDS* and *transparentClockPortDS* respectively. Note that these datasets are optional (section 8.3.1).

Virtual PTP Ports

A virtual PTP port on a PTP clock is a connection to an external phase/time input reference (such as a GNSS receiver), but to the clock, it looks like a PTP slave port. This allows the external interface to participate in the BMCA and can be selected as a source of time/ phase. This feature is not included in the IEEE 1588-2008 standard but is covered in Annex B of ITU-T G.8275, "Inclusion of an external phase/time interface on a PTP clock."

This feature allows a lot of flexibility when deploying a mixed solution of PTP combined with GNSS receivers. Figure 7-5 shows a basic outline of a virtual port in an example topology.

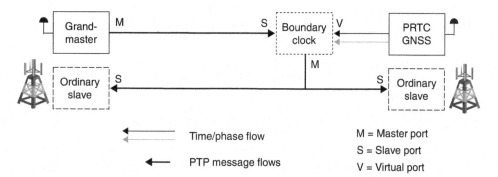

Figure 7-5   *Example of a Virtual Port in a Timing Distribution Network*

When this feature is used in a boundary clock, it allows the BMCA to select the local time reference as a primary source of time/phase but still have the possibility to select another port (synced in slave state to a remote grandmaster) should the local source fail.

The physical interface to the local time source is a normal time/phase interface, such as the 1 pulse per second (1PPS) and time of day (ToD); there may also be a frequency source, such as 10 MHz. No PTP messages cross this interface. The data fields that the BMCA requires for the virtual port to determine the order of selection of slave port are configured on the boundary clock. For example, fields like stepsRemoved, domain number, and priority1/priority2 are defined by the network engineer in the configuration. Other fields, such as the grandmaster identity or stepsRemoved, must be determined by the implementation on the BC (since there is no real PTP grandmaster here).

Negotiation

The IEEE 1588-2008 standard covers optional components. For example, the *unicast discovery* option (section 17.5) allows PTP to contact the master in a network that does not offer multicast (which might be the case in many IP networks). It is used in combination with another optional component, *negotiation* (section 16.1), to discover the best master available over IP unicast. Therefore, the following discussion is of most interest to those working with unicast PTPoIP deployments.

Firstly, the engineer configures the desired slave port with the IP addresses of all possible masters. Using the negotiation process, that PTP port then requests synchronization service from those IP addresses by a transmission of a unicast PTP *signaling message*. Simply put, the slave requests the master IP for the unicast transmission of Announce, Sync, and Delay_Resp (or Pdelay_Resp) messages using the *REQUEST_UNICAST_TRANSMISSION* type, length, value (TLV) entity. The port being asked for service will be referred to as the *grant port*.

When negotiating for a unicast transmission, the slave requests a message type, a message rate (*logInterMessagePeriod*), and the length of time (*durationField*) for which the message flow is desired. As is normal in IEEE 1588, the rate is expressed as the log (base 2) of the interval between messages, so that, for example, −4 is 16 packets per second, while +1 is one packet every 2 seconds. The mechanism behind the use of logarithm for inter-message period is explained in the section "Message Rates" earlier in the chapter.

The grant port that receives this request TLV shall respond with a signaling message containing a *GRANT_UNICAST_TRANSMISSION* TLV. This transmitted grant TLV signals to the requesting slave whether it grants or denies the request.

These are the possible operations available for the negotiation process:

- **REQUEST UNICAST TRANSMISSION:** The slave uses this TLV to request a flow of one of the timing messages (for example, Announce messages to see how good the configured master is). This message TLV contains fields that specify the log of the desired message interval as well as the duration of the grant—the default being 300 seconds. Before the grant expires, the slave must issue another request if it wants to continue receiving the service.

- **GRANT UNICAST TRANSMISSION:** Any grant port that receives a request TLV responds with a grant TLV confirming the grant or denying it. The TLV contains fields that specify the message type, the log of the granted message interval, and the duration of the grant. If the grant is denied, the duration will be zero seconds.

- **CANCEL UNICAST TRANSMISSION:** A slave that has requested a grant may inform the grant port that it no longer needs the service it was granted. It does this by transmitting this cancel TLV message with the name of the message type that it wishes to cancel. A grant port providing the service may also decide to inform the grant port that it can no longer provide the message flow and send the same TLV to the slave port.

- **ACKNOWLEDGE CANCEL UNICAST TRANSMISSION:** The grant port receiving the cancel TLV responds with this acknowledge cancel TLV and then immediately stops transmitting the messages the slave wants to cancel. Similarly, any slave port that receives the cancel TLV should also send this acknowledge cancel TLV message and cease using the service.

■ **UNICAST NEGOTIATION ENABLE:** This message TLV is used to enable or disable the unicast negotiation mechanism. A grant port receiving this TLV message will cancel all outstanding grants and disable the mechanism.

Note   PTP ports in the following states are not allowed to participate in the negotiation process: initializing, faulty, or disabled.

According to Annex A.9.4.2 of IEEE 1588-2008, for unicast implementations, the recommended and allowed values for the message intervals of the request TLV are outlined in Table 7-6.

Table 7-6   *Recommended Message Interval Rates for Unicast Negotiation*

Message	Default Message Interval	Minimum Message Interval	Maximum Message Interval
Announce	1 (1 per 2 sec)	–3 (8 per sec)	3 (1 per 8 sec)
Sync	–4 (16 per 1 sec)	–7 (128 per sec)	1 (1 per 2 sec)
Delay_Resp	–4 (16 per 1 sec)	–7 (128 per sec)	6 (1 per 64 sec)

According to the same recommendation, the duration field should have a configurable range of 10–1000 seconds (default is 300 seconds).

Figure 7-6 illustrates how the negotiation process between two PTP ports would work.

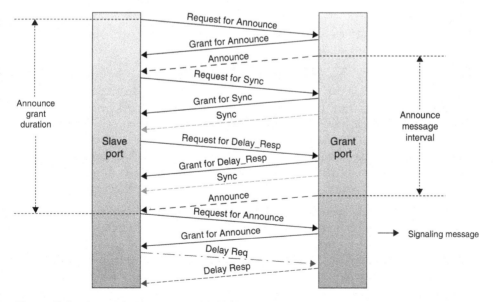

Figure 7-6   *Example Negotiation Process*

The slave begins by asking for a service of Announce messages to determine the quality of the upstream clock. The grant port responds with the acknowledgement of the grant and commences the transmission of (unicast) Announce messages. The BMCA on the slave determines (via the clock quality and priority) that this grant port is an acceptable master clock and then requests a flow of Sync event messages. The grant port acknowledges the grant of Sync service and immediately starts sending the Sync messages at the requested rate (for a two-step clock, each Sync message will be followed by a Follow_Up message).

If this is a two-way mechanism, the slave port will now ask for a grant of Delay_Resp messages. Again, the grant port responds with the acknowledgement of the grant of Delay_Resp service. The Announce messages continue to arrive at the requested rate until the duration of the grant has expired, and the slave will then have to again request a flow of Announce messages (granted by the grant port). Remember that, by default, the grant interval requested is 300 seconds, so the slave port needs to request service for each message type only once every five minutes.

Now, because the slave has already received a grant for the Delay_Resp, the slave is then free to send Delay_Req messages to the grant port. The grant port answers each Delay_Req with a Delay_Resp to determine the final timestamp needed for the calculations.

PTP Clocks

There are five basic types of PTP devices:

- Ordinary clock (OC)
- Boundary clock (BC)
- End-to-end transparent clock (E2E TC)
- Peer-to-peer transparent clock (P2P TC)
- Management node

In the definition from IEEE 1588-2008, the OC is simply a PTP clock that has only one single PTP port, which under normal conditions is in either slave or master state (there are other states possible). Obviously, this means that an OC is located either at the beginning (see grandmaster in the next section) or end (ordinary slave) of a clock hierarchy in a PTP domain. Because the OC can handle time transfer only in one direction from a single PTP port, it cannot be in the middle of the chain.

A BC is a clock that has more than one PTP port, some combination of ports in either master, slave (or passive) states. In simple terms, it receives timing information from an upstream master, recovers the clock, and distributes it downstream via its master ports. Given that it terminates one PTP flow and reinitiates a new flow between the master port and the slave port, a BC "resets" the PDV of the PTP messages. In modern telecom networks, the BC is normally the most common type of clock.

A TC is a clock that assists the accuracy of time transport by determining an accurate *residence time* for PTP event messages. The device measures this time (meaning the time taken for it to receive, process, and transmit the PTP event message) and writes this residence time into the correctionField of the PTP message. For example, the E2E TC does not terminate or interrupt the Sync and Delay_Req message flows between the master and slave ports but updates a field in the message that helps the slave clock to recover the time with increased accuracy.

A management node is included for completeness because it is not involved in the processing of PTP time information (and therefore is not really a clock) but is a device used to manage, configure, and monitor the clocks throughout the network. However, it is mentioned here because it is defined as a clock type in IEEE 1588-2008, and it processes PTP management messages, although PTP management messages are not used in the telecom profiles.

For further reference, the specification for ordinary and boundary clocks is covered in Chapter 9 of IEEE 1588-2008, and transparent clocks are covered in Chapter 10.

Grandmaster (Ordinary) Clocks

A grandmaster clock is the ultimate source of time for all the clocks in a PTP domain, whereas a master clock is a clock that is the source of time to which other clocks on a single PTP communications path synchronize. As shown in Figure 7-7, PTP clocks are organized into a master-slave synchronization hierarchy, but the clock at the top of the hierarchy of the entire system is the grandmaster clock. Remember that because an ordinary clock can have only one port, any ordinary clock must either be the grandmaster clock in a domain hierarchy or be an ordinary slave clock.

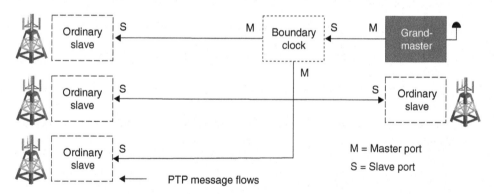

Figure 7-7 *Master-Slave and Grandmaster Hierarchy*

Note from Figure 7-7 that the grandmaster has a mushroom-shaped icon protruding from the right side of the box, representing the connection to a GNSS antenna (the antennas are frequently shaped somewhat like a mushroom). This icon indicates the presence of a GNSS receiver as either a source of frequency or a source of frequency, phase, and time—a PRC or a PRTC, respectively.

Because 5G mobile most often requires phase synchronization, you can assume this source of synchronization is most often a PRTC. The antenna icon indicates that the PRTC is a GNSS receiver (being the most common source), but it could also be some other device—such as an atomic clock. Just view it as a symbol of a valid source of time or frequency.

One option for deployment is that the GM has an external GNSS receiver (or atomic clock) and the GM receives frequency, phase, and time by way of three input signals, namely 10 MHz, 1PPS, and ToD. The GM (being a PTP entity) is therefore a separate device from the PRTC. Chapter 3 provides more details on the PRTC and input signals from sources of synchronization.

However, another deployment option is that the GM could have a built-in GNSS receiver and the device recovers the sources of synchronization internally. In this case, the grand-master is a combined GM+PRTC, and the only connector required is one to attach the external GNSS antenna (because the antenna is nearly always external to the device).

In the specifications for the telecom profiles for phase synchronization, the grandmaster is known as the *Telecom Grandmaster* (T-GM).

Slave (Ordinary) Clocks

Besides the grandmaster, the other form of ordinary clock is the slave clock. Again, the slave clock has only a single PTP port with a PTP connection to an upstream master port.

The slave may either listen for the upstream PTP flows (via some layer 2 or layer 3 multicast mechanism) or request unicast service of PTP messages via the *negotiation* process. In the case of unicast negotiation (over PTPoIP), the slave is configured with the IP address of the possible masters.

Specifically, for the telecom profiles, the slave clocks have two allowed transport mecha-nisms (depending on the profile—see details in the following sections):

■ PTP over layer 2 (Ethernet) multicast

■ PTP over IP (either IPv4 or IPv6)

With the L2 mechanism, the PTP port on an interface simply listens to one of the two standard multicast MAC addresses for PTP messages. There is no specific configuration to contact any master port on another PTP clock. The port simply subscribes to the traf-fic flow, and BMCA decides what state the port will enter (in this case, slave) and the port will start to recover the clock from the PTP messages.

In the L3 (IP) profiles, the transport is unicast (note that other non-telecom profiles may use multicast), so there is no ability to "subscribe" to a multicast address. In this case, the operator must configure the slave port with the IP address of a remote master port to contact. The slave port will then contact the PTP port on that IP address and request the packet flows it wants to receive. It would first ask for Announce messages to determine whether that master port is advertising itself as traceable to a grandmaster. If so, then it can ask to also receive timing messages and start to send Delay_Req messages to the master. See the earlier section "Negotiation" in this chapter for details of how this process works.

For redundancy, the operator would normally configure the slave port with several IP addresses so that the slave can always find a valid master to synchronize with should one or more of the connections fail.

Note that on layer 3 ports, you might not be able to predict the ingress (or egress) port of the PTP packets because that might be determined dynamically by the IP routing protocols. This can be a problem because the interface selected for sending and receiving PTP messages might not be activated for timestamping. This is especially so for devices either in the middle of the network or situated in rings, but for a router at a remote cell site, you would expect perhaps only a single link to the backhaul network.

Boundary Clocks

According to the official PTP definition, a BC is a clock with more than one PTP port. At the minimum, it has one slave port (recovering the clock from a remote master port) and one master port (passing the clock downstream to another slave port). Of course, it could have more than one master port because it is able to send PTP traffic in many directions.

Figure 7-8 illustrates how a BC fits into a timing distribution network. The PTP message flow from the master port (in this case on a GM) will be forwarded to the slave port of the BC. At the ingress port, those messages are timestamped, processed, and discarded. Similarly, messages in the reverse direction, such as Delay_Req, are created, the time of transmission is recorded, and the message is transmitted to the master.

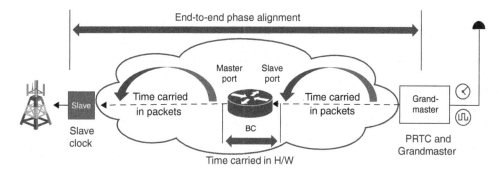

Figure 7-8   *PTP Boundary Clocks*

At the egress port, the mirror image of this process occurs. PTP messages are created, timestamped, and transmitted to the downstream slave. Similarly, traffic in the reverse direction, such as incoming Delay_Req, will be timestamped, answered (with a Delay_Resp), and discarded. Note that because the PTP message flow has been re-created, the PDV of the PTP message flow has been "reset" to zero.

However, internally to the BC, there are hardware connections to distribute time and frequency around the node from the ingress port, to/from the timing hardware to the egress port. This means that the transfer of time through a properly designed boundary clock is *predictable* because the transfer of time is not compromised by either other traffic, forwarding plane events or asymmetrical delays on the node.

The section "Hardware and Software Solution Requirements" in Chapter 12, "Operating and Verifying Timing Solutions," covers the hardware aspects of design, including more details of what is required to build an accurate clock. But in summary, when a properly designed boundary clock transfers time, it should add only a very small time error (in the order of tens of nanoseconds) to the end-to-end phase offset. Chapter 9 and Chapter 12 cover more details of the performance of telecom profile boundary clocks.

Transparent Clocks

There are two basic forms of transparent clocks: E2E TCs and P2P TCs. Those deployed in the telecom profiles are E2E TCs, but some industry profiles, such as that supporting the power industry, use P2P TCs. This section examines both in some detail, but understand that depending on the profile and industry use case, one or the other (or both) may not be allowed.

The idea behind using a TC is that by counting the amount of time the PTP event message was delayed, the slave can reduce the error in the recovered clock. Table 7-7 shows a simplified example of four messages that flow between a master and slave—they flow over the same path with two TCs between the master and slave.

Table 7-7   *Example Sync Messages*

Sync Message Time in Flight	Residence Time in TC-1	Residence Time in TC-2	Corrected Time in Flight
250 µs	20 µs	30 µs	250 − (20 + 30) = 200 µs
350 µs	100 µs	45 µs	350 − (100 + 45) = 205 µs
210 µs	5 µs	5 µs	210 − (5 + 5) = 200 µs
400 µs	50 µs	155 µs	400 − (50 + 155) = 195 µs

Given the four Sync messages were in flight for quite different times (between 210 and 400 µs), then most of those Sync messages were delayed in transit (as captured in the TC transit times columns). Therefore, the timestamps on the event messages have "aged" and at least three of them would be useless to help calculate accurate values for clock offset and mean path delay. But if we correct the timestamps for the residency time inside the two TCs, then one can see that the corrected time in flight is now in the range of 195–205 µs, which is much more useful.

So, how is this done?

End-to-End Transparent Clocks

An E2E TC allows PTP traffic to pass through but will update a field (the *correctionField*) in PTP event messages to adjust for the time that the message was resident in the TC. It does this by timestamping the message on ingress and again on egress and subtracts the ingress time from the egress time. That result of that subtraction is the time that the message is resident in the clock, and that value is then added to the *correctionField* in the PTP message. The correctionField in the Delay_Req is calculated and updated similarly.

Figure 7-9 outlines how the PTP messages flow through the various clocks in the E2E delay mechanism. Note that the BCs terminate the PTP flow, while the TCs retransmit the incoming PTP messages. Note that IEEE 1588-2019 now says that the PTP message in the TC is retransmitted as a new packet, although conceptually, it seems to pass through with only a modification of the correctionField.

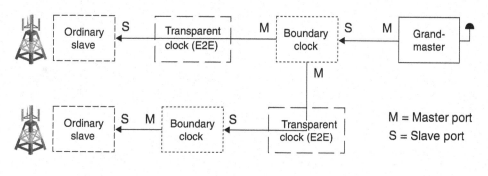

Figure 7-9 *PTP Clock Types—End-to-End Mechanism*

For the case of two-step masters (where the Sync message has the *twoStepFlag* set), the E2E TC will not update the correctionField in the Sync message but will instead update the same field in the associated Follow_Up message with the residence time for the Sync message. So, with one-step clocks, the t_1 timestamp and correctionField are both in the Sync message, but with two-step, the t_1 timestamp and correctionField would be both in the Follow_Up (but the timestamp it contains refers to the Sync message, not the Follow_Up).

That correctionField value is then read on the downstream slave port so that it can take residence time into account when recovering the time and phase. Remember that E2E is the only method used with telecom profiles. The details of this mechanism are explained in section 11.3 of IEEE 1588-2008.

Recall from the discussion of one-step and two-step clocks that there is a distinction between one-step and two-step TCs. As one example, if a one-step E2E TC receives a Sync message, it will update the correctionField with the residence time as normal. However, a two-step TC will only update the Sync message to indicate that there is a Follow_Up message associated with this Sync message (by setting the twoStepFlag). It will then generate and send a Follow_Up with the timestamp and correctionField filled in.

> **Note**   There are many detailed examples of the use of the correctionField with TCs, for each of the E2E and P2P cases with both one-step and two-step clocks, in IEEE 1588-2008 Annex C, "Examples of residence and asymmetry corrections."

Peer-to-Peer Transparent Clocks

The P2P TC treats the Sync and Follow_Up messages the same as the E2E case but drops any Delay_Req and Delay_Resp messages. Therefore, unlike the E2E case, the calculation of the mean path delay between the master and the slave ports cannot be done using the Delay_Req/Delay_Resp mechanism between the master and slave ports. Instead, the mean path delay is calculated between neighboring peers using the P2P mechanism. This value is added to Sync or Follow_Up messages passing through the P2P TC. Figure 7-10 shows how the P2P TC interacts with the boundary and ordinary clocks.

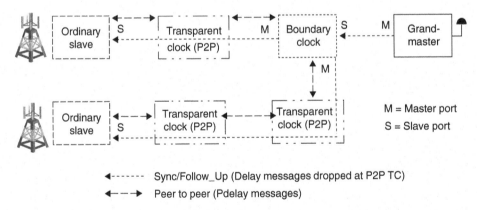

Figure 7-10   *PTP Peer Delay Topology and Message Flow*

Note in Figure 7-10 that PTP ports in master and slave state also support the P2P delay mechanism. The P2P TC treats the Sync and Follow_Up messages (short-dash arrows) the same as the E2E case but drops any Delay_Req and Delay_Resp messages.

With a P2P TC in the path, the calculation of the E2E mean path delay between the master and the slave ports is not possible because the Delay_Req and Delay_Resp are dropped by a P2P TC. Instead, the P2P mechanism calculates the link delay between the neighboring peers (long-dash arrows) using Pdelay messages, and the E2E TC adds this value to the correction field of the passing Sync or Follow_Up message.

The computation of link delay is based on an exchange of Pdelay_Req, Pdelay_Resp, and possibly Pdelay_Resp_Follow_Up messages with the peer, as shown in Figure 7-11. As a result of these exchanges, the link delay is known for each port of the P2P TC.

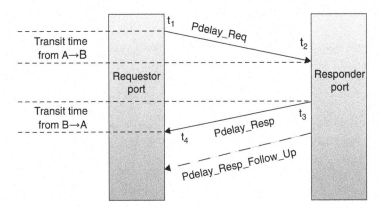

Figure 7-11   *PTP Peer-to-Peer Delay Mechanism*

The P2P method uses the timestamps in the same way as the Sync message case in the E2E method. So, the one-way link delay can be calculated by using the timestamps in the following formula:

$$meanPathDelay = (t_2 - t_1 + t_4 - t_3) / 2$$

Note that devices that do not support peer-to-peer are not allowed between P2P TCs (because this would skew the results of the calculation). It therefore follows that P2P and E2E TCs should not coexist in the same PTP domain. Also, the P2P delay mechanism is restricted to topologies where each P2P port communicates PTP messages with at most one other such port (point to multipoint does not work). One way to think of it is that it is purely a "hop-by-hop" mechanism.

So, in summary, the difference between the two mechanisms can be summarized as shown in Figure 7-12.

Note   There are many detailed examples of these message flows, for each of the E2E and P2P cases with both one-step and two-step clocks, in IEEE 1588-2008 Annex C, "Examples of residence and asymmetry corrections."

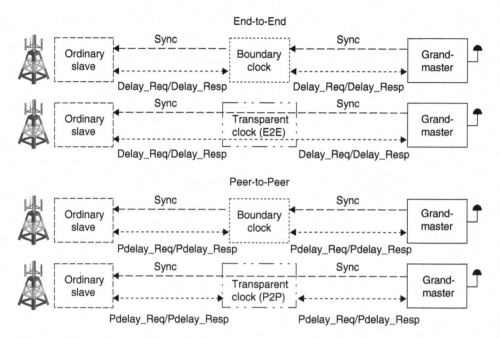

Figure 7-12 *End-to-End Versus Peer-to-Peer*

Management Nodes

PTP management nodes are mainly concerned with the processing of PTP management messages, which are used between the management nodes and clocks. These messages are used to query and update the PTP datasets maintained by clocks as well as configure PTP clocks and perform monitoring and fault management.

In summary, these messages perform actions such as GET and SET on a range of any one of about 50 TLV entities. Some of these TLVs apply to the whole clock and others apply only to a port on the clock. An example of a clock TLV is DEFAULT_DATA_SET, which is used to retrieve (GET) one of the datasets on the clock (see the earlier section "PTP Datasets"). An example of a port TLV is ANNOUNCE_RECEIPT_TIMEOUT, which is used to retrieve (GET) or update (SET) the value of the Announce timeout on a port.

This book does not cover management nodes in any more depth, as they are not widely used and their use is not allowed in the telecom profiles. See Chapter 15 of IEEE 1588-2008 for further details on management nodes and management messages.

Profiles

By now, you should realize that the IEEE 1588-2008 specification is like a Swiss Army knife in that it contains any number of tools that can be applied to a multitude of tasks. The earlier section "General Overview" mentioned that PTPv2 introduced the concept of profiles to PTP. Profiles allow an organization to design a specific implementation of PTP to apply to a specialized use case or application.

The multitude of possible options means there is no point requiring that a network element should "support 1588-2008" with no further elaboration of the details. The preceding sections should demonstrate how many different methods are available in the 1588 standard, so getting two PTP clocks to talk to each other means agreeing on a range of implementation options. There are domain numbers, TLV and dataset values, message rates, encapsulation, transport, and much more. To say a device "supports 1588" is like saying that it "supports routing." There are many different routing protocols, and they do not readily talk to each other.

Of course, there are many elements of 1588-2008 that must be followed, but no vendor will implement, validate, and support all the 1588 options under all circumstances. It makes sense that a smaller subset of the available options is chosen to implement a specific solution. It also makes predicting the performance more straightforward and vastly simplifies interoperability.

This section covers the most important profiles, although there is an emphasis on the telecom profiles, because they are the ones you are most likely to encounter in WAN topologies.

Default Profiles

When IEEE 1588-2008 introduced the concept of profiles, it planted a seed by proposing several "default profiles" that are described in Annexes A and J of the standard. These annexes contain a (small) number of parameter values to help implement a "default" PTP solution. Most parameters are given a value that should be the default (for example, the default domain number should be 0) and minimum supported range (for example, the log of Announce message rate should be in the range between 0 to 4). Vendors can support wider ranges if desired.

There are two choices of default profile:

- Annex J.3 Delay Request-Response Default PTP profile

- Annex J.4 Peer-to-Peer Default PTP profile

Obviously, J.3 uses the E2E delay request-response mechanism by default, whereas the Annex J.4 profile uses the P2P mechanism by default. Even though these two mechanisms are quite different, the standard allows links performing either mechanism in both profiles. Beyond that basic choice of mechanism and a few values and message rates, nothing else is stipulated. Note that Annexes J.3 and J.4 have moved to I.3 and I.4 in the 2019 edition of IEEE 1588.

So, as far the specification of transport and encapsulation, L2 or L3, Ethernet, IPv4 or IPv6, multicast or unicast, or some combination—nothing is proposed. Obviously, there needs to be significant further agreement to ensure two clocks, both supposedly supporting the default profile, could ever interoperate.

Note that sometimes you may run across the profile in some equipment known as *Telecom 2008*. This is a pre-standardized version of the Telecom profiles based loosely on the default profile that very closely resembles G.8265.1 (or G.8275.2). The authors frequently encounter it in mobile operators that are doing mostly frequency synchronization over packet, although it can work in the phase scenarios. Details are in Annex A of IEEE 1588-2008.

Also be aware that default profiles give very little guarantee as to the accuracy of the time recovered by the slave or the performance of the clocks (network elements) carrying time. A good way to think of default profile performance is that, after implementation, you may find that "your mileage may vary." A number of other profiles are designed with the aim to rectify that situation.

Telecom Profiles

There are three telecom profiles:

■ **G.8265.1**: Telecom profile for frequency synchronization

■ **G.8275.1**: Telecom profile for phase/time synchronization with full timing support from the network

■ **G.8275.2**: Telecom profile for phase/time synchronization with partial timing support from the network

The selection of the appropriate profile is based on the use case and network topology. Figure 7-13 gives an illustration of the different profiles. The first profile is for the distribution of frequency synchronization, so it shows only the master clock receiving a frequency signal (from a PRC/PRS), while the phase/time profiles are shown receiving phase and time, as well as frequency, from a PRTC.

The first profile specifically disallows PTP assistance from the network (meaning it does not allow boundary or transparent clocks). The second profile mandates that every node must be a PTP clock (boundary or transparent), and the third profile allows PTP assistance where it is possible.

Note how the bottom two phase/time profiles have a new set of names for their clocks, generally constructed by adding the word "Telecom" to the front of each IEEE 1588 term. These clock types are defined elsewhere in G.8275 and their features and performance are specified in the G.8273.*x* family of recommendations—see Chapter 8, "ITU-T Timing Recommendations," for more details of how the ITU-T recommendations are structured.

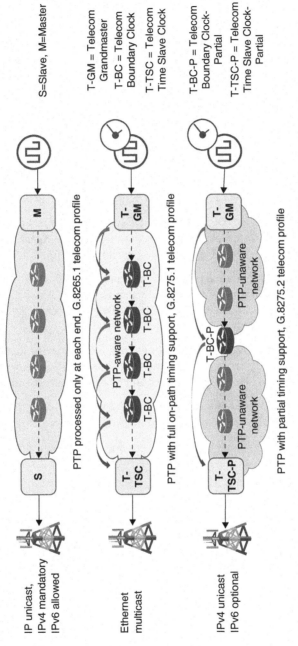

S=Slave, M=Master

T-GM = Telecom Grandmaster
T-BC = Telecom Boundary Clock
T-TSC = Telecom Time Slave Clock

T-BC-P = Telecom Boundary Clock-Partial
T-TSC-P = Telecom Time Slave Clock-Partial

PTP processed only at each end, G.8265.1 telecom profile

IP unicast, IPv4 mandatory IPv6 allowed

PTP with full on-path timing support, G.8275.1 telecom profile

Ethernet multicast

PTP with partial timing support, G.8275.2 telecom profile

IPv4 unicast IPv6 optional

Figure 7-13   *The Three Telecom Profiles*

G.8265.1—Telecom Profile for Frequency Synchronization

This recommendation defines a profile intended to be used to carry frequency across a packet network and where it is not an option to do that with SyncE. As shown in Figure 7-13, this profile applies where the network elements have no support for PTP in any intermediate node between the PTP master and the PTP slave. Of course, given that limitation, using this profile over such a network does not guarantee that the performance requirements of a given application will be met.

The main features of this profile are as follows:

■ Domain number should be within a range of 4–23 with a default value of 4.

■ Unicast UDP over IPv4 (Annex D of IEEE 1588-2008), and optionally PTP over IPv6 (Annex E of 1588-2008), can be used to transport the PTP messages.

■ Message rates for Sync, Delay_Req, and Delay_Resp are from a minimum of 1 packet per 16 seconds to 128 per second. The message rate for Announce is from a minimum of 1 packet per 16 seconds to 8 per second with a default of 1 every 2 seconds.

■ Unicast message negotiation is used—a mechanism allowing slaves to request timing service from a master in a unicast environment (see the earlier section "Negotiation" in this chapter and clause 16.1 of IEEE 1588-2008). The default interval to request service is 300 seconds, with a range of 60–1000 seconds.

■ Defines an alternate BMCA (aBMCA), with the main feature being that the clock state is always static, so a master is always a master and a slave is always a slave— there is no dynamic changing of the port state between master and slave.

■ One-step and two-step clocks are both allowed, with the master port free to use whatever method it prefers, and the slave required to accept either.

■ One-way operation to recover frequency is allowed, although some implementations continue to use a two-way system (see the section "One-Way Versus Two-Way PTP" earlier in this chapter).

■ Packet timing signal fail (PTSF) is defined, a condition where either the Announce messages from the master are lost (PTSF-lossAnnounce); or the timing messages are lost (PTSF-lossSync); or the messages are outside the input tolerance of the slave (PTSF-unusable). PTSF-unusable usually comes about because of excessive PDV of the Sync messages. These signal failure events trigger the algorithm to select another reference with the highest quality level that is not experiencing PTSF.

■ Quality (clockClass) values with a range of 80–110 (quite a different range from most other profiles) are supported. These values are used to carry clock quality level information from the master to the slave. See the following bullet point.

■ Allows interoperability with other synchronization networks (such as SyncE and SDH). The profile does this by mapping QL values (SSM or ESMC) to the clockClass field of the Announce message. Doing this allows full traceability of the clock back to the PRC/PRS, even though the timing signal may have crossed other networks such as SDH and SyncE.

These last two points are the primary benefit of deploying G.8265.1 for frequency. Table 1 of G.8265.1, reproduced in this book in Table 9-2, shows the mapping between the various methods of transporting QL information: the synchronization status message (SSM), G.781 (Option I, II and III) and PTP G.8265.1 telecom profile.

G.8265.1 is deployed as PTPoIP over an Ethernet, IP, and MPLS network with the nodes between the master and slave simply switching the frames like any other. Of course, the network should be conditioned with quality of service (QoS) to try and reduce the PDV of these time-sensitive messages.

The slave is provisioned with the IP address of any potential master clocks and the slave sets up the unicast negotiation, asking for a stream of Announce messages and then timing messages. This is all statically defined and configured; there is no discovery mechanism. In the case where the slave is provisioned with multiple master IP addresses, the following parameters contribute to the master selection process:

- Clock quality level (clockClass)

- PTSF status

- Local priority configured on the slave

This profile is widely used in two use cases. The first is circuit emulation (CEM), whereby the legacy synchronous networks are replaced by packet transport and SyncE is not available. The second is the 3G and 4G mobile networks that need frequency synchronization but, again, SyncE is not an option.

G.8275.1—Telecom Profile for Phase/Time Synchronization with Full Timing Support from the Network

This recommendation defines a profile intended to be used to carry phase/time across a packet network in combination with a physical form of frequency (commonly SyncE). As shown in Figure 7-13, this profile only applies to the topology that features full timing support from the network (every network element is a PTP clock).

In this profile, the grandmaster clocks are known as Telecom Grandmasters (T-GM), the boundary clocks are known as Telecom Boundary Clocks (T-BC), and the slave clocks are known as Telecom Time Slave Clocks (T-TSC). There is also the ability to deploy (E2E) Telecom Transparent Clocks (T-TC).

Using this profile over a network containing only PTP clocks allows the accuracy of the distributed time to be tightly controlled. That is because the performance limits of the T-BC and T-TSC clocks are specified in another recommendation (G.8273.2) and every node is processing PTP. This is not the case in G.8265.1 and not necessarily true for G.8275.2.

It is not the profile itself that guarantees any performance because it merely describes the functions of the PTP. Other recommendations, such as those covering the clock characteristics and performance and the E2E network topology, are concerned with performance. Chapter 8 covers the interactions between these various documents in more detail.

The recommendation for this profile specifies the characteristics and behavior of the following clock types from G.8275:

- **Telecom Grandmaster (T-GM):** A 1588-2008 grandmaster ordinary clock with a single master-only port with characteristics from G.8272.

- **Telecom Grandmaster (T-GM):** Another grandmaster with the same characteristics, but it has multiple (master only) ports. Because of 1588-2008 rules, any clock with more than one port is a boundary clock, so according to that rule, this device is a 1588-2008 boundary clock acting as an (ordinary clock) T-GM.

- **Telecom Boundary Clock (T-BC):** A boundary clock that can be a master or a slave to another GM with performance characteristics from G.8273.2.

- **Telecom Transparent Clock (T-TC):** An E2E TC as defined in 1588-2008 and with performance characteristics in G.8273.3.

- **Telecom Time Slave Clock (T-TSC):** An ordinary slave clock with performance characteristics as defined in G.8273.2.

The main features of this profile are as follows:

- Domain number should be within a range of 24–43 with a default value of 24.

- Multicast Ethernet (Annex F of 1588-2008) is used to transport the PTP messages. Both the non-forwardable multicast address 01-80-C2-00-00-0E and the forwardable multicast address 01-1B-19-00-00-00 are supported. As outlined in Annex H of this profile, PTP can also be carried by the optical transport network (OTN).

- VLAN tagging is expressly forbidden and so there is no class of service (CoS) as there is no VLAN header to carry it, although there are some exceptions for T-TCs. Further information on QoS mechanisms are contained in the Chapter 9.

- Message rates for Sync, Delay_Req, and Delay_Resp are fixed at 16 messages per second, while the message rate for Announce is also fixed at 8 per second. There is no negotiation for timing service; the messages are constantly sent to whichever of the two multicast addresses is configured.

- Defines an alternate BCMA (aBMCA), allowing PTP ports to be dynamically switched between master and slave states, which enables the profile to support automatic configuration. This mechanism allows the aBMCA to establish an efficient PTP hierarchy based on the Announce messages, discovering the shortest path to the grandmaster. This is especially helpful in ring topologies, and during failure events and network rearrangements. See section 6.3.1, "Alternate BMCA," and Annex B of G.8275.1 for further details. Remember that a T-GM only has ports in master state (they cannot be slave) and T-TSC ordinary clocks can only have PTP ports in slave state.

- Requires a physical form of frequency (normally SyncE) in combination with PTP. This is commonly referred to as "hybrid mode" because it is a hybrid of packet and physical methods.

- One-step and two-step clocks are both allowed, with the master free to use whatever method it prefers, and the slave required to accept either.

- Required to use the two-way method, as this is needed to recover phase. Delay_Req and Delay_Resp messages are used, but the Pdelay message set is not used, so there are no P2P TCs allowed in this profile.

- Quality (clockClass) values have specific values within the range of 6–255 (depending on the clock type). When operating normally, a value of 6 is expected across the network.

- Local priority can be configured on each PTP port of a clock to act as a tiebreaker in the case where multiple paths to a grandmaster are available. This is used where the operator wishes to define a preferred path.

- Fixes the value of priority1 at 128 but allows the value of priority2 to be configured in the range of 0–255 for T-GM and T-BC clocks (default of 128) and fixed at 255 in T-TSC clocks. The aBMCA selects a port receiving the higher priority2 (lower value) to be the slave to a grandmaster if the clock quality is otherwise equal. Appendix IV of G.8275.1 describes possible use cases for the priority2 attribute.

G.8275.1 is very easy to configure and deploy—do not let the reference to multicast put you off. For Cisco routers running IOS-XR, PTP and SyncE are enabled on the physical interface with a few lines of configuration and that is all that needs to be done. The PTP messages are sent out that interface to the multicast MAC address. When the interface at the other end of the link is configured to run PTP, it will receive and process these messages—terminating the PTP flow.

This profile is designed to allow selection of the best reference source based on clock quality traceability (combined with an optional configurable local priority), which helps it automatically establish an optimal network topology for the time distribution. This process is rerun whenever there is a failure event or rearrangement in the network, in order to converge on a new topology.

The T-BC or T-TSC runs a single PTP clock that has one or more PTP ports. The aBMCA in the clock processes the Announce messages that it receives on its ports and decides what state each port should be in. Because this profile is only communicating over a single hop to its neighbor, every PTP port could be receiving different Announce information. Based on the Announce message and the configured local priority, the aBMCA will then place each port in either master, slave, or passive state.

After the aBMCA surveys the ports, the one selected to be a source of time will be placed in slave state, and the other candidates will be placed in passive state (there is only one port in slave state at any one time). Other ports are placed in the master state (to give downstream clocks the possibility to accept time). See the earlier sections "The Announce Message" and "Best Master Clock Algorithm" for more details.

The main use for this profile is to carry frequency, time, and phase across a WAN, and today it is most commonly deployed for mobile technologies that need phase synchronization (especially 5G and time division duplex (TDD) radios).

G.8275.2—Telecom Profile for Phase/Time Synchronization with Partial Timing Support from the Network

This recommendation defines a profile intended to be used to carry frequency and phase/time across a packet network and where SyncE is optional. This profile is based on a network with partial timing support (PTS) or the special case of assisted partial timing support (APTS).

Using this profile over a PTS network where the elements are PTP unaware, and are merely switching PTP messages as IP, does not allow the accuracy of the distributed time to be controlled. However, the more PTP-aware elements that are included in the network, the better the performance will be. Additional considerations and details on PTS versus APTS are covered in the G.8271.2 recommendation on network topology.

This recommendation specifies the characteristics and behavior of the following clock types from G.8275:

- **Telecom Grandmaster (T-GM):** A 1588-2008 grandmaster ordinary clock with a single master-only port.

- **Telecom Grandmaster (T-GM):** Another grandmaster with the same characteristics, but that has multiple (master only) ports. Because of 1588-2008 rules, any clock with more than one port is a boundary clock, so according to that rule, this device is a 1588-2008 boundary clock acting just like a (ordinary clock) T-GM.

- **Telecom Boundary Clock-Partial (T-BC-P):** A boundary clock with partial support from the network. The difference between the T-BC-P and the multi-port T-GM is that the T-BC-P can be a slave to another clock, whereas a T-GM can never be.

- **Telecom Boundary Clock-Assisted (T-BC-A):** A boundary clock with only partial support from the network but assisted by a local time reference (such as a GNSS receiver) as a primary source of time. A slave port recovering clock over the PTS network is only used as a reference upon failure of the local time source. Clock feature and performance characteristics of the T-BC-A are covered in G.8273.4.

- **Telecom Transparent Clock-Partial (T-TC-P):** An E2E TC as defined in 1588-2008 with only partial support from the network.

- **Telecom Time Slave Clock-Partial (T-TSC-P):** An ordinary slave clock (and therefore only one PTP port) with partial support from the network.

- **Telecom Time Slave Clock-Assisted (T-TSC-A):** An ordinary slave clock with partial support from the network but assisted by a local time reference (such as a GNSS receiver) as a primary source of time. A slave port recovering clock over the PTS network is only used upon failure of the local time source. Performance characteristics of the T-TSC-A are covered in recommendation G.8273.4.

Note that when the clock is either a T-TSC-P or T-TSC-A, it is an ordinary clock with either a single slave port or alternatively with multiple PTP ports, which according to

IEEE 1588 definitions makes it a boundary clock. On a T-TSC with multiple PTP ports, only one port can be in a slave state at any one time, and no port can go to master state.

The main features of this profile are as follows:

- Domain number should be within a range of 44–63 with a default value of 44.

- Unicast UDP over IPv4 (Annex D of 1588-2008), and optionally PTP over IPv6 (Annex E of 1588-2008), is used to transport the PTP messages.

- Message rates for Sync, Delay_Req, and Delay_Resp are from a minimum of 1 packet per second to 128 per second. The message rate for Announce is from a minimum of 1 packet per second to a maximum of 8 per second.

- Unicast message negotiation—a mechanism allowing slaves to request timing service from a master in a unicast environment (see the earlier section "Negotiation" in this chapter and clause 16.1 of 1588-2008). The default interval to request service is 300 seconds, with a range of 60–1000 seconds.

- Defines an alternate BCMA (aBMCA), allowing PTP ports to be dynamically switched between master and slave states, which enables the profile to support automatic configuration. See section 6.7.1, "Alternate BMCA," and Annex B of G.8275.2 for further details on using an aBMCA to establish the PTP hierarchy. The mechanisms of G.8275.1 and G.8275.2 are quite similar; however, the effect of using a PTS network to transport messages means that the outcome of this selection process might not be optimal, since not every upstream node is processing PTP (and the aBMCA has no visibility of that topology).

- One-step and two-step clocks are both allowed, with the master free to use whatever method it prefers, and the slave required to accept either.

- The slave port can decide to request traffic for either one-way or two-way traffic from the remote master port based on its requirements. The master must be able to support both methods. For two-way operation, the Delay_Req and Delay_Resp messages are used, but the Pdelay message set is not used, so there are no P2P TCs allowed in this profile.

- Quality (clockClass) values have specific values within the range of 6–255 (depending on the clock type). When operating normally, a value of 6 is expected across the network.

- Local priority can be configured on each PTP port of a clock to act as a tiebreaker in the case where multiple paths to a grandmaster are available. This is used where the operator wishes to define a preferred path.

- Packet timing signal fail (PTSF) whereby the timing messages from the master are either lost (PTSF-lossSync) or outside the input tolerance of the slave (PTSF-unusable). There is a new port dataset member (portDS.SF) that indicates that the port (if the value is TRUE) has suffered a signal failure and is no longer capable of being considered as a candidate port in the decision to select the best master.

- Fixes the value of priority1 at 128 but allows the value of priority2 to be configured in the range of 0–255 for T-GM, T-BC-P, and T-BC-A clocks (default of 128) and fixed at 255 in T-TSC-A and T-TSC-P clocks. The aBMCA selects a port receiving the higher priority2 (lower value) to be the slave to a grandmaster if the clock quality is otherwise equal. Appendix I of G.8275.2 describes possible use cases for the priority2 attribute.

G.8275.2 is deployed as PTPoIP over an Ethernet, IP, and MPLS network with the unaware nodes between the master and slave port simply switching the frames like any other. Of course, the network should be designed to minimize any impact that this unaware switching has on the delivered timing accuracy—more on that in Chapter 9.

G.8275.2 is straightforward to configure and deploy, although in a large network, the manual provisioning of the addresses of the closest masters and management of the IP addresses can be complicated and burdensome. For Cisco routers running IOS-XR, PTP is enabled on the L3 interface with a few lines of configuration (and optional SyncE on the physical interface).

Like G.8265.1, the topology is all statically defined and configured, and there is no discovery mechanism. The slave port is provisioned with the IP addresses of any potential master clocks, and it sets up the unicast negotiation, asking for a stream of Announce messages and then timing messages. However, unlike G.8265.1, this profile allows boundary clocks, and their ports can change state, just like G.8275.1. Based on the Announce message and the configured local priority, the aBMCA will place each port in either master, slave, or passive state. See the earlier sections "The Announce Message" and "Best Master Clock Algorithm" for more details.

Again, like G.8275.1, this profile is designed to allow selection of the best reference source based on clock quality traceability (combined with an optional configurable local priority), which helps it establish an optimal network hierarchy for the time distribution. This process is rerun whenever there is a failure event or rearrangement in the network, in order to converge on a new topology. The major difference here is that the best master cannot be "discovered"; it must belong to one of the IP addresses provisioned on the slave port by the operator.

Although the process is basically the same as G.8275.1, the outcome can be quite different and not optimal. Take the topology illustrated in Figure 7-14. In this case, the operator configures the slave with the IP addresses of two masters. One of those takes a path that is well supported by BCs and the other goes through a network of indeterminate quality.

The slave is provisioned to communicate with two separate T-GMs, each with its own IP address. One expects that the network, fully provisioned with boundary clocks between the T-GM (IP address of 10.2.2.2) and the slave, would deliver the most accurate clock. The unaware network between the slave and the T-GM (IP address of 10.1.1.1) is likely to be producing bad PDV with substantial asymmetry.

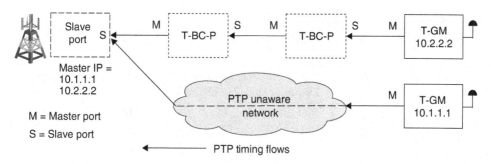

Figure 7-14 *Example Network with G.8275.2 and PTPoIP*

However, with all else being equal, it is very likely the aBMCA will select the GM across the unaware network because it is "closer" to the T-GM when comparing the steps Removed parameter. This situation would not arise with a deployment of the G.8275.1 profile. The operator must estimate and provision the best path manually using local priorities, or some other mechanism, because PTP does not know anything about the network path beyond an IP address. Managing the configuration on the slave to select the best path to the closest master is a significant operational burden in a large network.

The other major difference between G.8275.1 and G.8275.2 is the treatment of the PTSF event, which is a means to signal the failure of the PTP messages on the slave. Although G.8275.1 relies on the Announce timeout mechanism to indicate loss of packet timing, the G.8275.2 mechanism is more thorough, which is not surprising considering the use of a wide-area routed network as a transport. Note that the very latest revisions to G.8275.1 are introducing the PTSF concept to this profile as well.

The use case for G.8275.2 is those circumstances where an operator needs frequency and phase timing and yet does not have the ability to have a network offering full on-path support. This situation arises when an operator may not own their own transport network and is relying on circuits carried over third-party networks. A very common scenario is to use G.8275.2 as a backup to a GNSS receiver at the cell site using the APTS approach. Chapter 9 provides much more information on this topic.

Profiles for Other Industries

This chapter has covered the original IEEE 1588-2008 and the telecom profiles in some depth; however, there are other profiles specifically designed to address additional use cases or unique industry specifications. This section covers these profiles at a high level to help you understand the differences compared to what has already been covered. There is ample material available on the Internet for each of these profiles, and given the understanding of profiles that you have gained up to this point, understanding them should be straightforward.

Table 7-8 gives a summary of the major profiles, the supporting SDO, their adoption status, and the basic characteristics of each.

Table 7-8 *Characteristics of Major Profiles*

Industry	SDO	Status	Profile	Transport/ Type	Clocks
1588*	IEEE	2019	Annex I.3 Delay Request-Response Default PTP profile	Undefined E2E	OC, BC, TC
1588*	IEEE	2019	Annex I.4 Peer-to-Peer Default PTP profile	Undefined P2P	OC, BC, TC
Telecom	ITU-T	2014	G.8265.1 telecom profile for frequency synchronization	IPv4, IPv6 E2E	OC
Telecom	ITU-T	2020	G.8275.1 telecom profile for phase/time (full support)	Ethernet E2E	OC, BC, TC
Telecom	ITU-T	2020	G.8275.2 telecom profile for phase/time (partial support)	IPv4, IPv6 E2E	OC, BC, TC
Enterprise	IETF	Draft	TICTOC enterprise PTP profile	IPv4, IPv6 E2E	OC, BC, TC
Industrial	IEC	2016	IEC 62439-3 PTP Industry Profile (Annex C) layer 3 E2E	IPv4 E2E	OC, BC, TC
Industrial	IEC	2016	IEC 62439-3 PTP Industry Profile (Annex C) layer 2 P2P	Ethernet P2P	OC, BC, TC
Industrial	IEC	2016	IEC 62439-3 PTP Industry Profile (Annex B) with HSR and PRP	Ethernet P2P	OC, BC, TC
Power	IEC	2016	IEC 61850-9-3 Power Utility Automation (optional HSR and PRP)	Ethernet P2P	OC, BC, TC
Power	IEEE	2017	C37.238-2017 "Power profile" Revision of C37.238-2011	Ethernet P2P	OC, BC, TC
AudioVideo	IEEE	2020	802.1AS TSN generalized PTP Profile (gPTP)	Ethernet P2P	OC, BC
Audio	AES	2018	AES67 Media Profile	IPv4 E2E/(P2P)	OC, BC
Video	SMPTE	2015	ST-2059-2 Broadcast Profile	IPv4/IPv6 E2E/(P2P)	OC, BC, TC
Timing	CERN	2011	White Rabbit v2.0	Undefined E2E	OC, BC

*Note that Annexes J.3 and J.4 have moved to I.3 and I.4 in the 2019 edition of IEEE 1588.

Here is a list of the major PTP profiles and their background:

- IEEE 802.1AS-2020: Timing and Synchronization for Time-Sensitive Applications. Generalized PTP (gPTP) is a profile covered by the Time-Sensitive Networking (TSN) Task Group from the IEEE (TSN was formerly known as Audio Video Bridging

[AVB]). In 2020, the TSN Task Group of IEEE 802.1 published an approved revision to the original 2011 standard.

■ PTP Industry Profile (PIP) from the International Electrotechnical Commission (IEC) for Layer 2 Peer-to-Peer (L2P2P) and Layer 3 End-to-End (L3E2E) transport. The IEC 62439-3 (2016) standard outlines the two profiles in Annex C. It also contains Annexes A and B that outline how to use Parallel Redundancy Protocol (PRP) and High-availability Seamless Redundancy (HSR) to create doubly attached (redundant) PTP clocks.

■ IEC 61850-9-3 (2016) Power Utility Automation Profile (PUP). This profile was adapted from the L2P2P profile from IEC 62439-3 (2016) Annex C and included in this joint IEC/IEEE standard. The use of doubly attached clocks is optional in this profile.

■ IEEE C37.238-2017. Known as the "Power Profile," this profile takes the preceding PUP and extends it to define the use of PTP in the power system, electrical substations, and across the distribution grid.

■ SMPTE ST-2059-2 and AES67 Media Profiles. This pair of profiles from two different industry organizations (the Society of Motion Picture and Television Engineers and the Audio Engineering Society) support the dissemination of audio and video content in a professional and broadcast setting.

■ PTP Enterprise Profile, RFC draft-ietf-tictoc-ptp-enterprise-profile-18 (might be a later draft version depending on when you are reading this). This Enterprise Profile is intended to work in large enterprise networks, for example, in financial corporations, where the time accuracy needed is tighter than available from NTP.

■ White Rabbit (WR). This is an extension to PTP that was developed to synchronize nodes in a packet-based network with sub-nanosecond accuracy. It makes use of PTP 1588-2008 combined with SyncE and a link setup and calibration method to address link asymmetry (that requires specialized hardware).

IEEE 802.1AS-2020: Timing and Synchronization for Time-Sensitive Applications: Generalized PTP (gPTP)

Time-Sensitive Networking (formerly Audio Video Bridging) is a set of standards from the IEEE that deliver deterministic behavior (for example, by controlling latency) on an Ethernet network to support time-sensitive applications. Applications can include audio/video streaming, low-latency industrial sensors, and real-time process control. TSN encapsulates a set of standards to improve the deterministic nature of the packet network in the following categories:

■ Time synchronization between the network elements (this profile)

■ Scheduling and traffic shaping across the E2E path

■ Path selection, reservation, and failover

The time section of these standards is contained in IEEE 802.1AS-2020, which includes a profile called "IEEE 802.1AS PTP profile for transport of timing over full-duplex, point-to-point links." This profile defines the generalized PTP (gPTP) profile as a tightly defined subset of the complete set of options in the IEEE 1588-2008 PTP document. You can obtain the 2020 version of gPTP, as well as the previous (now revised) 2011 version, and the corrigenda from the https://standards.ieee.org website.

From the introduction of IEEE 802.1AS-2000:

> This standard specifies the protocol and procedures used to ensure that the synchronization requirements are met for time-sensitive applications, such as audio and video, across bridged and virtual bridged local area networks consisting of LAN media where the transmission delays are fixed and symmetrical.

This standard requires that any two time-aware systems separated by six or fewer time-aware systems (meaning up to seven hops) will be synchronized to within 1 μs (peak-to-peak) of each other during normal steady-state operation. Message rates are fixed with Announce and Pdelay_Req at 1 per second, while Sync is 8 per second. The domain number is fixed at 0 and the values of priority1 are limited to only a few values (non-GM nodes are fixed at 255).

The differences between the gPTP profile and IEEE 1588-2008 include the following points:

- gPTP defines performance characteristics (in Annex B), while 1588-2008 does not. There are requirements for the local (free-running) clocks, the time-aware systems (clocks), and the E2E performance.

- gPTP requires PTP transport with layer 2 (1588 Annex F), whereas 1588-2008 allows higher-layer transport (including IPv4/IPv6 in Annexes D and E). The 802.1AS document is addressed to an audience comfortable with 802 standards and covers PTP from that aspect (meaning it is not easy reading otherwise).

- gPTP defines PTP clocks as time-aware systems; ordinary clocks as time-aware "end stations"; and boundary clocks as time-aware "bridges." These time-aware nodes need to support the P2P delay mechanism and act much like the P2P TC from 1588-2008 (but participate in the BMCA). Therefore, with this P2P requirement, there is no possibility to mix unaware nodes into a timing distribution network.

- gPTP requires the use of two-step processing (with Follow_Up and the Follow_Up from the Pdelay_Resp), while 1588-2008 allows both one-step and two-step processing.

- The BMCA is the default BMCA from 1588-2008 with some small extensions.

Ports, at the direction of the BMCA, can adopt one of four roles:

- **Master:** The port closest to the source of time from any downstream time-aware system

- **Slave:** The port closest to the root time-aware system

- **Passive:** Any port that is not master, slave, or disabled

- **Disabled:** A port that is disabled or not capable of 802.1AS processing

Figure 7-15 illustrates the hierarchy of a timing network built using the BMCA from 802.1AS-2020 in combination with the Announce messages (a good way to think of it is that it is much like the Rapid Spanning-Tree Protocol).

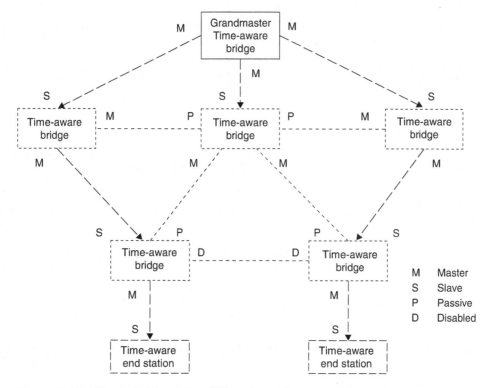

Figure 7-15 *Example Hierarchy of Time-Aware Systems with Port States*

TSN becomes a topic again when discussing the timing requirements for the 5G radio access network (RAN) fronthaul network. The IEEE TSN Task Group has developed the IEEE 802.1CM-2018 standard "Time-Sensitive Networking for Fronthaul," which enables the transport of time-sensitive fronthaul traffic in the RAN by Ethernet networks. See the section "IEEE TSN" in Chapter 4, "Standards Development Organizations," for more details on TSN and TSN profiles.

The IEEE 802.1CM standard defines performance profiles that specify features, options, configurations, defaults, protocols, and procedures of Ethernet devices that are necessary to build networks capable of transporting radio traffic. Note that the RAN fronthaul network uses TSN technologies in 802.1CM, but it does not use the 802.1AS PTP profile.

The TSN Task Group is also working on two profiles where precise time synchronization plays an exceptional role:

- **IEC/IEEE 60802**: TSN Profile for Industrial Automation

- **IEEE P802.1DG**: TSN Profile for Automotive In-Vehicle Ethernet Communications

The use cases and strict requirements in IEC/IEEE 60802 have triggered new work on an amendment for gPTP hot-standby services, meaning the ability to use independent backup GMs with multiple domains.

IEC 62439-3 (2016) PTP Industry Profile (PIP)

IEC 62439-3:2016 specifies two redundancy protocols (HSR and PRP) designed to provide seamless recovery in case of a single failure in a network, either of a link or a network element. This redundancy is based on the parallel transmission of duplicated information using "doubly attached" network elements connected by one of these two methods (or a combination of them).

Besides defining the PRP and HSR methods, on the topic of timing, the standard defines a PTP Industry Profile (PIP) to synchronize an industrial (Ethernet) network to sub-microsecond-level accuracy. Two variants of the PIP profile are specified in Annex C of the standard:

- L3E2E (layer 3, end-to-end) for clocks operating on L3 networks with E2E delay measurement (from 1588-2008 Annex J.3)

- L2P2P (layer 2, peer-to-peer) for clocks operating on L2 networks with P2P delay measurement (from 1588-2008 Annex J.4)

The IEC, in cooperation with the IEEE, has adopted the L2P2P version as the Power Utility Automation Profile (PUP) and published it jointly as IEC/IEEE 61850-9-3 (see the following section). The SDOs have agreed to keep these two documents aligned. Note that the IEC standards refer to IEC 61588:2009 (edition 2.0) when referring to PTP, which is just the IEC's adopted version of IEEE 1588-2008. The IEC is adopting 1588-2019 as edition 3 without changes.

Annex A specifies how PTP clocks are attached over redundant, simultaneously active paths using PRP or HSR with no single point of failure at the network level. It is quite a detailed annex (about 45 pages) showing how PTP is implemented in combination with these protocols (it is not a profile). The idea is for PTP to run on top of HSR/PRP to make use of its redundancy features.

However, Annex A extends the default BMCA because of the need to support doubly connected ports synchronized by the same master. Because messages from the same grandmaster appear on two different interfaces, there needs to be an application-specific process to select the port with the best clock quality. Any PTP clock that is doubly connected needs to support this extension.

Annex B defines a new profile for doubly attached clocks named "IEC/IEEE 61850-9-3 Precision Time Protocol profile for Power Utility Automation." It is simply the same L2P2P profile from Annex C that has an additional requirement that the clocks must be doubly attached (using the methods in Annex A). The difference between 61850-9-3 (also copied from Annex C) allows the PTP clocks to be doubly attached (redundancy clause 9), but does not require it, whereas the 62439-3 Annex B profile requires it.

IEC 62439-3:2016 also acts as the foundation for a range of other standards used to deal with packet-based time synchronization for the power industry. Figure 7-16 illustrates the high-level interrelationships between these various PTP standards. Note that the IEC uses its own published PTP standard (IEC 61588) as the PTP reference rather than IEEE 1588.

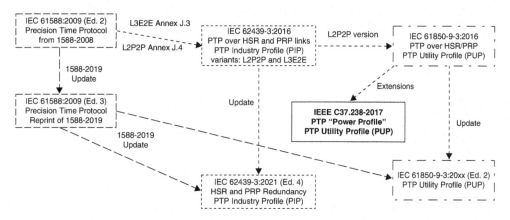

Figure 7-16   *Relationship Between PTP Standards for the Power Industry*

There have been three "editions" of the IEC 62439-3 standard, in 2010, 2012, and 2016. The 2016 edition is a technical revision that replaces the other editions, and there is expected to be a fourth edition published in 2021 that aligns with edition 2 of the 61850-9-3 profile and addresses changes from the new IEEE 1588-2019 version.

These profiles make a lot of use of transparent clocks and have a stricter requirement on their performance than that of boundary clocks. This profile defines numerous PTP parameters to be specific values, such as restricting the message interval for Announce, Sync, and Pdelay_Req to be 1 per second.

IEC 61850-9-3 (2016) Power Utility Automation Profile (PUP)

As mentioned in the preceding section on IEC 62439-3, this profile is just a replication of the L2P2P profile from Annex C of that standard, with the optional requirement to support HSR/PRP redundancy outlined in Annex A of 62439-3. The goal of this profile is to achieve network time inaccuracy better than ±1 µs after crossing approximately 15 TCs or 3 BCs.

This PUP profile utilizes layer 2 multicast communication according to IEEE 1588-2008 (Annex F) and uses the P2P delay measurement method derived from the default P2P profile in IEEE 1588-2008, Annex J.4—although with a restricted range of values. Like its parent, this profile sets numerous PTP parameters to specific values, such as the message interval for Announce, Sync, and Pdelay_Req is 1 per second.

With singly attached clocks, this is a somewhat restricted subset of the 1588-2008 default profile with the use of the default BMCA. When clocks incorporate the optional double attachment, then the profile makes use of the extended BMCA as specified in IEC 62439-3:2016, Annex A.

These PIP and PUP profiles make substantial use of TCs and have stricter requirements on TC performance than that of boundary clocks. In these power topologies, a boundary clock is a clock that has ports in two or more domains. The boundary clock synchronizes to a grandmaster in one domain and acts as a grandmaster in another domain.

The BC is used to time-synchronize two or more separate networks infrastructures to one grandmaster without the need of bridging data packets between the networks. One way to think of it is that they act somewhat like a router connecting two LAN broadcast domains together.

The IEC/IEEE published PUP in 2016 and it has a stability date of 2020, which means that a new version is expected in 2021. As this profile is a copy of the IEC 62439-3 PIP profile, the preparatory work has been undertaken by that group in the IEC, and it is reflected in a new version of 61850-9-3 (edition 2). Among the updates is that it aligns with the new PTPv2.1 version of IEEE 1588-2019.

Devices implementing the PUP profile tend to be equipment specialized for substation and power grid monitoring like intelligent electronic devices (IED) and phasor measurement units (PMU). On the other hand, the profile is not so widely implemented in general-purpose routers deployed by service providers.

Every 2 years, the industry gets together to perform IEC 61850 interoperability (IOP) testing activities, with (as of writing) the last taking place in October 2019. Part of that testing is to test the performance and capability of PTP implementations of the substation equipment. At the 2019 event, there were 13 companies that provided IEC 61850-9-3/IEEE C37.238-compatible equipment, and 13 companies also provided compatible applications or IEDs.

IEEE C37.238-2011 and 2017 Power Profile

The Power Profile is defined in IEEE C37.238-2017, "IEEE Standard Profile for Use of IEEE 1588 Precision Time Protocol in Power System Applications," which is a revision to the previous C37.238-2011 standard (Annex B of C37.238-2017 outlines the differences between the 2011 and 2017 versions). The profile version number has been bumped from 1.0 to 2.0.

C37.238 specifies a profile that extends the capabilities of the PUP (IEC 61850-9-3:2016) profile to define the use of PTP for power system protection, control, automation, and data communication applications utilizing a packet transport. An example of an extension is the introduction of a new TLV in the Announce message to track grandmaster and network time inaccuracy.

Another extension in the 2011 version over 61850-9-3 was the requirement of PTP messages to be sent with the (CoS) priority value of 4, so they must include an 802.1Q VLAN tag with a default VLAN ID of 0. This VLAN tag was known to cause interoperability issues between the 2011 edition and 61850-9-3.

Subsequently, C37.238-2011 was divided into a base profile specified in the PUP (IEC 61850-9-3) profile and the 2017 version that concentrates on the extended capabilities beyond the PUP profile. The goal of this change was to increase interoperability between different implementations. This is the same reason that many options and values available in IEEE 1588-2008 have been set to fixed values in these profiles. As with the two preceding profiles, the message interval for Announce, Sync, and Pdelay_Req is fixed at 1 per second.

Not only variations in the message rates but some other mechanisms are specifically disallowed with this profile. They include the E2E path delay mechanism, unicast transmission, and negotiation.

The devices in the transport network need to distribute time from a grandmaster clock to the end devices over 16 network hops (TC nodes). The time error between the time recovered by the end device and the standard time source (UTC) should be less than ±1 µs. This applies even for network loads up to 80% wire-speed (line rate) on each link.

SMPTE ST-2059-2 and AES67 Media Profiles

The SMPTE profile is part of the SMPTE ST-2059 standard and was created specifically to synchronize video equipment in a professional broadcast environment. This profile helps keep multiple video sources in synchronization across numerous pieces of equipment by providing time and frequency synchronization to all devices. It was designed with the following goals:

- For a slave to achieve synchronization within 5 seconds of being connected to an operational PTP network

- Upon phase alignment, to maintain time accuracy between any two slave devices with respect to the master reference within 1 µs

- To transport synchronization metadata (SM) required for synchronization and time labeling of audio/video signals

That last requirement is supported by the master port in a grandmaster clock sending a Management message with an SM TLV appended once per second (something that, for instance, the telecom profiles will not support). Regarding frequency accuracy, the PTP grandmaster clock shall also maintain a frequency with a maximum deviation within 5 parts per million (ppm) from the SI second.

The AES67 Media Profile is part of the AES67 standard and supports professional-quality audio applications for high-performance streaming in media networks. The requirement for accuracy and frequency stability is not quite as strict as ST-2059-2 (for example, frequency to within 10 ppm rather than 5 ppm for the video standard).

Although sourced from different organizations, most of the major parameters in the two profiles, AES67 and SMPTE ST2059-2, have minimal differences. The selected transport type for both is IPv4 (multicast) and based on Internet Group Management Protocol (IGMP) version 2 (IGMPv3 optional), although SMPTE allows IPv6 as well. Each requires the delay-request/response mechanism (Annex J.3 from IEEE 1588-2008) as the path delay measurement mechanism, although both SMPTE and AES optionally support the P2P delay measurement mechanism (Annex J.4 from 1588-2008) as well. They both support ordinary and boundary clocks and the default 1588 BMCA algorithm.

Both profiles support a different range of message rates, with SMPTE requiring a higher rate for Sync and Announce, but there is substantial overlap in the ranges allowed. Because of these similarities, it is possible, by careful selection of the profile parameters, to combine them on the same network.

At the agreed point that fits within both profiles, media streams of different types can be synchronized by using a single, common PTP clock distribution system (see AES-R16-2016, "PTP parameters for AES67 and SMPTE ST 2059-2 interoperability").

Note that a new edition of the SMPTE ST 2059-2 was due for publication at the end of 2020 but at the time of writing is not yet published. It is expected to contain the following changes:

- New defaults to better align SMPTE and AES67 profiles

- Clarified communications modes of multicast, unicast, and mixed

- Makes use of the traceable flag

- Removes some inappropriate terms.

These two standards (along with others relating to the audio and video itself) are a very important component of industries related to broadcasting.

PTP Enterprise Profile (Draft RFC)

The Enterprise Profile is a proposed standard by the Timing over IP Connection and Transfer of Clock (TICTOC) Working Group of the IETF (https://tools.ietf.org/wg/tictoc/), which is responsible for about four published RFCs related to PTP and timing. This draft document defines a PTP profile targeted for large enterprise networks, such as those found in large financial organizations. See Chapter 2 for some examples of these use cases.

There are several requirements driving the adoption of accurate timing in the enterprise:

- Regulation, such as MiFID-II from the European Union, requiring accurate time-stamping of business transactions and traceability of those timestamps (think of high-frequency and low-latency trading)

- Requirements to measure one-way latency and cumulative delays across multiple computers not only locally but across data centers

- Increased scale and speed requiring more accurate time-tagged measurements

The Enterprise Profile does not detail any specifics on timing performance, but an increasing number of enterprise timing requirements cannot be achieved using more traditional methods such as NTP (see the section "PTP Versus NTP" toward the beginning of the chapter). NTP is normally a software-only solution with limited accuracy and could not improve this performance without adding significant hardware support, support that is already available for PTP.

Because IEEE 1588-2008 includes many options, this profile is designed to limit the available choices, values, and parameters to increase interoperability between different implementations.

The Enterprise Profile uses PTP transport with UDP over both IPv4 (1588-2008 Annex D) and IPv6 (Annex E) with an E2E delay mechanism. The P2P mechanism is specifically not allowed. The profile supports OC, TC, and BC clocks with a mixture of unicast and multicast transmission, as follows:

- Announce messages are multicast (fixed at 1 per second).

- Delay_Req can be either unicast or multicast, and the master port must reply with the Delay_Resp carried by the same method as the Delay_Req.

- Sync messages are multicast.

The options defined for the Enterprise Profile allow for interoperability for clocks running the IEEE 1588-2008 E2E default profile (Annex J.3) when transported with PTPoIP.

One interesting feature of this profile is that clocks should support different domains. This allows the operator to run multiple, independent, and redundant sources of time within the enterprise network. End devices can use timing information from multiple masters by combining information from multiple instantiations of a PTP stack, each operating in a different domain.

If the network is well designed, and each domain takes different paths, then this results in a very robust and resilient distribution of time. The slave ports can cross-validate the different values of timing they are receiving in different domains and discard any rogue results. So, the default version of BCMA is supported within a domain on a clock, but the clock may be tracking multiple grandmasters if they are in different domains.

Message rates for this profile are a default of 1 per second with a range from 128 per second to one every 128 seconds. Only the Announce message is fixed at 1 per second. Negotiation of message rates is not allowed.

White Rabbit

White Rabbit (WR) is a project that originated within the European Organization for Nuclear Research (CERN) facility and has now spread out to other research institutions. The initial requirement was to support scientific experiments in particle physics, but it has since been applied to many applications requiring highly accurate time. The project aimed at creating a low-latency network based on Ethernet that could provide high-accuracy timing distribution. The goal was to achieve synchronization between thousands of nodes, spread across tens of kilometers to sub-nanosecond accuracy.

WR makes use of the following technology:

- Ethernet-based transport (over fiber, which is typically a single BiDi fiber to reduce any asymmetry)

- PTP for phase/time transfer

- SyncE for frequency synchronization

- Methods to reduce link asymmetry and increase clock accuracy

Several techniques were used to increase the accuracy:

- Use SyncE to deliver frequency synchronization in combination with PTP (also available in other implementations, such as the G.8275.1 telecom profile).

- Increase the granularity and accuracy of the timestamping (which was commonly 8 ns in previous implementations, limiting the possible accuracy).

- Measure and calibrate any asymmetry in the hardware implementation of time distribution and the circuit board layout within the WR network element (assumed to be mostly a fixed value and so can be compensated for).

- White Rabbit extension to PTP (WRPTP), a PTP profile that includes a WR link setup process with extra messages and TLVs.

- Measurement and correction of asymmetry errors caused by the physical medium that are fixed and can be measured and corrected for. An example would be the difference in propagation velocity for two different lambdas in a single BiDi fiber.

- Calibration and correction of asymmetry in the physical link between nodes, which is normally considered to be dynamic (such as cyclical temperature changes in the fiber over the day and seasons causing changes in the propagation speed).

The process for measuring the two-way delay relies upon the E2E delay mechanism from PTP 1588-2008 (Annex J.3) with its associated accuracy to give a first-order estimate. Then, another process, using a technique called Digital Dual Mixer Time Difference (DDMTD), is used to measure/control phase difference between each end of the link at a much higher level of precision. The details of this method are beyond the scope of this book, but there are references at the end of the chapter (see reference with the term "DDMTD").

When building a WR-enabled network, it uses the Ethernet 802.1 bridge (switch) functionality with some WR extensions to ensure more deterministic performance with lower latency (much like TSN; see the earlier section dedicated to IEEE 802.1AS-2020).

For highest precision, WR uses a single BiDi fiber for two-way communication with a different lambda in each direction. In this case, most of the asymmetry is caused by the difference in propagation velocity in each direction, as the velocity in fiber is dependent on the frequency of the laser light. This can be accurately determined for each type of fiber (although that difference can change with temperature, which may also be modelled and corrected).

The WRPTP profile specifies a *modified* BMCA (mBMCA), with the major difference being that a WR clock can have multiple ports in slave state, with each of them communicating with a different grandmaster clock—a form of "hot standby." This allows the WR clock to maintain the desired accuracy during an event that requires switching to a new grandmaster.

Given the hardware support to assist in the time distribution, the actual PTP traffic is generally not as high in WR as one might expect. The number of Sync messages has a default of 1 per second with a possible range of 2 per second to 1 per 64 seconds. The domain number is fixed at 0. Generally, the specification beyond that is mostly undefined (for example, transport type).

WR is an open standard, for both hardware and software implementations, so information and products supporting it are readily available. The current version as of writing is 2.0 from 2011. Compatibility with existing standards enables mixed networks, where time-critical end nodes are connected directly to a WR network core, while other less-critical devices use standard switches to connect to it.

Note that the WR specification is likely to be replaced by the *High Accuracy Delay Request-Response Default PTP Profile* (Annex I.5) from IEEE 1588-2019.

PTP Security

It is clear that timing and synchronization is a critical component of our modern infrastructure, so like all essential services, it will attract bad actors with malicious intentions, looking for any open vulnerabilities. Therefore, security of the timing distribution network is an important consideration for operators deploying it. From a security aspect, the timing distribution network has one advantage when it comes to security and has a couple of other disadvantages that set it apart.

The advantage is that the packet-based timing runs over essentially the same network as all the rest of the important data. This means that the normal security measures that network engineers perform for their data network now apply equally to the packet-based timing distribution network. It is beyond the scope of this book to cover security and trust of the packet network in general; there are ample references on that topic. One source you might find especially helpful for background is IETF RFC 7384, which covers the security requirements of both PTP and NTP.

The security disadvantages of using the transport network for timing fall into two categories:

- The end-to-end timing solution may require a significant number of GNSS receivers or other radio equipment as sources of time. This equipment can be quite vulnerable to jamming, not only by state actors with military means, but even to an unsophisticated attack. This aspect will be covered in Chapter 11, "5G Timing Solutions," when looking at deployment tradeoffs.

- PTP traffic is not like any other packet traffic because it is time sensitive at the sub-microsecond level. The time sensitivity means that it is vulnerable to any form of delay in transmission, especially if the delay is asymmetric (different length of delay in each direction).

A bad actor only needs to delay the packet stream in one direction by a few microseconds to take all downstream devices out of alignment. There is very little that can be done to mitigate this (or even detect it) beyond the basic access control measures applicable to all other packet networks.

Beyond that, the vulnerability of PTP is not as great. The data in most PTP messages is primarily timestamps, so the data is not sufficiently sensitive to require strong encryption. Some propose the use of encryption to "hide" from any attacker the fact that sensitive PTP packets are in the data stream. That method does not help, as the innate characteristics of PTP traffic make it easy to detect, even when encrypted, allowing a simple delay attack. See the white paper "Encryption is Futile" (listed in the "References" section) for further details.

The other aspect is verification of the timestamps, such as the need to provide source authentication, message integrity protection (to ensure it is unaltered), and defense against replay attacks. There is a mechanism to do this described in Annex K of IEEE 1588-2008, "Security Protocol (experimental)." It is based on standard cryptographic techniques and makes use of shared keys and keyed-hash message authentication codes (HMAC).

This feature, as suggested by the name of the annex, is not mandatory and is not widely adopted. Furthermore, many of the profiles based off the 1588-2008 standard do not make use of the mechanism either. For reasons of interoperability and operational simplicity, it is not widely adopted. Despite that, it is undeniable that security is an aspect of PTP that needs further development. It is not surprising, then, that although Annex K has been dropped from the new version of PTP, security is getting much greater attention in 1588-2019 with its Annex P, "Security."

This new integrated security mechanism from 1588-2019 provides source authentication, message integrity, and replay attack protection for PTP messages. Like the now defunct

Annex K from 1588-2008, the mechanism relies upon standard cryptographic techniques, shared keys, and HMAC integrity checks, but it is a different implementation. See 1588-2019 section 16.14, "PTP integrated security mechanism," and the informative Annex P. These sections define two basic methods:

- An AUTHENTICATION TLV, which allows source authentication and message integrity. This TLV will contain various fields, including the calculated Integrity Check Value (ICV) of the PTP message used to ensure it has not been altered by an unauthorized party.

- A key-exchange mechanism to enable (secure) distribution of security parameters (such as the algorithm type and shared key) needed to construct and verify the AUTHENTICATION TLV. The standard does not specify the method for this key-distribution process.

The ICV is calculated by a hashing algorithm, and HMAC-SHA256-128 is suggested as one that should be supported by an implementation. The mechanism envisages two basic integrity check methods, one where the message is immediately verified before additional processing, and another where verification can take place after processing. There are different considerations for each case. One involves the use of TCs.

Because TCs need to update the correctionField of the Sync message during transmission from master to slave port, this obviously invalidates the existing integrity check, so either TCs must have a copy of the shared key to regenerate the ICV or else the correctionField must be excluded from the hash calculation (which could defeat the whole point of the exercise).

For the delayed verification case, the key used to generate the ICV can be disclosed for later integrity checking, but this does not help a TC that needs the key to recalculate the ICV after updating the correctionField. However, to allow this, there is a mechanism in the standard to exclude the correctionField from the ICV hash calculation.

Annex P of 1588-2019 is a good source of information on the details of the key-distribution problem and multipronged approaches to securing the PTP data transport. Note that there are additional projects approved by the IEEE (see the upcoming section "Next Steps for IEEE 1588") to further define some of these aspects following the publication of IEEE 1588-2019.

Looking beyond the security of PTP itself, diversity in clock sources and correlation and cross-checking between them is the best way to make an application based on precise time synchronization more robust. This is true for power automation (substation security), and it is also commonly used in the financial services industry. Chapter 9 provides more information on the practical security aspects for real-world deployments.

IEEE 1588-2019 (PTPv2.1)

As mentioned in preceding sections, IEEE 1588-2008 (PTPv2) has been updated with a new revision, formally approved by the IEEE as 1588-2019, but known colloquially as

PTPv2.1. The idea behind choosing this version number is to indicate that a major goal of v2.1 was to remain as close to backward compatible with PTPv2 as possible.

In short, the two versions are compatible with each other, so long as

- No option or feature present in only one of the versions is used.

- Features common to both versions are used and similarly configured.

From the preceding section, you might understand that an example of an option that might have compatibility issues between editions is the optional security mechanism.

The new standard goes to great lengths to point out any differences and specifically addresses compatibility. Section 19.4 of the new standard gives details of the areas where compatibility between former and newer implementations can arise, while Table 139 specifically cautions on the compatibility of options.

However, there is a lot more material written into the new version than previously, with the page count dramatically increasing. As a short summary, the following main areas have been addressed by the new version, 1588-2019:

- **High accuracy:** Improved time sync performance with high accuracy techniques with an optional new profile to support it (which might likely replace the White Rabbit specification).

- **Management:** Performance monitoring and data information models. Methods and options for monitoring PTP implementations are covered in 16.11, 16.12, and Annex J and adoption of an information model for the PTP datasets (to allow remote management).

- **Architecture update and clarifications:** Methods for profile isolation; PTP redundancy; and restructuring of the standard to separate the media-dependent functions from the media-independent functions, allowing the creation of special PTP ports. One other architectural change is the adoption of the concept of the PTP Instance as an instantiation of the PTP protocol, operating in a single device, within exactly one domain (separating the concept of a running PTP instance from the PTP domain).

- **Security:** A PTP integrated security mechanism to provide a security layer to PTP encompassing message integrity and authentication. IEEE added a new Annex P to aid implementation of a secure timing distribution network.

Of course, 1588-2019 includes the usual clarifications, small amendments, and typographical corrections. The following two sections cover the details of the more substantial changes between the versions and the new features added to PTPv2.1.

Changes from PTPv2 to PTPv2.1

The following major changes have been made between PTPv2 and PTPv2.1:

- The 2008 version allowed for the concept of multi-domain PTP networks, but the specification focused on networks with a single PTP domain. As a result, there were some inconsistencies in some parts of the standard that could lead to confusion. Changes to the 2019 edition (especially to section 6.1 on the general requirements) address those concerns. A new informative Annex O gives examples of inter-domain interactions.

- The specification of TCs has changed in two areas. The 2008 specification of TC allowed an (optional) single copy of default and port datasets. The 2019 version deprecates the old transparentClockDefaultDS and transparentClockPortDS and instead allows attributes in the default and port datasets for each PTP instance. Previously, the specifications for TC (concerning the datasets) were independent of the domain, but since the new specification for datasets are per PTP Instance, they are domain specific.

- New rules have been added for assignment of the clockIdentity to reduce the possibility of assignment of duplicate clockIdentity values to different clocks. This is because the rules for constructing a EUI-64 value were changed by the IEEE Registration Authority after the publication of the 2008 edition.

- The common header of the PTP packet was modified to allow a reserved field to become a PTP minor version number, or *minorVersionPTP*. For this new edition of the standard, the versionPTP attribute has the value 2, and the minorVersionPTP attribute has the value 1. This allows the protocol version of the new edition to be expressed as 2.1 rather than 3. Messages with equal or different values for this minorVersionPTP field, but equal values for the versionPTP field, are accepted by this version of PTP.

- The section on PTP datasets has been changed to enable the management of PTP datasets with external management tools beyond the PTP management mechanism in (optional) section 15. It suggests the specification of the datasets in section 8 can be mapped to other data modelling languages, including MIB (for SNMP) and YANG. So, section 8 has expanded from the definition of data used in the operation of the PTP protocol to an information model.

New Features in v2.1

The following is a list of the most important new features and interesting additions to v2.1:

- **Improved accuracy:** Accuracy is improved by making use of layer 1 frequency synchronization (Annex L) and asymmetry calibrations (16.7, 16.8), and a new High Accuracy Delay Request-Response Default PTP Profile in Annex I.5. Two new informative annexes on enhanced performance, namely M and N, were added, with Annex M including the use of a DDMTD-based phase offset detector (as used by White Rabbit).

- **Special PTP ports:** Special ports enable the inclusion of network links based on technologies that provide inherent timing support as opposed to the use of PTP timing messages, for example, IEEE 802.11 Wi-Fi and passive optical networking (PON) (see section 19.4.6). From the point of view of a PTP network, a special PTP port can only be connected to a compatible special port.

- **Mixed multicast/unicast model of operation:** The new edition includes new TLVs to provide stronger support for mixing multicast and unicast messages in a PTP session (see section 16.9, "Mixed multicast/unicast operation"). This optional feature was available but not well described in the 2008 version.

- **External configuration:** A new (optional) feature allows the external configuration of the port state under management control (see section 17.6). The management method to do this is not defined but could include a standard mechanism such as SNMP or YANG. Previously port state could only be determined resulting from the BMCA process. The (optional) section 17.7, "Reduced state sets and use of the <foreignMasterList> feature," also allows the use of a restricted set of port states.

- **Security enhancements:**The 1588-2008 (optional) Annex K on security has been removed, and 1588-2019 introduces an (optional) section 16.14 on the PTP integrated security mechanism and introduces an informative Annex P on security. See the earlier section "PTP Security" for more details on changes to the security mechanisms.

- **Profile isolation:** New (optional) section (16.5) on the isolation of PTP Instances running under profiles conceived by different SDOs when they are both running over the same network. Profile isolation in 2008 was very rudimentary (mainly relying on domain number) but is now expanded to include a PTP header field *sdoId* (repurposed from the transportSpecific field). The sdoId is an organizational ID value that an SDO can obtain from the IEEE Registration Authority. By requiring the use of this value in their profiles, SDOs can isolate messages belonging to their profiles from those of other SDOs. See section 16.5, "Isolation of PTP Instances running under profiles specified by different standards organizations."

- **Slave event monitoring:** This (optional) feature allows for monitoring timing information from a PTP port in slave state. Some networks have deployed various application-specific methods that allow a similar capability. However, this feature defines an application-independent mechanism. It includes the definition of three new TLVs to carry information away from the slave. SLAVE_RX_SYNC_ COMPUTED_DATA, one of these TLVs, allows the sharing of computed fields like the offsetFromMaster and meanPathDelay. Additional datasets to support this feature are also defined. This simplifies the implementation and deployment of monitoring timing information from the slave. See section 16.11, "Slave Event Monitoring," for details.

- **Performance monitoring options:** Annex J introduces another (optional) mechanism to report metrics from the PTP instances running on the node. It defines new datasets, including the performance monitoring dataset (*performanceMonitoringDS*) and performance monitoring port dataset (*performanceMonitoringPortDS*). This allows management access to statistics and metrics from the PTP instance. including clock and port parameters. See Annex J, "Performance monitoring options," for details.

■ **On-path performance metrics:** There is also an additional (optional) mechanism to report synchronization accuracy metrics from various points along the PTP timing path. Each node on the path implementing this feature (including TCs) updates the accuracy metrics based on its contribution to the expected degradation in time accuracy. A set of metrics can be included in the ENHANCED_ACCURACY_METRICS TLV to determine an overall expectation of time inaccuracy. This TLV is propagated down the timing path for nodes to use as required. For example, a PTP profile could define an aBMCA that makes use of this data to allow the slave to select the most accurate grandmaster. See section 16.12, "Enhanced synchronization accuracy metrics (optional)," for details.

It is evident that v2.1 contains numerous new, interesting, and helpful options. In the interests of backward compatibility, many of them are optional, but it is clear there will be increased adoption of some of these options over the next several years.

Next Steps for IEEE 1588

The publishing of the 2019 edition is not the end of the evolution of the standard. The IEEE has already begun projects to address topics that will be included in future amendments of IEEE 1588. These projects are described in what the IEEE calls project authorization requests (PAR) and are known by the titles P1588a through P1588g. These projects are expected to complete through the end of 2022. These projects will address the following topics over the next few years:

■ **Security:** Define a default security profile and proceed on aspects of security (for example, key management)

■ **BMCA enhancements:** Enhancing the data made available to support the BMCA; provides information to allow the adoption and specification of alternative BMCAs; corrections and clarifications

■ **Management:** Define YANG and MIB data models

■ **Transport:** Specification for mapping PTP to OTN

■ **Calibration:** Enhances support for latency and asymmetry calibration

■ **Additional edits and clarifications:** To fix errors and clarify unclear text, including developing inclusive names as an alternative for some terms in the standard

Of course, the SDOs writing profiles now will begin to adopt the desired new features from v2.1 into the definition of their profiles. On the other hand, they will have to be somewhat cautious so as not to cause issues with backward compatibility.

As outlined in the section "IEEE PTP" in Chapter 4, there are several PARs now approved as follow-on work for IEEE 1588. Please see the references at the end of Chapter 4 for links to more information on these projects.

Summary

This chapter covered the full range of PTP technologies, both broadly and in depth. The earlier half covered the details of PTP based on the IEEE 1588 standards, whereas the latter part of the chapter (starting at "Profiles") outlined the different flavors of PTP based on the founding 1588 standard. Each of the major profiles was covered in the context of the industry segment that it is designed around along with a summary of their basic features. The chapter finished with a discussion of the security aspects, the changes in the IEEE 1588-2019 edition, and the next steps for the IEEE with 1588.

Now that PTP has been covered in some depth, the next task is to look at the substantial number of ITU-T recommendations used to define many aspects of using PTP as well as physical methods to support time transfer.

Chapter 8 looks in detail at the ITU-T recommendations that are linked to SyncE and the telecom profiles for PTP. The goal behind providing this information is to allow the network engineer to gain an understanding of what document applies to which problem area as well as an outline of what is contained in each recommendation.

References in This Chapter

3GPP. "Evolved Universal Terrestrial Radio Access (E-UTRA); Base Station (BS) radio transmission and reception." *3GPP*, 36.104, Release 16, 2021. https://www.3gpp.org/DynaReport/36104.htm

Annessi, R, J. Fabini, F. Iglesias, and T. Zseby. "Encryption is Futile: Delay Attacks on High-Precision Clock Synchronization." *arXiv*, 2018. https://arxiv.org/pdf/1811.08569.pdf

Audio Engineering Society (AES)

"PTP parameters for AES67 and SMPTE ST 2059-2 interoperability." *AES*, AES-R16-2016, 2016. http://www.aes.org/publications/standards/search.cfm?docID=105

"AES standard for audio applications of networks – High-performance streaming audio-over-IP interoperability." *AES*, AES67-2018, 2018. http://www.aes.org/publications/standards/search.cfm?docID=96

Cota, E., M. Lipinksi, T. Wlostowski, E. van der Bij, and J. Serrano. "White Rabbit Specification: Draft for Comments." *Open Hardware Repository*, 2011. https://ohwr.org/project/wr-std/wikis/Documents/White-Rabbit-Specification-(latest-version)

IEEE Standards Association

"Active projects of the P1588 Working Group." *IEEE P1588*, 2020. https://sagroups.ieee.org/1588/active-projects/

"IEC/IEEE International Standard – Communication networks and systems for power utility automation – Part 9-3: Precision time protocol profile for power utility automation." *IEC/IEEE Std 61850-9-3:2016*, 2016. https://standards.ieee.org/standard/61850-9-3-2016.html

"IEEE Standard for a Precision Synchronization Protocol for Networked Measurement and Control Systems." *IEEE Std 1588:2002*, 2002. https://standards.ieee.org/standard/1588-2002.html

"IEEE Standard for a Standard for Precision Synchronization Protocol for Networked Measurement and Control Systems." *IEEE Std 1588:2008*, 2008. https://standards.ieee.org/standard/1588-2008.html

"IEEE Standard for a Standard for Precision Synchronization Protocol for Networked Measurement and Control Systems." *IEEE Std 1588:2019*, 2019. https://standards.ieee.org/standard/1588-2019.html

"IEEE Standard for Local and Metropolitan Area Networks – Timing and Synchronization for Time-Sensitive Applications in Bridged Local Area Networks." *IEEE Std 802.1AS-2011*, 2011. https://standards.ieee.org/standard/802_1AS-2011.html

"IEEE Standard for Local and Metropolitan Area Networks – Timing and Synchronization for Time-Sensitive Applications." *IEEE Std 802.1AS-2020*, 2020. https://standards.ieee.org/standard/802_1AS-2020.html

"IEEE Standard for Local and Metropolitan Area Networks – Timing and Synchronization for Time-Sensitive Applications." *IEEE 802.1AS-Rev, Draft 8.3*, 2019.

"IEEE Standard for Local and Metropolitan Area Networks – Time-Sensitive Networking for Fronthaul." *IEEE Std 802.1CM-2018*, 2018. https://standards.ieee.org/standard/802_1CM-2018.html

"IEEE Standard Profile for Use of IEEE 1588 Precision Time Protocol in Power System Applications." *IEEE Std C37.238-2011*, 2011. https://standards.ieee.org/standard/C37_238-2011.html

"IEEE Standard Profile for Use of IEEE 1588 Precision Time Protocol in Power System Applications." *IEEE Std C37.238-2017*, 2017. https://standards.ieee.org/standard/C37_238-2017.html

"Standard for a Precision Clock Synchronization Protocol for Networked Measurement and Control Systems Amendment: Enhancements for Best Master Clock Algorithm (BMCA) Mechanisms." *IEEE Amendment P1588a*, 2020. https://standards.ieee.org/project/1588a.html

"Standard for a Precision Clock Synchronization Protocol for Networked Measurement and Control Systems Amendment: Addition of Precision Time Protocol (PTP) mapping for transport over Optical Transport Network (OTN)." *IEEE Amendment P1588b*, 2020. https://standards.ieee.org/project/1588b.html

"Standard for a Precision Clock Synchronization Protocol for Networked Measurement and Control Systems Amendment: Clarification of Terminology." *IEEE Amendment P1588c*, 2020. https://standards.ieee.org/project/1588c.html

"Standard for a Precision Clock Synchronization Protocol for Networked Measurement and Control Systems Amendment: Guidelines for selecting and operating a Key Management System." *IEEE Amendment P1588d*, 2020. https://standards.ieee.org/project/1588d.html

"Standard for a Precision Clock Synchronization Protocol for Networked Measurement and Control Systems Amendment: MIB and YANG Data Models." *IEEE Amendment P1588e*, 2020. https://standards.ieee.org/project/1588e.html

"Standard for a Precision Clock Synchronization Protocol for Networked Measurement and Control Systems Amendment: Enhancements for latency and/or asymmetry calibration." *IEEE Amendment P1588f*, 2020. https://standards.ieee.org/project/1588f.html

"Standard for a Precision Clock Synchronization Protocol for Networked Measurement and Control Systems Amendment: Master-slave optional alternative terminology." *IEEE Amendment P1588g*, 2020. https://standards.ieee.org/project/1588g.html

International Electrotechnical Commission (IEC). "Industrial communication networks – High availability automation networks – Part 3: Parallel Redundancy Protocol (PRP) and High-availability Seamless Redundancy (HSR)." *IEC Std 62439-3:2016*, Ed. 3, 2016. https://webstore.iec.ch/publication/24447

European Broadcasting Union (EBU). "Technology Pyramid for Media Nodes." *EBU*, Tech 3371 v2, 2020. https://tech.ebu.ch/publications/technology_pyramid_for_media_nodes

International Telecommunication Union Telecommunication Standardization Sector (ITU-T)

Arnold, D. "Changes to IEEE 1588 in the 2019 edition." *ITU-T SG 15 (Study Period 2017) Contribution 1966*, 2020. https://www.itu.int/md/T17-SG15-C-1966/en

"G.8275: Architecture and requirements for packet-based time and phase distribution." *ITU-T Recommendation*, 2020. http://handle.itu.int/11.1002/1000/14509

"Q13/15 – Network synchronization and time distribution performance." *ITU-T Study Groups*, Study Period 2017-2020. https://www.itu.int/en/ITU-T/studygroups/ 2017-2020/15/Pages/q13.aspx

Kirrmann, H. and W. Dickerson. "Precision Time Protocol Profile for power utility automation application and technical specifications." *PAC World Magazine*, 2016. http://www.solutil.ch/kirrmann/PrecisionTime/PACworld_2016-09_038_043_IEC_IEEE_61850-9-3.pdf

Internet Engineering Task Force (IETF)

Arnold, D. and H. Gerstung. "Enterprise Profile for the Precision Time Protocol with Mixed Multicast and Unicast Messages." *IETF*, draft-ietf-tictoc-ptp-enterprise-profile-18, 2020. https://tools.ietf.org/html/draft-ietf-tictoc-ptp-enterprise-profile-18

Mills, D. "Network Time Protocol (NTP)". *IETF*, RFC 958, 1985. https://tools.ietf.org/html/rfc958

Mills, D. "Network Time Protocol Version 3: Specification, Implementation and Analysis." *IETF*, RFC 1305, 1992. https://tools.ietf.org/html/rfc1305

Mills, D. "Simple Network Time Protocol (SNTP) Version 4 for IPv4, IPv6 and OSI." *IETF*, RFC 4330, 2006. https://tools.ietf.org/html/rfc4330

Mills, D., J. Martin, J. Burbank, and W. Kasch. "Network Time Protocol Version 4: Protocol and Algorithms Specification." *IETF*, RFC 5905, 2010. https://tools.ietf.org/html/rfc5905

Mills, D. "Security Requirements of Time Protocols in Packet Switched Networks." *IETF*, RFC 7384, 2014. https://tools.ietf.org/html/rfc7384

Mills, D. "Computer Network Time Synchronization: the Network Time Protocol on Earth and in Space," Second Ed., CRC Press 2011

Mills, David L. et al, "Network Time Synchronization Research Project." https://www.eecis.udel.edu/~mills/ntp.html

Moreira, P., P. Alvarez, J. Serrano, I. Darwezeh, and T. Wlostowski. "Digital Dual Mixer Time Difference [DDMTD] for Sub-Nanosecond Time Synchronization in Ethernet." *2010 IEEE International Frequency Control Symposium*, 2010. https://ieeexplore.ieee.org/document/5556289

SMPTE. "SMPTE Profile for Use of IEEE-1588 Precision Time Protocol in Professional Broadcast Applications." *SMPTE ST 2059-2:2015*, 2015. https://ieeexplore.ieee.org/document/7291608

UCA International Users Group (UCAIug). Precision Time Protocol Profile IEC 61850 Interoperability Testing (IOP), "IEC 61850 2019 Interoperability Testing (IOP) – Final Test Report." *UCAIug*, 2019. http://www.ucaiug.org/IOP_Registration/IOP%20Reports/IEC%2061850%202019%20IOP%20Final%20Report%2020200122.pdf

Chapter 7 Acronyms Key

The following table expands the key acronyms used in this chapter.

Term	Value
1PPS	1 pulse per second
2DM	two-way delay measurement
3GPP	3rd Generation Partnership Project (UMTS)
3G	3rd generation (mobile telecommunications system)
4G	4th generation (mobile telecommunications system)
5G	5th generation (mobile telecommunications system)
aBMCA	alternate best master clock algorithm
AES	Audio Engineering Society
APTS	assisted partial timing support
ARB	arbitrary (timescale)
AVB	Audio Video Bridging
BC	boundary clock

Term	Value
BiDi	bidirectional
BITS	building integrated timing supply (SONET)
BMCA	best master clock algorithm
CEM	circuit emulation
CERN	Conseil Européen pour la Recherche Nucléaire
CERN	European Organization for Nuclear Research
CLI	command-line interface
CoS	class of service
CPU	central processing unit
DDMTD	Digital Dual Mixer Time Difference
E2E	end-to-end
E2E TC	end-to-end transparent clock
ESMC	Ethernet synchronization messaging channel
EUI	extended unique identifier
GM	grandmaster
GNSS	global navigation satellite system
GPS	Global Positioning System
gPTP	generalized precision time protocol (IEEE 802.1AS)
HMAC	keyed-hash message authentication codes
HSR	High-availability Seamless Redundancy
ICV	Integrity Check Value
IEEE	Institute of Electrical and Electronics Engineers
IETF	Internet Engineering Task Force
IEC	International Electrotechnical Commission
IED	intelligent electronic devices
IERS	International Earth Rotation and Reference Systems
IGMP	Internet Group Management Protocol
IGMPv3	Internet Group Management Protocol version 3
IOP	interoperability
IOS-XR	Internet Operating System—XR edition
IP	Internet protocol

Term	Value
IPv4, IPv6	Internet protocol version 4, Internet protocol version 6
ITU	International Telecommunication Union
ITU-T	ITU Telecommunication Standardization Sector
L2	layer 2 (of the OSI model)
L2P2P	layer 2 peer-to-peer
L3	layer 3 (of the OSI model)
L3E2E	layer 3 end-to-end
LAN	local-area network
NTP	Network Time Protocol
NTPv4	Network Time Protocol version 4
MAC	media access control (address)
mBMCA	modified best master clock algorithm
MIB	Management Information Base
MiFID-II	Markets in Financial Instruments Directive (2014/65/EU)
MPLS	Multiprotocol Label Switching
OC	ordinary clock
OTN	optical transport network (optical transport technology)
P2P	peer-to-peer
P2P TC	peer-to-peer transparent clock
PAR	project authorization request
PDV	packet delay variation
PHY	PHYsical Layer (of OSI Reference Model)—an electronic device
PIP	PTP Industry Profile
PMU	phasor measurement unit
PON	passive optical networking (optical broadband)
POSIX	Portable Operating System Interface
PRC	primary reference clock (SDH)
PRP	Parallel Redundancy Protocol
PRS	primary reference source (SONET)
PRTC	primary reference time clock
PTP	precision time protocol

Term	Value
PTPv1	precision time protocol according to 1588-2002
PTPv2	precision time protocol according to 1588-2008
PTPv2.1	precision time protocol according to 1588-2019
PTPoIP	PTP over Internet protocol
PTS	partial timing support
PTSF	packet timing signal fail
PUP	Power Utility Automation Profile
QoS	quality of service
RAN	radio access network
RFC	Request for Comments (IETF document)
SDO	standards development organization
SDH	synchronous digital hierarchy (optical transport technology)
SI	International System of Units
SM	synchronization metadata
SMPTE	Society of Motion Picture and Television Engineers
SNMP	Simple Network Management Protocol
SNTP	Simple Network Time Protocol
SNTPv4	Simple Network Time Protocol version 4
SSM	synchronization status message
SSU	synchronization supply unit (SDH)
SyncE	synchronous Ethernet—a set of ITU-T standards
TAI	Temps Atomique International
T-BC	Telecom Boundary Clock
T-BC-A	Telecom Boundary Clock—Assisted
T-BC-P	Telecom Boundary Clock—Partial support
T-GM	Telecom Grandmaster
TC	transparent clock
TDD	time division duplex
TICTOC	Timing over IP Connection and Transfer of Clock (Working Group)
TLV	type, length, value

Term	Value
ToD	time of day
TSN	time-sensitive networking
T-TC	Telecom Transparent Clock
T-TC-P	Telecom Transparent Clock—Partial support
TTL	Time to Live (from IP)
T-TSC	Telecom Time Slave Clock
T-TSC-A	Telecom Time Slave Clock—Assisted
T-TSC-P	Telecom Time Slave Clock—Partial support
TV	television
UDP	User Datagram Protocol
UDP/IP	User Datagram Protocol/Internet protocol
UTC	Coordinated Universal Time
VLAN	virtual local-area network
WAN	wide-area network
WR	White Rabbit
WRPTP	White Rabbit (extension to) PTP
YANG	Yet Another Next Generation (data modelling language)

Chapter 8

ITU-T Timing Recommendations

In this chapter, you will learn the following:

- **Overview of the ITU:** Describes the general structure of the International Telecommunication Union (ITU), and the part of the organization (the ITU-T) responsible both for standardization and for timing and synchronization. It also categorizes the large number of recommendations and provides detail about how to read and use them.

- **ITU-T physical and TDM timing recommendations:** Provides an overview of the ITU-T recommendations that concern frequency timing using physical and time-division multiplexing (TDM) techniques such as the synchronous digital hierarchy (SDH).

- **ITU-T recommendations for frequency in packet networks:** Provides an overview of the ITU-T recommendations that concern frequency timing using physical techniques in packet-based networks, such as Ethernet.

- **ITU-T packet-based timing recommendations:** Provides an overview of the ITU-T recommendations that concern frequency, time, and phase timing using packet-based techniques, such as precision time protocol (PTP) in Ethernet and IP networks.

- **Possible future changes in recommendations:** Gives a short overview of what is currently being worked on, and therefore what changes might occur over the next few years.

The most accurate reference material for timing and mobile comes from numerous standards development organizations (introduced in Chapter 4, "Standards Development Organizations"). SDOs produce large amounts of written information on each specification as well as the need and usage of timing in numerous deployment scenarios. The problem for those wishing to start educating themselves is that these standards and recommendation documents are not exactly easy to find, read, or understand.

This chapter introduces the most important and relevant of these recommendations from just one organization, the ITU-T. You will see the various specifications placed into context so that you can decide what material is of value for your application. This chapter does not attempt to be the ultimate authority on each recommendation, just an orientation for those unfamiliar with the material in the public domain. For the fine details beyond that, readers who wish to dive down to the next level are pointed to the relevant standards.

This chapter aims to orientate the reader as a guide to navigating the many standards.

Overview of the ITU

The ITU is a multilateral organization headquartered in Geneva, Switzerland. It is an agency of the United Nations that is specialized in information and communication technologies. The ITU website (https://www.itu.int) gives the following summary of its role:

> Founded in 1865 to facilitate international connectivity in communications networks, we allocate global radio spectrum and satellite orbits, develop the technical standards that ensure networks and technologies seamlessly interconnect, and strive to improve access to ICTs [Information and Communication Technologies] to underserved communities worldwide. Every time you make a phone call via the mobile, access the Internet or send an email, you are benefitting from the work of ITU.

The ITU consists of three main areas of activity organized into the following *sectors*:

- **ITU Telecommunication Standardization Sector (ITU-T):** Mainly responsible for the development of standards to support telecommunications and other technical areas such as audio and video compression

- **ITU Radiocommunications Sector (ITU-R):** Mainly responsible for harmonized wireless and radio standards, as well as allocation of spectrum and satellite orbits

- **ITU Telecommunication Development Sector (ITU-D):** Responsible for numerous programs centered on improving access to modern telecommunications services for people living in underserved regions and emerging markets

For the purposes of this book, the most important sector is the ITU-T, which supports the telecommunications industry by developing a very large number of standards—although the ITU-T prefers to call them recommendations. For the mobile, broadcasting, and satellite industries, the ITU-R also plays a very important role.

The ITU-T sector is broken up into numerous *study groups* (about 11), and each of these study groups has various subgroups that the ITU-T calls *questions* (up to about 20 per study group). These subgroups focus on specific areas based on the goals of the current study period (at the time of this writing, the 2017–2020 study period is currently ending). With a new study period can come a change in priorities and focus.

As outlined in Chapter 4, the specific section of the ITU-T that is concerned with timing is known as Question 13 (Q13), which is part of Study Group 15 (SG15). Collectively, this group is referred to as Q13/15, and so the following section deals with this group in more detail.

ITU-T Study Group 15 and Question 13

Study Group 15 develops recommendations that define technical specifications for *networks, technologies and infrastructures for transport, access, and home.* The 19 individual questions within SG15 encompass a wide range of communications technologies such as passive optical networks (PON), optical fibers, Dense Wavelength Division Multiplexing (DWDM), Course Wavelength Division Multiplexing (CWDM), and digital subscriber line (DSL). The questions also cover various other optical systems, such as optical transport network (OTN), packet transport network (PTN), and metro transport network (MTN).

Question 13 of SG15 (abbreviated to Q13 or Q13/15) is responsible for *network synchronization and time distribution performance* and the recommendations that flow from this work. As a component of its tasks, Q13/15 also cooperates with several other SDOs, such as the following:

- Institute of Electrical and Electronics Engineers (IEEE) 1588, for PTP

- IEEE 802.1 and 802.3, for Ethernet

- Bellcore/Telcordia Technologies (now a division of Ericsson) and the American National Standards Institute (ANSI), which developed the General Requirements (GR) standards for synchronous optical network (SONET)

- Alliance for Telecommunications Industry Solutions (ATIS), accredited by ANSI and also a sector member and major contributor to the ITU and an organizational partner of the 3GPP

- 3rd Generation Partnership Project (3GPP), which is the principal organization driving the Long Term Evolution (LTE) and 5G mobile standards

- MEF Forum (MEF), developing services across automated networks

- O-RAN Alliance, for radio access network (RAN) fronthaul network specifications

- Internet Engineering Task Force (IETF), for Internet-related standards

There is a lot more information on each of these organizations and their contributions in Chapter 4. Within the ITU-T, Q13/15 is responsible for the following major recommendations related to timing:

- **Definitions, architecture, and functional models:** G.781, G.781.1, G.810, G.8260, G.8265, G.8273, G.8275

- **Network performance:** G.823, G.824, G.825, G.8261, G.8261.1, G.8271, G.8271.1, G.8271.2

- **Clocks:** G.811, G.811.1, G.812, G.813, G.8262, G.8262.1, G.8263, G.8266, G.8272, G.8272.1, G.8273.1, G.8273.2, G.8273.3, G.8273.4

- **PTP profiles:** G.8265.1, G.8275.1, G.8275.2

■ **Other documents and recommendations:** G.703, G.8264, GSTR-GNSS, Supplement 68—Synchronization Operations, Administration and Maintenance (OAM) Requirements

This chapter will go through each of these recommendations and documents in turn. Q13/15 also covers several other recommendations, which are not relevant to this topic and so will not be covered.

The three-digit recommendations, G.7xx and G.8xx, are primarily relevant for TDM and physical methods of timing transport, whereas the four-digit recommendations tend to be for packet-based transport, including synchronous Ethernet (SyncE). This is where the two definitions overlap a little, as SyncE uses a packet-based transport but is still a physical method of frequency transfer.

Most of the recent development effort of Q13/15 is in packet-based recommendations, although it has been involved in other efforts—for example, the enhanced PRC (ePRC) definition in G.811.1. Table 8-1 shows the categorization of only the packet-based recommendations and the role of each recommendation in the overall solution architecture. Further information on the non-packet (physical) recommendations is provided in the upcoming section "Physical and TDM Versus Packet Recommendations."

Table 8-1 *ITU-T SG15 Q13 Packet-Based Recommendations*

Role	Frequency	Phase/Time Partial Support	Phase/Time Full Support
Definitions and Terminology		G.810	
		G.8260	
Basic/network requirements	G.8261	G.8271	
	G.8261.1	G.8271.1	G.8271.2
Clock models and performance limits	G.811 (PRC)	G.8272 (PRTC)	
	G.811.1 (ePRC)	G.8272.1 (ePRTC)	
	G.8263 (packet slave)	G.8273.1 (T-GM) (not yet published)	
	G.8266 (packet master)	G.8273.3 (T-TC)	
	G.8262 (SyncE EEC)	G.8273.2 (T-BC, T-BC-P, T-TSC, and so on)	
	G.8262.1 (SyncE eEEC)		G.8273.4 (APTS)
Methods and framework	G.8264 (ESMC)	G.8273 (framework)	
	G.8265 (architecture)	G.8275 (architecture)	
Telecom profiles	G.8265.1	G.8275.1	G.8275.2
Synchronization layer functions	G.781	G.781.1 (not yet published)	

Role	Frequency	Phase/Time Partial Support	Phase/Time Full Support
Other documents and recommendations	G.703 (interface definitions)		
	GSTR-GNSS (Considerations on the use of GNSS as a primary time reference in telecommunications)		
	Supplement 68—Synchronization OAM Requirements		

Note Use of the terms "master" and "slave" is ONLY in association with the official terminology used in industry specifications and standards, and in no way diminishes Pearson's commitment to promoting diversity, equity, and inclusion, and challenging, countering, and/or combating bias and stereotyping in the global population of the learners we serve.

Note that there are several recommendations for PRC and PRTC clocks that are not strictly for packet networks, but they are necessary components of a packet-based timing distribution network, so they are included. For packet-based transport, the recommendations numbered in the G.826x range (to the left) are concerned with frequency distribution and those within the G.827x range (the right two columns) involve phase and time distribution.

How Recommendations Come About

The ITU-T has many *sector members*, which include a wide range of private and public companies, organizations, educational institutions, service providers, regulators, and governments. About once every 9 months, the complete Study Group 15 is hosted for two weeks at the ITU building complex in Geneva for a consolidated meeting known as a *plenary*. Member organizations of the ITU-T send *delegates* to the plenary to represent the organization, and for SG15, several hundred people attend.

During this two-week process, there are many meetings, most of them among delegates concerned with a specific question, but there are other, larger gatherings and several joint meetings between questions when the delegates need to work together. Q13/15 normally welcomes about 35–45 delegates at the plenary. The meetings are chaired by a *rapporteur* and assisted by an *associate rapporteur*.

The delegates to the plenary submit *contributions* to the meeting that contain ideas and suggestions and make *proposals* about new material to insert or changes that need to be made to a recommendation. From the number of contributions that are submitted, Q13/15 is one of the most active questions, with well over 100 contributions not unusual at every meeting. During the (very long) meetings, each contribution is taken in turn, and the proposing delegate presents the contribution to the meeting. At that point, following open discussion, the delegates in the room arrive at a consensus to either accept or reject the proposal for inclusion—or may suggest further study and consideration.

Sometimes the delegates need to work together to refine some proposals before consensus is reached. This consensus will be captured in a *working document*, and vigorous discussion continues over the lunch table, during the coffee breaks, and at restaurants during dinner.

The text from the proposals and working documents are then worked into the latest draft by the *editor* of the recommendation (each recommendation has one of the delegates as an editor). Once all the contributions have been considered, the process of *drafting* takes place. For each recommendation, the editor will go through the changes made to the text and the delegates in the room will approve, by consensus, the appropriate language for inclusion. Given the number of contributions and recommendations, it should not surprise you how many hours this process takes.

At the end of the drafting process, several recommendations may be proposed for *consent*, which means they are finished (for now) and the document is in a state fit for publication. This document can be an amendment, a correction, a complete revision, or a totally new publication. Following consent by the wider study group, a formal approval process takes place within ITU-T officialdom, after which there is some further professional editing and then publication. This process may take several months following the consent at a plenary. At the last plenary (September 2020), Q13/15 consented 8 new editions of recommendations, while at the one before that (February 2020), it was 14 new documents, 12 of them being recommendations. At time of this writing, mid-April 2021, there is a virtual plenary under way that proposes to consent 6 documents, consisting of revisions and amendments to existing recommendations.

That is the process for making recommendations. The only additional detail is that some questions have so much material to consider that they have *interim* meetings (where just delegates from a single question assemble) scheduled between the plenary meetings. These interim meetings are usually hosted by a member organization at some facility, located pretty much anywhere. For Q13/15, they occur 3–5 months before or after a plenary, run for one week, and typically attract around 30–40 delegates. No recommendations can be consented at an interim meeting, but contributions and drafting take place as normal.

Of course, in 2020 with the COVID pandemic, this process evolved to a slightly different approach, changed to use online, virtual meetings. One advantage is that the lack of travel allows many more delegates to participate, but time zone differences make it difficult for many.

Notes on the Recommendations

Be aware that the recommendations are being constantly reviewed and updated by Q13/15, so various versions of any given document are available for download at any one time. Even though a single version (usually the latest) might be shown to be *in force*, that does not mean that the in-force version is reflected in the devices sitting in your network.

Generally, the implementation of these standards in everyday products tends to lag by a significant period—up to several years. This is because it takes time for the component vendors who make oscillators, phase-locked loop (PLL), servos, stacks, clocks, physical layer (PHY) devices, Global Navigation Satellite System (GNSS) chipsets, and so on to update their designs and then test, certify, and manufacture newer versions of their products.

Then the equipment vendors need to incorporate these new devices into later generations of their products, which may require a complete redesign or re-spin of the current hardware design. Once the hardware is done, it is followed by engineering of the required support in software, performing validation testing, performance tuning, and finally shipping to the operator. Just be aware that there is some significant lag in adoption of new features and measures.

Over the past few years, there has also been change in the way the ITU-T distributes its recommendations. Historically, each recommendation had to be read in combination with several amendments, corrections (corrigenda), errata, and revisions. This made it difficult for the casual reader to consume, as there was rarely a single document that contained the complete text of a single recommendation.

Recently, however, the ITU-T has moved to a system whereby it re-releases updated recommendations in full, with change bars indicating alterations to the text. In this way, an amendment to a recommendation now contains the full text, whereas previously it contained just the changes to the original text. This approach makes it much easier for the reader.

Additionally, many recommendations contain annexes and appendixes. The difference between them is that an annex is *normative* (meaning it is part of the specification) and an appendix is *informative* (only for information and educational purposes). To indicate the importance of material in an annex, the annex has a subheading that states "This annex forms an integral part of this Recommendation."

To claim compliance with a recommendation, an implementor must address the annexes but not necessarily the appendixes. Despite this, the annexes to some recommendations may be an optional component. On the other hand, the appendix is for any information that is considered as merely illustrative or supplementary to the recommendation.

This description of those recommendations is current to those editions published resulting from the ITU-T (virtual) Plenary Meeting in September 2020. As outlined in the section on the development process, the consent happens at that plenary meeting, but the formal process of approval and publication takes several months.

Physical and TDM Versus Packet Recommendations

The ITU has developed many recommendations related to timing in the areas of SDH, SONET, and SyncE. For the purposes of this book, the treatment of these recommendations is split into physical (non-packet) and packet-based timing, but of course, there is overlap. Obviously, there is no one ideal method for categorization, but this layout should make it as clear as possible. Just remember that the boundaries between the categories can be somewhat blurred. For example, there is merging of aspects of the packet and physical approach to frequency synchronization in some of the recommendations such as G.8261. In addition, the implementation of packet-based phase and time synchronization may be implemented in combination with a physical frequency distribution over a packet network, such as SyncE.

Table 8-2 illustrates the method used to group and categorize the recommendations for this chapter.

Table 8-2 *Layout and Categories of the Recommendations for This Chapter*

Methods	Frequency	Phase and Time
Physical methods	Physical and TDM use cases: ■ Legacy TDM and Circuit Emulation ■ Frequency-only 2G/3G/4G	Legacy synchronization (not covered here)
Packet methods	Frequency in packet networks: ■ SyncE and ESMC ■ Frequency-only 2G/3G/4G ■ Frequency with packet methods	Packet-based time/phase: ■ PTP phase/time ■ Time/phase 4G/5G mobile
PTP telecom profiles	PTP G.8265.1 telecom profile	G.8275.1 and G.8275.2 profiles

Types of Recommendations

There are different types of recommendations written for different purposes, so it is important to know what sort of recommendation you need to consult and for what purpose. They are layered in approach, starting with the top-level recommendations that define an end-to-end problem to be addressed and the assumptions and requirements to use when solving it. Subsequent recommendations address different levels and facets of the synchronization solution. Figure 8-1 illustrates how these recommendations work to support each other.

The architecture documents set up the assumptions and define the problem. The network limits documents then define the end-to-end performance budget. The equipment limits documents define the performance of each component in the chain. Again, there is some overlap, and the boundaries are sometimes fuzzy, but this is the general approach.

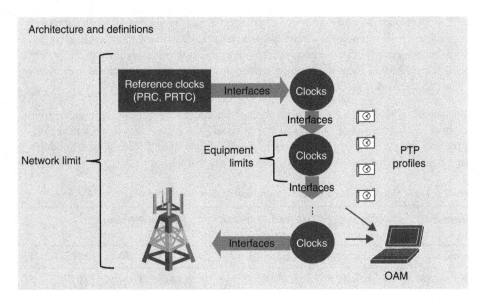

Figure 8-1 *Categories of ITU-T Recommendations on Synchronization and Timing*

The following list details the types of recommendations within each category of synchronization method:

■ *Functional architectures* can contain definitions, terminology, and abbreviations as well as the architecture, framework, and requirements of the timing scenario.

■ *Synchronization layer functions* contains atomic functions or function blocks for the scenario. These recommendations may also include fundamental logic flow and state diagrams for the control-plane functions. Many engineers regard them as describing the control-plane aspects of the solution.

■ *Network limits and solution requirements* outline the end-to-end performance and requirements that need to be met across the complete timing distribution network. These documents define the expected quality of the timing signal that emerges from the end of the timing chain and are extensively modelled by sophisticated mathematical simulations.

■ *Equipment clock specifications and limits* define the performance and functional requirements that each component of the end-to-end timing distribution needs to meet, so that the network limits will not be breached. This includes many recommendations that define the specifications and performance for reference clocks such as a PRTC.

■ *Telecom profiles* define the protocol, message flows, fields, and values as well as the clock selection mechanisms that will run on the previously defined clocks. These profiles define the exact mechanism and messages used to transfer frequency, phase, and time.

- *Other documents and recommendations* exist that may be harder to characterize, such as those defining the electrical specifications of a timing interface (such as a 10-MHz signal) or the method used to carry quality-level traceability in the network (such as ESMC). There may also be documents concerning some related area of interest to the scenario or an operational aspect such as OAM.

Table 8-3 puts the Q13/15 recommendations into context within these categories, grouping them via the method of time transfer. You should understand that it is not necessary to try and understand all these recommendations, or even a significant subset. So, don't lose hope if it seems somewhat overwhelming at this point.

Table 8-3 *TDM Versus Packet-Based Synchronization Methods*

Requirements	TDM and Physical Network	Packet Network
Functional architecture	G.810	G.8260, G.8265, G.8273, G.8275
Synchronization layer functions	G.781, G.707 (SSM)	G.8264 (ESMC), G.781.1
Network limits and solution requirements	G.823, G.824, G.825	G.8261, G.8261.1, G.8271, G.8271.1 (FTS), G.8271.2 (PTS)
Reference clock specifications	G.811 (PRC), G.811.1 (ePRC)	G.8272 (PRTC), G.8272.1 (ePRTC)
Equipment clock specifications for frequency	G.812, G.813, G.8262 (OEC), G.8262.1 (eOEC)	G.8262 (EEC), G.8262.1 (eEEC), G.8263, G.8266 (PEC-M)
Equipment clock specifications for phase/time	—	G.8273.1 (T-GM), G.8273.2 (T-BC/T-TSC), G.8273.3 (T-TC), G.8273.4 (APTS)
Telecom profile for frequency	—	G.8265.1
Telecom profiles for phase/time	—	G.8275.1, G.8275.2
Characteristics of interfaces	G.703	—

However, there is another way to distill this down to a more reasonable subset of recommendations if you are only concerned with using PTP as the timing transport. This distillation makes sense for many readers because it aligns with a use case that only needs a smaller subset of recommendations to implement. Table 8-4 gives a short overview of common use cases and where to get started when deploying a PTP-based solution for either frequency or phase/time.

For example, for an engineer implementing timing for a mobile network, the place to start might be the chosen profile, say G.8275.1. Then the next would be the recommendation for the associated Telecom Boundary Clock (T-BC) and Telecom Time Slave Clock (T-TSC) in G.8273.2. Then, the network architecture associated with that profile (for full on-path timing support) would be the G.8271.1 recommendation for full timing support.

Table 8-4   *Commonly Grouped Recommendations for PTP Deployments*

Network Architecture	Architecture	Network Limits	Clocks	Profile
Packet frequency	G.8265	G.8261.1	G.8263	G.8265.1
Full timing support	G.8275	G.8271.1	G.8273.2	G.8275.1
Partial timing support	G.8275	G.8271.2	G.8273.4	G.8275.2
Assisted partial timing support	G.8275	G.8271.2	G.8273.4	G.8275.2

Just understanding what is contained in those few recommendations covers most of what the engineer would need. Perhaps questions around SyncE or enhanced SyncE might come up, which might bring in G.8262, G.8262.1, and G.8264. Whatever the case, understand that you don't need to consult these documents in day-to-day operations—most often only when initially deploying, doing acceptance testing, fixing interoperability issues, or something goes awry (hopefully rarely).

Reading the Recommendations

As outlined in Table 8-1, the following sections group each of the recommendations into the following major categories:

- Physical and TDM timing
- Frequency in packet networks (SyncE)
- Packet-based timing distribution (of frequency, phase, and time)

The section for each category includes a summary outline of each of the important recommendations belonging to that category. The goal for these sections is to enable the reader to navigate to where further information can be found on topics of interest and to provide a summary of what can be found there.

Each of these recommendations is available for download (for free!) in PDF format from the website https://www.itu.int. With a decent understanding of the concepts contained in this book, you should be able to understand these documents and put them into the overall solution context. This is not to say that you will understand all of it, but you should understand it well enough for 99% of what you need to do.

ITU-T Physical and TDM Timing Recommendations

The first section of the recommendations to address is that related to frequency transfer using physical and TDM methods. These recommendations include many that specify the historical methods of frequency synchronization, although these techniques were (and still are) widely used in the older 2G/3G radio base stations for mobile networks. They also are still relevant to more modern techniques, such as SyncE, because it was designed with similar goals to the existing legacy methods that came from SDH and SONET.

The first section will categorize the different types of recommendations and how they fit into the overall picture. Subsequent sections will address each category in turn and outline a summary of the recommendations that apply for that category.

Types of Standards for Physical Synchronization

The frequency timing recommendations using physical and TDM methods can be broadly grouped into these following categories:

- **Definitions, architecture, and requirements:** G.781, G.810

- **End-to-end solution and network performance:** G.823, G.824, G.825

- **Node and clock performance:** G.811, G.811.1, G.812, G.813

- **Other recommendation:** G.703

Many readers might decide that this section on frequency transfer via physical means is not of great interest to them. Others, such as those interested in circuit emulation (of TDM circuits over packet networks) will recognize that they are important and relevant. But this section is also relevant for those interested in mobile synchronization because even the packet-based methods for frequency transfer (and SyncE) are underpinned by these physical recommendations.

Another reason for coverage of these recommendations in the book becomes apparent in the section "Testing Timing" in Chapter 12, "Operating and Verifying Timing Solutions." These recommendations define the frequency limits for both performance of standalone node clocks (of various quality levels) and end-to-end performance in a frequency distribution chain. When testing either standalone network elements or network performance with lab equipment, the engineer will need to know what performance *mask* to apply to the test results to determine a pass or fail. Those masks are defined by these recommendations.

Definitions, Architecture, and Requirements

These high-level recommendations are the basis for all the following physical frequency recommendations. They contain the definitions and terminologies as well as the architecture and requirement for distributing frequency synchronization using physical methods. This is not the entire set but includes those directly applicable to physical frequency timing.

G.781: Synchronization Layer Functions

ITU-T G.781 defines sets of atomic functions that are part of what it defines in three layers: synchronization, network, and transport. These functions describe the synchronization of SDH, Ethernet, and OTN network elements and how these elements are involved in network synchronization. From the scope of the recommendation:

This Recommendation specifies a library of basic synchronization distribution building blocks, referred to as "atomic functions" and a set of rules by which they are combined in order to describe a digital transmission equipment's synchronization functionality.

Although G.781 is quite a large document and somewhat dense for the new reader, it does contain quite a deal of useful information around the basic concepts. This recommendation breaks the available TDM network technologies into the following *options*:

- **Option I:** Based on the 2048-kbps hierarchy (for example, E1). It includes dedicated timing circuits (without traffic) of 2048 kHz and 2048 kbps.

- **Option II:** Based on the 1544-kbps hierarchy that includes the rates 1544 kbps, 6312 kbps, and 44,736 kbps (for example, T1). It includes dedicated timing circuits (without traffic) of 64 kHz and 1544 kbps.

- **Option III:** Based on the 1544-kbps hierarchy that includes the rates 1544 kbps, 6312 kbps, 33,064 kbps, 44,736 kbps, and 97,728 kbps. It includes dedicated timing circuits (without traffic) of 64 kHz and 6312 kHz.

G.781 includes the following main topics, annex, and appendixes:

- Synchronization basics, including an overview of synchronization interfaces, definitions of quality levels (QL), synchronization status message (SSM), and type, length, value (TLV) for the various quality levels. Also covers concepts of the hold-off and wait-to-restore timers, source priorities, automatic source selection, and timing loop prevention.

- Atomic function of the various layers of synchronization.

- Annex A, "Synchronization selection process." This is a very important annex because it details the process flow for a clock to select the best frequency source under numerous conditions.

- Appendix IV, "Interworking of option II equipment supporting second-generation SSM and first-generation SSM using a translation function."

- Appendix VI, "Clarification of the use of the acronym 'SEC'." This appendix clarifies the use of the term SEC to mean synchronous equipment clock as well as SDH equipment clock and how it relates to the quality level QL-SEC. This point is also made throughout this chapter.

- Appendix VII, "Use cases of mixing SSM and eSSM."

G.781 defines a very important function for the frequency synchronization approach, which is defining the process to select the best available frequency source (in clause 5 in combination with Annex A). See the section "Clock Selection Process" in Chapter 6, "Physical Frequency Synchronization," for an explanation of how this process works.

Another important point in G.781 also deals with the new meaning of the acronym SEC, which is explained in Appendix VI and clause 3.6. Originally the term "SEC" was used to refer to an *SDH* equipment clock, but now the term has been changing (from about 2018) to mean *synchronous* equipment clock. There is also a note explaining this in the section on G.8261 later in the chapter.

This new SEC is a generic term used to represent all the following: the SDH equipment clock (G.813), the Ethernet equipment clock (EEC in G.8262), and the OTN equipment clock (OEC in G.8262). However, some of the older, TDM-based recommendations (for example, G.813) still use SEC to refer to the specific SDH equipment clock.

Note   Link to recommendation: https://handle.itu.int/11.1002/1000/14240

G.810: Definitions and Terminology for Synchronization Networks

ITU-T G.810 is quite an old recommendation, and it has hardly been touched for more than 20 years. It gives definitions and abbreviations used in the ITU-T timing and synchronization recommendations. From the scope of the recommendation:

> This Recommendation provides definitions and abbreviations used in timing and synchronization Recommendations. It also provides background information on the need to limit phase variation and the impairments on digital systems.

This recommendation includes the following main topics and appendix:

- Definitions relating to clock equipment, synchronization networks, clock modes of operation, clock characterization, and some SDH-specific terms

- Phase variation and impairments, and the specification of them

- Clock measurement

- Appendix II, "Definitions and properties of frequency and time stability quantities," which covers Allan deviation (ADEV), modified ADEV (MDEV), time deviation (TDEV), time interval error (TIE), and maximum time interval error (MTIE)

This recommendation is good background reading to get a different explanation of the common terms and concepts used in synchronization and timing. The fact that it is not very hard to read (at least most of it) makes it a handy aid to the learning process.

Note   Link to recommendation: https://www.itu.int/rec/T-REC-G.810/en

End-to-End Network Performance

These recommendations belong to a collection that defines the end-to-end network performance of a group of clocks and network elements carrying frequency with physical methods. They also go back a long way in time and have barely been updated for the last 20 years.

G.823: The Control of Jitter and Wander Within Digital Networks Which Are Based on the 2048 kbit/s Hierarchy

ITU-T G.823 covers the network limits for jitter and wander for traffic and synchronization networks based on the 2048-kbps hierarchy. The electrical characteristics of these interfaces are defined in G.703. From the introduction to the recommendation:

> An excessive amount of jitter and wander can adversely affect both digital (e.g. by generation of bit errors, slips and other abnormalities) and analogue signals (e.g. by unwanted phase modulation of the transmitted signal). ...

> It is therefore necessary to set limits on the maximum magnitude of jitter and wander, and the corresponding minimum jitter and wander tolerance at network interfaces, in order to guarantee a proper quality of the transmitted signals and a proper design of the equipment.

This recommendation includes the following main topics, annexes, and appendixes:

- Network limits for traffic interfaces (output jitter, output wander)

- Network limits for synchronization interfaces (output jitter, output wander) for clocks of different quality, such as primary reference clocks (PRC), synchronization supply units (SSU), SDH equipment clocks (SEC), and plesiochronous digital hierarchy (PDH)

- Jitter and wander (input) tolerance of traffic and synchronization interfaces

- Annex A, "Network model underlying the synchronization network limit"

- Annex B, "Network wander reference model and parameters"

- Appendix I and II: Cover consideration and measurement methods for wander

G.823 is more than 20 years old. To read the complete recommendation, you only need to read the latest recommendation from 2000. Since this applies to circuits based on the 2048-kbps system, it interests only those regions using European type circuits, which is most places except for the United States, Canada, and Japan.

> **Note** Link to recommendation: https://www.itu.int/rec/T-REC-G.823/en

G.824: The Control of Jitter and Wander Within Digital Networks Which Are Based on the 1544 kbit/s Hierarchy

ITU-T G.824 covers the network limits for jitter and wander for traffic and synchronization networks based on the 1544-kbps hierarchy. The electrical characteristics of these interfaces are defined in G.703. This is basically the T1 version of G.823. From the scope of the recommendation:

> An excessive amount of jitter and wander can adversely affect both digital (generation of bit errors, uncontrolled slips) and analogue signals (unwanted phase

modulation of the transmitted signal). It is therefore necessary to set limits to the presence of jitter and wander at the network interfaces, in order to guarantee a proper quality of the transmitted signals.

This recommendation includes the following main topics and annex:

- Network limits for traffic interfaces (output jitter, output wander)

- Network limits for synchronization interfaces (output jitter, output wander) for PRC clocks and 1544 reference interfaces

- Jitter and wander tolerance of traffic (input) and clock (input) interfaces

- Annex A, "Wander reference models and wander budgets"

Although G.824 is more than 20 years old, a correction was published in 2015. To read the complete recommendation, you will need both documents, which are accessible at the following link. Because this recommendation applies to circuits based on the 1544-kbps system, it interests only those using North American–type circuits, which includes the United States, Canada, and Japan.

> **Note**   Link to recommendation: https://handle.itu.int/11.1002/1000/12560

G.825: The Control of Jitter and Wander Within Digital Networks Which Are Based on the Synchronous Digital Hierarchy (SDH)

ITU-T G.825 covers the network limits for jitter and wander for traffic and synchronization networks based on the SDH hierarchy. The architectural aspects of the SDH networks are defined in G.803 with optical, electrical, and other information across numerous other recommendations. From the scope of the recommendation:

> The scope of this ITU-T Recommendation is to define the parameters and the relevant values that are able to control satisfactorily the amount of jitter and wander present at the SDH network-network interfaces (NNI).

As seen previously, jitter and wander requirements for PDH and synchronization networks based on the first-level bit rate of 2048 kbps are specified in G.823. Networks based on the first-level bit rate of 1544 kbps are covered in G.824.

This recommendation includes the following main topics and appendixes:

- Network limits for synchronous transport module (STM-N) interfaces (output jitter, output wander)

- Jitter and wander tolerance of STM-N input interfaces

- Jitter and wander generation and transfer

- Appendix I, "Relationship between network interface jitter requirements and input jitter tolerance"

■ Appendix II, "Measurement methodology for output wander of synchronous interfaces"

G.825 is also more than 20 years old, but an erratum was published in 2001, and an amendment was published in 2008. The amendment added information on STM-256 (40 Gbps) interfaces, which may be of interest. However, to read the complete recommendation, you will need all three documents, which are accessible at the following link: https://www.itu.int/rec/T-REC-G.825/en.

Node and Clock Performance

These recommendations cover the timing requirements for a hierarchy of clocks in the physical timing distribution network. The network builds a hierarchy of clocks using a master-slave relationship between each level of the hierarchy. Each clock is synchronized to a higher level over a synchronization distribution network, with the highest level being the PRC.

The hierarchical levels and their associated recommendations are

■ PRC in G.811

■ Slave clock (transit node) in G.812

■ Slave clock (local node) in G.812

■ SDH network element clock in G.813.

Figure 8-2 illustrates the hierarchy.

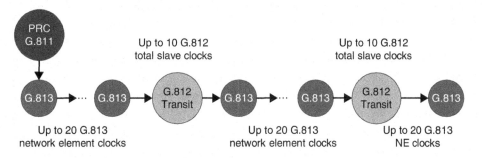

Figure 8-2 *Synchronization Network Reference Chain (Based on Figure 8-5 of ITU-T G.803)*

The further restriction on that topology is that the total number of SDH network element clocks in a chain is limited to 60. This does not mean a limit of 60 network elements, simply a limit of 60 in a single chain. For further information on many aspects of the synchronization chain, see clause 8, "Architecture of synchronization networks," in G.803. The recommendations for these different clock types are each addressed in the following sections.

Because traditional frequency timing is not the focus of this book, this is the limit of the coverage of the topic. There are many other resources available that cover PDH, SDH, and SONET timing in far more detail. So, the following section covers the different TDM clocks types, and then the coverage moves on to packet timing.

G.811: Timing Characteristics of Primary Reference Clocks

ITU-T G.811 defines the (timing) characteristics of PRC clocks as sources of frequency for a synchronization network. From the scope of the recommendation:

> This Recommendation outlines the requirements for primary reference clocks (PRCs) suitable for synchronization supply to digital networks. These requirements apply under the normal environmental conditions specified for digital equipment.

This recommendation includes the following main topics:

- Frequency accuracy requirements (1 part in 10^{11} over one week)

- Noise generation (wander and jitter at the output interface)

- Output interfaces defined

Even though the original recommendation goes back to the 1970s, the latest version of G.811 is "only" 25 years old, although there was an amendment published in 2016. The amendment covers some changes for the enhanced primary reference time clock (ePRTC) and for 10-MHz interfaces. Thankfully, the amendment was published with full text, so only that single document is needed to read the complete recommendation.

> **Note**   Link to recommendation: https://handle.itu.int/11.1002/1000/12792

G.811.1: Timing Characteristics of Enhanced Primary Reference Clocks

ITU-T G.811.1 defines the (timing) characteristics of enhanced PRC clocks as sources of frequency for a synchronization network. The frequency accuracy of an ePRC is ten times that of a PRC. From the scope of the recommendation:

> This Recommendation outlines the requirements for enhanced primary reference clocks (ePRCs) suitable for frequency synchronization. These requirements apply under normal environmental conditions specified for the equipment.

This recommendation includes the following main topics:

- Frequency accuracy requirements (1 part in 10^{12} over one week)

- Noise generation (wander and jitter at output)

- Output interfaces defined

G.811.1 is a recent recommendation and covers the latest ePRC clocks. Like G.811, it is a relatively short and straightforward document.

Note   Link to recommendation: https://handle.itu.int/11.1002/1000/13301

G.812: Timing Requirements of Slave Clocks Suitable for Use as Node Clocks in Synchronization Networks

ITU-T G.812 outlines minimum requirements of timing devices used as nodes in synchronization networks. This covers aspects such as frequency deviation; pull-in, hold-in, and pullout range; noise generation, tolerance, and transfer; and transient response and holdover performance. From the scope of the recommendation:

> The function of a node clock is to select one of the external synchronization links coming into a telecommunication station as the active synchronization reference, to attenuate its jitter and wander and subsequently to distribute the reference to the telecommunication equipment in the station.

These node clocks are basically timing devices that are better than network element clocks and can be used in a frequency distribution network to provide better performance (especially during holdover when the connection to the PRC fails). So, they are better than network elements but not as good as PRC clocks.

G.812 defines six separate clock types:

- **Type I**: For use in networks optimized for the 2048-kbps hierarchy

- **Type II**: For use in networks optimized for the 1544-kbps hierarchy with more stringent holdover requirements

- **Type III**: For use in networks optimized for the 1544-kbps hierarchy with less stringent holdover requirements—for use in end offices

- **Type IV**: For use in networks optimized for the 1544-kbps hierarchy

- **Type V**: Transit node clock from the 1988 version of this recommendation (historical)

- **Type VI**: Local node clock from the 1988 version of this recommendation (historical)

This recommendation includes the following main topics, annex, and appendix:

- Frequency accuracy

- Pull-in, hold-in, and pull-out ranges

- Noise generation (locked wander, non-locked wander, jitter)

- Noise tolerance (wander and jitter)

- Noise transfer

- Short-term and long-term (holdover) phase transient response

- Interface definitions

- Annex A, "Specifications for Types IV, V and VI clocks"

- Appendix II, "Measurement method for noise transfer"

Note   Link to recommendation: https://www.itu.int/rec/T-REC-G.812/en

G.813: Timing Characteristics of SDH Equipment Slave Clocks (SEC)

ITU-T G.813 outlines the minimum requirements of SDH equipment clocks. From the scope of the recommendation:

> This Recommendation contains two options for the SEC. The first option, referred to as "Option 1," applies to SDH networks optimized for the 2048 kbit/s hierarchy. These networks allow the worst-case synchronization reference chain as specified in Figure 8-5 [of] G.803. The second option, referred to as "Option 2," applies to SDH networks optimized for the particular 1544 kbit/s hierarchy that includes the rates 1544 kbit/s, 6312 kbit/s, and 44 736 kbit/s.

This recommendation includes the following main topics and appendixes:

- Frequency accuracy for Option 1 and Option 2 SEC nodes

- Pull-in, hold-in, and pull-out ranges for both options

- Noise generation, noise tolerance, and noise transfer

- Transient response and holdover performance

- Appendix I, "Guidance on the relationship between network limits and input noise tolerances"

- Appendix II, "Considerations on bandwidth requirements, noise accumulation and payload wander accumulation"

The appendixes are very good reading if the topic is of interest.

The main reason for coverage of this recommendation in this book is that it defines the frequency limits for the Option I and Option II SEC nodes (Ethernet, TDM, and optical) when you want to test the timing performance of a clock.

Note   G.813 is approaching 20 years old, although two corrigenda (corrections) have been published, in 2005 and 2016. To read the complete recommendation, you will need all three documents, which are accessible at the following link: https://handle.itu.int/ 11.1002/1000/13084.

Other Documents

There is one document that does not quite fit well with the others in this physical frequency section, but it is quite important. G.703 is a recommendation that specifies the physical and electrical characteristics of interfaces. This is an important recommendation to help achieve interoperability between vendors and implementations. Of course, many of the interfaces defined in G.703 are not especially interesting for timing, but there are several defined that are highly relevant for carrying physical timing signals.

G.703: Physical/Electrical Characteristics of Hierarchical Digital Interfaces

ITU-T G.703 defines the physical and electrical characteristics of a whole range of interfaces applicable to PDH and SDH. From the scope of the recommendation:

> Recommendation ITU-T G.703 specifies the recommended physical and electrical characteristics of the interfaces at hierarchical bit rates as described in Recommendations ITU-T G.702 (PDH) and ITU-T G.707 (SDH). The interfaces are defined in terms of general characteristics, specifications at the output ports and input ports and/or cross-connect points, earthing of outer conductor or screen and coding rules.

This recommendation includes the following main topics:

- Many interfaces from 64 kbps up to 139,264 kbps

- Timing interfaces, including 2048 kHz, 1 pulse per second (1PPS), 10 MHz, and time of day (ToD)

- Definitions of modified alternate mark inversion codes such as High Density Bipolar of Order 3 (HDB3)

The main reason for including G.703 in this chapter is that it defines the electrical characteristics of three interfaces used expressly to carry timing signals, for example, between a GNSS receiver and a packet grandmaster. These three interfaces are defined in the following G.703 clauses:

- Clause 15, "2048 kHz synchronization interface (T12)": Defines the 2048-kHz synchronization signal (for frequency)

- Clause 19, "Time synchronization interfaces defined in ITU-T G.8271/Y.1366": Defines the V.11-based (using RJ-45) time/phase distribution interface to carry 1PPS and ToD signals

- Clause 20, "10 MHz synchronization interface": Defines the 10-MHz interface (for frequency)

This recommendation is very old, but the latest version from 2016 was released with full text, so only that version of the document is required to get the complete recommendation.

> **Note**   Link to recommendation: https://handle.itu.int/11.1002/1000/12788

ITU-T Recommendations for Frequency in Packet Networks

The ITU-T frequency timing recommendations for packet networks are divided into two groups, but there is some overlap, so precise categorization is not always easy. One group includes the recommendations that apply to physical timing in a network designed to transfer packets (namely SyncE). The other group includes the recommendations that apply when transporting frequency using the packets themselves (namely PTP and other techniques used with circuit emulation). However, some recommendations, such as G.8261, apply more generally to both cases.

The recommendations covered in this section include

- Packet-based frequency and circuit emulation network limits: G.8261, G.8261.1

- Synchronous Ethernet clocks: G.8262, G.8262.1

- Carrying timing quality in packet networks (ESMC): G.8264

Packet-Based Frequency and Circuit Emulation

This section is still concerned with the transport of frequency, but this time the recommendations are for when a frequency signal is carried using packet-based techniques rather than a physical method.

The most common use case for this section is for networks implementing circuit emulation (CEM) as a replacement for end-to-end TDM circuits. But many mobile networks also use these methods to provide frequency synchronization for 2G/3G networks in those circumstances where a physical method (such as SyncE) is either not available or too expensive.

G.8261: Timing and Synchronization Aspects in Packet Networks

ITU-T G.8261 defines numerous characteristics of frequency synchronization in packet networks. It specifies the limits of network jitter and wander that should not be exceeded. It also stipulates the minimum tolerance of the network node to jitter and wander at the TDM and synchronization interfaces. From the scope of the recommendation:

> This Recommendation defines frequency synchronization aspects in packet networks. It specifies the maximum network limits of jitter and wander that shall not be exceeded. It specifies the minimum equipment tolerance to jitter and wander that shall be provided at the boundary of these packet networks at TDM and synchronization interfaces. It also outlines the minimum requirements for the synchronization function of network elements.

This means that it defines the specific limits (using MTIE and TDEV masks) for the jitter and wander in several scenarios. For that reason, when testing frequency performance, either network-wide or standalone node performance, G.8261 contains the reference performance metrics to meet.

Common terms you may encounter in reading this recommendation are

- **Packet-based equipment clock (PEC):** A clock supporting packet-based methods of time distribution

- **Synchronous equipment clock (SEC):** A clock supporting synchronous methods of time distribution, such as

   - **SDH equipment slave clock (SEC):** A clock that uses SDH TDM circuits to transfer time

   - **OTN equipment clock (OEC):** A clock that uses synchronous methods, such as SyncO, over OTN

   - **Ethernet equipment clock (EEC):** A clock that uses synchronous methods such as SyncE

As mentioned earlier in the section "G.781," in the original recommendations, such as G.813, "SDH" was used to refer to the SDH equipment clock alone. But since about 2018, the Q13/15 is transitioning to using "SEC" as a more generic term for any clock that supports synchronous methods of frequency distribution. (See Appendix VI of G.781 for more details.) Therefore, many recommendations are now adopting this nomenclature to avoid having a different name for every clock using a different form of transport technology. That is the reason "SEC" is used in several different ways in G.8261 (and elsewhere).

The G.8261 recommendation covers two main use cases:

- When carrying a reference timing signal for frequency synchronization, what is referred to in the recommendation as packet network timing (PNT) domain.

- When using circuit emulation, and where the circuit has its own service clock that is being transported using packet techniques such as adaptive clock recovery (ACR). This is referred to in the recommendation as the circuit emulation service (CES) domain.

Some of the information for this second case is for further study. This recommendation includes the following main topics, annexes, and appendixes:

- Packet network synchronization and TDM timing requirements (SDH and PDH). Included are aspects of synchronization network engineering in packet networks including the PNT domain and CES domain.

- Reference timing signal distribution over packet networks (PNT domain).

- Timing recovery for constant bit rate services transported over packet networks (CES domain), including synchronous, differential, and adaptive techniques.

- Network limits for CES and PNT using several deployment cases.

- Lots of information on the impact of impairments to the transport of frequency.

- Annex A, "Proposed network architecture for synchronous Ethernet."

- Annex B, "IWF functional partitioning into CES and PNT IWF and network examples" (describes interworking between the PNT and CES domains, including adaptive and differential clocks).

- Appendix IV, "Applications and use cases" (provides examples of applications and use cases for frequency synchronization for various mobile technologies).

- Appendix VI, "Measurement guidelines for packet-based methods" (gives test cases and scenarios to confirm the required performance metrics).

- Appendix XI, "Relationship between requirements contained in this Recommendation and other key synchronization related recommendations" (this is a very good reference appendix).

- Appendix XII, "Basic principles of timing over packet networks."

Appendix VI contains a set of 17 test cases with different topologies, traffic conditions, and fault scenarios, including models for background traffic. Many operators use these test cases to confirm performance of packet clocks in the lab. It is very important to be aware of the existence of Appendix VI in G.8261, especially when you start to explore testing of frequency synchronization.

Note Link to recommendation: https://handle.itu.int/11.1002/1000/14526

G.8261.1: Packet Delay Variation Network Limits Applicable to Packet-Based Methods (Frequency Synchronization)

ITU-T G.8261.1 covers the packet delay variation (PDV) network limits applicable when frequency synchronization is carried via packets. It also specifies several hypothetical reference models (HRM) that correspond to the worst-case models for most of the mobile backhaul networks. From the scope of the recommendation:

> Two main applications are addressed in this Recommendation: the distribution of a synchronization network clock signal via a packet-based method (e.g., using PTP or NTP packets and using an adaptive approach), and the distribution of a service clock signal over a packet network according to an adaptive clock recovery method...

This recommendation also covers two main use cases:

- When carrying a reference timing signal for frequency synchronization, there must be limits on the PDV generated in the network in order to meet acceptable performance requirements when trying to recover frequency from the packet stream at the slave.

- When using circuit emulation, and where the circuit has its own service clock that is being transported using ACR. Most of the information for this second case is for further study.

This recommendation includes the following topics:

- Several network HRMs.

- Reference points for network limits in packet networks, which are the points at which network limits are applicable. Examples are at the output of the PRC, the output of the master packet clock, and the input and output of the slave packet clock.

- PDV network limits, which are defined as the maximum permissible levels of PDV at the input to the packet slave clock. These limits can differ based on the HRM.

G.8261.1 really needs to be read in conjunction with G.8260, but it gives an idea of what network topologies and performance aspects are required to carry frequency over packet networks.

Note   G.8261.1 has not yet been consolidated into a single document, so both the February 2012 version and the amendment 1 from May 2014 need to be read together. Both documents are available at https://handle.itu.int/11.1002/1000/12190.

Synchronous Ethernet

The recommendations for SyncE build upon the existing ones for TDM and physical frequency transfer as well as those that apply to frequency transfer using packet techniques. Therefore, in this section, you will see the recommendations for clocks that deal with the physical transfer of frequency via an Ethernet (packet) link. This section covers the standard SyncE clock and the newer, more accurate *enhanced* SyncE clock.

These clocks used to be described as the Ethernet equipment clock (EEC) and the enhanced Ethernet equipment clock (eEEC). But since about 2018, Q13/15 is beginning to refer to the SDH equipment clock (SEC) from G.813, the EEC (G.8262), and the OTN equipment clock (OEC) as the generic synchronous equipment clock (SEC). Similarly, the eEEC (and enhanced eOEC) are now referred to as the enhanced SEC (eSEC).

It is not helpful that the acronym SEC is reused in that way, but you need to be aware of it. Some older recommendations, such as G.813, will likely stay with the definition of SEC meaning the SDH equipment clock, whereas others (G.781 and G.8261, for example) use the new term because they apply across the different transport types.

G.8262: Timing Characteristics of a Synchronous Equipment Slave Clock

ITU-T G.8262 defines the Ethernet equipment clock for SyncE (or in newer language, the Ethernet version of the SEC). From the scope of the recommendation:

> Recommendation ITU-T G.8262 outlines requirements for timing devices used in synchronizing network equipment that uses the physical layer to deliver frequency synchronization. This Recommendation defines the requirements for clocks, e.g., bandwidth, frequency accuracy, holdover and noise generation.

This is the network clock in a packet network (for example, in a router that supports SyncE) that is equivalent to the SEC clock of G.813 (for physical/TDM). You should see that the limits are equivalent, at least for Option I networks (based on E1/SDH).

This recommendation includes the following main topics and appendixes:

- Frequency accuracy for an EEC in Option I and Option II networks

- Pull-in, hold-in, and pull-out ranges for both options

- Noise generation, noise tolerance, and noise transfer

- Transient response and holdover performance

- Synchronization input and output interfaces

- Appendix II, "Relationship between requirements contained in this Recommendation and other key synchronization-related Recommendations"

- Appendix III, "List of Ethernet interfaces applicable to synchronous Ethernet"

- Appendix IV, "Considerations related to synchronous Ethernet over 1000BASE-T and 10GBASE-T"

The appendixes are quite important to understand, especially III and IV. If you are using older and slower Ethernet technologies in your network, do not assume your network will automatically support SyncE. You may also need to address some issues when deploying SyncE over copper-based Ethernet (see Chapter 9, "PTP Deployment Considerations"). G.8262 was republished in full, with an amendment, in March 2020.

Note   Link to recommendation: https://handle.itu.int/11.1002/1000/14208

G.8262.1: Timing Characteristics of an Enhanced Synchronous Equipment Slave Clock

ITU-T G.8262.1 defines the *enhanced* Ethernet equipment clock for *enhanced* SyncE. From the scope of the recommendation:

> This Recommendation outlines new requirements for timing devices used in synchronizing network equipment that supports synchronous clocks, involved in time and phase transport. It supports clock distribution based on network-synchronous

line-code methods (e.g., synchronous Ethernet, synchronous optical transport network (OTN) to deliver frequency synchronization).

This Recommendation focuses on the requirements for the enhanced synchronous Ethernet equipment clock (eEEC) and the enhanced synchronous OTN equipment clock (eOEC).

This eEEC (or in newer language, the Ethernet version of the eSEC) is the newer, more accurate alternative to the EEC clock of G.8262. Equipment supporting it is starting to become generally available. One additional note is that the eEEC also needs to support the extended range of quality level values (eESMC) described in the latest amendments for G.8264.

This recommendation includes the following main topics and appendixes:

- Frequency accuracy

- Pull-in, hold-in, and pull-out ranges

- Noise generation, noise tolerance, and noise transfer

- Transient response and holdover performance

- Synchronization input and output interfaces

- Appendix I, "Relationship between requirements contained in this Recommendation and other key synchronization-related Recommendations"

- Other appendixes from G.8262 are still valid for G.8262.1

Note   Link to recommendation: https://handle.itu.int/11.1002/1000/14011

Ethernet Synchronization Messaging Channel (ESMC)

The concept of carrying information about the source and traceability of a timing signal was covered in previous chapters such as the section "Clock Traceability" in Chapter 6. In another section "Ethernet Synchronization Messaging Channel" in Chapter 6, the mechanism to carry QL information using Ethernet frames is explained. This recommendation details that mechanism.

This recommendation is slightly unusual (meaning that it is hard to classify) as it relates to the transport of physical clock quality levels across a packet network. This is the direct equivalent of the SSM from the SDH/SONET world as laid out in G.781. The main difference is that the information is carried in frames using a method native to Ethernet. And like G.781, G.8264 can be thought of as a recommendation that specifies control-plane functions for synchronization.

G.8264: Distribution of Timing Information Through Packet Networks

ITU-T G.8264 defines the mechanism that allows clock quality traceability in a packet network and initially focuses on Ethernet networks as opposed to other packet networks.

This recommendation also details the required architecture in formal modelling language and uses timing flows to describe where and how timing flows through the architecture. From the scope of the recommendation:

> It specifies the synchronization status message (SSM) transport channel namely the Ethernet synchronization messaging channel (ESMC), protocol behaviour and message format.

> The physical layer that is relevant to this Recommendation is the Ethernet media types defined in [IEEE 802.3].

Therefore, G.8264 defines ESMC and is designed to support SyncE. This recommendation includes the following main topics and appendixes:

- Packet network architectures, timing flows, and functional blocks

- Next-generation timing architectures

- Frequency transfer using SyncE

- SSM for SyncE

- Use of SyncE in a multi-operator context

- Appendix I, "Examples of timing flows"

- Appendix II, "Functional models based on ITU-T G.805 and ITU-T G.809"

Amendment 1, released in 2018, makes changes to incorporate the ePRC into the tables of enhanced SSM codes for SyncE. Previous amendments added the enhanced QL values, which are carried in an extended TLV. This extended QL TLV mechanism was developed for use with the eEEC values for QL. For further information on these changes, see the sections "Ethernet Synchronization Messaging Channel" and "Enhanced ESMC" in Chapter 6.

Therefore, one important component to understand is the interworking between different SyncE generations. It is possible to have network nodes such as an eEEC supporting extended QL TLV; an EEC supporting standard QL TLV; and an EEC supporting extended QL TLV in the one network. These questions are addressed in the section "Interworking between different SyncE generations" in G.8264.

Note   Link to recommendation: https://handle.itu.int/11.1002/1000/13547

ITU-T Packet-Based Timing Recommendations

The ITU-T develops its synchronization recommendations based on a solution-led approach. The process starts where other standards organizations define a problem to be solved. For example, 3GPP requires a solution to deliver accurate phase synchronization to 5G mobile radios.

Taking that problem, the ITU-T puts together a range of recommendations to solve the problem, end to end. The ITU-T specifies all aspects of the problem, such as solution architecture, clock performance, network topology, time error budgets, and, finally, the profiles to implement it all. See the section "Types of Recommendations" earlier in the chapter for more detail on these different categories.

The end-to-end assumptions are then validated and confirmed by very thorough and sophisticated simulation. It is not just about PTP profiles; they arise as a result of the process, not the other way around.

This approach is clearly demonstrated in this section on packet-based timing for both frequency and time and phase. It is for this reason that the authors strongly recommend going with the standards-based approach for implementation of timing. Following the path already laid out delivers predictable outcomes, whereas doing it in your own way may not produce the results desired. There are many very large networks already built and in operation that serves as a testament to this approach.

Types of Standards for Packet-Based Synchronization

The packet-based timing recommendations are broadly grouped, and within these categories, the following are key to packet-based timing:

- **Definitions, architecture, and requirements:** G.8260, G.8265, G.8273, G.8275

- **End-to-end solution and network performance:** G.8271, G.8271.1, G.8271.2

- **Node and clock performance:** G.8272, G.8272.1, G.8273.1, G.8273.2, G.8273.3, G.8273.4

- **PTP profiles:** G.8265.1, G.8275.1, G.8275.2

- **Other documentation:** G.8264, GSTR-GNSS, Supplement 68—Synchronization OAM Requirements

The following sections summarize each of the recommendations within these categories. There is included an outline for each of the most important recommendations related to packet-based timing. The goal of this section for the reader is twofold:

- To point to where the recommendations define and describe the end-to-end solution

- To find where further information on implementation can be found, along with a summary of what is contained there

As mentioned previously, each of these recommendations is available for download in PDF format from the https://www.itu.int website.

Definitions, Architecture, and Requirements

These recommendations are the basis for all the other recommendations following. They contain the definitions and terminologies as well as the architecture and requirements for distributing frequency, time, and phase for packet-based synchronization.

G.781.1: Synchronization Layer Functions for Packet-Based Networks

G.781.1 specifies a functional architecture model and the corresponding atomic functions for the transport of time and frequency synchronization via packet-based methods (using PTP). From the scope of the recommendation:

> The functional architecture contains two synchronization layers, the network synchronization layer and the synchronization distribution layer. In addition, some of the synchronization-related atomic functions defined in this recommendation are part of the transport layer.

Note that this recommendation is still being worked on and is not yet published, so it is not yet available for public download.

G.8260: Definitions and Terminology for Synchronization in Packet Networks

ITU-T G.8260 provides the definitions, terminology, and abbreviations used across the recommendations on timing and synchronization in packet networks. From the scope of the recommendation:

> It includes mathematical definitions for various synchronization stability and quality metrics for packet networks, and also provides background information on the nature of packet timing systems and the impairments created by packet networks.

This recommendation contains definitions of numerous concepts that are covered in this book, although G.8260 addresses the topic somewhat more mathematically than here. Although it is not a very long document, it goes to the next level of complexity in explaining concepts. This recommendation includes the following main topics:

- Definition of terms, such as different forms of time error
- Nature of packet timing systems, including the definition of *significant instants*
- Differences between packet-based and physical layer timing systems
- Classes of packet clocks, including packet master and packet slaves
- Two-way timing protocols and the flow of timing from master to slave
- Packet timing signal equipment interface characterization

Most importantly, G.8260 defines the following terms: packet master clock, packet slave clock, and packet timing signal. There is also an appendix outlining packet measurement metrics such as MTIE and TDEV as well as information on filtering and packet selection.

This recommendation is a source for further information on topics covered in Chapter 5, "Clocks, Time Error, and Noise." A revision (with full text including all amendments) was published in 2020.

G.8265: Architecture and Requirements for Packet-Based Frequency Delivery

ITU-T G.8265 describes the architecture and requirements for packet-based frequency distribution in telecom networks. It briefly describes examples of packet-based frequency distribution such as NTP and PTP. From the scope of the recommendation:

> This Recommendation describes the general architecture of frequency distribution using packet-based methods. The requirements and architecture form a base for the specification of other functionality needed to achieve packet-based frequency distribution in a carrier environment. The architecture described covers the case where protocol interaction is at the end points of the network only, between a packet master clock and a packet slave clock.

This recommendation therefore does not cover details of architectures where devices between the packet master and packet slave clocks participate in the timing solution. Included is discussion on these topics:

- Requirements for packet timing
- Architecture of packet-based frequency distribution, including redundancy and network partitioning
- Packet-based protocols for frequency distribution: PTP and NTP
- Security aspects

G.8265 is informative and easy to read. More details on frequency distribution using packet timing are also provided in Appendix XII, "Basic principles of timing over packet networks" of G.8261, although it is a more mathematical treatment the subject.

G.8273: Framework of Phase and Time Clocks

ITU-T G.8273 is a framework recommendation for phase and time clocks using packet-based methods for transferring time/phase. This means that it puts the clock recommendations (the G.8273.x series) in perspective but does not contain a large amount of content about them. From the scope of the recommendation:

> This Recommendation is a framework Recommendation for phase and time clocks for devices used in synchronizing network equipment that operate in the network architecture defined in [ITU T G.8271], [ITU-T G.8275] and the ITU-T G.8271.x-series of Recommendations.

This Recommendation serves as a framework for phase and time clocks defined in the ITU T G.8273.x-series. It includes annexes with detailed testing and measurement methods of phase and time clocks.

This recommendation includes the following main topic, annexes, and appendixes:

- General introduction of phase and time clocks that outlines the clock types covered by the G.8273.x set of recommendations

- Annex A, "Testing and measurement of time/phase clocks"

- Annex B, "Phase/time clock equipment specification related measurement methods"

- Several appendixes on aspects of testing clock performance

The annexes and appendixes contain details on testing and measurement of phase and time clocks that are worthwhile for studying.

There is a correction to this recommendation consented at the October 2020 plenary, but it includes a full reprint of the text, so only a single document is needed.

Note   Link to recommendation: https://handle.itu.int/11.1002/1000/14528

G.8275: Architecture and Requirements for Packet-Based Time and Phase Distribution

ITU-T G.8275 describes the architecture and requirements for packet-based time and phase distribution in telecom networks using PTP. From the scope of the recommendation:

> The requirements and architecture form a base for the specification of other functionalities that are needed to achieve packet-based time and phase distribution in a carrier environment. The architecture described covers the case where protocol interaction is at all nodes, between a packet master clock and a packet slave clock or only a subset of the nodes between a packet master clock and a packet slave clock.

This scope mentions that every node between a master and a slave clock will be involved in "protocol interaction," which basically means that the nodes are understanding and processing PTP (and not simply switching the packets containing it). This is another way of saying that every node needs to be a boundary clock or a transparent clock. Despite this sentence, if you read section 7 or Appendix I, you'll notice that they cover scenarios where that is not the case.

Comparing the scope to that of G.8265, you will notice they are basically the same, except G.8265 is for frequency and G.8275 is for time and phase. Many of the terms used in G.8275 build on the concepts introduced in G.8260 (and G.810).

This recommendation includes the following main topics, annexes, and appendixes:

- General introduction and requirements for packet-based time/phase distribution

- Architecture of packet-based time/phase distribution (including redundancy)

- Security aspects

- Annex A, "Time/phase models based on ITU-T G.805"

- Annex B, "Inclusion of a virtual PTP port on a PTP clock" (which was moved to G.8275 from the G.8275.x profile documents as it was material applicable to both profiles; same approach was taken for Annex C, Annex D, and Appendix VIII)

- Annex C, "Options to establish the PTP topology with the alternate BMCA"

- Annex D, "Synchronization uncertain indication (optional)"

- Appendix I, "Architecture for time and phase distribution over a packet network providing PTS [partial timing support] at the protocol level" (meaning, nodes that do not process PTP)

- Appendix VIII, "Description of PTP clock modes and associated contents of Announce messages"

Appendix VIII is very interesting reading because it describes the linkage between the state of the clock ports, the clock state as a whole, and the contents of the Announce message.

G.8275 clause 7, "Architecture of packet-based time/phase distribution," covers the different cases for the design of the packet network and is especially interesting reading. The architecture describes two cases where

- Support for timing is provided by all nodes in the network (by being boundary clocks, for example) combined with physical layer frequency support (normally SyncE).

- Intermediate nodes do not provide timing support, but timing support is provided by GNSS at the network edge, with PTP acting as a backup. This is termed assisted partial timing support (APTS).

See the earlier sections on G.8271.1 and G.8271.2 for detailed treatment of these two different network topologies. Finally, G.8275 also examines the different options around the positioning of the sources of time and how to address redundancy concerns. A new version of G.8275 was consented at the October 2020 plenary.

Note Link to recommendation: https://handle.itu.int/11.1002/1000/14509

End-to-End Solution and Network Performance

The packet-based timing recommendations covered previously are centered on terms, definitions, requirements, and architecture. These form the basis for this next category, those relating to end-to-end network topology, design, and performance.

Because the ITU-T designs a set of recommendations as a total timing solution, one of the first tasks is to define the characteristics of the packet network that will carry the timing. These recommendations also define the performance limits that must be met by the time signal distributed via this network. These values are known as the *network limits* and include the timing error budget for the end-to-end network.

The ITU-T covers this for the case of packet networks carrying frequency (G.8261 and G.8261.1) and then for the case of packet networks carrying time and phase (G.8271, G.8271.1, and G.8271.2). The first recommendation in each series covers the general aspects, while the .1 and .2 versions go into the specific limits and error budgets of a specific network topology.

G.8271: Time and Phase Synchronization Aspects of Telecommunication Networks

ITU-T G.8271 provides the basis for phase and time synchronization. Therefore, it is the equivalent of what G.8261 is to frequency distribution. From the scope of the recommendation:

> This Recommendation defines time and phase synchronization aspects in telecommunication networks. It specifies the suitable methods to distribute the reference timing signals that can be used to recover the phase synchronization and/or time synchronization according to the required quality. It also specifies the relevant time and phase synchronization interfaces and related performance.

This recommendation includes the following main topics, annex, and appendixes:

- The need for time and phase synchronization, including the phase accuracy requirements for various types of mobile radio technologies

- Methods to distribute phase and time either using a distribution of GNSS receivers or packet-based approaches

- Network reference model and time and phase synchronization interfaces

- Annex A, "One pulse-per-second (1PPS) time and phase synchronization interface specification" (for 1PPS and ToD timing ports)

- Appendix I, "Time and phase noise sources in time distribution chains" (characteristics of timing noise for PRTC, masters, and slaves)

- Appendix II, "Time and phase end application synchronization requirements" (gives more details of the requirements for various types of mobile radio technologies, and includes references to source documents from other standards organizations)

- Appendixes III, IV, and V: Collectively cover asymmetry compensation in the network and transport links

- Appendix VI, "Time synchronization aspects in TDD based mobile communication systems (issues to note from mobile radio systems based on time division duplex techniques)

- Appendix VII, "Time scales" (information on various time scales)

Appendix I is particularly interesting, even if you avoid some of the mathematics. It outlines all the various places where and the reasons why time error can creep in. A new complete reprint of G.8271 was consented in March 2020.

Note   Link to recommendation: https://handle.itu.int/11.1002/1000/14209

G.8271.1: Network Limits for Time Synchronization in Packet Networks with Full Timing Support from the Network

ITU-T G.8271.1 specifies the network limits for time and phase synchronization in packet networks with full timing support from the network (or "with full on-path support"). This case *with* full on-path support is directly equivalent to the G.8271.2 recommendation that covers the case *without* the on-path support.

Note that because the network defined here has full timing support, every node is required to process PTP, so the network is implemented as a "hop-by-hop" topology. In this case, PTP is carried by layer 2 (Ethernet) multicast and is implemented according to the G.8275.1 telecom profile.

This recommendation specifies those limits that apply to the network model for packet-based distribution outlined in G.8271. The limits covered here are the maximum time and phase error generated in the network as well as the minimum phase and time error that the equipment needs to tolerate. From the scope of the recommendation:

> This Recommendation specifies the maximum network limits of phase and time error that shall not be exceeded. It specifies the minimum equipment tolerance to phase and time error that shall be provided at the boundary of packet networks at phase and time synchronization interfaces. It also outlines the minimum requirements for the synchronization function of network elements.

> This Recommendation addresses the case of time and phase distribution across a network with packet-based method with full timing support (FTS) to the protocol level from the network.

G.8271.1 treats two deployment cases, one where the slave clock is embedded in the end application (which for mobile means that the radio has a PTP slave clock), and one where the slave clock is external to the end application. In this second case, the time signal is passed to the final application via some other method (an example would be a timing signal, such as 1PPS, carrying phase).

For each of these two deployment cases, the standard defines limits at various points, A, B, C, D, and E. Point D is hidden in deployment case 1, as it is embedded in the end application. Figure 8-3 shows the deployment case 2 with the non-embedded clock.

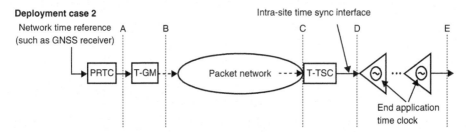

Figure 8-3   *Deployment Case 2 from G.8271.1 with Non-embedded Slave Clock (Based on Figure 7-1 of ITU-T Recommendation G.8271.1)*

This recommendation includes the following main topic and appendixes:

- Network limits at the various points in the network from A through E

- Appendix I, "Clock models for noise accumulation simulations" (noise accumulation models for T-BC and T-TC)

- Appendix II, "Hypothetical reference models used to derive the network limits"

- Appendix III, "Network limit considerations" (how to measure the network limits in the embedded slave case)

- Appendix V, "Example of design options" (includes TE budgeting and failure scenarios)

- Appendix VII, "Maximum relative time error" (covers the difference between absolute time error and relative time error, with Figure VII.1 demonstrating the difference; a very important concept for engineers deploying time in the fronthaul)

- Appendixes VIII, "Models for budgeting in a chain of microwave devices"

- Appendix IX, "Models for budgeting in a chain of xPON or xDSL devices"

- Appendix XII, "Examples of design options for fronthaul and clusters of base stations"

G.8271.1 is one of the most important of the recommendations, because it defines the behavior of the end-to-end network that is used to transport the packet-based synchronization. This is the "best in class" model for deployment of packet-based transport of time. It includes a wealth of detailed information on the more detailed design decisions for a timing synchronization network.

Two versions of this recommendation were consented in 2020, with both being a complete reprint.

> **Note** Link to recommendation: https://handle.itu.int/11.1002/1000/14527

G.8271.2: Network Limits for Time Synchronization in Packet Networks with Partial Timing Support for the Network

ITU-T G.8271.2 specifies the network limits for time and phase synchronization in packet networks with partial timing support from the network (or "without full on-path support"). This case of partial on-path support is directly equivalent to the G.8271.1 recommendation that covers the case with full on-path support.

This recommendation covers two use cases with partial support:

- Assisted partial timing support (APTS), where PTP is used as a backup timing source to a local GNSS primary source. This backup is only meant to be used in the case of a local outage of GNSS and for a limited period (commonly 72 hours).

- Partial timing support (PTS) from the network, where PTP (carried over the transport network) is the primary source of synchronization, but elements in the network do not process PTP to reduce the time error.

From the scope of the recommendation:

> This Recommendation addresses the distribution of time and phase across a network, using the packet-based method with partial timing support to the protocol level from the network. In particular, it applies to the assisted partial timing support (APTS) and partial timing support (PTS) architectures described in [ITU-T G.8275] and the precision time protocol (PTP) profile defined in [ITU-T G.8275.2].

Note that because the network defined here has only partial timing support, PTP must be carried by IP (either IPv4 or IPv6) and therefore is closely associated with the G.8275.2 telecom profile that uses PTP over IP (PTPoIP).

It is this recommendation that covers three new PTP clock types:

- T-BC-P Telecom Boundary Clock with partial support

- T-TSC-A Telecom Time Slave Clock with assisted partial support

- T-TSC-P Telecom Time Slave Clock with partial support

The -P clocks are boundary and slave clocks embedded in a network with PTP timing support available only from a subset of the network elements. The T-TSC-A is a PTP slave clock assisted by a local time reference (normally a GNSS receiver) in the same partially aware topology. Figure 8-4 shows the topology of a network terminated in a T-TSC-A.

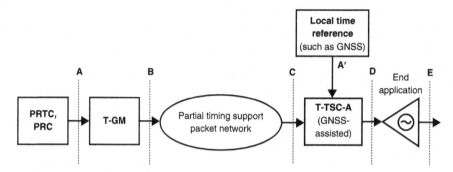

Figure 8-4 *Network Terminating in a T-TSC-A Assisted Partial Slave Clock (Based on Figure 1 of ITU-T Recommendation G.8271.2)*

This recommendation includes the following main topic and appendixes:

- Network limits at the various points in the network at A′ and A through E

- Appendix I, "Deployment scenarios for partial timing support networks"

- Appendix II, "Considerations for handling precision time protocol traffic in networks with partial timing support" (because there is no assistance for PTP in the network elements, the engineers must very carefully design the network to minimize the sources of time error)

- Appendix III, "The use of frequency to maintain precise time"

- Appendix IV, "Noise accumulation model in partially aware networks"

G.8271.2 is also a very important recommendation because it applies to those situations where an operator is not able to provide full timing support on their network.

> **Note**   Link to recommendation: https://handle.itu.int/11.1002/1000/13768

Node and Clock Performance Recommendations

Following on from the end-to-end network topologies, limits, and budgets, the next step is to address the individual elements and clocks within that network. Once the end-to-end budget is allocated, the performance of the individual network elements needs to be characterized. In ITU-T terms, these node values are known as *equipment limits* and include several types of timing error for each of the individual clock types (grandmaster clocks, boundary clocks, transparent clocks, and slave clocks).

These clock recommendations include parameters such as *constant time error, noise generation, noise tolerance, noise transfer, phase error, jitter* and *wander*, and so on. Please see Chapter 5 for more details on these parameters.

The following sections start with the case of network elements carrying frequency with PTP. The recommendations covering the source of frequency (PRC or ePRC) for this case is in the preceding section covering G.811 and G.811.1. Since there is no on-path support with this architecture, there is only a single recommendation (G.8263) covering master and slave packet clocks.

Following that section is coverage of the case of network elements carrying time and phase starting from the PRTC source of time (G.8272, G.8272.1). After that is coverage of the PTP grandmaster clock (G.8273.1), the T-BC boundary and T-TSC slave clock (G.8273.2), the T-TC transparent clock (G.8273.3), and the APTS clock (G.8273.4).

G.8263: Timing Characteristics of Packet-Based Equipment Clocks

ITU-T G.8263 outlines the minimum requirements for the timing functions of the packet-based equipment clock-slave (PEC-S, from G.8265), when timed from a packet-based equipment clock-master (PEC-M, from G.8266). It deals with the packet-based transport of frequency synchronization as opposed to some physical method. From the scope of the recommendation:

> This Recommendation focuses on mobile applications, and in particular on the delivery of frequency synchronization for end applications, such as mobile base stations. It supports the architecture defined in [ITU-T G.8265]. Other applications are for further study.

> This Recommendation includes clock accuracy, packet delay variation (PDV) noise tolerance, holdover performance and noise generation.

G.8263 clocks are typically deployed on the types of networks defined in G.8261.1 and carried using the G.8265.1 PTP profiles. Figure 8-5 shows a simplified view of one of the deployment cases.

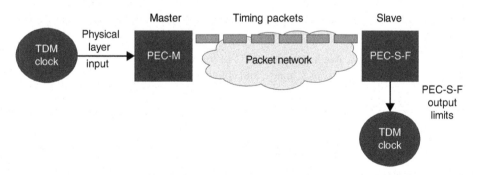

Figure 8-5 *Topology for G.8263 Packet Equipment Slave Clocks for Frequency*

The input of the PEC-M is connected to the output from a physical-layer frequency clock (SEC, SSU, PRC). The master generates the packets from this frequency source and sends it to the packet-based equipment clock-slave-frequency (PEC-S-F) that recovers the frequency.

G.8263 defines the applicable limits at all points in this network. This recommendation includes the following main topics:

- Hypothetical reference models

- Reference points for network limits in packet networks

- PEC-M network limits

- PEC-S-F network limits

- PDV network limits

Note   Link to recommendation: https://handle.itu.int/11.1002/1000/13320

G.8266: Timing Characteristics of Telecom Grandmaster Clocks for Frequency Synchronization

ITU-T G.8266 outlines the minimum requirements for the timing functions of the packet-based equipment clock-master (PEC-M), when timed from a physical frequency source traceable to a PRC/PRS or PRTC. It deals with the packet-based transport of frequency synchronization as opposed to some physical method. From the scope of the recommendation:

> This Recommendation defines the minimum requirements for the timing functions of the telecom grandmaster clocks that operate in the network architecture as defined in [ITU-T G.8265] and the related profile defined in [ITU-T G.8265.1]. It supports frequency synchronization distribution when using packet-based methods.

This recommendation includes the standard metrics of clock accuracy, noise tolerance, noise transfer, noise generation, and holdover specifications for T-GM clocks when used for frequency synchronization. Basically, it is a packet version of G.812 (G.8262 for Ethernet interfaces) with many of the same limits depending on whether the clock is Type I, II, or III.

G.8266 defines the applicable limits at the PEC-M for frequency. This recommendation includes the following main topics, annex, and appendix:

- Frequency accuracy

- Pull-in, hold-in, and pull-out ranges

- Noise generation (wander and jitter)

- Noise tolerance and transfer

- Transient response and holdover performance

- Annex A, "Telecom grandmaster clock functional model" (includes a functional model for a T-GM/PEC-M)

- Appendix I, "Measurement method for wander transfer"

> **Note**   Link to recommendation: https://handle.itu.int/11.1002/1000/13548

G.8272: Timing Characteristics of Primary Reference Time Clocks (PRTC)

ITU-T G.8272 provides the requirements for PRTC clocks suitable as a source of time, phase, and frequency synchronization in packet networks. From the scope of the recommendation:

> This Recommendation defines the PRTC output requirements. The accuracy of the PRTC should be maintained as specified in this Recommendation.
>
> This Recommendation also covers the case where a PRTC is integrated with a T-GM clock. In this case it defines the performance at the output of the combined PRTC and T-GM function, i.e., the precision time protocol (PTP) messages.

So, G.8272 basically specifies the accuracy with which the PRTC provides a reference time signal traceable to a recognized time standard, normally Coordinated Universal Time (UTC). As mentioned in the scope, there are two cases:

- Where the PRTC provides only physical signals for frequency, phase, and time—commonly the 1PPS, ToD, and 10-MHz signals outlined in Chapter 3, "Synchronization and Timing Concepts," and Chapter 7, "Precision Time Protocol"

- Where the PRTC is combined with a PTP grandmaster and the frequency, phase, and time is distributed with PTP packets from the device directly

The requirements for accuracy include both cases, although in the second case, one is obviously measuring the error of the T-GM as well as that of the PRTC itself. But the recommendation makes no allowance for that; the limits are the same, although in the combined case, the time error samples are measured through a moving-average low-pass filter.

There are two classes of device covered in this recommendation: PRTC-A and PRTC-B. The time output of PRTC-B is more accurate than that of PRTC-A, so it is better suited to those applications where more accurate time is required. This is covered in much more depth in Chapter 10, "Mobile Timing Requirements," and Chapter 11, "5G Timing Solutions," when addressing timing in the mobile fronthaul network. Generally, PRTC-B is achieved by using GNSS receivers with enhanced capabilities as well as being located where there is better control of the environment, especially temperature.

This recommendation includes the following main topics and appendix:

- Constant time offset compared to the reference time (say, UTC) when locked

- Phase error (wander and jitter) produced at the output of the device, which is specified (for wander) as MTIE and TDEV masks for both classes of device

- Phase and time as well as frequency interfaces, defining the characteristics of the interfaces used to carry physical timing signals, such as 1PPS or a 2-MHz signal

■ Appendix I, "Measuring the performance of a PRTC or a PRTC combined with T-GM" (a very useful source of information when you are considering testing a PRTC in a lab)

Note that that there is currently no requirement for holdover from a PRTC.

If you are interested in the PRTC, you may also find very helpful the technical report on GNSS systems, "GSTR-GNSS Considerations on the use of GNSS as a primary time reference in telecommunications." It is covered later in the chapter in the "Other Documents" section, which includes a link for its location.

Note   Link to recommendation: https://handle.itu.int/11.1002/1000/14211

G.8272.1: Timing Characteristics of Enhanced Primary Reference Time Clocks

ITU-T G.8272.1 provides the requirements for higher accuracy ePRTC clocks suitable as a source of time, phase, and frequency synchronization in packet networks. From the scope of the recommendation:

The ePRTC provides a reference time signal traceable to a recognized time standard (e.g., coordinated universal time (UTC)) and also a frequency reference. Compared to the primary reference time clock (PRTC) as defined in [ITU-T G.8272], the ePRTC is subject to more stringent output performance requirements and includes a frequency input directly from an autonomous primary reference clock.

This Recommendation defines the ePRTC output requirements, including the ePRTC integrated with a T-GM clock.

A difference between the PRTC and ePRTC cases is that the ePRTC has a (stringent) holdover requirement that requires the stable source of frequency. This means that the ePRTC basically requires an atomic clock (PRC) as an oscillator to meet its performance requirements. The case for a combined ePRTC + T-GM is also covered—it allows no additional allowance for time error or wander when the ePRTC has a T-GM function embedded within it.

There are also two classes of device covered in this recommendation: ePRTC-A and ePRTC-B, with the ePRTC-B expected to be higher performance than that of ePRTC-A, although the values for its performance have not yet been agreed.

This recommendation includes the following main topics and annex:

■ Constant time offset compared to the reference time (say, UTC) when locked.

■ Phase error (wander and jitter) produced at the output of the device, which is specified (for wander) as MTIE and TDEV masks.

■ Transient response and phase/time holdover performance, based on the local frequency reference (atomic clock) during loss of phase/time input (from GNSS).

- Phase and time as well as frequency interfaces, defining the characteristics of the interfaces used to carry physical timing signals. There are performance requirements on the V.11 1PPS interface to deliver higher accuracy.

- Annex A, "ePRTC autonomous primary reference clock requirements."

As is the case with the PRTC, if you are interested in the ePRTC, you may also find very helpful the technical report on GNSS systems, "GSTR-GNSS Considerations on the use of GNSS as a primary time reference in telecommunications." It is covered later in the "Other Documents" section, along with a link for its location.

Note Link to recommendation: https://handle.itu.int/11.1002/1000/14014

G.8273.1: Timing Characteristics of Telecom Grandmaster Clocks for Time Synchronization (T-GM)

ITU-T G.8273.1 outlines the timing characteristics for the T-GM clocks. From the scope:

> This Recommendation defines the minimum requirements for the timing functions of the telecom grandmaster clocks that operate in the network architecture as defined in [ITU-T G.8275] and the related profiles defined in [ITU-T G.8275.1] and [ITU-T G.8275.2].

Note that this recommendation is still being developed and is not yet published, so it is not yet available for public download. There are aspects of this document that are covered in other recommendations.

G.8273.2: Timing Characteristics of Telecom Boundary Clocks and Telecom Time Slave Clocks for Use with Full Timing Support from the Network

ITU-T G.8273.2 is one of the most important recommendations for many deployments of accurate phase and time synchronization carried over a packet network. It specifies the minimum requirements for T-BC and T-TSC time and phase clocks when deployed with full timing support from the network. From the scope of the recommendation:

> This Recommendation defines the minimum requirements for telecom boundary clocks and telecom time slave clocks in network elements. These requirements apply under the normal environmental conditions specified for the equipment. The current version of this Recommendation focuses on the case of physical layer frequency support. Requirements related to the case without physical layer frequency support (i.e., the PTP only case) are for further study.
>
> This Recommendation includes noise generation, noise tolerance, noise transfer and transient response for telecom boundary clocks and telecom time slave clocks.

Note that the recommendation currently only applies to the case with "physical frequency support," meaning where PTP is deployed in combination with some physical method (commonly SyncE) for frequency transport.

This recommendation includes the following main topics and appendixes:

- Physical layer frequency performance requirements.

- T-BC performance requirements with full timing support. It includes parameters for time error noise generation, constant time error (cTE), filtered dynamic time error (dTE), relative time error (rTE), noise tolerance, and noise transfer.

- Transient response and holdover performance.

- Phase and time as well as frequency interfaces, defining the characteristics of the interfaces used to carry physical timing signals. There are some increased requirements for class C and D clocks.

- Appendix III, "Background to performance requirements of the T-BC and T-TSC"

- Appendix V, "Performance estimation for cascaded media converters acting as T-BCs and for T-BC chains"

- Appendix VIII, "Measurement of Relative Time Error between two T-BC output ports" (defines two new relative time error generation limits between any two outputs of T-BC)

This recommendation defines T-BC and T-TSC clocks into four classes based on their performance, known as Class A through D (in order of increasing accuracy). Class A and B clocks are appropriate to deploy together with SyncE (G.8262) or some equivalent TDM-based method. For Class C and D clocks, the values in the recommendation were developed based on simulations that assume it is being deployed with eSyncE (G.8262.1).

The importance of these classes of clocks will become apparent in the sections on deployment and time error budgeting. Obviously, the more accurately that a clock can transport time, the more clocks the time can pass through without exceeding the end-to-end budget limits. Similarly, the time error delivered to an application at the end of a chain of clocks can be minimized by utilizing clocks with a lower contribution to the overall time error.

There are two versions of this recommendation consented in March and October 2020.

> **Note**   Link to recommendation: https://handle.itu.int/11.1002/1000/14507

G.8273.3: Timing Characteristics of Telecom Transparent Clocks for Use with Full Timing Support from the Network

ITU-T G.8273.3 defines the minimum requirements for telecom transparent clocks (T-TCs) when used to transport time and phase over a packet network with full timing support. From the scope of the recommendation:

> This Recommendation specifies minimum requirements for time and phase synchronization devices used in synchronizing network equipment... It supports time and/or phase synchronization distribution for packet-based networks.

This Recommendation defines the minimum requirements for transparent clocks. These requirements apply under normal environmental conditions specified for the equipment.

This recommendation includes the following main topics and appendixes:

- Physical layer frequency performance requirements.

- T-TC performance requirements. It includes parameters for maximum absolute time error noise generation (max|TE|, constant time error (cTE), filtered dynamic time error (dTE), relative time error, noise tolerance, and noise transfer.

- Transient response requirements are for further study and there is no holdover capability supported in a transparent clock.

- Appendix I, "Traffic load test patterns" (describes some guidelines for testing transparent clocks)

- Appendix III, "Performance estimation for cascaded media converters acting as T-TCs"

This recommendation defines the noise generation performance of T-TC clocks into three classes, Classes A, B, and C (in order of increasing performance). The requirements for these three classes mirror those from the T-BC and T-TSC cases.

Note that the recommendation focuses on syntonized T-TC clocks, meaning that some physical method (commonly SyncE) is used for frequency synchronization of the T-TC clocks. Therefore, it also requires "physical frequency support" and applies only to transparent clocks running in end-to-end transport clock mode (see the section "End-to-End Transparent Clocks" in Chapter 7). It also assumes that the deployment meets the end-to-end network limits for full timing support (from G.8271.1) and that it uses the G.8275.1 PTP telecom profile.

A new version of G.8273.3 was consented at the October 2020 plenary.

> **Note**   Link to recommendation: https://handle.itu.int/11.1002/1000/14508

G.8273.4: Timing Characteristics of Telecom Boundary Clocks and Telecom Time Slave Clocks for Use with Partial Timing Support from the Network

ITU-T G.8273.4 specifies the minimum requirements for synchronization clocks used in a time and phase synchronization network operating in either APTS or PTS architectures. From the scope of the recommendation:

This Recommendation allows for proper network operation for phase/time synchronization distribution when a network equipment embedding an APTS or a PTS clock is timed from another telecom boundary clock (T-BC) or a telecom grandmaster (T-GM). This Recommendation specifies the minimum requirements for APTS and

PTS clocks in network elements (NEs). These requirements apply under the normal environmental conditions specified for the equipment.

This Recommendation includes noise generation, noise tolerance, noise transfer, and transient response for APTS and PTS clocks.

The APTS and PTS architectures are defined in G.8271.2, and the resulting synchronization deployment assumes the use of the G.8275.2 PTP telecom profile. Physical sources of frequency (such as SyncE) are optional for PTS in this recommendation, and they are not used for APTS.

This recommendation (in combination with G.8271.2) defines the requirements for both boundary and slave clocks in both the FTS and APTS architectures. So, it specifies these clock types:

- **T-BC-A**: Telecom Boundary Clock with assisted partial support

- **T-BC-P**: Telecom Boundary Clock with partial support

- **T-TSC-A**: Telecom Time Slave Clock with assisted partial support

- **T-TSC-P**: Telecom Time Slave Clock with partial support

This recommendation includes the following main topics and appendix:

- Performance requirements for T-BC-A and T-TSC-A, which includes maximum absolute time error (max|TE|), constant time error (cTE), filtered dynamic time error (dTE), noise tolerance, noise transfer, transient response, and holdover

- Performance requirements for T-BC-P and T-TSC-P with a similar range of values to the assist clock case

- Appendix III, "Consideration on the use of synchronous equipment clock to maintain time holdover for partial timing support networks"

APTS is specifically designed for short-term holdover scenarios when the primary GNSS source is lost and a backup PTP source is temporarily unavailable. It is not appropriate to assume that the APTS will provide accurate long-term holdover or deliver accurate timing under double failure conditions (such as subsequent rearrangements in the unaware network). A new version of G.8273.4 was consented at the March 2020 plenary.

Note Link to recommendation: https://handle.itu.int/11.1002/1000/14214

Telecom Profiles

The most frequently consulted recommendations are those that cover the three main telecom profiles, namely G.8265.1, G.8275.1, and G.8275.2. Although the previously covered recommendations describe the solution, the network, and the clocks, it is the profiles that are implemented on the elements that make up the nodes of the world's networks.

If there are any questions about the implementation or interoperability of these protocols during testing, acceptance, initial deployment, or day-to-day operations, then the recommendations on the individual profiles are the place to answer any questions or resolve any disagreements.

Commonly, there is only a single profile deployed in a network, but some mobile operators may run a combination of frequency support using G.8265.1 alongside one of the two phase profiles: G.8275.1 and G.8275.2. Note, however, that many network elements only contain enough hardware to implement a single clock, so it is not possible for them to be a PTP clock in multiple clock domains at the same time. See Chapter 12, "Operating and Verifying Timing Solutions," for details on the hardware design of clocks.

G.8265.1: Precision Time Protocol Telecom Profile for Frequency Synchronization

ITU-T G.8265.1 is the PTPoIP profile used to transport frequency in a packet network. From the scope of the recommendation:

> The profile specifies the [IEEE 1588] functions that are necessary to ensure network element interoperability for the delivery of frequency only. The profile is based on the architecture described in [ITU-T G.8265] and definitions described in [ITU-T G.8260].

> This Recommendation also specifies some aspects necessary for use in a telecom environment which are outside the scope of, and complement the PTP profile.

The profile completes the picture, as it is the protocol implementation that delivers the solution outlined in the architecture and network topology recommendations. This recommendation includes the following main topics, annex, and appendix:

- Details on the use of PTP for frequency distribution, including selection of some PTP options available in IEEE 1588-2008. This includes the PTP mapping (IP) and message rates and which form of the best master clock algorithm (BMCA) is allowed. The profile requires support for IPv4 (IEEE 1588-2008 Annex D), with IPv6 (IEEE 1588-2008 Annex E) being an optional permitted transport mechanism.

- Annex A, "ITU-T PTP profile for frequency distribution without timing support from the network (unicast mode)." This annex contains the main details such as allowed values for datasets, TLVs, flag values, and so on.

- Appendix I, "Use of mixed unicast/multicast mode for PTP messages." This appendix contains information on deployment of PTP, especially considering that PTP was originally designed to be used in a multicast environment. Some implementations of PTP in a telecom environment used a "mixed mode" whereby some messages were multicast and others were unicast.

This recommendation outlines the implementation and deployment details for G.8265.1. This profile is widely used in the circuit emulation application as well as frequency synchronization for older versions of the mobile telecommunications networks. See Chapter 2,

"Usage of Synchronization and Timing," for an overview of the application of frequency synchronization to the mobile space.

A new full-text version of G.8265.1 with an amendment was published in 2019. This amendment included a couple of new notes in a table mapping between SSM quality levels and PTP clock class values.

> **Note**   Link to recommendation: https://handle.itu.int/11.1002/1000/14012

G.8275.1: Precision Time Protocol Telecom Profile for Phase/Time Synchronization with Full Timing Support from the Network

The ITU-T G.8275.1 profile is one of the most important for modern networks and for those operators transitioning to 5G. It utilizes the layer 2 version of PTP for those networks that fully support PTP at every hop from the master to the slave. From the scope and summary of the March 2020 edition of the recommendation:

> Recommendation ITU-T G.8275.1 contains the ITU-T precision time protocol (PTP) profile for phase and time distribution with full timing support from the network. It provides the necessary details to utilize IEEE 1588 in a manner consistent with the architecture described in Recommendation ITU-T G.8275.
>
> The parameters defined in this version of the profile are chosen based on the case where physical layer frequency support is provided, and the case without physical layer frequency support (i.e., PTP only) is for further study.

The profile is the final piece that completes the puzzle, as it is the protocol implementation that delivers the solution outlined in the architecture and network topology recommendations. G.8275.1 is the recommended profile to deploy phase timing in a transport network for best-in-class performance. It requires full timing support from the network (which may be difficult for some operators who do not own their own transport links), but it will deliver to expectations.

This recommendation includes the following main topics, annexes, and appendixes:

- Details on the use of PTP for phase/time distribution, including selection of some PTP options available in IEEE 1588-2008. This includes the PTP mapping (Ethernet multicast), message types, BMCA, and which types of clocks are supported.

- Annex A, "ITU-T PTP profile for phase/time distribution with full timing support from the network." This annex contains the main details such as allowed values for datasets, TLVs, flag values, and so on.

- Annex B, "Options to establish the PTP topology using the Alternate BMCA" (moved to G.8275 along with Annex C, Annex E, and Appendix V).

- Annex H, "Transport of PTP over OTN."

- Appendix III, "Considerations on the choice of the PTP Ethernet multicast destination address." Discusses the issues around the choice of MAC address for the multicast traffic (normally configurable in the clock).

- Appendix V, "Description of PTP clock modes and associated contents of Announce messages" (moved to G.8275).

- Appendix VII, "Relationship between clockClass and holdover specification."

Section 6.2.7 has a special note on the use of the virtual local-area network (VLAN) tag from IEEE 802.1Q for PTP messages when using G.8275.1. In short, VLAN tags are *not* allowed, and if a T-GM, T-BC, or T-TSC receives a PTP message with a VLAN tag, the message is dropped. There may be some exceptions when using a T-TC Transparent clock, but a PTP message with a VLAN tag should never appear at a PTP port on any other clock type.

There are numerous appendixes and annexes contained in this recommendation that are very helpful to understanding a wide-scale deployment of the technology. Only the main ones are listed here, but it is strongly suggested that you download the document and familiarize yourself with them if you are deploying this profile.

G.8275.1 is one of the most active recommendations at the ITU-T, meaning that it receives many contributions from the delegates, so it is updated rather regularly. There was a revision in March 2020 that includes several amendments and an amendment consented at the October 2020 plenary.

Note   Link to recommendation: https://handle.itu.int/11.1002/1000/14543

G.8275.2: Precision Time Protocol Telecom Profile for Phase/Time Synchronization with Partial Timing Support from the Network

The ITU-T G.8275.2 profile is another of the important profiles for those operators transitioning to 5G, most especially in North America. It covers the layer 3 (unicast) version of PTP for those networks that cannot fully support PTP at every hop from the master to the slave. From the scope of the recommendation:

> The profile specifies the IEEE 1588 functions that are necessary to ensure network element interoperability for the delivery of accurate phase/time (and frequency) synchronization. The profile is based on the use of partial timing support (PTS) or assisted partial timing support (APTS) from the network architecture as described in [ITU-T G.8275] and definitions described in [ITU-T G.8260].

> It is assumed that this profile will be used in well-planned cases where network behaviour and performance can be constrained within well-defined limits, including limits on static asymmetry. Control of static asymmetries can be achieved in case of assisted partial timing support. Use of this profile in unassisted mode would require careful considerations on how to control static asymmetries.

G.8275.2 is the recommended profile where full timing support from the network is not an option for operators. However, as mentioned in the preceding scope, and throughout this book, the engineer must be careful how it is deployed, as switching PTP over unaware, non-PTP-aware network elements can lead to significant degradation in time accuracy.

This recommendation includes the following main topics, annexes, and appendixes:

- Details on the use of PTP for phase/time distribution, including selection of some PTP options available in IEEE 1588-2008. This includes the PTP mapping (IP unicast), message types, BMCA, unicast negotiation, and which types of clocks are supported.

- Annex A, "ITU-T PTP profile for time distribution with partial timing support from the network (unicast mode)" (contains the main details such as allowed values for datasets, TLVs, flag values, and so on).

- Annex B, "Options to establish the PTP topology using the Alternate BMCA" (moved to G.8275 along with Annex C, Annex E, and Appendix IV).

- Annex C, "Inclusion of an external phase/time input interface on a PTP clock" (moved to G.8275).

- Annex F, "Mapping from PTP clockClass values to quality levels." This annex covers the quality levels that a PTP clock will output on its frequency interfaces.

- Appendix IV, "Description of PTP clock modes and associated contents of Announce messages" (moved to G.8275).

- Appendix VI, "Considerations of PTP over IP transport in ring topologies" (discusses the serious issues that can arise when deploying PTPoIP in a ring topology).

- Appendix VIII, "Operations over link aggregation" (discusses the issues that arise with the transfer of PTPoIP in a topology that includes Link Aggregation Group [LAG] or "Ethernet/port bundles").

As with G.8275.1, there are numerous appendixes and annexes contained in this recommendation that are very helpful to understanding a wide-scale deployment of the technology. Only the main ones are listed here, but it is strongly suggested that you download the document and familiarize yourself with them if you are deploying this profile.

G.8275.2 is also a very active recommendations at the ITU-T, receiving numerous contributions from delegates, and so it is also updated rather regularly. There was a revision in March 2020 that includes several amendments and an amendment consented at the October 2020 plenary.

Note   Link to recommendation: https://handle.itu.int/11.1002/1000/14544

Other Documents

There are a couple of other documents that you might find interesting. These are not recommendations, but rather are a source of extra information that may be helpful.

GSTR-GNSS Considerations on the Use of GNSS as a Primary Time Reference in Telecommunications

ITU-T GSTR-GNSS is a very useful technical report that explains much of the technology underpinning the use of GNSS receivers as a time source. It provides pertinent information on designing and operating telecom GNSS-based clocks for those applications where highly accurate time recovery is critical. From the scope of the document:

> This technical report is addressed to telecom operators, manufacturers, silicon vendors, and test equipment vendors who are interested in a high-level view of the information and issues associated with GNSS reception. This includes general information; basic variables, parameters, and equations; modes of operation; and the nature of the challenges and the methodology to mitigate them.

> This document does not aim to collect or summarize the large amount of material continually published on the topic of GNSS reception, which includes scientific publications, doctoral theses, experimental test reports, and articles in online encyclopedias.

This report includes the following main topics and appendixes:

- High-level description of GNSS systems

- Factors influencing the performance of a GNSS-based PRTC

- Sources of time error in GNSS time distribution

- Mitigation of time error in a GNSS-based PRTC

- Operational schemes for mitigation of time error in GNSS time distribution, which covers multi-constellation and multi-band receivers, augmentation systems, and techniques like differential GNSS and post-processing

- Appendix I, "Cable delay effects and correction in a GNSS receiver"

- Appendix II, "Ionospheric delay, its effect on GNSS receivers, and mitigation of these effects"

- Appendix V, "The effect of multiple reflections within the antenna cable"

- Appendix VI, "Satellite common-view" (whereby accurate time comparisons are made possible by the observation of the same satellites at the same time in different locations)

- Appendix VII, "The effect of multipath within the receiver signal processing"

This report is quite easy to read and full of interesting information.

> **Note**   Link to technical report: http://handle.itu.int/11.1002/pub/815052de-en

Series G Supplement 68—Synchronization OAM Requirements

ITU-T Supplement 68—Synchronization OAM Requirements is a supplement that gives an overview of OAM for a deployed timing solution. From the scope of the recommendation:

> This Supplement provides an overview of Synchronization operations, administration and maintenance (OAM) and includes fault management, performance monitoring, alarms and events.

This document includes the following main topics:

■ Fault management

■ Performance monitoring

■ Alarms and events

This supplement covers three use cases:

■ Frequency synchronization over the physical layer

■ Frequency synchronization via packet-based transport

■ Time synchronization

> **Note**   Link to supplement: https://handle.itu.int/11.1002/1000/14232

Possible Future Changes in Recommendations

The last few years have seen some amazing developments in the standardization process for synchronization. This is being driven mostly by the innovation coming from 5G and the mobile telecommunications industry. Predicting in which direction the standards are likely to move is a risky undertaking, but some changes are apparent.

Perhaps the most likely change to the Q13/15 recommendations will come because of the publication of the new edition of IEEE 1588-2019 with v2.1 of the PTP protocol. According to clause 19 of IEEE 1588-2019, PTP v2.1 is designed to be interoperable with the current PTP v2.0. Of course, it will not be long before operators start to deploy PTP-capable nodes that support elements and features from PTP v2.1. If that network clock does not use any functionality introduced in v2.1, then it should be compatible with the existing deployments using v2.0.

However, at some point, the ITU-T recommendations will have to align with the new version and start adopting some of the increased functionality available in the new 2019 edition of the IEEE 1588 standard. An example of this might be elements of the security features introduced with the new PTP edition.

There will also be ongoing work on the details and results of the mathematical simulations that support G.8271.1, G.8273.2, and G.8273.3. There is also work already underway to define the relative time error (rTE) between the ports of a boundary clock (in updates to G.8273.2). This limits the amount of time error that arises between different interface ports on a network element or between the PTP and 1PPS outputs.

The only other change that looks likely will be the further adoption of new network topologies and limits (in G.8271.1) to specifically support the case of fronthaul in the RAN using a network of highly accurate Class C T-BC boundary clocks. There is also an ongoing effort to support the concept of cluster-based synchronization in 5G RAN networks (see Appendix XII of G.8271.1 for an idea of what that looks like).

Much of this adoption for fronthaul is coming from 5G evolution in the RAN as well as industry initiatives around standardization in the radio segment of the mobile network (for example, the O-RAN Alliance). See Chapter 4 for more information on the fronthaul standardization efforts.

Summary

The ITU-T organization and the material that it (and specifically Q13/15) produces is essential to the timing and synchronization community—especially for the network operator and the mobile service provider. One problem for people coming to this topic is to understand how this material is developed, organized, and produced. This chapter laid out a mechanism to categorize the relevant recommendations and provided a synopsis of each one within each category. There is no ideal way to present this material, but the authors selected this approach because it aligns more closely with the layout and topics of the book.

To make the best use of this chapter, you should use the rest of the book to understand the synchronization needs for your use case and the likely alternatives for defining a solution. Then this chapter serves as a source of the documents most likely to apply to that scenario.

Now that all the pieces are in place, Chapter 9 will look at the factors that a designer needs to consider when designing a timing solution based on PTP technologies. The subsequent chapters will then address the use case of mobile communications in some detail and show how to apply these lessons to that specific situation.

References in This Chapter

IEEE Standards Association. "1588 Standard for Precision Synchronization Protocol for Networked Measurement and Control Systems." *IEEE Std. 1588:2008*, 2008. https://standards.ieee.org/standard/1588-2008.html

IEEE Standards Association. "1588 Standard for Precision Synchronization Protocol for Networked Measurement and Control Systems." *IEEE Std 1588:2019*, 2019. https://standards.ieee.org/standard/1588-2019.html

ITU-T. "Q13/15 – Network synchronization and time distribution performance." *ITU-T Study Groups*, Study Period 2017-2020. https://www.itu.int/en/ITU-T/studygroups/2017-2020/15/Pages/q13.aspx

ITU-T. "Transmission systems and media, digital systems and networks." *ITU-T Recommendations*, G Series. https://www.itu.int/rec/T-REC-G/en

Telcordia Technologies. "Synchronous Optical Network (SONET) Transport Systems: Common Generic Criteria." *GR-253-CORE*, Issue 5, 2009. https://telecom-info.njdepot.ericsson.net/site-cgi/ido/docs.cgi?ID=SEARCH&DOCUMENT=GR-253&

Chapter 8 Acronyms Key

The following table expands the key acronyms used in this chapter.

Term	Value
1PPS	1 pulse per second
3GPP	3rd Generation Partnership Project (SDO)
2G	2nd generation (mobile telecommunications system)
3G	3rd generation (mobile telecommunications system)
4G	4th generation (mobile telecommunications system)
5G	5th generation (mobile telecommunications system)
ACR	adaptive clock recovery
ADEV	Allan deviation
ANSI	American National Standards Institute
APTS	assisted partial timing support
ATIS	Alliance for Telecommunications Industry Solutions
BMCA	best master clock algorithm
CEM	circuit emulation
CES	circuit emulation service
cTE	constant time error
CWDM	Course Wavelength Division Multiplexing (optical technology)
DSL	digital subscriber line (a broadband technology)
dTE	dynamic time error
DWDM	Dense Wavelength Division Multiplexing (optical technology)

Term	Value
E1	2-Mbps (2048-kbps) signal (SDH)
EEC	Ethernet equipment clock
eEEC	enhanced Ethernet equipment clock
eOEC	enhanced OTN equipment clock
ePRC	enhanced primary reference clock
ePRTC	enhanced primary reference time clock
ePRTC-A	enhanced primary reference time clock—class A
ePRTC-B	enhanced primary reference time clock—class B
eSEC	enhanced synchronous equipment clock
eESMC	enhanced Ethernet synchronization messaging channel
eSSM	enhanced synchronization status message
eSyncE	enhanced synchronous Ethernet
ESMC	Ethernet synchronization messaging channel
FTS	full timing support (for PTP from the network)
GNSS	global navigation satellite system
GR	Generic Requirement
HDB3	High Density Bipolar of Order 3
HRM	hypothetical reference model
IEEE	Institute of Electrical and Electronics Engineers
IETF	Internet Engineering Task Force
IP	Internet protocol
IPv4, IPv6	Internet protocol version 4, Internet protocol version 6
ITU	International Telecommunication Union
ITU-D	ITU Telecommunication Development Sector
ITU-R	ITU Radiocommunication Sector
ITU-T	ITU Telecommunication Standardization Sector
IWF	interworking function
LAG	Link Aggregation Group
LTE	Long Term Evolution (mobile communications standard)
max|TE|	maximum absolute time error
MDEV	modified Allan deviation

Term	Value
MEF	MEF Forum (formerly Metro Ethernet Forum)
MTIE	maximum time interval error
MTN	metro transport network
NNI	network-network interface
NTP	Network Time Protocol
O-RAN	Open Radio Access Network
OAM	Operations, Administration, and Maintenance
OEC	OTN equipment clock
OTN	optical transport network (optical transport technology)
PDF	Portable Document Format
PDH	plesiochronous digital hierarchy
PDV	packet delay variation
PEC	packet-based equipment clock
PEC-M	packet-based equipment clock—master
PEC-S	packet-based equipment clock—slave
PEC-S-F	packet-based equipment clock—slave—frequency
PHY	PHYsical Layer (of OSI Reference Model)—an electronic device
PLL	phase-locked loop
PNT	packet network timing
PON	passive optical networking (optical broadband)
PRC	primary reference clock (SDH)
PRS	primary reference source (SONET)
PRTC	primary reference time clock
PRTC-A	primary reference time clock—class A
PRTC-B	primary reference time clock—class B
PTN	packet transport network
PTP	precision time protocol
PTS	partial timing support
PTPoIP	PTP over Internet protocol
Q13 or Q13/15	Question 13 of Study Group 15 of the ITU-T (concerned with sync)
QL	quality level

Term	Value
RAN	radio access network
RJ-45	registered jack 45
rTE	relative time error
SEC	synchronous equipment clock
SEC	SDH equipment clock
SDH	synchronous digital hierarchy (optical transport technology)
SDO	standards development organization
SG15	Study Group 15 (of the ITU-T)
SONET	synchronous optical network (optical transport technology)
SSM	synchronization status message
SSU	synchronization supply unit (SDH)
STM-N	synchronous transport module level "N" (N can be 1, 4, 16, etc.)
SyncE	synchronous Ethernet—a set of ITU-T standards
T1	1.544-Mbps signal (SONET)
T-BC	Telecom Boundary Clock
T-BC-A	Telecom Boundary Clock—Assisted
T-BC-P	Telecom Boundary Clock—Partial support
T-GM	Telecom Grandmaster
T-TC	Telecom Transparent Clock
T-TSC	Telecom Time Slave Clock
T-TSC-A	Telecom Time Slave Clock—Assisted
T-TSC-P	Telecom Time Slave Clock—Partial support
TDD	time division duplex
TDEV	time deviation
TDM	time-division multiplexing
TIE	time interval error
TLV	type, length, value
ToD	time of day
UTC	Coordinated Universal Time
V.11	ITU-T specification for differential data communications
VLAN	virtual local-area network

Further Reading

Pointers to the recommendations in each section are listed throughout the chapter and are available from the ITU website. Refer to the following recommended sources for further information about some of the topics covered in this chapter.

Alliance for Telecommunications Industry Solutions (ATIS): https://www.atis.org/

ITU-T Study Group 15 – Networks, Technologies and Infrastructures for Transport, Access and Home: https://www.itu.int/en/ITU-T/about/groups/Pages/sg15.aspx

O-RAN Alliance: https://www.o-ran.org/

Telcordia, The Generic Requirement (GRs) Development Process (now a division of Ericsson). https://telecom-info.njdepot.ericsson.net/site-cgi/ido/docs2.pl?ID=199178067&page=home

Chapter 9

PTP Deployment Considerations

In this chapter, you will learn the following:

■ **Deployment and usage:** Covers the specifics of sourcing timing signals and carrying them across a packet network for both the frequency and phase/time use cases. For the case of frequency, it covers using both PTP and non-PTP methods and points out the appropriate solution to use in different scenarios. For the phase/time scenario, it explains a range of issues that one must consider, including capabilities of the network.

■ **Factors impacting timing performance:** Covers the issues that must be dealt with to produce an accurate and robust solution. This includes the timing performance of the PTP clocks and the influence of packet delay variation (PDV) and asymmetry on the phase/time recovered at the end of the timing chain.

■ **Parameters for timing performance:** Builds upon the introductory theory in Chapter 5, "Clocks, Time Error, and Noise," to show how the performance parameters must be controlled to deliver an accurate timing signal.

■ **Clock performance:** Extends the coverage in previous chapters about the characteristics of several types of frequency and phase/time clocks. Deals with each category of performance and shows how it affects the design and deployment.

■ **Budgeting end-to-end time error:** Covers the factors that apply to the end-to-end performance, and specifically examines how to budget end-to-end time error.

■ **Network holdover:** Shows what holdover can do for a synchronization design and what it is not able to do.

■ **Packet network topologies:** Discusses the influence that many common packet technologies can have on the timing design and how to mitigate the consequences of how they function.

■ **Packet transport:** Examines the effect of different transport types on the transport of timing signals and what must be done to mitigate and minimize their ability to alter the time signal and degrade the accuracy.

■ **Non-mobile deployments:** Examines a couple of alternative deployment cases such as the cable industry and other industrial uses such as the power industry.

This chapter examines the main issues that need to be addressed when using a packet-based transport to carry a timing signal across a wide-area network (WAN). In most cases, this means using some form of PTP to carry that signal, although there are other packet-based techniques that are covered.

This chapter is where all the theory from the previous chapters comes together so that you can see how to design and implement a timing solution using the transport network. This chapter is not specific to the mobile or 5G use case, and the lessons largely apply to any deployment with similar characteristics and performance targets.

Chapter 11, "5G Timing Solutions," covers the more specific aspects of the 4G and 5G synchronization deployments, but the basic tenets are covered here first. In some areas where there is repetition between the mobile and non-mobile case, this chapter may refer to more details in other chapters (especially Chapter 11).

Some portions of this chapter are more quantitative, but others are much more qualitative. If the mathematics is not important to you, then feel free to breeze over it—you should not feel it will prevent you from being able to successfully deploy a timing transport network.

Deployment and Usage

This chapter deals with the aspects of designing a network to carry timing using a packet-based method. Mostly this means using some form of PTP, although there are also other packet-based methods that can carry frequency, phase, and time. For the case of phase/time, this chapter focuses on PTP, but it covers other options to carry frequency via packets.

The engineer should consider the options available to deploy a timing solution as summarized in Table 9-1. The section "Solution Options" in Chapter 11 covers the trade-offs between these options in a little more detail and details the advantages of each (in a mobile setting).

Table 9-1   *Timing Methods for Different Synchronization Types*

Method	FrequencySync	PhaseSync	TimeSync
Precision time protocol (1588, PTP)	✓	✓	✓
Network Time Protocol (NTP)	✓	✓	✓
Other packet methods	✓		
Synchronous Ethernet (SyncE)	✓		
Global navigation satellite (GNSS) receiver	✓	✓	✓
Atomic clock	✓	Needs calibration (GNSS)	Needs calibration (GNSS)

This chapter deals with timing transport using packet methods, so it concentrates on PTP, although it briefly covers NTP and some other packet methods. An additional resource is Annex A, "Using PTP," of IEEE 1588-2008, which covers the highlights of issues that may arise during PTP deployments.

The flow of this chapter starts from the head of the timing network and proceeds down toward the application using the timing signal. Therefore, the place to start is at the head of the timing distribution network, and that is with the source of the timing signal to be carried.

Physical Inputs and Output Signals

This section begins with the source of timing signals that are to be transported across the network. Chapter 3, "Synchronization and Timing Concepts," introduces the various forms of primary reference clock (PRC), primary reference source (PRS), and primary reference time clock (PRTC) and their associated signals. This section expands on Chapter 3 with more specific details on their usage and implementation.

Frequency Inputs and Outputs (10 MHz, BITS, and so on)

Here is a summary of the major factors to consider when examining the source of timing signals as well as some common pitfalls when doing so:

- Ensure the cabling and connectors are correct and that they properly coexist (for instance, by matching signal characteristics such as impedance and gain/loss).

- Choose signal types and understand the differences in how they are configured.

- Provide quality level (QL) traceability and ensure the configuration matches the type of quality levels being passed. If the signal has no QL values, then configure the receiving node to assume an expected QL to use whenever the signal is present.

At the start of a packet-based frequency distribution network, there will be a master packet-based equipment clock (PEC-M) or PTP grandmaster (GM) with a frequency input. This input reference is usually a signal such as 10 MHz, 2 MHz, synchronous Ethernet (SyncE), or some E1/T1 time-division multiplexing (TDM)–based circuit from a synchronization supply unit (SSU) or building integrated timing supply (BITS). The most basic thing to ensure is that the cable and connectors are correct, and the following section looks at them in some detail.

> **Note** Use of the terms "master" and "slave" is ONLY in association with the official terminology used in industry specifications and standards, and in no way diminishes Pearson's commitment to promoting diversity, equity, and inclusion, and challenging, countering, and/or combating bias and stereotyping in the global population of the learners we serve.

There are many, many types of connectors, and several look very similar to each other. In fact, it is quite easy to plug one connector into the other and damage them both, since they are "almost" compatible, but not quite. Avoiding this damage requires the engineer to be very diligent about understanding what connections are needed and what type they are—the authors cannot stress this enough. About 95% of the time, the first big issue with any new timing project is finding the correct connectors and cables for the timing input/output signals.

The next consideration is to confirm whether the connector is male or female. The timing engineer is advised to always carry a few *gender bender* adapters that allow male-to-male and female-to-female connections in case the cables do not match gender.

The following are the major connector types commonly found for these forms of input:

- **Bayonet Neill-Concelman (BNC):** BNC connectors are very common on PRC/PRS devices and equipment that tests timing or anything that needs a frequency input. BNC connectors are often used to carry 10-MHz/2-MHz signals and can be used to carry *unbalanced* 75-ohm E1/T1 signals (more on that in a moment).

  BNC connectors come in 50-ohm and 75-ohm versions, with the related cable type matching the impedance. You should check the impedance, because unbalanced E1/T1 are 75 ohm, whereas most 10-MHz timing interfaces are 50 ohm, although for short cable lengths, the unbalanced E1/T1 may still function adequately well. There is also a smaller version of BNC called mini-BNC ("mini" is relative; they are still somewhat large).

- **SubMiniature version B (SMB) connector:** This is a pluggable version of the connector commonly found as the GNSS antenna connectors. The major difference is that SubMiniature version A (SMA) is a screw-on connector, whereas SMB is a click-on pluggable. They look like some other types of connectors, so be careful when connecting them to equipment you are unfamiliar with. SMB is used by some vendors to help reduce the size taken up by input/output ports on their equipment.

- **DIN 1.0/2.3:** The large size of BNC is not a major issue in a large desktop device, but for the front faceplate on a pizza box router, real estate is at a premium. Therefore, Cisco and some other equipment manufacturers use DIN 1.0/2.3 connectors for 10-MHz input/output. Although many people mix them up or assume that they are the same thing, DIN 1.0/2.3 connectors are not the same as SMB connectors. DIN 1.0/2.3 connectors are widely used in the audio/visual industry and available from vendors serving that space. They are click-on and quick-disconnect pluggable devices much like SMB, but they are *not* compatible.

- **BITS and E1/T1:** There are two basic forms of the traditional E1/T1 circuit commonly seen in devices. Unbalanced 75 ohm used to be widespread (two cables, one for transmit, one for receive, with 75-ohm coaxial cable and commonly BNC connectors), but nowadays, the BITS interface on the faceplate is mostly an RJ-48c port generating a *balanced* signal over CAT-5 unshielded twisted-pair cable. The impedance is 120 ohm for E1 and 100 ohm for T1, and the signal is carried on pins 1, 2, 4, and 5. This system can be used to carry unframed 2-MHz signals as well.

BNC Connectors

The PRC/PRS is almost always equipped with several (female) BNC connectors (called *sockets* or *jacks*). Most of the time, but not always, the equipment that receives the signal also has a female connector. So, the normal cable used to connect them together is a male-to-male cable with a *plug* on each end. Figure 9-1 shows a female BNC jack connector front and center with a male-to-male cable (containing a plug on each end) behind it.

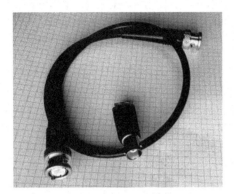

Figure 9-1   *Female BNC Connector and Male-to-Male Coaxial Cable*

Be aware that these cables are expected to maintain a quality signal only over a very short run, and you should not expect to use cables of more than a few meters. In fact, using a cable of more than 10–15 m might mean that a signal will not be detected on the far end. There are many more details in ITU-T G.703 on the electrical and other standards behind these interface types.

Of course, if the interface on the PRC/PRS does not match the interface on the router or timing equipment, then you need to get an adapter to change the gender and/or connector type. If one interface is unbalanced coax (50/75 ohm) and the other is a balanced UTP, then you'll need a *balun* to convert the impedance. This device will also convert the media type, allowing a UTP RJ-48c to be plugged into one end and the BNC coaxial cables into the other. Figure 9-2 shows a typical balun that performs both roles of media convertor and impedance matcher.

Figure 9-2   *Balun to Convert Between 120-Ohm Balanced UTP and 75-Ohm Unbalanced Coax*

DIN 1.0/2.3 Connectors

When performing timing projects using Cisco routers, the next most common connector on a timing interface is the DIN 1.0/2.3. The connectors on the front faceplate are female jacks, so the connecting cable has a male plug. Figure 9-3 shows the front panel of a Cisco ASR-900 series router on the left, a female jack in the middle, and the male plug on the right-hand side.

Figure 9-3   *DIN 1.0/2.3 Jacks with the Corresponding Male Plug*

Because many other timing devices (particularly large equipment with plenty of real estate) have BNC connectors, one of the handiest cables to have readily available is the BNC male-to-DIN 1.0/2.3 male adapter cable.

Note   As DIN is an acronym referring to a German standards development organization, there are many readily available items that are "DIN standard." One of those is a very popular (well, it was 30 years ago) audio connector, so do not make the mistake of getting the wrong one. In fact, they were the default connector for the early generations of mouse cable, so old desktop personal computers have them on the rear of every machine. So, you need to be specific about which DIN connector you need. See https://www.din.de/en for further information on the DIN organization.

RJ-45 UTP BITS Interface

The last major connector to consider is the RJ-45 type connector (in fact, it is an RJ-48c, but most people refer to it as RJ-45, as it looks like an Ethernet port). Typically, this is used where a router has a BITS port to provide either a 2-MHz, 2-Mbps E1, or 1.544-Mbps T1 signal. Note that 2 MHz refers to a frequency of 2048 kHz as per G.703, but the terms 2 MHz and 2.048 MHz are used somewhat interchangeably.

This interface looks a lot like an Ethernet port and cable, but its specific pinouts (1, 2, 4, and 5) means that most Ethernet cables will probably not match your intended usage (at this point it is better to avoid the rabbit hole of UTP wiring standards). The short lesson is that you should be prepared to change the pinouts on the cable connectors unless you use UTP cables specifically made for E1/T1. Just because the RJ-45 connector on the UTP cable fits the RJ-48c port does not mean the wiring connection is correctly made.

Note that the BITS port is frequently used by engineers and mobile service providers (MSP) to quickly check the frequency alignment of a router. A piece of test equipment (that has a frequency reference signal available) is plugged into the BITS port and the frequency of the clock in the network element can be quickly measured against the test device (more in Chapter 12, "Operating and Verifying Timing Solutions," on test equipment). One could do something similar by measuring the frequency of an Ethernet port configured with SyncE, but that introduces security issues because it requires someone to configure an active port to measure it.

The BITS port is currently being phased out on devices from most vendors. There are several reasons for this:

- Real estate (space) constraints on the front panel means that hardware engineers want to remove it and use the space saved for something else (like more Ethernet ports).

- The electronics industry that produces components is removing the products (such as framers) needed to build BITS interfaces. There are several reasons for this, including low volumes and expensive production processes to manufacture them.

- Router vendors are seeing less demand for BITS interfaces in customer requests, and the higher cost of the components to make them leads to a price disadvantage compared to routers from other vendors.

The BITS port is quickly becoming a legacy device, so do not assume that they are always present, especially on smaller platforms where faceplate space is at a premium and device cost is a constant consideration. The alternative is to use a 10-MHz interface (which may also support 2 MHz, but not E1/T1).

Frequency Inputs and Outputs (SyncE)

Devices such as a PRC/PRS that provide a reference frequency are increasingly also equipped with an Ethernet port (as well as the commonly available 10 MHz). Any modern PRTC delivering phase is almost certainly Ethernet enabled with support for SyncE output. This means that one of the best sources of frequency for a device is to use SyncE, mainly because it is simply Ethernet.

There are several conditions that you must ensure are met to confirm a node correctly supports SyncE. Generally, fiber Ethernet interfaces can be designed to support SyncE without too much issue. Of course, the device should state in its documentation that it supports SyncE, but also be sure the individual port is also SyncE capable (some ports may have limitations, such as supporting output only).

There can be issues in some situations involving older interface types. For example, low-speed copper interfaces (1 GE or less) might not support SyncE or might have restrictions on the support that it offers (for example, the interface may not support a change in direction of the SyncE flow without a port up/down event). A rule of thumb would be that the slower and older a port is, the less likely it is to implement SyncE or have some caveat.

But one issue to be acutely aware of is that SyncE is generally not supported on small form-factor pluggable (SFP) devices for copper cables. These SFPs are devices that plug into a 1-gigabit Ethernet (GE) SFP port on a router and allow the device to be connected to UTP (1000BASE-T) copper cables. This is because pluggable ports compliant with the standards do not have enough pins to connect with the copper SFP to carry SyncE. Note that this is a problem with the SFP port and not a normal copper port, which does not have this limitation.

Some proprietary SFPs are available that partially solve this issue, although they have some shortcomings. One of the major drawbacks is that they come as two separate devices: one is a SyncE master and the other is a SyncE slave port, and they cannot change state. This means they have limited use in rings because the direction of SyncE in the ring cannot change without swapping the direction of the signal (and maybe the SFPs themselves if their role is hardwired at the factory).

The best solution is to be sure that if your solution requires SyncE to run over a copper cable, be sure your network element contains native 1-GE RJ-45 ports for copper 1000BASE-T because copper SFPs will almost certainly not work.

Frequency Inputs and Outputs (Quality Levels)

Receiving a stable and accurate reference frequency signal as an input is important, but an equally important component of that signal is to receive information about the QL and traceability to the signal source. For further information, see the section "Clock Quality Traceability" in Chapter 3.

Not all types of frequency transport support the ability to carry information about the clock quality. For example, the pure frequency signals such as 10 MHz and 2 MHz do not have any method of signaling clock quality or traceability. To solve this shortcoming, the engineer can configure the network elements with a QL to assume for the input signal whenever it is available.

That way, if the engineer configures a 10-MHz input to have a QL of QL-PRC (primary reference clock), the router will assume that the signal is PRC quality while the link is up. Note that in North America, one would configure it to be QL-PRS (primary reference source) for an Option II network because QL-PRC is for an Option I network (see following point on options).

When the PRC/PRS frequency source loses its input reference (such as the GNSS signal from space), it must signal this loss to the downstream node, but again, for signals like 10 MHz, there is no QL available. So to signal the loss of quality, the PRC/PRS takes down (squelches) the frequency signal it is transmitting. The downstream network element relying on that reference sees the signal disappear and therefore rules that frequency source (and the QL configured for it) out of contention as a source. It then selects a new source of frequency out of the signals that it can see, based on the QL levels.

Some other signal types support QL information only when configured to use a specific mode of transmission. This means that it is critical that the engineer configure both the PRC/PRS and the router to use a form of transmission that supports carrying the QL. By doing that, the router will be able to receive the incoming QL signals and use that information to select the best frequency source available. Of course, the received QL on any interface can be overridden by configuration on the router.

A good example of where this is the case is with the synchronization status message (SSM) bits for E1/T1 circuits. For E1 to send the SSM bits to signal its quality level, it must be configured as using the framing type of Cyclic Redundancy Check 4 (CRC-4). Other framing will not support SSM. There is a similar situation for T1 circuits, because SSM signaling is only supported when configured to use the framing method known as the extended super frame (ESF).

The last trap that is easy to fall into is where the devices processing QL information are configured to use incompatible sets of QL values: some using those from the SDH E1 world and others from the SONET T1 world. By default, Cisco routers interpret SSM and QL values according to the ITU-T SDH E1 scheme (*Option I*), so engineers in North America should change it immediately to use the scheme appropriate for their own geography (*Option II*).

If the PRC/PRS is putting out a set of SSM values that are aligned with the Option II markets and the router is still at the default of Option I, the node will misinterpret the values it receives and make wrong decisions about which sources it should use. This is a very common mistake—the whole network must be running the same option value for QL scheme (unless you are running international TDM links between North America and other continents).

Phase/Time Inputs/Outputs (1PPS, ToD)

The phase input signals of 1 pulse per second (PPS) and time of day (ToD) are somewhat distinct and different from each other, although they both work together to transmit phase and time. In its simplest form, the 1PPS is just a simple pulse that cycles at a rate of once per second. On the other hand, the ToD is a method of transmission that carries a representation of the time of day from the PRTC to the router. See "1PPS" and "Time of Day" in Chapter 3 for more details.

As seen in Chapter 3, the length of the positive pulse of the 1PPS signal must lie in a range between 100 ns and 0.5 s (according to clause 19.2.1 of ITU-T G.703). This is often configurable, so the engineer can choose a value that the receiving device might require or prefer. Sometimes the receiving device is required to be calibrated to whatever length of pulse is chosen. In most cases, the 1PPS signal is carried over 50-ohm coax cable (coverage of other options follows). Remember that the length of coax cable that can carry a 1PPS pulse is quite limited, certainly much less than 10 m; however, for high-accuracy applications, the coax cable should be of very good quality and be short (less than 2 m).

The 1PPS connector often is the same type as the frequency connectors—on timing equipment, it is very common to see BNC, whereas other devices might support SMB. Cisco routers support 1PPS signals with a DIN 1.0/2.3 jack, the same as the 10-MHz frequency connector (refer to the left-hand side of Figure 9-3, earlier in the chapter).

Some network elements also support a form of connector (like SMB and DIN 1.0/2.3) that is known as micro coaxial connector (MCX) or micro-miniature coaxial (MMCX). These are popular connectors in Europe (they are specified in an IEC standard) and are commonly used as a GPS antenna connector on devices too small to house an SMA connector (such as handheld navigation devices).

With all these types of connectors out there, you can see that it is vitally important to understand exactly what is needed so that you don't arrive onsite with the wrong cable or adapter.

A second option for 1PPS is to carry it alongside the ToD running on a different set of pins on a type of serial interface over 100-ohm UTP (see clause 19.1 of ITU-T G.703 for specifications and pinouts). In an RJ-45 form factor, this is a balanced signal (such as ITU-T V.11) with two pins carrying the positive and negative of the ToD and another two pins carrying the 1PPS.

Of course, using the UTP/RJ-45 method to carry a combined 1PPS and ToD is easier from a cabling and deployment point of view than using separate connectors for each signal. But, be prepared to change the pinouts if you are trying to use standard Ethernet cables.

But perhaps the biggest issue is with the format of the ToD. There is now a standardized format for ToD available from the ITU-T (G.8271 Annex A), but the adoption across the industry is not yet universal. There are a couple of de facto standards (NTP, UBX, NMEA), but nothing is supported by 100% of the industry. There is also a format known as "Cisco" format (see Chapter 3 for more information on the formats).

Virtual Port (1PPS, ToD, and Frequency)

The concept of a virtual port was introduced in the ITU-T G.8275.1 recommendation as Annex C, so the feature is sometimes referred to as *Annex C* support. Because it applies to both G.8275.1 and G.8275.2, the specification was moved into Annex B of ITU-T G.8275, so the original name is now confusing. It is better to refer to it as *virtual port support*.

The idea of a virtual port is that the timing ports (1PPS, ToD, and 10 MHz) or GNSS inputs are made to look like a "virtual" PTP slave port. The normal PTP slave port has its port dataset and real transfer of PTP messages. On the other hand, the virtual port connects to an external PRTC source (no PTP) and has a virtual dataset so that it looks like other slave ports to the state machine.

Most of the parameters normally associated with a port, such as the values in the Announce message received on that port, are either left as default or configured by the operator. As there is no actual PTP Announce message coming into the virtual port, the engineer must select and configure values that apply to that network and do not conflict with the real GM. These might include priority2 and the clock quality parameters, including clockClass, from the Announce message.

One thing to note is that the virtual port can only be configured on a Telecom Boundary Clock (T-BC) because this extra port causes it to exceed the port count (of one) normally allowed with any other ordinary clock such as the Telecom Grandmaster (T-GM) or Telecom Time Slave Clock (T-TSC). Figure 9-4 illustrates the T-BC with a virtual port and the difference when compared to a T-GM.

One small but important detail is that a T-GM can never have a remote backup because one of the cast-iron rules in 1588 is that the T-GM can never take a clock input from another clock source. It can take frequency, phase, and time in from a local source, but it can never have a slave port synchronized to another PTP clock.

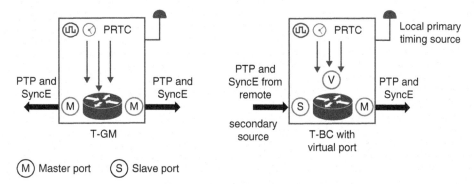

Figure 9-4 *T-GM with Embedded GNSS and T-BC with Virtual Port*

For that reason, if you have a clock that uses GNSS as its primary source, but you require a form of backup from another (remote) PTP T-GM, then you *must* configure the device as a T-BC with a virtual port. Making it a T-BC allows you to configure a slave port to use as a backup in case of an outage in the local GNSS.

For details of the virtual port, please refer to Annex B of ITU-T G.8275.

GNSS

Chapter 11 offers a lot more details on the use of GNSS-based systems to build a distributed timing solution for 5G synchronization, so refer to that chapter for a discussion of the overall solution design alternatives. The following section covers several implementation considerations for GNSS deployments.

The source of the signal for a T-BC with a virtual port (and T-GM) can be an embedded GNSS receiver with some internal connectors to route the frequency and phase/time signals to the timing subsystem. It can also be an external GNSS receiver (a separate device) and connected to the T-GM/T-BC by external physical cables (1PPS, ToD, and 10 MHz/ 2 MHz/SyncE).

Either way works, although the embedded (single box) solution allows for a much tighter integration between the two functions, especially for maintenance and operational reasons. Using separate systems works better if the operator wants to choose a more "full-featured" GNSS receiver than one that might be embedded in a router. That would allow the operator to select a receiver that offers optional extras such as a rubidium holdover oscillator or multiple output connectors for physical signals.

For the antenna, many devices in the timing world use an SMA jack as the default connector for an external GNSS antenna (SMA is the screw-on alternative to the SMB snap-on version). Other types of connectors are sometimes used, such as BNC and Threaded Neill-Concelman (TNC), which is a threaded version of BNC, but SMA is very common. SMA (along with MCX) is found on many of the cheap consumer "patch" antennas available for use with devices like portable receivers for car navigation. For some informal testing and proof of concept in a lab environment, one of these with an SMA can be pressed into service.

Most expert antenna installations using long runs of cable (>50 m) normally use a very heavy and thick low-loss coaxial cable, so the SMA connector is too weak to hold the weight and strain of the cable. Typically, this cable is terminated at a junction box (see the bullet point on splitters in the following list) with much larger and stronger connectors—such as a large connector known as the *Type N connector*. Then a lighter Type N–to–SMA patch cable is used to complete the connection to the receiver.

This raises the point of how to select an antenna for production usage or for usage in a formal laboratory setting. This question is often posed to the authors, and the short answer is that the antenna selection and installation should be performed by local con-tractors who have the required expertise and experience. Many local requirements need to be met to install an antenna safely and accurately (and legally), and an off-the-shelf antenna installed by an amateur may meet none of those.

Here are a few items to consider when planning a GNSS antenna installation:

- The antenna must be chosen based on the (signal) gain it produces, which is then consumed by signal loss caused by attenuation in the cable. The higher the loss in the (lower quality or cheaper) cable, the more gain must come from the antenna to overcome it. The installer must balance this gain/loss to allow a sufficiently strong signal to arrive at the receiver given the length and quality of the antenna cable.

- To boost the very weak signal from space, the GNSS antenna contains a pre-amplifier in its base. This pre-amplifier is powered by a voltage (typically less than 5-V DC) that is applied to the coaxial antenna cable by the GNSS receiver. Again, this gain from the amplification is added to the calculation when determining the signal strength that reaches the antenna port on the receiver.

- Many times, when an antenna is installed for a GNSS application, there are many other receivers in the same location that can also share the antenna. For this rea-son, many installations incorporate a *splitter*, which allows the signal to be shared between many receivers. This splitter must be capable of moderating all the DC volt-ages coming from the receivers without over-supplying power to the pre-amplifier. The splitting of signals into multiple paths also reduces the signal strength, adding another factor to the gain/loss calculation.

- The antenna will detect a wide band of unwanted radio signals around the narrow spectrum carrying the desired GNSS signal. For this reason, antenna installations have filters that only allow a very narrow range of frequencies to pass, those that the receiver wishes to detect. The installer must make sure that the antenna is sensitive to the band that the GNSS receiver can process, and that the correct narrow-band filters are used so that the sought-after signals pass through to the receiver.

  It is pointless to have a multiband and multi-constellation receiver when either the antenna does not detect the transmission or the filters block most of the signals.

■ Local (fire) authorities may have regulations about what sort of cable can be run through the ducting of a building and specifically the rating on the outside insulation layer of the cable. For example, there is a cable type called low smoke and fume (LSF) that may need to be used to comply with various legal requirements. There might also be similar regulations for the segment of the cable that must run outdoors.

■ The antenna must be located on a high position with a clear (as possible) view of the sky. Furthermore, it must be placed at a distance away from other antennas and especially away from other sources of radiated energy. The antenna may even require some form of shielding to protect it from the ingress of electromagnetic interference (EMI). Finding the best location for GNSS signal reception is not as obvious as one might think; experience is required.

■ In many buildings, an installer needs to be certified (occupational health and safety) to get access to restricted spaces, like internal cable ducting and rooftop areas. There can be somewhat dangerous equipment on rooftops, such as microwave transmitters.

■ The antenna installation needs to incorporate surge protection to protect people and equipment in the building from the ingress of dangerous sources of energy, such as lightning. This is a serious safety-of-life issue if not done correctly. For similar reasons, when playing around with an off-the-shelf consumer antenna, do not place it up high or leave it outside, especially in stormy weather conditions.

■ Finally, the antenna installation should be surveyed for accurate position and height to ensure the antenna is receiving a valid signal and not a multipath signal reflected off a nearby structure. Similarly, the length of antenna cable should be determined and the GNSS receiver calibrated to compensate for that cable length.

For all those reasons, the authors recommend operators of GNSS equipment that need a reliable and accurate source use a local professional to design and install GNSS antennas. For further information on deploying a solution based on GNSS versus other approaches, refer to Chapter 11. There is also an ITU-T document, "GSTR-GNSS Considerations on the use of GNSS as a primary time reference in telecommunications," that contains useful information on the topic. Go to https://www.itu.int/pub/T-TUT-HOME-2020 to download the report.

Frequency Distribution in a Packet Network

Chapter 6, "Physical Frequency Synchronization," covered many critical aspects of carrying a frequency signal across a transport network using a physical method (the signal is carried in a wire or fiber). But there are many cases where the operator wishes to decommission those types of networks, or the equipment to transport those signals is no longer cost effective.

Some service providers (SP) who provide legacy transport services (such as E1/T1 and lower speed optical) are similarly moving away from TDM-based networks and adopting packet-based transport. Many times, the SP leaves the customer side of the circuit as TDM-based but migrates the core of the network delivering that service to a packet-based transport. At the point where the customer TDM circuit is terminated in the nearest point of presence (POP), the TDM signal is sampled, turned into packets, and forwarded to the other end. At the other end, the TDM circuit is regenerated for the terminating link to the customer. The industry generally refers to this approach as *circuit emulation* (CEM), which is illustrated in Figure 9-5.

Replacing a portion of an end-to-end synchronous network link obviously means that although each end of the link might be running a synchronous protocol, there is a "gap" in the middle where that synchronous scheme is not present. Therefore, the SP must somehow build a timing network to replace the one that it removed with the decommissioned TDM core. For simple financial reasons, most SPs decide that this new timing signal should be transferred using the transport network carrying the services.

As described in Chapter 6, for Ethernet-type networks, the obvious choice in a packet network for frequency transport is to use SyncE (a slight diversion from packet-based frequency for a moment). Figure 9-5 illustrates the approach using SyncE to provide a timing signal to the customer TDM circuits (known in CEM specifications as *attachment circuits*).

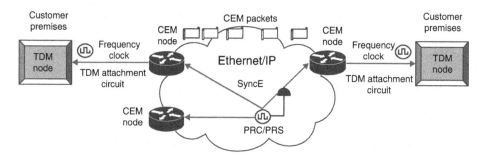

Figure 9-5 *TDM Circuits Emulated over an Ethernet Network, Timed with SyncE*

All the CEM routers at every POP in the SP network receive a frequency signal via SyncE from a reference PRC/PRS. Alongside SyncE, the CEM router also receives a QL signal, via the Ethernet synchronization messaging channel (ESMC), that indicates this signal is traceable to a high-quality source. The CEM uses these signals to both clock the TDM attachment circuit and set the SSM bits to signal the QL to the customer device. At each of the customer premises, the TDM device sees that the circuit is traceable to a high-quality source. So, it recovers the frequency from the line and uses it as the device clock and applies that frequency to receive and transmit its signal to the POP.

Mobile operators first encountered this problem in the 3G era when many mobile radio devices moved from E1/T1 transport to Ethernet, and yet the radio still required frequency synchronization. At that time, large segments of the operators' backhaul network did not support SyncE (the preferred solution) with the network elements they had installed. Given these constraints, many adopted a packet-based approach to frequency distribution, most commonly using the PTP telecom profile for frequency synchronization, G.8265.1. For more details on the mobile use cases for frequency, please see Chapter 11.

One of the constraints of the preceding model is the following question: What if the SP's customer has its own source of frequency that it wants the SP to carry transparently for the customer—meaning, the customer wants its own frequency on both ACs and not the frequency from the SP? See Figure 9-6 in the upcoming section "Adaptive Clock Recovery" for an illustration of this problem.

Differences Between Physical and Packet Methods

This section provides a quick summary of the differences between the physical and packet methods to carry frequency and the impact of the choice on the quality of the frequency recovered at the end of the timing chain. Although the metrics have a lot in common, they have different characteristics. They fall into three major categories:

- **Bandwidth of the signal:** The packet rate used by frequency over packet is in the range of 1–128 messages per second or 128 Hz (for PTP). For an E1 timing signal, it would be 2 MHz, whereas for SyncE it could be many megahertz. Therefore, the rate at which timing data can be transferred varies widely between the two methods.

- **Regularity:** The physical signal is very regular at short time scales, whereas the packet rate may vary considerably over short intervals. Yes, a packet stream may have a very consistent average rate over time, but PDV will cause the intervals between messages to be irregular.

- **Noise characteristics:** For packet timing, the main cause of timing noise is PDV, which has a larger amplitude and distribution than the jitter and wander in a physical signal. Packet timing relies on not just the network elements but also the transport and the topology. All these factors make the noise more course-grained, complex, and difficult to predict and counteract.

- **Traffic insensitivity:** Changing traffic patterns will affect the PDV generated in the packet timing signal, whereas the SyncE signal is unaffected in any way by the flow of packets.

The timing signals from the physical layer are usually mostly static with a Gaussian (bell curve) shaped distribution (G.8260); therefore, it makes sense to use every significant instant (data point) available to filter out as much noise as possible and achieve the best stability.

In contrast, packet-based timing signals are much more dynamic and do not have a Gaussian distribution (for an example, see Figure 9-16 later in the chapter, in the section "Packet-Based Frequency Distribution Performance"), meaning that there are far more outlying data points. Therefore, the recovery mechanism usually needs to select only a subset of the entire packet population by some form of filtering. It is a similar problem with measurement methods when trying to estimate the likely performance of a packet stream.

Despite those good reasons, SyncE still might not be the most appropriate solution for an operator. Before talking about the deployment of PTP as a packet method to carry frequency, you should understand some of the other popular methods that can solve some of these problems. There is additional reference information on these techniques in G.8261, specifically clause 8.

Adaptive Clock Recovery

One way in which it is possible to carry a different clock over a piece of shared infrastructure is to use a technique called *adaptive clock recovery* (ACR). This is used in the case where the clock coming from the customer must be carried transparently and regenerated at the AC at the other end of the service. One common name for this external clock is called the *service clock*—as opposed to the *network clock*, the name given to the clock used by the SP network.

An example of the use of ACR might be where the customer has its own private telephone network in headquarters but requires a link to a small business telephone exchange (PABX) at a remote office. This circuit is now being emulated with packets, so the service clock needs to be transparently carried over the CEM circuit between the two locations.

Before discussing the basics of how this system works, you will have to understand a little bit about CEM itself. Obviously, this book cannot include a full treatment of CEM, so consult the references (mainly from the IETF) at the end of the chapter for a source of further reading.

The basic idea of CEM is that the interworking router terminates the TDM circuit and an electronic circuit regularly samples the incoming signal (for an E1, it would be 2048 kHz) and recovers the data. It then bundles that data into an IP packet (for example, once every millisecond) and forwards it to the other end of the emulated circuit. Figure 9-6 shows an illustration of a CEM service with timing based on ACR. It is quite common that this functionality is implemented using a field-programmable gate array (FPGA).

The CEM router at the other end of the service starts to receive the packets from the source but does not immediately start to process and forward those packets. The receiver always lags a little behind the transmitter so that there is always a packet available to process. With a synchronous protocol, there needs to be data available when the TDM frame is ready to be filled, or blank space must be inserted.

To guard against this blank space, the receiver builds up a small buffer of packets in case some packets arrive too late. This buffer is called the *dejitter buffer*. For emulation of E1/T1 circuits, this normally a few milliseconds of data, and is configurable so that it can be tuned to the amount of PDV/jitter experienced in the transport network.

So, after the dejitter buffer fills up about 50%, the circuitry on the CEM router starts to drain the dejitter buffer (first-in, first-out, obviously). This gives it a few milliseconds of buffer in case of the late arrival of a packet. It reads the first packet in the queue and pushes it into the TDM framer on the AC for transmission to the customer site. Note that there is an equivalent process going on in the opposite direction, as each of these circuits is bidirectional. But the timing can only flow in one direction (left to right in Figure 9-6).

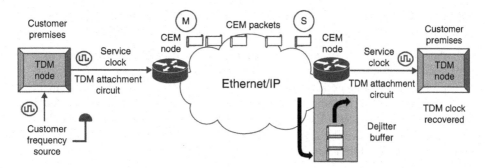

Figure 9-6 *TDM Circuit Emulation, Timed with Adaptive Clock Recovery*

Note in Figure 9-6 that the timing flow has subtly changed from Figure 9-5. On the left-hand side, the CEM node is no longer providing clock to the TDM node in the customer premises; rather, it is receiving it. The attachment circuit interface on that CEM node is using the TDM as a line clock to recover the frequency from the customer. It then uses that recovered frequency to transmit data back over the attachment circuit. The CEM node itself might still (optionally) have a network clock (such as SyncE), but that does not matter, because it does not use it for that circuit. That is all fine, but this creates a problem at the downstream end.

The issue is that the CEM on the right-hand side does not have that frequency signal from the master AC available to it, so it cannot drive the slave AC at the correct frequency. Of course, the CEM node could take a frequency signal from the network clock (with SyncE, for example). But this frequency will be different from that on the input AC, which will cause data integrity issues between the two ends of the customer circuit.

If the CEM (ACR slave clock) node on the right-hand side is running the attachment circuit more quickly than the left-hand (master) side, then it will eventually run out of data in the dejitter buffer to send (causing a buffer *underflow*). If it pushes data out too slowly, then the dejitter buffer will fill up and generate a buffer *overflow*—thereby losing data.

So, the right-hand CEM slave node uses the ACR method to regulate the frequency of the attachment circuit—based on the percentage fill of the dejitter buffer. If the dejitter buffer is filling up, then it is driving the attachment circuit too slowly, and it must speed up the frequency. If the dejitter buffer is emptying, then it is driving the circuit too quickly, and it must reduce the frequency. Thus, it is adapting the frequency that it sends to the downstream customer device based on the packet arrival rate, which is derived from the master clock on the upstream side.

Note that because the timing only passes in one direction (in Figure 9-6, from the clock master on the left to the slave on the right), this ACR clock recovery only happens on the right-hand slave node. The left-hand CEM node takes its clock from the customer circuit and uses that to become the ACR clock master.

Remember, this is a service clock on only a single interface; the CEM node would never use the customer frequency to drive the (network) clock on the entire CEM node. Another AC on the same router could use a different clock from another customer.

Still other ACs could reverse the clock direction and use the network clock—which means the customer equipment would have to receive the clock as is shown in Figure 9-5.

Regarding performance, the recovery of accurate frequency by ACR is susceptible to PDV. If the delay caused by the packet network is constant, the frequency of arrival of packets at the destination node is not affected by the network. There will be a lag in the phase of the recovered clock due to the delay through the network, but it will not suffer frequency or phase wander at the slave end. On the other hand, if the delay varies, the clock recovery process on the slave may interpret that as a change in phase or frequency of the master clock (which it is not).

The big advantage of ACR is that *every* interface can run a different (service) clock. Therefore, ACR is a very good tool for a third-party circuit provider to transparently carry a different service clock from many customers over their shared infrastructure that is still using its own network clock.

But ACR does have some limitations, not the least being that it might not scale so well for very large and dense installations of TDM circuits. Also, the mechanism for locking and getting the clock to settle is somewhat looser and less deterministic than when compared to any physical method. This is because ACR is fundamentally an averaging process over time.

For this reason, after an ACR clock is provisioned and starts recovering the clock, a waiting period of 15–20 minutes is recommended before measuring the maximum time interval error (MTIE) for the recovered clock. This behavior is documented in Appendix 2 of ITU-T G.8261, "Timing and synchronization aspects in packet networks." Also, maintaining a stable frequency output on the slave clock places increased requirements on the quality of the oscillator, especially if the packet network is suffering high PDV.

For those (and other reasons) there is an alternative used in large installations, called differential clock recovery (DCR), which brings the network clock back into the solution.

Differential Clock Recovery

DCR is designed to solve the same problem as ACR, which is for a network to carry a service clock transparently without the AC having to align with the network clock of the transport network. As seen in the preceding section, with ACR, the downstream node recovers the clock from the upstream end of the CEM circuit by aligning its frequency with the packet arrival rate.

With DCR, the upstream (master) clock inserts timestamps in the real-time transport protocol (RTP) header included in the CEM data packets. The slave clock recovers the clock from those differential timestamps. Including RTP is normally an optional component when configuring CEM circuits, but when using DCR, it is mandatory because the differential clock information is transferred in the RTP. See IETF RFC 3550 for more details on RTP.

Figure 9-7 illustrates how the DCR method works. Both the master and slave end of the CEM circuit have a network clock available, traceable to a common frequency source (normally it would be traceable to a PRC/PRS, but it could be just a common clock of different quality). The upstream master AC interface on that CEM node is using the TDM as a line clock to recover the frequency from the customer. Like the ACR case, it uses that frequency to both receive the incoming AC data and transmit the output data.

But with DCR the electronic circuitry (such as an FPGA) on the master side generates a timestamp that captures the difference in frequency between the network clock and the (customer) service clock. It includes that timestamp in the RTP header along with the data from the attachment circuit. On the (right-hand) slave side, the CEM node uses the timestamps to re-create the TDM clock frequency from the master end by replaying the timestamps against its own network clock reference.

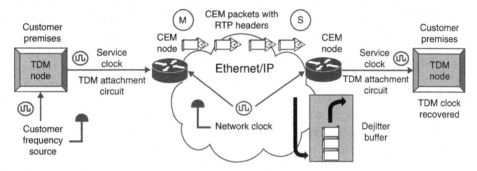

Figure 9-7 *TDM Circuit Emulation, Timed with Differential Clock Recovery*

The major difference of DCR compared to ACR is that both ends of the emulated circuit must have traceability to a shared source of frequency for DCR to work. Operators and service providers tend to prefer DCR over ACR because it scales better for large, carrier-scale deployments. Additionally, DCR is not sensitive to PDV, certainly when compared to ACR.

NTP for Frequency

Most network engineers use NTP only to keep the system clocks of nodes in the network roughly aligned with each other, most of the time for use in operational tasks such as fault isolation and event correlation. When network alarms and syslog messages are captured with inaccurate timestamps, it becomes very difficult to try and determine the order of events in a complex situation. However, in some specific circumstances, NTP has been deployed to transmit accurate frequency across a packet network.

The most common case that the authors are aware of is Ericsson using NTP to give frequency synchronization to its base stations (eNodeB/eNB), mostly for 3G/LTE mobile networks. Mobile radio requires 50-ppb stability at the air interface, and engineers accept that this maps to a requirement of 16 ppb at the input to the radio equipment. For small cells (see Chapter 11), it might be somewhat looser.

In theory, NTP and PTP provide basically the same performance, assuming that one is using the same algorithm, same clock, and the same network conditions. Also, both implementations must be supported by hardware timestamping (not so common for NTP). One difference in performance comes from the fact that NTP does not support the higher packet rates available when using PTP. The reality is that PTP has much wider support in the timing and networking communities and much better hardware support.

This deployment option is mentioned here only for completeness because it might be an option that you come across, particularly in a pre-5G mobile network with Ericsson equipment. Otherwise, there is no reason to expect you would have to deal with it because the rest of the industry prefers other methods, most particularly PTP.

PTP for Frequency

The most popular alternative to carrying frequency over a network using packet methods is to use PTP technologies. For most scenarios, this will mean using the ITU-T G.8265.1 telecom profile for frequency synchronization. There are two main reasons:

- It has strong support from industry and network elements.

- It is designed to interoperate with the timing aspects of SyncE, PDH, and SDH/SONET.

Figure 9-8 illustrates the same CEM application as shown previously, but in this case all the CEM nodes are timed using a PTP clock rather than SyncE (compare it to Figure 9-5 where SyncE is used). The operator positions a PRC/PRS somewhere in the network combined with a GM running the G.8265.1 profile. Each of the nodes in the network that requires frequency runs as a PTP slave to recover the clock. The transport uses IPv4 (optionally IPv6), and there are no transparent or boundary clocks because on-path support is expressly forbidden in the defining standard.

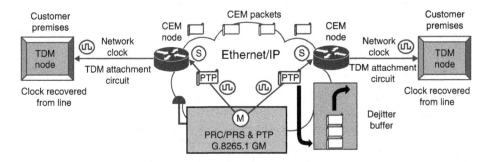

Figure 9-8 *TDM Circuit Emulation, Timed with PTP Telecom Profile G.8265.1*

Note that because G.8265.1 is a method to transfer only frequency, it does not need to be a two-way protocol, meaning that only the messages from the master to the slave are required—namely the Sync and optional Follow_Up messages. For more details, see the sections "One-Way Versus Two-Way PTP" and "G.8265.1—Telecom Profile for Frequency Synchronization" in Chapter 7, "Precision Time Protocol." For this reason, asymmetry is not an issue with G.8265.1 deployments, but excessive PDV is a serious problem that must be controlled (more on PDV in a following section).

Probably the two most deployed use cases for G.8265.1 are circuit emulation and mobile networks where phase is not yet necessary (for example, 3G/4G frequency division duplex [FDD] radios). As the mobile use case begins to require phase/time, the chief remaining use for this solution is CEM. And again, that applies only where other alternatives are not available—because SyncE would be the preferred option.

Figure 9-9 shows how the flexibility of G.8265.1 can be used to provide a form of "glue" to bind different frequency domains together, especially in those areas where no physical methods to carry frequency are available. In this case, the right-hand Ethernet network between router C and D does not support SyncE, which would normally be a problem.

The frequency is sourced from a PRC/PRS on the far left and used as a frequency source in router A. The QL information indicating the source of frequency would be carried using the SSM bits (if supported by the BITS signal). Router A uses that frequency as a source for the frequency on the outgoing SyncE link to router B. The QL information is carried over that same link in ESMC packets.

Router B uses the ESMC QL information to help select the SyncE input from A as a network clock. That frequency is then used as the clock source for a TDM link to router C. As there is no CEM in this case, the TDM link between B and C is solely an IP data transport for the application. Again, the QL information could be carried in SSM bits or an equivalent for the transport type. Router C will recover the frequency from the line clock as it can see it is traceable to a PRC and use it as the network clock for the node.

However, the network from router C to router D does not support SyncE, but router C supports the G.8265.1 packet master function. Router C uses the reference frequency recovered from the TDM circuit as a source for the PTP G.8265.1 master. The QL information, signaling traceability to a PRC/PRS, is mapped onto the appropriate PTP clock quality information such as the clockClass attribute. This information is carried in the Announce message forwarded to router D.

Router D supports the PTP slave function and sees the Announce message indicating traceability to a PRC/PRS source of frequency. It then initiates a PTP message flow with the master port on router C (see the section "Negotiation" in Chapter 7). Router D uses the flow of PTP timestamps to recover the frequency and derive the network clock from it. That clock is then used as a source of frequency for other circuits—such as SyncE— onto other nodes further downstream.

For all this interoperability to work effectively, there must be an understanding of how to map the different clock QL levels between the different systems of carrying frequency. Table 9-2 shows the mapping of QL information between the different mechanisms, which is outlined in Table 1 of G.8265.1. To take the example in Figure 9-9 and assuming it is an Option I network, if router A gets an SSM value of 0x2, it interprets it as QL-PRC. The ESMC value would then also be QL-PRC, and router B would turn it back to an SSM value of 0x2. Router C would recover that SSM and map it to clockClass of 84 for the PTP link. Router D would then interpret that clockClass as QL-PRC for the downstream SyncE/ESMC.

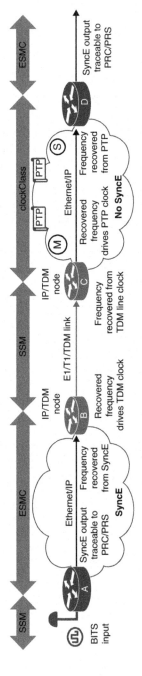

Figure 9-9   *Frequency Synchronization Between E1/T1, SyncE, and G.8265.1 PTP Profile*

Table 9-2 *Mapping of QL Values Between Different Frequency Transport Technologies*

SSM QL	ITU-T G.781 Option		PTP clockClass
	Option I	**Option II**	
0001		QL-PRS	80
0000		QL-STU	82
0010	QL-PRC		84
0111		QL-ST2	86
0011			88
0100	QL-SSU-A	QL-TNC	90
0101			92
0110			94
1000	QL-SSU-B		96
1001			98
1101		QL-ST3E	100
1010		QL-ST3/ QL-EEC2	102
1011	QL-SEC/QL-EEC1		104
1100		QL-SMC	106
1110		QL-PROV	108
1111	QL-DNU	QL-DNU	110

Source: ITU-T G.8265.1, Table 1

This example is given to show the flexibility in working with the standards to ensure that difficult network problems can be addressed simply with the tools available to you. Of course, this only demonstrates that the signal can pass from end to end and that the mapping of QL will be faithfully reproduced. But it does not mean that the frequency signal will meet the application requirements when it arrives at the desired location. There will be time error introduced at every step of the path, so it is the responsibility of the timing engineer to ensure that the frequency quality metrics all meet the requirements.

Note that sometimes you may run across the profile in some equipment known as *Telecom 2008*. This is a pre-standardized version of the telecom profiles based loosely on the default profiles, but which very closely resembles G.8265.1 (or G.8275.2). The authors frequently encounter it in use by mobile operators that are doing mostly frequency synchronization over packet, although it can work in the phase scenarios. See Annex A of IEEE 1588-2008 for details.

Packet-Based Phase Distribution

It is now time to move on from distributing frequency based on packet transport to phase/time distribution using the same techniques. In fact, a lot of the details have some commonality; if anything, the phase/time case is a superset of the frequency case, meaning that it is much the same, but harder.

There are a limited number of options available to run accurate phase/time across a packet-based transport network, and just about all of them involve some flavor of PTP. There might be slightly different implementations of PTP; for example, a specific industry may mandate a profile that uses a purpose-built type of communication method. That requirement might be based on a situation unique to that industry or because the messages must be carried by a transport method that is already widely deployed. The Power Profile is one good example. See details of the Power Profile and other profiles in Chapter 7.

For the following section, the focus is on the telecom profiles, but the principles apply to other cases and profiles trying to achieve accurate phase/time synchronization.

PTP for Phase/Time

Figure 9-10 illustrates the architecture for the carriage of phase/time over packet. The approach is for frequency, phase, and time to be taken from a time source (PRTC) and used by a grandmaster to send time information across a packet network. These signals are then recovered on the slave clock and used to synchronize the slave node and any associated application hosted there (represented here as a radio in a cell tower).

As you can see, the concept is simple, but there is a lot of detail involved in making it work well. The following sections of this chapter go through the major factors that contribute to a successful deployment. Many of the topics have been quickly covered in previous chapters, but this chapter will tie that background to specific choices when faced with a real implementation.

The choice for a network timing engineer is very clear. If possible, a packet timing network should use the following approaches to distribute time:

■ Utilize a full timing-aware network with timing awareness in every node (and the transport) and using the Layer 2 ITU-T G.8275.1 telecom profile.

■ Distribute the timing infrastructure as much as possible. It is not recommended to deploy a couple of PRTC+T-GM in the center of the network and then attempt to run timing out to the furthest reach of the network over many hops.

■ Use T-BC or Telecom Transparent Clocks (T-TCs) with G.8275.1 support embedded in every network element, with a certified performance level (certainly for noise generation) from G.8273.2 (for T-BC or T-TSC) and/or G.8273.3 (for T-TC).

■ Combine PTP phase/time with SyncE or, if possible, enhanced SyncE (G.8262.1) to carry frequency in what is called *hybrid mode*.

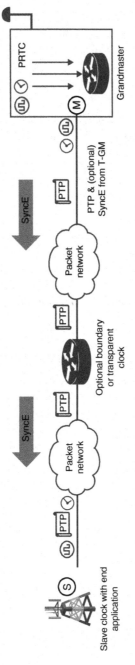

Figure 9-10 *Phase/Time Synchronization over Packet Using PTP with Optional SyncE*

- Use a PTP-aware transport system, which aims to reduce asymmetry as far as possible and controls any generation of PDV.

- Avoid (unless necessary) the use of G.8275.2 telecom profile over unaware nodes. If its use cannot be prevented, then its use must be very limited in scope because there is no control over how well even a single *unaware* hop will treat the timing signal as it passes through. The best approach in these circumstances is to run G.8275.1 as far as possible and then use profile interworking to convert it to G.8275.2 at the last hop (see Chapter 11 on profile interworking).

- Design a timing solution around budgeting time error, not as a networking exercise. There is much more on budgeting in the section "Budgeting End-to-End Time Error" later in the chapter.

  The engineer needs to know how accurate the phase and frequency must be when needed at the application, and then figure out how to get time delivered to it within that time error budget.

For details of each of these factors, see the following three sections. Chapter 11 will further expand on this chapter with a specific focus on the 4G and (especially) 5G radio applications. To avoid repetition, the details of some aspects are more thoroughly covered in Chapter 11. Even if mobile synchronization is not your problem, there are lessons to be learned from an industry that needs microsecond-level timing accuracy across the whole country, even on remote mountaintops.

Full On-Path Timing Support Versus Partial Timing Support

Chapter 8, "ITU-T Timing Recommendations," deals with the ITU-T recommendations that cover the network topology and requirements, detailing the form and topology of network suitable for a PTP deployment. One case is a network that is fully "timing aware" and aids the timing signals at every point of the network between the source of time and the consumption of it—referred to as full timing support (FTS).

The alternative is where there might be assistance at some points but not at every point—referred to as partial timing support (PTS). See the Chapter 8 sections on G.8271.1 (FTS) and G.8271.2 (PTS) for information on these two different approaches.

The expectation is that the FTS case is implemented using the ITU-T G.8275.1 telecom profile in combination with SyncE to carry frequency, while the PTS case is implemented using the ITU-T G.8275.2 telecom profile with PTP over IP (PTPoIP) and where SyncE for frequency is optional. The clear difference between them comes down to a single critical factor: the FTS case can guarantee a level of deterministic performance, whereas the PTS case cannot.

Figure 9-11 illustrates the differences between the two options.

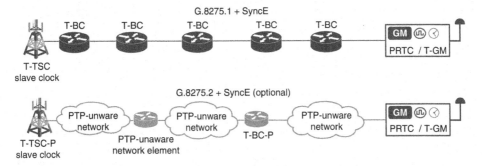

Figure 9-11 *Full Timing Support Versus Partial Timing Support*

One point that is worth reinforcing very strongly is also repeated quite plainly in some of the ITU-T recommendations, such as G.8271.2:

> A network that is based on the architecture defined in [ITU-T G.8265], and is designed to meet the network limits for frequency defined in [ITU-T G.8261.1], is not designed for the delivery of accurate time or phase, which requires much lower levels of packet delay variation (PDV) and asymmetry. The network limits specified in this document are for small, well-controlled networks (e.g., in-building or last-mile network segments), which can guarantee that the stringent PDV and asymmetry network limits are met.

This means that if you have a network already (successfully) carrying frequency using G.8265.1, you should not assume that this same network will accurately carry phase/time using G.8275.2. Although they are quite similar in many respects, the carriage of phase/time will probably not succeed without significant mitigation and refresh. This is especially the case for asymmetry because it is not a significant factor when recovering only frequency, but it needs to be very tightly controlled for phase/time.

The second point from the preceding quote is that using PTPoIP over unaware PTP nodes must be limited to very well-controlled cases with very good packet performance and only a small number of hops.

Hybrid Mode Versus Packet Only

At every opportunity, this book proposes that the deployment of PTP (to carry phase/time) should occur in combination with SyncE (to carry frequency). When it comes to deployment, the timing engineer needs to understand whether this is possible, and what are the trade-offs for either approach.

There is a somewhat semitechnical way to describe the reasons for the difference in performance when using a physical method for frequency. When setting out to synchronize a slave clock, the timing system needs to perform two separate functions. First, the slave must get the local oscillator running as close to its (correct) nominal frequency as possible (20 MHz has been used as a typical reference oscillator in earlier examples).

To achieve this goal with a packet-only method, the slave must watch the clock for a while and figure out if the time is going too quickly or slowly, then adjust the local oscillator, wait for a while, and retest. This is necessary because the data from the master is irregular (non-Gaussian) and of low bandwidth because of the limited message rate. The slave has only the periodic timestamp to guide it and, given the variability in the packet network, that data can move around somewhat.

But with a physical frequency signal, such as the 125-MHz signal from gigabit Ethernet, the slave can recover the frequency from the incoming SyncE, and dividing it by the appropriate factor gives the timing subsystem a very clean reference for the phase-locked loop (PLL) to feed to the local oscillator. This means that the slave quickly attains a stable frequency lock because of the (relatively) rock-solid reference signal.

Once the clock has attained frequency lock and the oscillator is running at the correct rate, the second task is that the PTP servo must align the phase with the master, using the standard PTP timestamp mechanism (Chapter 7). This is straightforward because the oscillators are now already syntonized by the physical signal; it is just a matter of figuring out the offset from the master and the path delay to calculate the phase/time.

The packet method for recovering frequency is somewhat slow and cumbersome, mainly because the amount of data it is receiving is limited to 16 sets of PTP timestamps a second (for G.8275.1). Meanwhile, purely from the bandwidth point of view, in the case of the physical signal, there are millions of reference data points (significant instances) per second. See Chapter 12 for details of the timing subsystem and the physical components involved in selecting, down-sampling, and aligning input signals.

The second major benefit of using SyncE (or some other physical method) to carry frequency is that it is not affected by the nondeterministic behavior of packet networks. The SyncE signal is always available and stable, no matter what the traffic load on the network. The packet signal, on the other hand, suffers the consequences of whatever else is happening on the path.

The other great assistance from SyncE is its ability to help the slave clock maintain very good holdover performance. This holdover improvement arises because once a slave clock is aligned in phase, if both the master and slave clock are running at the same frequency, then the phase difference between them cannot change. Then when the PTP source of phase/time disappears, there is still a frequency signal to keep the oscillators aligned, so the phase cannot move out of specification. Without the frequency source, the oscillator is left to cope on its own and is at the mercy of its own physics. These cases (with and without SyncE) are covered in more detail later in the chapter in the section "Holdover."

SyncE clearly has real benefits, and for successful deployments, it is critical that the network elements and transport systems can transport SyncE (and ESMC) correctly and that the PTP clocks can operate in hybrid mode (physical frequency plus packet phase/time).

PTP-Aware Nodes Versus PTP-Unaware Nodes

The other aspect that makes the job of the network timing engineer easier is when the network, as far as is possible, consists of network elements that can function as PTP-aware devices (meaning some form of PTP clock). Of course, the most common of these is the boundary clock and the transparent clock. For the telecom profiles, these are known as the T-BC and the T-TC, respectively.

The transparent clock is a straightforward device in concept; it uses an accurate oscillator (ideally with a frequency reference such as SyncE) to measure the residence (transit) time of the PTP message as it passes through the node. This residence time is then added to the correctionField so that the slave can compensate for any delays that the message suffered in its path from master to slave. An example of how the correctionField is used is included in the section "Correction and Calibration" later in the chapter.

The boundary clock is quite a bit more involved. As illustrated in Figure 9-12, the concept is straightforward, although the hardware design is quite complex. There is a whole section in Chapter 12 on clock hardware design, so you are referred to that for a more detailed description on how a PTP-aware T-BC is designed and implemented.

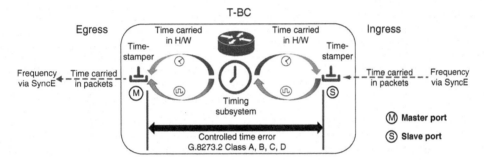

Figure 9-12 *Functionality of a PTP Telecom Boundary Clock*

Look first at the frequency subsystem, as SyncE on the (right-hand) ingress port is being used as the frequency source for the system. The SyncE is recovered and routed over a hardware signal to the PLL dealing with frequency. The output frequency of the PLL is then eventually routed out to all the egress ports (including the one on the left).

For phase/time, the ingress port is configured as a slave port and incoming PTP messages are timestamped on arrival. The source of clock for the timestamping comes over a hardware path from the timing subsystem. Following timestamping, the incoming messages and timestamps are forwarded to that component of the timing subsystem dealing with packet processing, and the timestamps are used to calculate the time offset and mean path delay.

Outgoing PTP messages are created in the packet processing component of the timing subsystem and then sent for forwarding out the egress interface designated as a PTP master port. As the messages exit the interface, they are timestamped using a reference time signal coming over a hardware path from the timing subsystem.

The whole point of a timing-aware clock is that the system is designed to eliminate many of the likely causes of timing error. Using hardware signals for time transport dispenses with many possible problems caused by the normal processing of a packet switch, such as queuing, scheduling, and so on. This means that the clock will, depending on the quality of the design and the components, have a readily predictable and reliable performance, which is not impacted by the traffic load. This performance is characterized by numerous metrics in the ITU-T G.8273.2 recommendation where the class of clock operation is specified as Class A, B, C, or D. Engineers are most concerned with the noise generation parameters, which is the amount of timing error (noise) introduced by time passing through the node. The section "Clock Performance," later in this chapter, provides more details on this topic, including noise generation. You should refer to Chapter 5 for details of the metrics themselves.

Assisted Partial Timing Support

The advantages of FTS topologies over PTS topologies in combination with PTP-aware nodes are plain to see, so there is a strong motivation to use that approach. When that approach is not possible, G.8275.2 PTPoIP can be switched through PTP-unaware nodes, with the disadvantage that the performance will be less predictable and be degraded by traffic. Which raises the question, given those constraints, what is the point of this profile?

The original use case for G.8275.2 PTPoIP was based on the need of operators (particularly in North America), with many deployed GPS receivers to have a backup in case of localized GPS outages. MSPs were also handicapped by the fact that many relied on third-party SP circuits to provide the majority of their backhaul links between the cell site and mobile core. This meant that they wanted to be able to have a transport solution for timing but had no opportunity to use on-path PTP support.

So, the operators, working with the ITU-T, agreed upon an option to carry PTP over the third-party circuit using IPv4 (with IPv6 optional) without any on-path assistance. Another motivation is that operators prefer to have IP-based negotiation because they like control over the configuration of devices on their networks. The PTPoIP model offers that, whereas the G.8275.1 profile is based more on an approach that automatically determines the best topology.

The downside to this approach was that any third-party circuit likely to be leased by the operator had no on-path timing support and yet probably contained a significant but unknown amount of asymmetry. The solution to this case was the assisted partial timing support (APTS). Figure 9-13 illustrates the major elements of this approach.

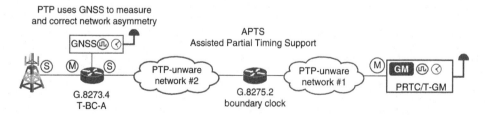

Figure 9-13 *Architecture of an Assisted Partial Timing Support Solution*

At the end of the third-party network segments, there is a BC or perhaps even a slave clock embedded in the end application (in this example, a cell site radio). By this point, there could be a significant amount of accumulated asymmetry because of the PTP messages have crossed numerous PTP-unaware network elements.

The final slave or BC has a local source of time (likely a GPS receiver in this case) that it uses as its primary source of synchronization. However, this clock also contains a slave port that is recovering time from a PTP time signal from a remote PRTC and T-GM combination. Because the PTP servo on that clock has a GNSS reference available, it can measure the difference between the phase delivered by the GNSS and the phase recovered from the PTP slave port. This measured difference is mostly made up of asymmetry, either dynamic or static (more on that in the following section "Asymmetry").

The amount of asymmetry is constantly measured and retained by the PTP slave. If the local GPS receiver ever becomes unavailable, then the PTP will select that PTP slave port as its source of phase/time. The extra step here is that the servo then applies the latest value for asymmetry as a correction to the incoming signal. This compensates for any asymmetry that was present when that measurement was taken.

Of course, any change in that error resulting from any subsequent rearrangements and transients in the network between this clock and the remote T-GM cannot be measured after the GPS reference disappears. It cannot be a long-term solution, but as a short-term emergency measure to add a level of resilience, it has merit.

There is more information on this solution in the section "Assisted Partial Timing Support with G.8275.2" in Chapter 11, as it is particularly aimed at the mobile operator.

Leap Seconds and Timescales

In talking with customers over the past year or so, most of whom are medium to very large MSPs, along with numerous cable operators, a couple of questions come up repeatedly:

- How do systems handle the leap second event, and what must the operator do?

- Why bother with the whole issue of leap seconds? The solution suggested by some is that it would be far simpler to just run some local wall-clock time and forget about all the complexities with different timescales.

The following sections deal with both these issues.

Leap Seconds

Both Chapter 1, "Introduction to Synchronization and Timing," and Chapter 7 covered the issue of leap seconds in some detail from the background point of view. But exactly how is the leap second treated in an operational environment?

There are two basic mechanisms available to distribute information on leap seconds:

- Use a source of Coordinated Universal Time (UTC) that also distributes leap second information

- Via configuration and operational procedures based on information made available by the *International Earth Rotation and Reference Systems* (IERS) service and its Bulletin C

GNSS systems carry leap second information in their timing and navigation messages to enable receivers to learn about and cope with leap second events. The GNSS systems signal that a leap second event is approaching and what sort of event it might be (the most common being the 61-second minute). This information allows the GNSS receiver to accurately translate the timescale native to the GNSS system into an accurate value for UTC. See Chapter 3 for more details on how the GNSS systems and their timescales work.

If the receiver supports the capability, and the ToD message format allows it, then the GNSS receiver may be able to pass that information to the PTP master function using the ToD link. That way, the PTP master can learn about the change and then use the Announce message to signal the downstream slaves that the leap event is coming.

The second approach is to use configuration on the PTP GM clocks to signal the leap events whenever the IERS announces an upcoming future event. Normally, bulletin C provides a warning of about five months before the event, allowing sufficient time to prepare for the impending change. The configuration on the GM indicates what the new (changed) time offset between TAI and UTC will be and the date and time from when it takes effect. So, a configuration might indicate that the leap second offset will change (from the current 37 seconds) to 38 seconds on 31-Dec-2021 at the second following 23:59:59 UTC.

Whatever method is used, the GM clocks will use that information to fill in the Announce message with the new leap second information when needed. Twelve hours before the event occurs, the PTP GM changes the advertised Announce messages to signal that a *Leap61* event is coming. This is done ahead of time so that if the slave is disconnected from the GM during the switchover time, then the slave clocks are still aware that the leap event is happening.

In normal operations, PTP uses the PTP timescale, which means that the timestamps reflect atomic clock time, Temps Atomique International (TAI). There are no leap events in the TAI timescale because it is *monotonic*. So, it is very important to understand that the leap second has *no effect* on the PTP timestamps. It affects only the contents of the Announce message, which is then reflected in the PTP datasets. The only scenario in which the UTC offset is needed is if a slave clock needs to convert the PTP timestamps into UTC (for any purpose requiring a human-readable timestamp or even as a source for an NTP server).

As an artificial example, the following shows what would happen on a PTP slave clock when the time for the leap second arrives. First, the slave receives the Announce message signaling that the leap second is coming in the next 12 hours. Then, 12 hours later, the Announce message is updated to clear the leap second flag, but increments the UTC offset from 37 to 38.

```
Dec 31 12:00:20.823: ptpd: Old leap61 Flag is 0  New leap61 Flag is 1
Jan 01 00:00:00.000: ptpd: Old leap61 Flag is 1  New leap61 Flag is 0
Jan 01 00:00:00.000: ptpd: OldUtcOffset is 37 , New UtcOffset is 38
```

See Annex B of IEEE 1588-2008 for detailed information on the usage of timescales and epochs in PTP.

Should the Clocks Run UTC or TAI?

As you read in the previous section, under normal conditions, PTP uses the PTP timescale, which means that the epoch is based on the PTP epoch (1 January 1970 00:00:00 TAI), and the timescale is TAI. See the section "Timestamps and Timescales" in Chapter 7 for more details.

Some engineers ask why they must deal with PTP timestamps based on TAI and all this leap second hassle, when the fact is that they want to see UTC. Some suggest they want to run their PTP installation on UTC, rather than TAI and ask what is wrong with that approach.

It is true that PTP has the capability of running a different timescale, what it calls the ARB (for arbitrary) timescale. This timescale could be any value that a user might desire, like the number of seconds since the last appearance of Halley's Comet. But it should be monotonic because running a non-monotonic timescale can cause some serious issues to other applications outside of PTP.

The main thing to consider is what would happen to the downstream systems when encountering timestamps that may go backward, repeat themselves, or just skip forward, as can happen to UTC time. You could even go further and implement a version of local time without using time zones, but this would cause chaos when summertime (daylight saving time) switchovers happen.

One of the authors remembers this implementation option being used on large IT mainframe systems that ran some form of local time. Because of that, all the machines would have to be taken down for a minimum of an hour on the autumn switch back from summer (daylight saving) to standard time. All of them had to be shut down before 3:00 a.m. (or whenever the switch back was scheduled), and the systems only restarted more than one hour later when the clocks again passed the time they were shut down.

This action was necessary to stop applications crashing and databases becoming corrupted because time had flowed backward (imagine a transaction that started at 2:50 a.m. and finished at 2:05 a.m.). A similar process was necessary during the skip forward to summertime (daylight saving) to ensure that the elapsed time for transactions was correct, although it was not necessary to wait one hour. Other machines that needed to calculate elapsed times, such as telephone call logging, always used GMT/UTC so that there were never any issues calculating the elapsed time for phone billing.

This anecdote is given as an example only to illustrate that there are many disadvantages when using a timescale that is not monotonic. The point is that there might be unintended consequences downstream of the PTP time clocks that are very hard to predict.

Factors Impacting Timing Performance

Chapter 6 covers many aspects of timing performance for the case where frequency is carried by a physical signal, and Chapter 5 covers many of the parameters and metrics that apply to the qualification of synchronization performance. On the other hand, Chapter 8 goes into a lot of detail about the various recommendations defining performance, both for clocks/nodes and end-to-end networks.

This section examines performance from the aspect of deploying packet timing for two different scenarios (both based on PTP):

- Packet-based frequency distribution

- Packet-based phase distribution

Separate from the packet performance of PTP, you also need to look at the time error and wander values in the signal after it is recovered by PTP. Not surprisingly, all the metrics used to measure the quality of frequency using physical methods also apply to packet-based methods. And because most of the needs for frequency in a packet network come from interworking with TDM networks, the metrics between physical and packet methods have a lot in common.

But the same goes for phase/time synchronization as well. The performance of the PTP packets and the quality of the frequency signal can have an impact on the time metrics for phase. Of course, it also depends on other things: the quality of the oscillator; the characteristics of the PLL; the bandwidth of the filtering; and the quality of the PTP servo. More details on the design of a PTP clock are provided in Chapter 12, but the next sections deal with the factors influencing timing performance.

Packet-Based Frequency Distribution Performance

As mentioned previously, one of the major differences between a physical frequency distribution and a packet-based one is the amount of data available to recover the frequency. When using PTP for frequency, a packet rate of about 64 packets per second is the commonly accepted rate (at least the rate for the Sync messages).

Even with only 64 packets per second, some of those arrive later than expected with data that was overly delayed and must be discarded. That is because excessive (and variable) delay in the PTP messages has allowed timestamps to become stale by aging more than other packets. To mitigate this, the packet selection mechanism will attempt to discard any "bad" data and concentrate on the "lucky" ones that were only minimally held up. See the section "Packet Delay Variation" later in the chapter to explore the reasons why PDV happens.

The question is then, what does the slave clock do with those lucky packets? Figure 9-14 illustrates a functional model for a slave clock recovering frequency from a packet timing signal. This model is loosely based on ITU-T G.8263 but is a common method of implementation.

The packet timing signal is first processed by a packet selection algorithm that selects the packets used to recover the clock. The time information from the selected packets is used as an input to the phase/time offset detector to compare the reference and local times. The difference between the two times is used to control the rate of the local oscillator that is driving the local time scale. The result is that the local time scale is adjusted to advance at close to the same rate as the master time scale.

Normally, this local reference is sourced from a (local) stable oscillator.

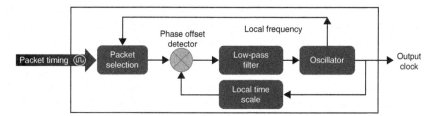

Figure 9-14 *Architecture of a Packet-Based Frequency Slave Clock*

The performance of the oscillator implemented in the slave equipment is the key factor that determines the frequency accuracy that the slave clock can achieve. From a packet point of view, the main impediment to recovering an accurate frequency is PDV. Therefore, the characteristics of the oscillator chosen in the clock design depends on the requirements of the recovered frequency, and on the level of PDV generated by the network.

For the performance limits for packet-based frequency, the following is a quote from G.8261 that provides a guide to the overall requirements for a typical deployment:

> As an example, the typical wander that is tolerated by end applications can be expressed in terms of [ITU-T G.823] and [ITU-T G.824] traffic masks. [ITU-T G.8261.1] defines the conditions under which a packet network is suitable to support packet-based methods, and [ITU-T G.8263] specifies the related clock characteristics.

This basically means that the frequency accuracy that comes out the end of a packet-based synchronization chain is aligned to the TDM traffic masks, such as those for E1/T1. It is ITU-T G.8261.1 that defines an output wander limit for frequency at the final clock in the network chain. That output wander limit is aligned with the G.823 limits (TDM version of a chain of frequency clocks) and aimed at meeting the frequency requirement of 16 ppb for the mobile cellular radio use case (see Figure 9-15).

Recommendation G.8261.1 has a second function that applies to packet-based frequency timing: it defines a model for PDV network limits. Of course, these network limits must be compatible with the minimum limit of PDV that the final clock in the chain is required to tolerate. The tolerance for clock PDV is specified in the recommendation on frequency clock performance, G.8263. Just be aware that for PTP, there is an amount of PDV that will prevent a slave clock aligning at all.

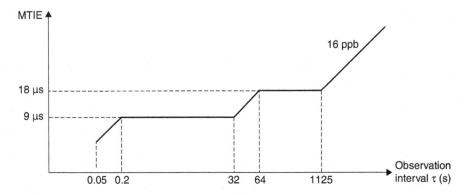

Figure 9-15 *Packet-Based Frequency Wander Limit MTIE Mask—Based on Figure 4 of G.8261.1*

Basically, the PDV must allow at least 1% of packets to arrive in a cluster close to the minimal possible transit time, or what is called the *floor delay*. This 1% is taken over a 200-second sliding window, and the cluster is defined in G.8261.1 as up to 150 μs from the floor delay. The method of selection is up to the implementation, and there is a lot of information on several approaches in G.8260.

Figure 9-16 illustrates what an ideal PDV profile might look like when the packet delay is plotted over time. It is obvious that there are many packets in this example that are close to the floor delay and therefore contain good data (fresh timestamps) to help the clock recover the frequency. There are also quite a few packets that have been seriously delayed, so the distribution is quite skewed. There are examples of PDV histogram charts in Figure I.2 of G.8260.

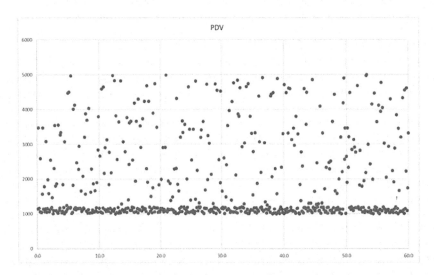

Figure 9-16 *Packet Delay Variation with Clearly Defined Floor Delay*

In summary, the major impact on performance for packet-based frequency is PDV and the only way that it can be mitigated is to remove/reduce it and to improve the oscillators in the slave clocks. Normally, using a PTP boundary clock between the master and the slave could reset the PDV, but the G.8265.1 telecom profile for frequency synchronization expressly forbids it.

Although this problem is tough enough, unfortunately the situation for the phase/time case is even more difficult and complex, and the extra dimension of asymmetry is added.

Packet-Based Phase Distribution Performance

Although the metrics might change, most of the lessons from the frequency situation also apply to the case for phase/time: PDV is still an issue, although the impact depends somewhat on how it is being deployed. Thankfully, there are techniques to help mitigate the PDV for the phase/time case that are not available when using the G.8265.1 profile. But the major additional headache for phase/time is asymmetry, a topic first covered in Chapter 7.

To refresh your memory, PTP on the slave needs to know how long a Sync message took to travel from the master, so that it can adjust for the propagation time when comparing the master's timestamp to its own clock. PTP determines that master-to-slave transfer time by assuming it is exactly half of the round-trip time. Anything that invalidates this assumption introduces an error in the calculation of correct time and results in a phase offset (error) on the slave. This is the asymmetry that causes problems with accuracy in the recovery of phase/time. See the section "The (Simple) Mathematics" in Chapter 7 and the section "Correction and Calibration" later in this chapter.

As a definition, *asymmetry* is any behavior or characteristic of the network that causes PTP messages travelling in the forward direction (from master to slave) to take a longer or shorter time than the trip in the reverse direction (slave to master). There is a lot more information on it in the upcoming section "Asymmetry," but before that, it is time to revisit some of the timing metrics first discussed in Chapter 5.

Parameters for Timing Performance

Besides the end-to-end network limits that define the performance of phase/time transfer with packets, there is a range of timing metrics that apply to the individual clocks themselves. Standards development organizations (mainly the ITU-T) have developed numerous recommendations to ensure that clocks can be assembled into a timing network without encountering issues. To do that, the SDOs have defined metrics for *equipment limits* that define the performance of individual network clocks.

Even the most basic problem could make a timing network unusable, such as where clock $N+1$ in the chain of clocks cannot tolerate (as input) what the preceding clock N is generating as output. When building the timing network, the engineer needs to procure network elements that meet defined equipment limits and criteria to ensure they can work with each other.

For phase/time distribution using a packet-based method, this covers factors such as

- **Noise generation:** The amount of time error produced at the output of a clock when it is supplied with an ideal input reference packet signal

- **Noise tolerance:** The minimum amount of time error that can be tolerated at the input of a clock without causing it to reject the input and generate errors

- **Noise transfer:** A clock property describing how time error from the input PTP interface can be detected in the PTP and 1PPS output interfaces

- **Transient response:** The response of a clock following the rearrangement of either the physical frequency (SyncE) or the PTP packet timing source signals

- **Holdover:** The maximum deviation in the PTP and 1PPS output signal during the loss of the PTP packet timing signal and/or physical layer frequency inputs

The section "Testing Timing" in Chapter 12 deals with the measurement and validation of these behaviors and limits by testing timing equipment and networks. This is not functional testing but testing to confirm that the equipment meets those specific timing metrics, so it involves using specialized time testing equipment. This testing is of value to the engineer who wishes to confirm that the chosen equipment is suitable for their deployment case.

Chapter 8 covered the recommendations that apply to measuring clock performance. For the PTP packet phase/time case, G.8273.2, which covers the boundary and slave clock performance, is one of the most important recommendations. And from that specification, timing designers consider noise generation to be the most important parameter when building a timing solution. For noise generation performance, G.8273.2 divides the levels of performance into classes (currently A to D) based (mostly) on the results of that test.

Table 9-3 shows the main noise generation time error (TE) values used in G.8273.2 to determine the class of performance for a T-BC or T-TSC clock (note that this is not the full set of values for noise generation).

Table 9-3   *G.8273.2 Performance Characteristics for Noise Generation*

Parameter	Class A	Class B	Class C	Class D
Max absolute TE (max|TE|)	100 ns	70 ns	30 ns	5 ns (low-pass)
Constant TE (cTE)	±50 ns	±20 ns	±10 ns	For further study
Dynamic TE (dTE_L)*	40 ns	40 ns	10 ns	For further study

Source: ITU-T G.8273.2, Tables 7-1, 7-2, 7-3, and 7-4

*dTE_L is the MTIE result measured through a first-order low-pass filter (LPF) with bandwidth of 0.1 Hz at constant temperature. The class D max\|TE\| is also specified after having passed through an LPF.

There is a lot more information on these parameters and what they mean in Chapter 5, so if you want to fully understand parts of this chapter, then you should be aware of the basic ideas from Chapter 5. But at a minimum, these are the main metrics you need to be concerned about, so the rest of the chapter explains them in more detail. The section "Clock Performance" later in the chapter outlines how these timing metrics are applied to each clock type.

Maximum Absolute Time Error

As described in Chapter 5, the maximum absolute time error, or max|TE|, is the maximum absolute value of the (unfiltered) time error (TE) observed over the course of a measurement. While TE is a measure of how far apart two clocks are from each other at any one instant, max|TE| is the maximum value that was reached over the period it was being observed or measured. In shorthand, TE indicates how bad it is now, whereas max|TE| says how bad it ever got.

The value for max|TE| is one of the most common values measured during testing for network end-to-end timing. Similarly, when talking about meeting a phase alignment requirement between two points in an application, max|TE| is the metric to be monitoring. But when applied to a standalone clock, max|TE| represents the maximum amount of noise that a clock generates during a test run when given ideal inputs.

Theoretically, the value of max|TE| could vary over time, and some unwanted noise might intervene to cause it to move more than expected. It is possible that if this state persists only for a short duration, it might not be an issue for the end application. Many times, this value is averaged over a longer period (typically 1000 s) to give an expected long-term average. This averaged value is known as the constant TE (cTE), and dynamic TE (dTE) is the short-term variation of the TE.

When the clock is functioning normally, the time output of the T-BC and the T-TSC should be accurate to within the max|TE| limits. This value includes all the noise components, which means the cTE and the dTE noise generation.

Constant Time Error

cTE is the most watched characteristic of a PTP clock, mainly because it is the principal metric used to select what class of boundary or slave clock it attains. The values for cTE and other metrics are outlined in the ITU-T Recommendation G.8273.2, which defines the performance characteristics for T-BC and T-TSC clocks. Chapter 12 contains a section on how G.8273.2 testing and measurement is performed.

Without spoiling Chapter 12 too much, cTE for a boundary clock is determined by giving an ideal PTP and SyncE input signal to the device slave port and measuring any variance in the timestamps returned from the master port. By doing this, the engineer measures how much noise has been generated as time transits through the node. These measurements are averaged over 1000 s to determine the cTE.

The cTE is very much a characteristic of a clock; this does not mean the cTE is a fixed value, but that it will not vary significantly, because a well-designed BC has only a small number of moving parts that might generate a different value for TE depending on the day (dynamic TE describes that). So, a PTP-aware network element generates a similar cTE value to that of other devices made the same way with the same components.

When building a network with several T-BC clocks, the cTE measures the TE that you could expect the T-BC to generate on an almost permanent basis. And that TE output from the first T-BC would be fed as an input to the next T-BC in the chain. Therefore, the effects of cTE are additive—over the length of the timing chain. The worst-case result could be when connecting ten Class B T-BC clocks together where every clock generated the maximum of +20-ns cTE. A chain of these clocks would generate up to 200 ns of cTE (10 × 20 ns) at the output of the final clock. The same would be true if they were all –20 ns of cTE, giving –200 ns total.

Of course, you would also need to add the cTE from the network links, which would arise from asymmetry in the transport. More on that in the following sections.

One additional point is that cTE captures the component of TE that is immune to filtering. cTE is the constant component of TE (referred to as *static offset*) and represents a measure of *accuracy*. The component of TE that can be somewhat controlled by filtering is the dTE.

Dynamic Time Error

It was shown previously that max|TE| measures the accuracy and stability that can be expected from the network or from a timing node. The max|TE| is decomposed into two subcomponents: cTE and dTE.

dTE is the dynamic component of TE that measures the variable portion of the TE. As shown in Figure 9-17, the dTE measures the *stability* of the TE and significantly contributes to max|TE|. So, to ensure a minimum of TE, it is important to keep the dTE low. In summary, cTE specifies the accuracy of the TE, whereas the dTE measures its stability.

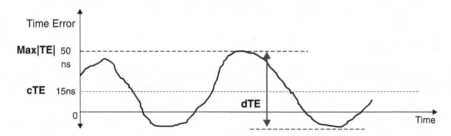

Figure 9-17   *Graph Showing cTE, dTE, and max|TE|*

As dTE can be such a major component of the total TE, what causes it? Any behavior of a network or timing node that contributes to TE in a nondeterministic way contributes to dTE. For example, one source of dTE for the end-to-end network case is the PDV induced by PTP-unaware nodes because the packet delays caused by unaware nodes are somewhat random, which adds significantly to dTE.

Another source of dTE comes from errors in timestamping. One of the characteristics of the timestamping feature on a PTP-aware node is the resolution (or granularity) of the timestamp. The resolution is simply the smallest value that the timestamp can represent, and the lower the resolution, the higher is the variability in timestamping accuracy.

Using the example of a timestamping engine with a resolution of 16 ns, the accuracy of any one timestamp can be anywhere in the range of −8 ns to +8 ns because the time-stamps are only available at values of 0, 16, 32, 48, 64, and so on. Because of this step function, the error is not fixed and varies within this 16-ns range for every timestamp. This variability becomes the major component of dTE generated within a PTP node. Note that the PTP messages themselves can carry sub-picosecond values for timestamps, but that does not help if the value of the timestamp is much less precise.

One of the most effective ways to minimize dTE because of this issue is to employ timing-aware nodes with a higher timestamping resolution. The details on the time-stamping engine and its accuracy in the design of a PTP clock are further covered in Chapter 12. For now, it is enough to know that this variability in timestamping is an important reason for dTE in a PTP clock.

It is also important to appreciate how time error accumulates and propagates in timing networks using hybrid mode, which uses SyncE for frequency combined with PTP for phase and time. To understand this, one should understand the relationship between the SyncE frequency domain and its impact on packet-based time synchronization process.

Figure 9-18 illustrates a model for T-BC and T-TSC that captures the relationship between the two timing domains very well (this is covered in Appendix III of G.8273.2). The figure shows a PTP clock with two separate clocks internally, a frequency clock locked to the physical frequency input and a time clock locked to the PTP input. There is a relationship between them, which is that the time clock uses the frequency clock to tick over the time. This also means that the TE from the frequency clock (after applying filters) is inherited by the time clock.

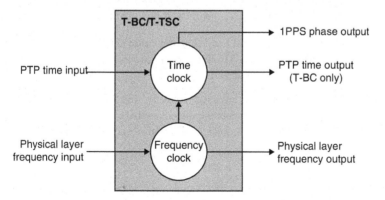

Figure 9-18 *Model of T-BC/T-TSC Showing Signal Flows—Based on Figure III.2 of ITU-T G.8273.2*

Recall that each timing node passes the physical layer frequency input through an LPF to filter out any jitter. When operating in hybrid mode, although the jitter is filtered out, the frequency clock injects the wander accumulated in the SyncE network into the time clock. This becomes yet another source of dTE for nodes operating in hybrid mode.

While discussing the mechanisms contributing to dTE, it is interesting to note that the rate of variation in TE will vary for the different sources. As discussed in Chapter 5, any variation in TE can be further classified as wander (occurring less frequently) and jitter (more frequently), based on the frequency of occurrence. Using the same convention, dTE is decomposed into two subcomponents: dTE_H and dTE_L, which represent the high and low frequency components of the dTE, respectively.

dTE_H can be referred to as the jitter portion of the dTE and dTE_L as the wander portion of dTE. These are delineated from each other based on the bandwidth (the line usually agreed upon is 0.1 Hz). These subcomponents of dTE can be filtered via low-pass and high-pass filters of a timing node. Refer to the concept of bandwidth and filters in Chapter 5 for more details.

Combining all three components of TE discussed until now, the max|TE| generated by a node can be understood to be a combination of cTE, dTE_H, and dTE_L values. Whereas cTE is measured and specified in nanoseconds (a positive or negative value), the dTE is the variation of TE, so it is not quite as straightforward. Also, in addition to the TE generated by the timing node, a method to determine the end-to-end network limits must be considered as well.

Instead of going into exact mathematical calculations, it is sufficient to understand that the calculation of max|TE| for network limits is done using a root mean square (RMS) method of averaging of dTE from all the nodes of the network. A simplified equation to represent the combination of the various TE components could be

$$\max|TE| \le SUM\left(cTE \ of \ all \ nodes\right) + RMS\left(dTE \ of \ all \ nodes\right)$$

Besides max|TE| for the network limits, the recommended limits of the two subcomponents of dTE are also measured and expressed as using the following measures:

- **dTE_L**: Maximum time interval error (MTIE) and time deviation (TDEV) masks are used to specify the dTE_L component of dTE. Refer to the section "Time Error" of Chapter 5 for further details on MTIE and TDEV.

- **dTE_H**: Peak-to-peak measurements of TE are used for measuring the dTE_H component of dTE. Note that "peak-to-peak" means the difference between the lowest peak (minimum observed value) and the highest peak (maximum observed value) of TE. This is specified in nanoseconds.

The limits for max|TE|, dTE_L, and dTE_H are specified in ITU-T G.8271.1 (FTS network), G.8273.2 (T-BC and T-TSC clocks), G.8273.3 (T-TC clock), and G.8273.4 (PTS/APTS clock) recommendations.

Asymmetry

The major problem with asymmetry is that it accumulates the further along the chain that a timing signal has traveled. Unless the asymmetry is known and constant, then once it is present, it cannot be undone or compensated for. This accumulation is not necessarily the same with PDV because the PTP master port on the BC regenerates the PTP message stream; therefore, the PDV is effectively "reset" to zero. Inserting BC clocks is a good solution available for phase/time profiles, not only for resetting the PDV but also for avoiding an excessive build up in asymmetry (since the PTP messages do not pass through the BC).

The options to mitigate against asymmetry are somewhat limited and include

- **Compensation:** Any asymmetry that is known to be fixed (an example is differences in cable length) can be measured and compensated for by configuration on the slave clock.

- **Avoidance:** Not allowing asymmetry to accumulate in the first place is a solution. This means selecting components and elements that are expressly designed to avoid it. The same applies to the transport technology because some transport types are inherently asymmetric.

- **PTP/timing awareness:** This really is a form of avoidance. When a network element allows time to transit via a non-packet hardware mechanism, then it avoids adding excessive asymmetry that it would suffer from the normal processes of packet switching inside the NE.

- **PDV avoidance:** When PDV arises in a node, that is a problem. But the bigger issue is that PDV happens to a different extent in both directions, which is asymmetric. Again, this is something that having PTP awareness helps avoid.

- **Avoid layer 3 transport:** L3 lends itself to useful techniques such as link bundling and path load-sharing that do not have mechanisms to ensure traffic is treated similarly in both directions, which will add asymmetry.

- **Non-dynamic routing:** A routing system can quite readily calculate preferred routing solutions between two end points that take different paths in each direction.

For more details of various methods that can cause asymmetry, especially in the L3 (PTPoIP) context with PTP-unaware nodes, consult the section "Partial Timing Support with G.8275.2 and GMs at the Edge" in Chapter 11.

Static Asymmetry

Static asymmetry is a form of asymmetry that the engineer can deal with; *static* means asymmetry that is fixed and unlikely to change without a significant event (more than reloading a router). The classic case of asymmetry is that arising from differences in the cable length between the transmit and receiver fiber in a fiber pair. Every extra meter of fiber requires about an extra 5 ns of time for the laser pulses to traverse. The case where an interface optic is patched with a 5-m cable in one direction and a 10-m cable in the other direction will introduce about 25 ns of asymmetry. This is one reason why engineers building extremely accurate time transfer use bidirectional fiber (the forward and reverse paths use different lambdas inside the same single fiber).

There are two basic approaches to dealing with static asymmetry:

- Estimate it from first principles

- Measure it against an external source of time, if possible

In the first approach, design engineers may understand that some optical and transport devices might have unique and fixed asymmetries. If they know those values and how many of each device is used in building the network link, then they can model an estimate of the asymmetry. Another example might be compensating for the different rates of propagation for different lambdas in a fiber, such as when using bidirectional fiber.

The second method, measuring static asymmetry, is the only method to discover the asymmetry when it cannot be discovered ahead of time. For example, the cable patching problem is a problem that may only be reasonably detected by measurement. The main disadvantage with measurement is that it can be operationally difficult and expensive. And then, when some change is made to the network, the measurement process may need to be repeated, which is an unwanted operational expense.

The upcoming section "Correction and Calibration" shows the details of the correction and calibration methods that PTP uses to treat and counteract static asymmetry.

Dynamic Asymmetry

Unlike the static kind, dynamic asymmetry does not appear to be predictable; it seems to vary with the phases of the moon, the temperature of the ice sheets, or just because it is Tuesday. Although dynamic asymmetry appears somewhat random, there are logical reasons for it (such as the laws of physics), but those reasons are too complex and chaotic to understand, model, and compensate for.

One of the most common sources of dynamic asymmetry comes from external traffic crossing the path between the master and the slave. This traffic is competing with the PTP message stream for access to the transport link, so the changing patterns of this traffic causes the PTP to suffer differing amounts of PDV (in each direction). Figure 9-19 illustrates how an unbalanced traffic pattern between the forward path and reverse path can affect PTP packets.

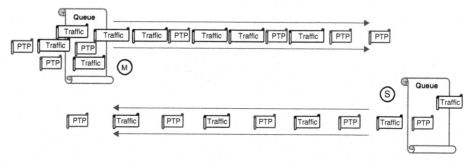

Figure 9-19 *Differing Traffic Patterns Causing PDV and Asymmetry*

Even with perfect quality of service (QoS) and priority queues, in a PTP-unaware node, it is still highly likely that a time-sensitive PTP message can be held up. This is because a large jumbo frame may have begun transmitting or passing through a congested pathway, and the PTP message must wait. While it is in a queue waiting, the time data it represents (or carries) is aging and becoming stale.

It should also be noted that this effect shrinks as the interface speeds increase, since a large packet takes much less time to pass out of a faster interface. The means that any queued packets will not have to wait as long to clear out the frame in front of them, reducing the PDV impact. To see more details of how the effects of queuing at an interface can be mitigated in unaware nodes, especially in low-latency situations, refer to Chapter 11.

However, for network elements, the best way to limit the effects of dynamic asymmetry is to use a well-designed PTP-aware clock, as it has been designed to either reduce dynamic asymmetry (boundary clock) or measure it (transparent clock). This is the main reason why a BC will only timestamp a PTP message as it is in the process of transmission, to minimize the aging while waiting in an output queue.

For the transport systems in the network, the only mitigation strategy for asymmetry is to avoid, limit, and reduce it before it reaches a level that causes problems. The main idea, then, is to filter out as much of it as possible and budget for the amount that cannot be removed through filtering. Because dynamic asymmetry changes rapidly, a low-pass filter will safely remove the high-frequency noise (the rapid changes). Where it becomes difficult to filter is for very low frequency effects, such as the daily peaks and troughs in traffic caused by human daily cycles.

So, the engineer accepts that some dynamic asymmetry will appear and manages it to be below a certain limit. Most end-to-end budgets allow for an amount of dynamic asymmetry in the network elements and network transport, so the idea is to manage the conditions to stay within those boundaries. See the previous section "Parameters for Timing Performance" and the later section "Budgeting End-to-End Time Error."

Correction and Calibration

The clear strategy for mitigating any static asymmetry is to either model or measure it, because anything that is known and fixed can be compensated for. The engineer can configure the PTP slave port with an estimate for the asymmetry, so that the servo corrects by that known value when solving the timing equations.

Similar mechanisms apply to building a PTP clock because electronic components will introduce delays to the signals inside the device. Signals within the node sometimes take alternate paths of different lengths, which can also impact the asymmetry. But if these values are not varying, the hardware designer can calibrate the signals to deal with the delay, latency, or resulting asymmetry. For more details on building an accurate PTP clock, refer to Chapter 12.

IEEE 1588-2008 has a very detailed Annex C, "Examples of residence and asymmetry corrections," that explains, with copious examples, the mechanisms for dealing with residence (transparent clock) and asymmetry corrections. Figure C.2 of 1588-2008 is a good figure to consult for the simple case of a one-step master, a one-step transparent clock, and a one-step slave port on a boundary clock. This example uses the end-to-end delay-response mechanism, which would be a common scenario for many readers. But at first glance it is a little intimidating, so it might be somewhat difficult to understand and absorb the important lessons. For that reason, this section explains the main mechanism using the same time values as in the C.2 example.

Figure 9-20 illustrates this example with a single PTP message exchange determining the four timestamps necessary to calculate the mean path delay and offset from master. For a primer on that process, see the section "The (Simple) Mathematics" in Chapter 7. What is being added to that simple example is the compensation for asymmetry and residence time. Both mechanisms rely on the use of the correctionField (CF) to carry a "running total" of the accumulated asymmetry.

The *originTimestamp* and *receiveTimestamp* in PTP event messages only contain fields for seconds and nanoseconds, so the sub-nanoseconds portion of a timestamp cannot be represented there. Besides carrying residence times, the other function of the CF is to represent the sub-nanosecond portion of the timestamps (any configured asymmetry gets reflected there as well). The timestamp fields are 48 bits for seconds and 32 bits for nanoseconds.

Some conventions: Although 1588-2008 does not describe it this way, you can consider the CF to have two subfields, nanoseconds, and sub-nanoseconds. Any nanosecond value (including sub-nanoseconds) pushed into this field gets multiplied by 2^{16}, so the lower 16 bits carry the sub-nanosecond portion of the residence times and timestamps. This means that the CF ends up as a 64-bit number with the top 48 bits representing the nanoseconds and the lower 16 bits representing the sub-nanoseconds.

In this example, as elsewhere in PTP documentation, timestamps are represented as *seconds:nanoseconds* (the *seconds* value in this example starts with 144, rather than seconds since the PTP epoch, to keep the number of digits reasonable).

Figure 9-20 warrants some explanation. First, the slave clock is 25.2 ns ahead of the master clock, so the timestamps from each clock have that offset built in between them. The transparent clock has no concept of time, but it calculates the residence time using its own oscillator or frequency reference. The mean transit time between GM and TC is 0.60 ns but there is 0.05 ns of asymmetry in the link, and similarly between TC and BC, the mean transit time is 0.70 ns with 0.20 ns of asymmetry. In contrast to the first hop, in the second hop the forward direction is shorter (0.70 compared with 0.90), which gives a negative value for asymmetry (−0.20).

These asymmetry values are known beforehand and are configured on the port facing the upstream clock. These configured values are applied to the CF on both reception and transmission.

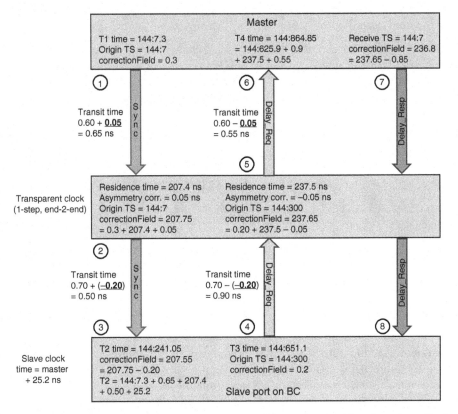

Figure 9-20 *Simplified Example of the Use of correctionField in One-Step Clocks*

Taking each step of the process in turn:

1. The timestamp for the origin of the Sync message is 144 s and 7.3 ns, which is represented as 144:7 in the originTimestamp field (which only holds seconds and nanoseconds) and 0.3 ns in the CF, which has the sub-nanosecond field. The combination of these two fields, when they arrive on the slave, will be used as t_1.

2. The TC receives the message 0.65 ns later (must be a quick link!) and adds the measured residence time of 207.4 ns to the 0.3 ns already in the CF. The TC is also configured with +0.05 ns of asymmetry for that link (positive being slower in the forward direction), which it also adds to the CF. So, the CF now contains 207.75 ns.

3. The BC receives the message 0.50 ns later, but that link has asymmetry (in the reverse direction), so the boundary clock adds that −0.20 ns of asymmetry to the incoming CF, so it is now 207.55 ns.

The t_2 timestamp, however, comes from the BC clock and its value is 144:241.05 when the Sync message arrives. One could determine this by adding the original time from the master (144:7.3), the two (real, unadjusted) transit times, the TC residence time, and the clock offset of 25.2 ns (since the slave is ahead of the master) to arrive at the same value.

4. The BC generates a Delay_Req and places an *estimate* of the t_3 timestamp in the originTimestamp field (it could also just place a zero in there). So, the t_3 value is 144:651.1 but the originTimestamp value written into the message is 144:300 (an estimation). The asymmetry correction of −0.20 ns is subtracted from the CF field (making it +0.20 ns) before transmission.

5. The TC receives the Delay_Req message 0.90 ns later (slightly slower in this direction) and adds the measured residence time of 237.5 to the CF. But it is also configured with −0.05 ns of asymmetry on the path to the GM, so it adds that value to the CF as well. Now the CF contains the value of 237.65 (0.20 + 237.5 + −0.05).

6. The GM receives the Delay_Req message 0.55 ns later and generates a timestamp of 144:864.85.

One could deduce that timestamp starting with the t_3 timestamp on the BC (144:651.1) and then subtracting the clock offset of 25.2 ns (the BC is fast by that much). Then to the remaining 144:625.9, add the TC residence time of 237.5 and the two (real, unadjusted) transit times of 0.90 ns and 0.55 ns.

7. The GM generates the Delay_Resp and places the seconds from the t_4 timestamp in the receiveTimestamp field (like originTimestamp, this field also carries no sub-nanoseconds). However, the GM must return the incoming CF value to the slave as well as the sub-nanoseconds of the t_4 timestamp. So, it subtracts the sub-nanosecond value of the t_4 timestamp from the CF value from the incoming Delay_Req.

So, the t_4 was 144:864.85 and the GM writes 144:864 into the receiveTimestamp field. But the incoming CF value was 237.65 (from the TC), and before copying that to the outgoing CF field, the GM subtracts the sub-nanosecond value of t_4, 0.85 ns, from the 237.65, giving 236.8 ns in the CF of the Delay_Resp.

8. Now the slave can calculate the mean path delay and offset from master using the four known timestamps. There is no correction applied to the Delay_Resp because it is not a time-sensitive message.

The four timestamps, as they are received on the slave port on the BC, are as follows:

1. 144:7 with CF of 207.55 ns that is added to the timestamp

2. 144:241.05

3. 144:651.10

4. 144:864 with CF of 236.80 that is subtracted from the timestamp

So, from the Chapter 7 section "The (Simple) Mathematics," you see that the calculation for mean path delay is $(t_2 - t_1 + t_4 - t_3) / 2$ and offset from master is $(t_2 - t_1 + t_3 - t_4) / 2$.

Mean path delay = [144:241.05 − (144:7 + 207.55) + (144:864 − 236.80) − 144:651.10] /2.

This gives 2.6 / 2 = 1.3 ns (the example shows that it is (0.65 + 0.5 + 0.9 + 0.55) /2 = 1.3 ns).

Offset from master = [144:241.05 − (144:7 + 207.55) + 144:651.10 − (144:864 − 236.80)] /2.

This gives 50.4 /2 = 25.2 ns (although you knew that at the beginning).

Despite the asymmetry of the links between the three participating nodes, the asymmetry was corrected by being measured, configured, and used by the slave in the determination of both the time values. In this case an asymmetry of +0.5 ns was configured on the TC for its incoming link from the GM, and an asymmetry of −0.20 ns was configured on the BC for its link to the TC.

Additionally, the asymmetry in the time spent in the TC was also eliminated by using the CF to receive and carry the residence time to the slave. The Sync message spent 204.7 ns passing through the TC, and the Delay_Req spent 237.5 ns in the other direction. This would have introduced 16.4 ns of phase error on the slave because the slave would have calculated the delay through the intervening node to be 221.1 ns (the mean/average of 204.7 and 237.5) rather than the real values.

Packet Delay Variation

The topic of PDV when discussing phase/time distribution covers two different scenarios. In the case where there is no physical frequency, the situation is much like the case with frequency-only packet. In some ways, that is what G.8265.1 and G.8275.2 have in common—they are both used to recover frequency with packets. Of course, the major difference is that G.8275.2 supports boundary clocks, so you have a tool to combat against an accumulation in PDV. Figure 9-21 illustrates how the addition of boundary clocks affects PDV.

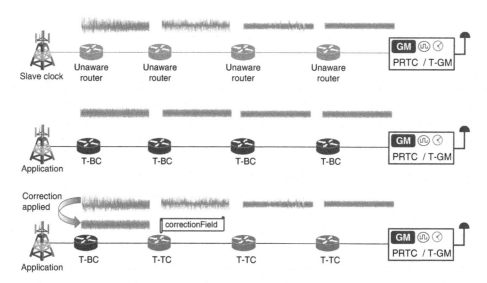

Figure 9-21 *PDV Accumulation Without and With Boundary Clock and Transparent Clocks*

The other case is where the PTP phase/time is used in combination with a physical frequency signal, most commonly SyncE. For that case, the frequency is carried using a physical method where PDV is not an issue. Note, however, that the SyncE will have wander, and that impacts the phase/time in the slave clock (see the earlier section "Dynamic Time Error").

In a packet-based transport, the phase/time signal is delivered over the network in packets or frames, and these packets become mixed in (multiplexed) with other traffic in the various network elements they pass through. Packet networks are not inherently synchronous, so there can be a difference in the rate at which packets are queued for transmission and the rate at which they are transmitted. This difference is handled by allowing time gaps between packets, buffering/queuing, or discarding packets.

The packets are also routed via intermediate switches and routers that cause delay due to processing, buffering, and queuing. Then, at a point of convergence, multiple packet streams may converge at a single choke point and are then queued until their turn to proceed comes. The resulting contention for a single resource introduces variable delay, and during times of congestion and oversubscription, packets will be dropped.

Because it is possible that individual packets may take different paths, a stream of messages from master to slave may display significant PDV when the packets arrive at the destination. In complex topologies, packets may even arrive out of sequence, resulting in devices holding packets until reordering is possible. A similar problem arises with packet fragmentation and reassembly. To help improve the service quality that otherwise would be affected by their nondeterministic nature, packet network elements have large buffers and deep queues of packets.

In summary, the following factors contribute to the PDV of packets:

- Queuing for output, especially when multiplexed with traffic containing very long frames (thousands of bytes), which cannot be interrupted once it starts being transmitted.

- Low-frequency noise, such as diurnal effects. Packet traffic load is quite dynamic and can be filtered to some extent, but long-term effects are almost impossible to filter out.

- Transmission reasons such as modulation and demodulation, waiting for transmission time slots, buffering, and the inherent delay characteristics of different transport technologies.

- Routing and rerouting.

- Path sharing techniques and load balancing such as interface bundling and multipath load sharing.

- Congestion or contention at some resource or interface and the associated QoS mechanisms to deal with it.

- Other errors such as packet error, reordering, fragmentation, and even packet loss.

For further reference material, G.8261.1 specifies the hypothetical reference model and the PDV network limits applicable when frequency synchronization is carried via packets. G.8263 provides a lot of mathematical treatment, especially Appendix I, "Packet delay variation noise tolerance—testing methodology." G.8263 clause 7, "Packet delay variation noise tolerance," has related coverage of PDV.

Packet Selection and Floor Delay

Even using SyncE to solve the PDV problems for frequency recovery, PDV is still a problem for the phase/time component because it leads to asymmetrical delay due to shifting traffic loads in unaware nodes. In fact, to recover an accurate phase and time, the discovery of a true delay floor in both directions is more critical than ever. For the frequency PTP case, it was just sufficient that the delay be somewhat constant and only in the master to slave direction.

The message that completes the journey in the delay floor time has traveled between master and slave in the minimum possible time. It means that there was no queuing, no buffering, and no resource contention in any component in the end-to-end path. On normally functioning networks with no congestion, a reasonable percentage of the packets will cross the network close to this floor time. Others might experience some delay, whereas a few others might experience much longer delay.

So, the distribution of the transit time, even on a lightly loaded network, trails off toward long delays, what is referred to as a *long tail*. For this reason, the methods used to select and qualify the message stream are an important part of the implementation of a PTP servo.

So, the engineer needs to know what contributes to PDV, and what can be done to limit it. At its most fundamental, PDV is a characteristic of a network element and how it is designed and implemented. The following is a list of items among others that can contribute to the delay floor value upon arrival at a PTP-unaware node:

- Input processing and packet classification delay

- Delay in making the forwarding decision and doing the next-hop lookup

- Time to forward the message across the switch fabric to the output line card

- Policy application delay (traffic policing that drops or remarks packets and traffic shaping that queues and delays them)

- Output queuing delay and head-of-line blocking

- Delays in the mapping and modulation in the transport

- Propagation delay through the transport

- Timestamp resolution

- Clock errors in the PHY and the backplane

Figure 9-22 illustrates the difference between PDV behavior that can be tolerated on the slave clock (left side) and that which cannot (right side). Both diagrams are a scatter plot of the transit time of a series of messages (y-axis) over an observation period of 60 seconds (x-axis).

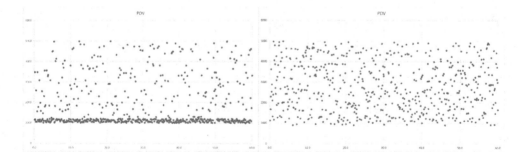

Figure 9-22 *PDV Accumulation With and Without a Clear Delay Floor*

A good reference for further information on packet selection and filtering is Appendix I of G.8260, specifically clause I.3, "Packet selection and filtering." Because this is outside the scope of what an engineer can conceivably influence for a timing design, it does not help to go into further depth. However, something that the operator can influence is PTP message rates.

Packet/Message Rates

One issue that provokes debate among engineers is the choice of the appropriate message rate needed to accurately carry time. The widespread adoption of the G.8275.1 profile has helped somewhat calm this debate because the rates in that profile are fixed and there is nothing to disagree about. In many implementations, the rates cannot be changed because, once the profile is configured, the rates are automatically set to the values in the recommendation. Some other, non-telecom profiles also fix the rates of some message types (see Chapter 7 for details on other profiles).

For the PTPoIP cases of G.8275.2 (and G.8265.1 for frequency), there is still flexibility because the engineer can still decide what is the desired rate for each of the messages. In the authors' experience, most customers tend to deploy those two profiles with a rate in the range of 32–64 messages per second, with 64 being the most common.

Increasing the message rate obviously effects the scalability of the clock, because exchanging more messages for every slave means that the master can support fewer slaves. Operators deploying PTP for use cases with lower accuracy requirements and more scalability tend to migrate toward lower rates (one example being the cable industry).

The main drivers behind increasing the traffic rate are as follows:

- Having more messages results in the slave having more data, which should theoretically improve accuracy, lock times, and general performance.

- Having more messages means a better selection and more chance of being able to receive a set of timestamps that represent the minimal transit time. If the messages suffer a lot of PDV, then having more to choose from gives a better chance of good timestamps arriving unhindered.

- If the operator is struggling to get a slave to lock to a master because of extremely high PDV, then increasing the packet rate is one way of attempting to receive enough good messages to lock. In practice, the experience of the authors is that this strategy is rarely successful.

Note that the choice of a packet rate may influence the requirements on the oscillator stability. This is because with less data arriving, the oscillator needs to be more stable so that it is impacted less by jitter. From experience with real deployments, increasing the message rate often does not fix any issues, but reducing it to a low level certainly starts to cause them.

Two-Way Time Error

As covered in the earlier section "Packet Selection and Floor Delay," only a subset of the arriving packets may have data fresh enough to accurately solve the time equations. When the SDOs design models for modeling clock behavior, they must specify metrics that (hopefully) reflect packet selection mechanisms used in real implementations. Appendix I of G.8260 provides quite a bit of background on these selection methods.

When the timestamped PTP Sync message arrives at the slave, it has been in transit for an interval, which is called the forward delay, d_{fwd}. This delay is composed of a fixed but unknown delay (switching time and propagation time) and a random variable component resulting from other factors such as queuing in intermediate equipment. This random component is the variation part of the PDV.

For frequency-only recovery (such as with G.8265.1), the fixed delay is not important, but, of course, the random delay (PDV) does cause error. On the other hand, for phase/time, both are important to the time error. In the case of phase/time using two-way traffic, the same effect occurs in the other direction with the Delay_Req message. This delay, d_{rev}, is also a combination of a fixed and dynamic error.

If the slave were to use the Sync t_1 timestamp from the master as a reference, the time error on the slave would be equal to $-d_{fwd}$ because the slave clock would be behind the master time by the amount of time the message was in transit. It is a similar approach for the reverse path; the error on the slave would be d_{rev} (the sign is reversed). This means that increasing the forward delay increases the negative TE in the slave, whereas increasing the reverse delay increases the positive TE.

Combining these two measures results in a two-way time error. However, the slave clock might need to select a good subset of packets before using the packets to characterize TE. There might also be optional filtering of the time data with the appropriate bandwidth and some mechanism to stabilize the recovered clock. The output of this process is called the *packet-selected two-way time error*, or *pktSelected2wayTE*. There are other versions of the two-way TE that represent the value of the metric after it has been filtered.

This sequence of measurements and its variations are used to directly characterize time error. For example, the specifications for noise tolerance use a form of pktSelected2wayTE (specifying the exact packet selection method) to define the input noise that a clock must tolerate. The specification uses the unfiltered version because it is a test for the tolerance to an input signal. This metric also turns up in the end-to-end network limits in G.8271.2.

Figure 9-23 illustrates an example of this measure over time and shows several of the values derived from it; namely, minimum, maximum, and peak-to-peak. The other interesting value is the absolute value of the maximum pktSelected2wayTE, or max|pktSelected2wayTE|, which caters for the fact that the important factor is the magnitude of the time error, and not the direction (values can be negative).

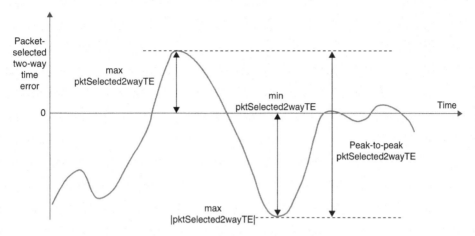

Figure 9-23 *Packet-Selected Two-Way Time Error*

So, for measuring an end-to-end network time error, the max|pktSelected2wayTE| and the filtered version of the same measure measures the accuracy of the recovered phase/time from the packet stream.

You will find more details on the measurement of two-way TE in Chapter 12. Refer to Appendix I of G.8260 for a more mathematical treatment of the metrics for packet measurement, although it is supplied in the recommendation for informational purposes only.

Clock Performance

One of the key aspects that determines the quality of the timing signal when it arrives at its destination is the performance of the clocks that it has passed through. Network elements that are not clocks have an unpredictable (but negative) influence on the quality of the timing signal, mostly because they have no capabilities to ensure that quality. But if the network element does support PTP at the protocol level and it is well implemented, then it can be relied upon to deliver a guaranteed level of performance under practically all circumstances. This section outlines these performance characteristics for the various types of clocks.

This section focuses chiefly on PTP performance because the frequency case using physical methods is covered already in Chapter 6 and in the preceding sections of this chapter when using packet methods. The clock performance measures for both frequency cases are very similar, the main difference being the influence that PDV and packet selection have on the performance in the case of packet transport.

The recommendations from the ITU-T that define the performance and characteristics of different types of clocks are based on the role of the clock and the timing network architecture. Table 9-4 lists the ITU-T recommendations for different clocks applied to each timing network architecture.

Table 9-4 *ITU-T Recommendations for Phase and Time Delivery Clocks*

ITU-T Recommendation	Clock Type	Network Type
G.8272	PRTC	—
G.8272.1	ePRTC	—
G.8273.2	T-BC / T-TSC	Full timing support
G.8273.3	T-TC	—
G.8273.4 (Clause 7)	T-BC-A / T-TSC-A	Assisted partial support
G.8273.4 (Clause 8)	T-BC-P / T-TSC-P	Partial support

There are at least three aspects to consider when trying to understand the performance characteristics recommended by ITU-T standards:

- **Performance metrics:** Performance metrics are specific to the clock type and are very precisely defined. If there is an MTIE or TDEV mask specified, usually it is different for each type of clock. For example, it would not serve any purpose if a T-BC were to fail the strict MTIE mask defined for a PRTC clock.

- **Measurement approach:** As the performance metrics are developed, there is a set of interfaces used to measure and qualify against the recommended metrics. It is necessary to understand at which position the data for measurement can be collected. For the case of phase/time delivery via packets, the data can be collected via PTP packets, at the output of the 1PPS interface, from frequency outputs of the clock, or at all of them.

- **Key elements of clock specifications:** As listed in the earlier section "Parameters for Timing Performance," the five key elements of ITU-T clock performance are *noise generation, noise tolerance, noise transfer, holdover,* and *transient response.* Limits and values for these metrics appear in almost all the clock specifications covering performance.

  In Chapter 5, the sections "Noise," "Holdover Performance," and "Transient Response" explain these terms in detail. Although Chapter 5 deals with the metrics for frequency delivery, the general approach for clock performance remains the same even for phase/time.

The following sections cover the key aspects of these performance metrics for each clock type. The section "Testing Timing" in Chapter 12 covers the process of testing and validating timing behavior and metrics.

Refer to the section "Node and Clock Performance Recommendations" in Chapter 8 for a list of the various ITU-T recommendations concerning the timing characteristics of clocks.

PRTC and ePRTC

Currently, the ITU-T defines two levels of PRTCs in G.8272, and new enhanced PRTCs (ePRTC) in G.8272.1. G.8272 retained the original PRTC specification of ±100 ns but renamed it to PRTC Class A (PRTC-A), and added a new Class B (PRTC-B), improving that accuracy to ±40 ns.

Refer to the section "Primary Reference Time Clock: PRTC" in Chapter 3 for details on the different classes of PRTC and the performance specifications. Table 9-5 summarizes the different PRTC classes along with their accuracy and holdover performance.

Table 9-5   *Classes of PRTC Performance*

Class of Clock	Accuracy	Holdover	ITU Recommendation
PRTC-A	±100 ns	None	G.8272
PRTC-B	±40 ns	None	G.8272
ePRTC-A	±30 ns	100 ns / 14 days	G.8272.1
ePRTC-B	±30 ns	For further study	G.8272.1

The TE (or noise generation) of a PRTC is characterized by the following:

- **Time of day error:** The ToD interface is used to determine the time accuracy at its output as compared to the reference. This refers to any error in the time of the day and time offset, if any, in comparison to UTC.

  If the PRTC is combined with a T-GM inside a single device, this time error can be measured using PTP packets received from an Ethernet interface.

- **Phase error:** The wander limits specified by recommendations are expressed using MTIE and TDEV masks. The signal output by the 1PPS interface is used to measure the phase error and plot the MTIE and TDEV graphs. Again, if the PRTC is combined with a T-GM, this phase error can be measured using PTP packets received from an Ethernet interface.

Note that the output for PRTC-B is expected to be accurate to ±40 ns when measured against the applicable primary time standard, such as UTC. However, the ToD interface does not produce an accurate time signal, so to measure the alignment to this level of accuracy, the ToD needs to be used in combination with the 1PPS.

As mentioned in Appendix I of G.8272, there are two aspects that engineers need to be cautious about:

- The ToD error is more difficult to measure because, unlike a *synthesized* frequency signal used as an input for many time-testing purposes, it needs a GNSS signal generator.

- The performance of a PRTC and T-GM depends on the local oscillator characteristics, which vary based on several factors, such as the environment and the *aging* characteristics of the oscillator. The measurements should be taken under similar environmental conditions to those expected at the final location of the PRTC/T-GM.

T-BC and T-TSC

ITU-T G.8273.2 specifies the minimum requirements for T-BC and T-TSC time and phase clocks when deployed with full timing support from the network. Refer to the section on G.8273.2 in Chapter 8 for an overview of this and related ITU-T recommendations.

Recall that a T-BC combines a slave port and one or more master ports. The slave port is where the input PTP message flow is terminated and used to recover the reference phase and time, whereas the master port is where a new PTP flow is generated from the recovered clock. This renewal of the PTP message flow stops the accumulation of end-to-end PDV in a network. The T-BC/T-TSC also uses physical layer frequency support (generally SyncE) to improve the stability and holdover period.

Figure 9-24 illustrates a simple model of a T-BC and T-TSC, indicating the points (refer to the numbers) that can introduce noise (either externally or from within the clock) and the points where the T-BC can transfer the noise to other downstream clocks in a clocking chain.

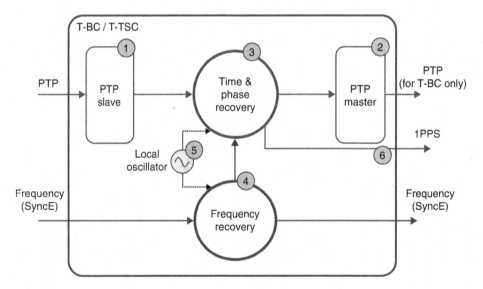

Figure 9-24   *Model of T-BC and T-TSC Showing Points Introducing Error*

These points indicate the following errors:

1. Timestamping noise (error) in the PTP slave port.

2. Timestamping noise (such as granularity) in the PTP master port (only the T-BC can have a master port). Obviously, this is a point where any noise introduced by the T-BC is transferred directly to downstream clocks.

3. Noise introduced within the time clock itself (for example, due to system issues such as noise in the hardware signals).

4. Any phase wander from the network that is introduced by the recovered physical frequency (such as SyncE). It is important to note that the recovered frequency would have been passed through a low-pass filter before being fed to the time clock.

5. Noise (such as wander) introduced by the local oscillator.

6. Noise introduced in the 1PPS output, which could be due to the hardware design of the clock (such as path delays or skewed rise times in the 1PPS signal).

The clock TE is a combination of all the mentioned possibilities, and the various performance limits (measured at the output ports) become a test of these specific points. For a T-BC, the performance is measured at the 1PPS interface, the PTP interface, and at any frequency outputs (such as SyncE). For a T-TSC, the 1PPS and frequency output interfaces are the only options because a T-TSC has no PTP master port.

Noise Generation

The noise generation of a clock is the noise (normally phase wander) that is measured at the output of the clock when supplied with an error-free (or ideal) reference at the input. Figure 9-25 illustrates this scenario for a T-BC and a T-TSC and indicates the points (shown as dots) that can introduce error within the T-BC/T-TSC.

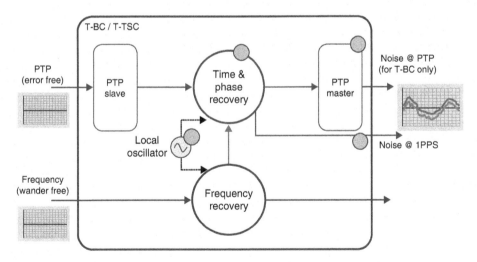

Figure 9-25 *Noise Generation Inside a T-BC and T-TSC*

The PTP and 1PPS signals will both output the cumulative phase and time noise generated from the different components within the T-BC (for T-TSC it is limited to the 1PPS signal). Note that the noise generated at the physical frequency output is affected only by the physical layer frequency input (covered in Chapter 5).

The noise generated is defined by three parameters, cTE, dTE, and max|TE|, and the maximum limits for T-BC/T-TSC at both PTP and 1PPS outputs are summarized in Table 9-6. Note that *for further study* means that the ITU-T is studying those details but has not produced agreement on the actual value. It can stay in that state for some time.

Table 9-6 *Maximum Noise Generation of T-BC/T-TSC*

Noise Type	Maximum Limits (as defined by ITU-T G.8273.2)			
	Class A	**Class B**	**Class C**	**Class D**
max|TE|	100 ns	70 ns	30 ns	5 ns (max|TE_L|)
cTE	±50 ns	±20 ns	±10 ns	For further study
dTE_L*	40 ns (MTIE)	40 ns (MTIE)	10 ns (MTIE)	For further study
	4 ns (TDEV)	4 ns (TDEV)	2 ns (TDEV)	
dTE_H	70 ns	70 ns	For further study	For further study

* The limits specified are for measurements at constant temperature.

Noise Tolerance

Noise tolerance defines how much noise a slave clock can receive (tolerate) on its input and still be able to maintain its output signal within the specified performance limits. Tolerance for noise is indicated by the clock continuing to function normally when receiving a noisy input. Conditions used to determine when the clock is continuing to behave normally include when the clock

- Is not raising any alarms

- Is not switching to a new reference input

- Does not move into holdover state

As shown in Figure 9-26, the input signals for T-BC/T-TSC are the PTP port and physical frequency input. To test tolerance, an input noise is generated on these interfaces (dots), and the clock is monitored to ensure it correctly tolerates the input noise.

For PTP input, there is no requirement for cTE tolerance because the PTP slave by itself is not able to identify or detect cTE. To do so requires additional information (such as external measurement against a reference clock) to understand that cTE is present and potentially correct it. Simply, the clock has no built-in mechanism to detect it, so it has no way to correct or reject it.

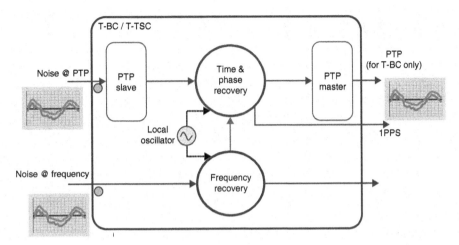

Figure 9-26 *Noise Tolerance of a T-BC and T-TSC*

For the solution to work, a T-BC/T-TSC (especially the last in the chain) must tolerate the maximum dTE that could accumulate over the complete chain of T-BCs. An MTIE mask is used to specify the dTE, and it is defined based on the network limit specified in G.8271.1. This makes sense because if the last node in the chain (which sees the most dTE) cannot tolerate the accumulated noise as an input, the chain is broken. Figure 9-27 illustrates the dTE network limit that a clock must be able to tolerate, as specified in G.8271.1.

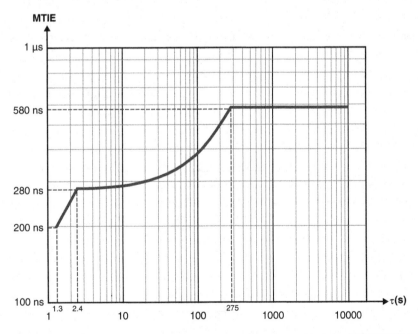

Figure 9-27 *Dynamic Time Error Network Limit (MTIE) for T-BC/T-TSC—Based on Figure 7-2 of ITU-T G.8271.1*

For frequency, the maximum noise that should be tolerated at the frequency input is described in ITU-T G.8262, G.813, and G.8262.1 (for Class C clocks based on eEEC).

Noise Transfer

As the name implies, the noise transfer of a clock describes how much of the noise present on the input of a clock gets passed through (transferred) to the output of the clock. This metric is usually expressed in terms of bandwidth because the clock is a filter to the input noise.

Bandwidth also describes the filter characteristics applied to the input signals before the recovery of frequency, phase, and time. Refer to the section "Noise Transfer" in Chapter 5 for more details on the noise transfer function of a clock and its impact on the timing network. That chapter extensively covers clock bandwidth in the section "Low-Pass and High-Pass Filters."

The characteristics of noise transfer are determined by the numerous timing paths that exist through the clock. These timing paths can transfer time error from the input to output, and each needs to be measured for its contribution to noise transfer. As illustrated in Figure 9-28, for a T-BC/T-TSC, there are three main timing flows:

- **PTP input to the PTP and 1PPS outputs:** Depicted by solid lines in Figure 9-28.

- **Frequency input to the PTP and 1PPS outputs:** Depicted by dashed lines in Figure 9-28. This path can transfer phase wander from the physical layer frequency interface to the PTP and 1PPS output interfaces.

- **Frequency input to frequency output:** Depicted by the line from left to right at the bottom of Figure 9-28.

Each of these flows illustrated in Figure 9-28 also has a bandwidth associated with it, which describes the characteristics of the filter in the path of the input signal through the clock.

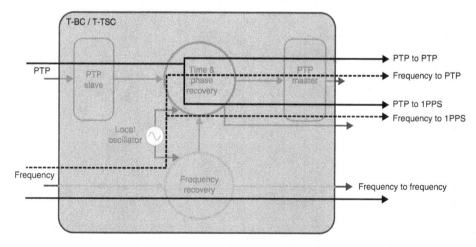

Figure 9-28 *Noise Transfer Paths Through a T-BC/T-TSC*

The frequency clock uses a phase-locked loop (PLL) (generally a hardware component) for frequency recovery. This PLL employs an LPF to filter the high-frequency noise from the input signal (remember that bandwidth also differentiates between jitter and wander). According to G.8262 this bandwidth is in the range of 1 Hz to 10 Hz for a (SyncE) Ethernet equipment clock (EEC).

In a similar way, the time clock also uses a PLL (albeit software based) to recover phase and time from the input signals. For a T-BC/T-TSC with full timing support, the inputs to this PLL are the PTP messages and frequency recovered from frequency clock. The concepts behind the use of PLLs in building time clocks is explained in detail in Chapter 12.

The time clock also uses an LPF to filter out jitter in the noise on the input from the PTP packets. However, according to G.8273.2, the bandwidth for this LPF must be between 0.05 Hz and 0.01 Hz, which is different to the bandwidth used by the frequency clock.

For the path from frequency to PTP (and 1PPS), the two filters described earlier combine their effect: the frequency input is first filtered by frequency clock and then gets filtered by the time clock but using a different bandwidth. And because the bandwidth is different for these filters, the net result of combining these filters look like a band-pass filter (see Chapter 5) with the following characteristics:

- A lower cut-off in a range of between 0.05 Hz to 0.1 Hz (from the time filter)

- An upper cut-off in the range of between 1 Hz to 10 Hz for T-BC/T-TSC Classes A and B, and range of 1 Hz to 3 Hz for Classes C and D (from the frequency filter)

Figure 9-29 illustrates this band-pass filter for the case of the Classes A and B clocks. Note that the figure is for illustration purposes only and is not strictly to scale.

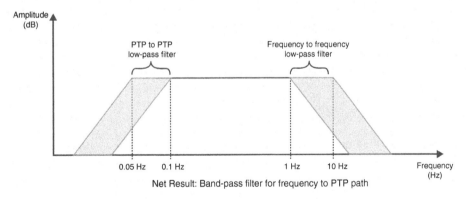

Figure 9-29 *Band-Pass Filter for Frequency to PTP Path for T-BC/T-TSC Classes A and B*

In summary, for a noise transfer test the following applies:

- **PTP to PTP (and 1PPS) path:** The phase and time noise presented on the PTP input of a T-BC/T-TSC will be transferred to the PTP (and 1PPS) output after the clock has filtered the noise using a bandwidth between 0.05 Hz and 0.1 Hz.

- **Physical layer frequency to PTP (and 1PPS) for Classes A and B clocks:** The noise presented on the physical frequency input will be transferred to the PTP (and 1PPS) output after the clock has filtered the noise through a band-pass filter. This filter has a frequency of between 0.05 Hz and 0.1 Hz at the lower corner and between 1 Hz and 10 Hz at the upper corner.

- **Physical layer frequency to PTP (and 1PPS) for Classes C and D clocks:** The noise presented on the physical frequency input will be transferred to the PTP (and 1PPS) output after the clock has filtered the noise through a band-pass filter. This filter has a frequency of between 0.05 Hz and 0.1 Hz at the lower corner and between 1 Hz and 3 Hz at the upper corner.

Transient Response

The transient response measures the reaction of a clock to some reorganization in its input reference signals. For the case of a T-BC/T-TSC with full timing support, a transient at the input results from a rearrangement of either the physical frequency signal (such as SyncE) or the PTP message flow.

A rearrangement of the PTP packet timing signal occurs when a T-BC/T-TSC loses its current PTP message stream and switches to an alternate flow. This switchover likely results in a transient at the PTP timing input of T-BC/T-TSC. Similarly, a rearrangement of the physical frequency transport occurs when the source reference changes (for example, triggered by changes in the QL values received).

Note that the rearrangement does not mean the input signal is lost; it is simply changed to a different path or source. The signal disappearing completely would not be a transient but would force the clock into holdover.

When using two different reference signals (one each for phase and frequency) on a T-BC/T-TSC with full timing support, there are four transient scenarios:

- Rearrangement of both the PTP timing signal and the physical frequency signal

- Rearrangement of only the PTP timing signal

- Rearrangement of only the physical frequency signal

- Rearrangement of physical frequency signal—for the "long term," the definition of which is that the PRC/PRS traceable frequency source is lost for more than 15 seconds

Currently, the ITU-T (in G.8273.2) has only defined a performance metric for point 3 (the rearrangement of physical frequency), and the expected response for all other scenarios is for further study. For physical rearrangement, the recommendation defines an acceptable response mask only for T-BC/T-TSC Classes A and B. Figure 9-30 shows this mask and indicates the maximum limits of the allowed response.

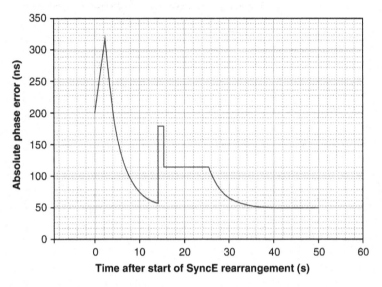

Figure 9-30   *Phase Error Limit for Physical Frequency Transport Rearrangement for T-BC/T-TSC Classes A and B—Based on Figure B.1 of ITU-T G.8273.2*

Holdover

Holdover defines the maximum deviation in the PTP and 1PPS output signal during the loss of the PTP packet timing signal and/or physical layer frequency inputs. So, there are two types of holdover available in a T-BC/T-TSC: one where only the PTP disappears and one where both signals are lost.

In the first case, a T-BC/T-TSC loses the PTP packet timing signal and yet the physical layer frequency reference remains traceable to a PRC/PRS. This is also referred to as frequency-assisted holdover, and it helps with holdover performance because the frequency reference is used to keep the time output "ticking" at close to the correct rate.

And so, it follows that the holdover depends on the quality of the frequency reference. If the frequency remains traceable to a PRC, it is likely to maintain very accurate phase/time for an extended period. A PRC-traceable frequency reference extends the period of holdover (at microsecond accuracy) for network elements from hours to more than a week.

The requirements for holdover are divided based on what class of T-BC/T-TSC that the device performance needs to meet. Table 9-7 summarizes the holdover requirements for Classes A and B for a maximum observation interval, *tau* (τ) of 1000 seconds with both constant and variable temperature. Note also that the holdover performance is a function of observation interval itself.

Table 9-7   *Holdover Limits During Loss of PTP Input for T-BC/T-TSC Classes A and B at Constant Temperature*

Temperature Condition	MTIE (ns)	Observation Interval τ (s)	Approximate Range (ns)
Constant	$22 + 40\ \tau^{0.1}$	$1 \le \tau \le 100$	62–85
	$22 + 25.25\ \tau^{0.2}$	$100 < \tau \le 1000$	85–123
Variable	$22 + 40\ \tau^{0.1} + 0.5\ \tau$	$1 \le \tau \le 100$	62.5–135
	$72 + 25.25\ \tau^{0.2}$	$100 < \tau \le 1000$	135–173
	For further study	$1000 < \tau \le 10\ 000$	—

Source: ITU-T G.8273.2, Tables 7-9 and 7-10

Figure 9-31 illustrates the permissible limits for phase error of T-BC/T-TSC clocks for Classes A and B holdover performance for 1000-s duration (starting immediately after the clock moves to holdover). Under constant temperature conditions (within ±1 K) the maximum observation interval is 1000 s.

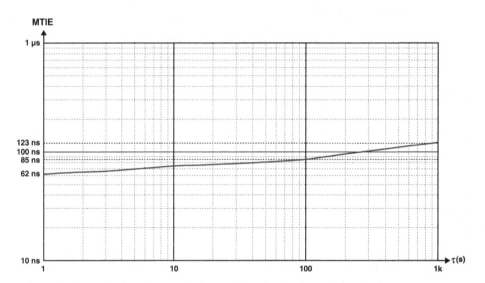

Figure 9-31   *Phase Error Limits for T-BC/T-TSC Classes A and B During Holdover with Constant Temperature—Based on Figure 7-1 of ITU-T G.8273.2*

In the second case of holdover in a T-BC/T-TSC, both the PTP packet timing signal and the physical layer frequency inputs are lost simultaneously, and the clock can rely only on the quality of its own internal oscillator. In that case, the holdover performance in a clock with an oven-controlled crystal oscillator (OCXO) is going to be much worse than one with a reference frequency input. See the section "Network Holdover" later in the chapter for more details.

T-TC

The function of a T-TC is to measure the residence time of transit PTP event messages (Sync and Delay_Req), so that the next slave in the chain can compensate for the PDV suffered by those messages. The T-TC is expected to measure the residence time and use it to update the CF of the PTP message. The performance of a T-TC is based on how accurately it can measure and reflect the delay experienced by the PTP event messages; in other words, how accurately it updates the CF.

ITU-T G.8273.3 specifies the limits for an end-to-end T-TC with a frequency input reference provided by the physical layer. It defines the maximum limit of error (or noise) that a T-TC adds to the CF. For a T-TC, this error in the CF is categorized as noise generation.

The frequency reference is needed to ensure that the oscillator is correctly calculating the residence interval. If the T-TC has no reference, then the accuracy of the measurement will be wrong if the oscillator is running too fast or slow. The performance of a T-TC without a frequency reference provided by the physical layer is for further study. Refer to the section on G.8273.3 in Chapter 8 for an overview of this ITU-T recommendation.

The types of inaccuracy (or noise) in the CF can be either of the following:

- **Fixed error (cTE):** When the time written into the CF is offset from the real residence time by a fixed value. The slave clock will incorporate this (erroneous) offset into its calculations to recover phase/time.

  However, this offset cancels out if it is equal in the forward and reverse direction through the T-TC. This means the Sync (forward) and Delay_Req (reverse) messages have equal offsets (in opposite directions). However, any differences between the offset in the forward and reverse direction will produce asymmetry and result in cTE on the input to the slave clock.

- **Variable error (dTE):** This is caused by packet-to-packet variation in CF accuracy. The classic example of dTE results from the resolution or granularity of the time-stamping unit updating the CF. Any dTE reflected in the CF results in the slave clock not being able to accurately compensate for the PDV.

ITU-T G.8273.3 defines these limits for a T-TC, and Table 9-8 summarizes the three main parameters for noise generation for a T-TC: max|TE|, cTE, and dTE.

Table 9-8 *Maximum Noise Generation of a T-TC*

Noise Type	Maximum Limits (as defined by ITU-T G.8273.3)		
	Class A	Class B	Class C
max|TE|	100 ns	70 ns	For further study
cTE	±50 ns	±20 ns	±10 ns
dTE_L $	40 ns (MTIE)	40 ns (MTIE)	10 ns (MTIE)
dTE_H	70 ns	70 ns	For further study

$ For constant as well as variable temperature except for Class C, which is constant temperature only

For the noise transfer behavior, the T-TC is not expected to amplify any input time error on its output. Both noise tolerance and transient response limits are for further study, and there is no holdover capability supported in a transparent clock.

T-BC-A, T-TSC-A

As explained in Chapter 7, a T-BC with assisted partial support (T-BC-A) is a boundary clock with only partial support from the network but assisted by having a local time reference (such as a GNSS receiver) as a primary source of time. The PTP slave port on this BC recovers the PTP clock over the PTS network and uses it as a reference only when the local time source fails. In the same fashion, a T-TSC with assisted partial support (T-TSC-A) is the slave-only equivalent of the T-BC-A boundary clock.

ITU-T G.8273.4 defines the performance limits for the T-BC-A and T-TSC-A. Refer to the section on G.8273.4 in Chapter 8 for an overview of this ITU-T recommendation.

Noise Generation

The noise generation of a T-BC-A/T-TSC-A represents the amount of noise produced at the output of the clock when locked to an ideal (wander-free) PTP signal as input. Like the T-BC case, the output of a T-BC-A is measured at either the 1PPS output or the PTP master port, whereas the output of a T-TSC-A is measured at the 1PPS output.

Table 9-9 summarizes the cTE and dTE that a T-BC-A/T-TSC-A is permitted to generate from within clock. The value of max|TE| for noise generation is for further study.

Table 9-9   *Maximum Noise Generation of T-BC-A/T-TSC-A*

Noise Type	Maximum Limits (as defined by ITU-T G.8273.4)	
	Class A	**Class B**
cTE	±50 ns	±20 ns
dTE_L	50 ns (peak-to-peak)	
dTE_H	For further study	

Source: ITU-T G.8273.4, Table 7.1

Noise Tolerance and Transfer

As the T-BC-A/T-TSC-A are taking an input from a local time reference (such as GNSS), obviously the clock must be able to tolerate whatever noise is output from a PRTC. Therefore, the noise that must be tolerated as an input to this clock is set at the same level as the maximum noise that is allowed to be generated at the output of a PRTC.

Therefore, G.8273.4 specifies that T-BC-A/T-TSC-A must be capable of tolerating input noise with the following specifications:

■ max|TE| ≤ 100 ns, the same as the TE limit at the output of a PRTC.

- Peak-to-peak pktSelected2wayTE < 1100 ns, with a selection window of 200 s and the selection percentage set at 0.25%. Refer to the section "Parameters for Timing Performance" earlier in this chapter for more details on pktSelected2wayTE and other parameters of packet selection. The peak-to-peak is used since the time error will be corrected by the APTS clock, so the actual value of the offset does not matter.

- For APTS networks, when the T-TSC-A is external (separate device) to the end application, and where the end application limit is 1500 ns, the T-TSC-A is also expected to maintain max|TE_L| below 1350 ns at its 1PPS output. This allows for a 150-ns budget in the end application (see the upcoming section "Budgeting End-to-End Time Error").

The metric max|TE_L| means that a low-pass filter with bandwidth of 0.1 Hz is applied to the TE measurement samples from the 1PPS output prior to evaluating the max|TE|.

The limits for noise transfer of T-BC-A/T-TSC-A focuses primarily on the maximum noise that the clock can transfer from the local time reference (1PPS) as input to the output at the 1PPS and PTP master port. This noise transfer is calculated and specified in terms of phase gain. According to G.8273.4, the phase gain should be smaller than 0.1 dB (1.1%).

Transient Response

For the T-BC-A/T-TSC-A, the transient is defined as the loss of the local time reference for a short period before it is restored. For the T-BC-A, the transient response is measured at the PTP and 1PPS output, whereas for T-TSC-A it is measured at the 1PPS output. The requirement covers the complete cycle from reference loss, through holdover, and back to reacquiring the signal and locking to it.

The clock undergoes a sequence of events during this transient. G.8273.4 specifies the transient response for the each of the main steps, which are as follows:

- The local time reference (such as a GNSS receiver) is lost. The transient response due to a loss of the local time reference shall be less than 22-ns MTIE.

- The clock enters holdover and maintains a holdover state for the short term. The performance requirement during this holdover is the same as the regular holdover of the clock. This is discussed in next section on holdover.

- The local time reference is restored, and the clock selects the local time reference. The transient response due to the selection of the local time reference (after restoration) shall be less than 22-ns MTIE.

The clock locks to the local time reference and the normal operation of the clock continues.

Holdover

A clock moves to holdover when it loses all its reference signals and must rely on its own devices to maintain phase and frequency. Because the T-BC-A/T-TSC-A also has a local time source as a primary input, it moves to holdover when both the local source and the backup PTP timing signal from the network are lost. The most likely cause of PTP loss is some network fault or misbehavior.

In this case, without any reference signal present, the holdover performance directly reflects the performance of the oscillator. This state is referred to as *holdover based on oscillator*, and good phase holdover cannot be achieved without including at least stratum 2 level (expensive) oscillators. Because network elements at the edge of a network are not normally so equipped, the phase holdover cannot stay in specification for very long. Network clocks commonly have stratum 3E clocks as a compromise (and may exceed 1 µs within a few hours).

Because holdover is important, good design ensures that having no PTP source available must only be allowed to occur for a short interval. So, holdover using the local oscillator is only expected to be a short-term temporary measure to maintain phase/time temporarily until the problem is rectified.

G.8273.4 specifies a mathematical function that limits the phase error of T-BC-A/T-TSC-A during a holdover based on oscillator. Without getting into mathematical details of the function itself, it is sufficient to understand that the permissible phase error is a time-based function that depends upon the following:

- The initial frequency offset

- Temperature variation, which has an adverse impact on the oscillator stability and hence on holdover performance

- The drift of the oscillator as the elapsed holdover time progresses

- A constant offset, which is specified as 22 ns

Figure 9-32 illustrates the G.8273.4 limit on phase error that a T-BC-A/T-TSC-A should display during holdover based on oscillator. The graph illustrates the limit for a clock at constant temperature and a maximum observation interval of 1000 s.

The graph in Figure 9-32 also illustrates how quickly the phase can degrade—starting from 22 ns at the start of holdover and crossing 1 µs within 1000 s. To get a true sense of comparison, it is a good idea to compare this holdover to that of T-BC/T-TSC assisted by physical layer frequency reference (shown in Figure 9-31 in the preceding "Holdover" section).

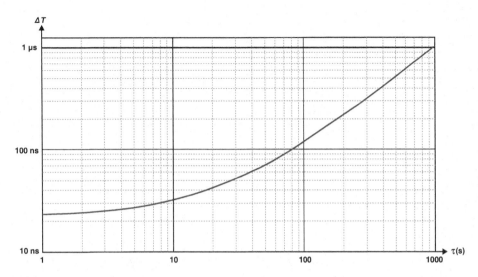

Figure 9-32 *Permissible Phase Error for TBC-A/T-TSC-A During Holdover—Based on Figure 7-1 of ITU-T Recommendation G.8273.4*

G.8273.4 defines one more holdover scenario for T-BC-A/T-TSC-A, one where the local time reference input is lost but the PTP input is valid and ideal. This is a case where the holdover is assisted by PTP and so is referred to as *holdover based on PTP*.

Under constant temperature conditions and for a maximum observation interval of 1000 s, the MTIE limit recommended is 222 ns.

T-BC-P, T-TSC-P

The T-BC with partial support (T-BC-P) is a BC with only partial timing support from the network and without any assistance from a local time reference. A physical frequency input signal is optional for this class of clock. The PTP slave port recovers clock over the PTS network, but this is its primary time reference signal, not a secondary/backup. In the same fashion, a T-TSC with partial support (T-TSC-P) is the slave-only equivalent of the T-BC-P boundary clock.

Not only for the assisted clocks, G.8273.4 (in clause 8) also defines the performance requirements for the T-BC-P and T-TSC-P partial clocks. Refer to the section on G.8273.4 in Chapter 8 for an overview of this ITU-T recommendation.

Noise Generation

The noise generation of a T-BC-P/T-TSC-P represents the amount of noise produced at the output of the clock when locked to an ideal (wander-free) PTP signal as input. As with the T-BC-A case, the output of the T-BC-P is measured at either the 1PPS output or the PTP master port, whereas the output of the T-TSC-P is measured at the 1PPS output.

Table 9-10 summarizes the cTE and dTE that a T-BC-P/T-TSC-P is permitted to generate from within the clock. The max|TE| for noise generation is for further study.

Table 9-10   *Maximum Noise Generation of T-BC-P/T-TSC-P*

Noise Type	Maximum Limits (as defined by ITU-T G.8273.4)	
	Class A	**Class B**
cTE	±50 ns	±20 ns
dTE_L	200 ns (peak-to-peak)	
dTE_H	For further study	

Source: ITU-T G.8273.4, Table 8-1

Noise Tolerance and Transfer

G.8273.4 specifies that the T-BC-P/T-TSC-P must be capable of tolerating input noise with the following specifications:

- max|pktSelected2wayTE| < 1100 ns, with a selection window of 200 s and selection percentage of 0.25%. Refer to the section "Parameters for Timing Performance" earlier in this chapter for more details on pktSelected2wayTE and other parameters of packet selection.

  The maximum value is used in the unassisted case because the TE is expected to be within the end-to-end timing budget of 1100-ns phase requirement from the mobile scenario—see Chapter 10, "Mobile Timing Requirements." This value is then used as the G.8271.2 network limit (meaning that the error at the end of the timing chain must be within this value to ensure that the requirements for the mobile use case are met). And the clock at the end of the timing chain must be able to tolerate it as input.

- As with the T-TSC-A case, for PTS networks when the T-TSC-P is external to the end application and where the end application limit is 1500 ns, the T-TSC-P is also expected to maintain max|TE_L| below 1350 ns at its 1PPS output. This allows for a 150-ns budget for the end application (see following section on budgeting).

The metric max|TE_L| means that a low-pass filter with a bandwidth of 0.1 Hz is applied to the TE measurement samples from the 1PPS output prior to evaluating the max|TE|.

The limits for noise transfer of a T-BC-P/T-TSC-P focuses primarily on the maximum noise that the clock can transfer from the input PTP timing signal to the output at the 1PPS and PTP master port. This noise transfer is calculated and specified in terms of phase gain. According to the recommendation, the phase gain should be smaller than 0.1 dB (1.1%).

Transient Response

For T-BC-P/T-TSC-P the transient is defined as any one of the following events:

1. The loss of the input PTP timing signal for a short period before it is restored

2. Switchover of the physical layer input while still maintaining the PTP input

3. Simultaneously switching both the physical layer and the PTP input

This section discusses only case 1, because at the time of writing cases 2 and 3 are for further study. For the T-BC-P, the transient response is measured at the PTP and 1PPS output, while for T-TSC-P it is measured at the 1PPS output.

Like the T-BC-A/T-TSC-A case, the clock undergoes a sequence of events during the short-term loss of the PTP timing signal in case 1. G.8273.4 specifies the transient response for each of the following main steps:

1. The input PTP timing signal reference is lost. The transient response due to a loss of the PTP reference shall be less than 22-ns MTIE.

2. The clock enters holdover and maintains a holdover state for the short term. The performance requirement during this holdover is the same as the performance requirement during the regular holdover of the clock. This is discussed in the next section on holdover.

3. The PTP time signal is restored, and the clock selects the PTP as a reference input. The transient response due to the selection (after restoration) shall be less than 22-ns MTIE.

The clock locks to the PTP reference and the normal operation of the clock continues.

Holdover

The T-BC-P/T-TSC-P moves to holdover when the PTP timing signal from the network is lost and it is not assisted by a physical layer frequency reference. In this case, without any reference signal present, the holdover is driven by the performance of the oscillator. The *holdover based on oscillator* requirements for the T-BC-P/T-TSC-P are exactly the same as those for the T-BC-A/T-TSC-A case. Refer to the preceding section on T-BC-A/T-TSC-A holdover.

The other type of holdover is when T-BC-P/T-TSC-P loses its PTP timing signal, but the physical layer frequency is valid and ideal. The performance limits for the *holdover based on physical layer frequency assistance* case is the same as that of T-BC/T-TSC. Refer to the preceding section on T-BC/T-TSC holdover.

Budgeting End-to-End Time Error

The performance of a timing signal recovered at the end of the timing chain depends on the combined effect of the links in the chain. Basically, it is the sum of the performance of the clocks in the path, the unaware nodes, and the transport infrastructure. The parameters for each of those components are covered in the preceding sections, including details of the clock performance. The sections following this cover the underlying packet transport itself.

The fundamental job of the designer of the timing solution is to ensure that the timing signal, on arrival at the application, meets the requirements of the application. The authors like to state that when you are required to solve a timing problem, you need a timing solution. All the networking details are very much secondary to the issue of time. The primary fundamental goal of the timing solution is to measure and control the timing error so that it arrives within the error budget.

One very specific example of a timing budget specific for the mobile use case is presented in G.8271.1. This use case is covered in detail in several other places, including Chapters 8 and 10. Figure 9-33 illustrates an end-to-end mobile budget using the FTS network with up to 20 T-BC clocks of Class B performance. Other industries have similar examples, in differing levels of detail.

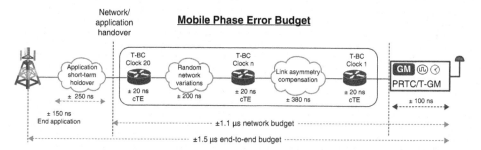

Figure 9-33 *End-to-End Time Error Budget Example for Mobile Networks*

This example includes the following points for apportioning and managing the time error (from left to right):

- There is an end-to-end budget of ±1.5 µs at the output of the end application. This is the hard limit that the timing solution must meet under all circumstances (this is a requirement for mobile radios).

- There is a budget limit of ±1.1 µs at the handover of the network to the application. The final T-TSC can be embedded in the final application (symbolized here by a radio tower), which makes this handover point difficult to measure. This means that the end application is allowed up to ±400 ns of time error.

- That ±400 ns in the end application is split into two components. The application is allowed up to ±150 ns of error and there is ±250 ns allowed for short-term holdover. This 250 ns gives the final application some space for the time to drift while the network is converging or recovering from a fault or some event.

- Each of the 20 T-BCs and the single T-TSC can have up to ±20 ns of cTE, giving a total of ±420 ns of error in the chain of nodes.

- The network is allowed up to ±200 ns to absorb random network variations, which can arise from any number of faults or events in the timing distribution network.

- The links between the network elements can contribute up to ±380 ns of time error owing to the asymmetry within the links.

- There can be up to ±100 ns of absolute time error at the output of the PRTC (this is PRTC-A from G.8272).

This budget is a standardized example from G.8271.1 and is included as guidance to allocate time error in the timing chain. These values have been settled on based on exhaustive simulations with the different types of error that arise in the components of the chain. Of course, for a different situation, different assumptions and conclusions apply.

The earlier section "Dynamic Time Error" explained that the calculation of max|TE| for network limits is done using a RMS method of averaging for dTE from all the nodes of the network. The following equation sums up the contribution of the TE components:

$$\max|TE| \leq SUM\left(cTE \ of \ all \ nodes\right) + RMS\left(dTE \ of \ all \ nodes\right)$$

Even if the mathematics is not your primary concern, at least you should understand that the three types of TE accumulate differently, because some of it is filtered and some is not. Basically, you do not just add them all together; the calculation depends on the type of noise.

The cTE component from previous clocks in the chain passes through unchanged and any local cTE is simply added to it. The low-pass filtering of the clock removes the high-frequency dTE, so the contribution to high-frequency dTE comes predominantly from the last network element in the timing chain. Meanwhile, the low-frequency dTE accumulates somewhat haphazardly because it cannot be filtered.

The high-band (or high-frequency) dTE is the jitter portion of the dTE. If all the network elements of the chain are adhering to standards for noise generation and transfer, this jitter portion is removed/filtered by each element in the path. And so, the high-band portion of the dTE is only contributed by the last network element.

Low-band (or low-frequency) dTE is the wander portion of the dTE. As clocks use low-pass filters, any accumulated wander would easily propagate through the chain of nodes and hence this dTE gets accumulated for all the clocks of the chain.

In summary:

- The cTE and any link asymmetry accumulates in a linear fashion (add it up).

- The high-frequency dTE (jitter) is mostly filtered from upstream and the main contribution comes from the last network element in the chain.

- The low-frequency dTE (wander) can arise haphazardly, but once it is present, it does accumulate linearly in the chain.

Network Holdover

If your application holdover performance relies on holdover, then you should also consult the section "Holdover" in Chapter 11 that is focused on the mobile use case. The section here, as in Chapter 11, starts by demonstrating how hard it is to maintain accurate holdover at reasonable cost. If holdover is important to you, then of course you need to consider it, but it comes attached with a considerable cost—holdover is not free.

First, you should have a clear understanding of what good holdover (at a reasonable cost) looks like. Table 9-11 gives an idea of the performance of a PRTC that meets the performance characteristics of a PRTC-B (the best PRTC class without having to buy a cesium atomic clock for tens of thousands of dollars). These characteristics are taken from a sample of currently available devices with the option of two different oscillators.

Table 9-11   *Typical Accuracy and Holdover Specifications for a PRTC Class B*

Metric	OCXO	Rubidium
10 MHz frequency accuracy over 24 hours (GNSS locked)	$< \pm 2 \times 10^{-12}$	$< \pm 1 \times 10^{-12}$
Frequency holdover at constant temp (per day)	4×10^{-10}	2×10^{-11}
1 PPS phase accuracy to UTC over 24 hours (GNSS locked)	±40 ns	±40 ns
Phase holdover at constant temp (after 8 hours)	5 μs	200 ns
Phase holdover at constant temp (after 48 hours)	10 μs	1 μs

The oscillators listed in Table 9-11 are using either a high-quality OCXO (better than stratum 3E) or a rubidium oscillator. All the holdover numbers are taken after a minimum of at least 48 hours stabilization time (locked to GNSS). Note that the specific numbers might vary a little (or even a lot) depending on the manufacturer, the cost, and the conditions of testing—but the table gives good estimates.

Another thing to consider is the accuracy of the PRTC itself. When going into holdover with a PRTC-A (for example), the initial error (when holdover starts) is quite likely to be significantly higher (100 ns versus 40 ns) than if it was starting with a PRTC-B source. There is a similar advantage to using eSyncE that is explained a bit later in this section.

So, if the scenario requires 1 microsecond of holdover, then it will require good equipment, such as PRTC-B devices with rubidium oscillators in a temperature-controlled environment. Normally, this level of performance is not found in normal network elements outside devices specifically manufactured for specialized timing applications.

There are also some operational factors to consider when comparing the performance of any clock in holdover:

- The rate of oscillator aging (even how new it is from the factory).

- The length of time the clock has already been running and aligned with the GNSS receiver, which impacts the quality and freshness of holdover data.

- Quality/stability of the reference signals received before holdover begins (PTP only, SyncE, eSyncE). This is a similar effect to that of the quality of receiver.

- Environmental factors, most especially temperature, sunlight/shadow, and airflow.

- Diurnal factors, although some of these are environmental, such as temperature cycles.

- Whether the clock in holdover has any assistance from a physical source of frequency (especially SyncE or eSyncE).

Looking beyond the design of an individual clock or network element, the topology and design of the network can have a significant impact on the quality and length of the hold-over. For most deployments the last bullet point is the most crucial. It has already been mentioned that SyncE is an important contributor to holdover performance, but now it needs an explanation.

In hybrid mode deployments, the physical frequency is driving the "tick" of the clock (refer to Figure 9-18 in the section "Dynamic Time Error"). If this signal continues to be available after the clock enters holdover, then the rate of progress of the clock continues to track the input reference. Given that the phase is taken from these oscillations, the phase does not have much opportunity to drift.

This is how SyncE helps achieve much improved holdover times. Once the slave is phase aligned, if the phase/time reference signal (meaning PTP) disappears but the SyncE is still available, then the holdover performance will be orders of magnitudes better than without the signal—because without it, the clock can only rely on the physics of its own oscillator combined with the holdover data learned during its locked time.

There is a similar advantage to using eSyncE over SyncE because of the improved performance of the physical signal that the holdover clock will receive. It has two benefits:

- Because the physical signal is better, the oscillator on the slave clock results ends up being more stable.

- When using eSyncE for assisted holdover, the reduced wander of the physical frequency improves the ability to maintain phase alignment.

Just look at some rough calculations. An oscillator in holdover might be accurate to (say) 10 ppb (at least for a short while). After 1 s, this oscillator could be 10 ns off from the nominal frequency. And after the end of 2 s, the error could be another 10 ns off from the error at the end of 1 s. Every second of error builds upon the previous second. Yes, it could average out (one up and one down) to some extent, but the point is that the error is additive on top of the previous frequency error—at least to the limits of the free-running performance.

But consider another oscillator that is also accurate (as an example) to 10 ppb but is locked to a SyncE reference. That means that after 1 second, this oscillator might be up to 10 ns off. But after 2 s, the oscillator will be still within 10 ns from the starting point (the required nominal frequency). After 1000 s, it will still be within 10 ppb of the (in this example) 20 MHz. In this case, the error is an absolute error compared to the requirement; in the other case, it is relative to the previous measurement.

In both cases, the frequency error induces some phase error, but the phase error is the frequency error integrated over time. You don't need to be mathematically inclined to understand that the longer the time interval the frequency is off from the nominal, the more the phase error accumulates. The holdover case without a frequency reference shows that there is nothing trying to return the frequency offset to the mean and so the phase error rapidly piles up.

Figure 9-34 illustrates the difference between the two cases and shows frequency error (the curved line) versus phase error (the shaded area under the frequency error curve).

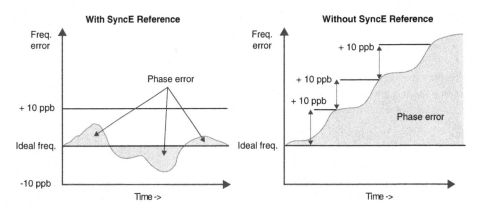

Figure 9-34 *Frequency and Phase Error With and Without a SyncE Signal Traceable to a Primary Reference*

In summary, holdover is a commodity with a price, and good holdover requires clocks with a combination of expensive oscillators, good design, and benign environmental conditions. For the designer of a timing solution, holdover is really a last resort, and the focus should be on a good network design to ensure that the application is always served with some reference signal.

Packet Network Topologies

Of course, when using the transport network to carry the timing signals, the critical factor is the characteristics of the transport system being used. But another aspect to consider is the topology and encapsulation of the transport. So, before moving onto the characteristics of the transport itself, there are a few items to understand about topology. Many of these issues have been covered in previous sections, so this section brings them together as a summary. Chapter 11 also provides more detail on these issues.

The following items must be guarded against in decisions on topology and deployment architecture:

■ Once again, PTPoIP over unaware nodes allows uncontrolled PDV, asymmetry, and time error. The use of a hop-by-hop topology reduces this risk, greatly improves end-to-end performance, and allows predictable budgeting of time error.

■ Using multiple stacked labels (for example, with MPLS) may lead to a situation where the PTP messages are not visible to the timestamping logic. In this case, it might be necessary to pop the MPLS labels at the last hop before a PTP-aware clock.

■ Forwarding timing information using pseudowires and tunnels is normally the wrong thing to do. With G.8275.1, the PTP message stream is hop-by-hop, configured on the physical interface, and contains no VLAN tags. This means it runs outside the service layer and so avoids these issues.

■ ESMC packets carrying clock traceability information similarly should not be tunneled. The whole point of ESMC is to carry QL information about the physical signal it is being carried over. Separating the QL information from the signal is probably not the right thing to do.

■ Difference in interface speeds between the master port and the slave port is not possible in the hop-by-hop scenario because they are connected to each other. With a PTPoIP model, an unaware node between the master and slave port can change the interface speed (for example, from 1 GE to 10 GE). The differences in the method and rate of serialization will introduce asymmetry (potentially hundreds of nanoseconds).

■ Routing protocols can cause issues that might not be visible to operations staff, the biggest risk being that the forward path and reverse path take different routes. That doesn't happen frequently, but there may be no hint that something is wrong, and it may be very hard to track down.

■ Rings are a good way to provide redundancy with the minimum number of links, but a ring must allow both SyncE and PTP to reverse direction based on the source of the best time signal. This places requirements on both SyncE and PTP. Not paying careful attention to configuration in these topologies can also cause timing loops (more so for SyncE and PTPoIP than PTP over Ethernet [PTPoE]).

■ Using bundled interfaces with PTPoIP can cause asymmetry issues. See the Chapter 11 section "Partial Timing Support with G.8275.2 and GMs at the Edge" for a detailed treatment of this topic because there are several aspects to consider.

That is a summary of the issues to consider concerning the topology when making decisions on the form of PTP to use. The next aspect to consider is the specific type of technology used to carry the timing signal.

Packet Transport

One of the major tasks in building a timing distribution network is to ensure that the underlying transport network is not working against what the designer is trying to achieve. It has been mentioned previously that major support is needed from the transport network, and the requirements can be summarized as follows:

- Ability to carry frequency in some physical form (SyncE/eSyncE for Ethernet)

- Not introduce excessive PDV once the PTP message is in transit

- Not introduce asymmetry between forward and reverse transmission

- Be timing aware, meaning the transport can either carry PTP preferentially or have a technique to carry phase/time natively within the transport

The best-in-class deployment for timing over the transport network is to use G.8275.1 in combination with a physical frequency signal because it gives the best result and predictable performance. For this to work, the first prerequisite is that the transport must be able to carry the physical frequency (SyncE/eSyncE for Ethernet) between the network elements.

The second issue is to understand what the transport does with the PTP message stream. It either can carry the PTP messages as frames or packets just like another other packet, or it can use the PTP to recover the phase/time and carry the signal using another method.

This section looks at both these situations in turn.

Carrying Frequency over the Transport System

Probably the most common form of interface on a network element is some form of Ethernet. For that interface to carry a traceable frequency signal, the network element, the interface, and the optic all need to support SyncE. Even if there is a specialized transport node between two locations, there is often a short-run patch cable based on the Ethernet connected to it.

If the node-to-node connection is Ethernet over a fiber link, the SyncE is an accurate signal that can be recovered at the other end and still be traceable back to a reference source. However, there are many other types of transport that do not allow that frequency to be carried transparently. The sorts of problems that might arise with transport that effectively blocks frequency are as follows:

- It cannot recover the frequency from the SyncE signal and uses its own internal clock for transmission and reception. An example might be a microwave system.

- An existing synchronous transport can have its own native clocking mechanism and therefore does not allow the transparent transport of an external frequency signal. This can be an issue with SDH/SONET optical, as one example. The section "Adaptive Clock Recovery" earlier in the chapter covered the case of carrying a service clock over a transport with a different clock.

- The path includes at least one link from a third-party service provider that will not allow customer frequency signals on its network or cannot transparently carry it.

- The transport features a hardware limitation whereby the frequency cannot pass a certain point because of a lack of some physical connection to complete the circuit. A good example is the case of using a 1000BASE-T copper SFP in a 1-GE interface port.

- Some feature is not implemented or some optional part of it is missing. An example would be an Ethernet switch without SyncE support because having one node without the feature breaks the chain of traceability.

The industries responsible for many different types of transport equipment are moving to reduce these problems and refresh their product portfolios to introduce systems allowing the transport of frequency (and phase/time). Of course, replacing the devices already deployed in service provider networks is a much slower process.

A good example of this migration are the microwave systems used for mobile backhaul networks. When mobile systems were connected to their core systems with TDM networks, the TDM frequency was used to synchronize the cellular radio as well as the radio in the microwave systems. The microwave would transmit its signal using this reference frequency, and this signal could be recovered at the far end.

When operators adopted mobile base stations based on packet technology, then hardware designers developed microwave systems supporting the carriage of packets, and the carriage of frequency on the Ethernet links was not necessary. A TDM source was still available to provide frequency to the base station (and the microwave). As those TDM links were decommissioned and SyncE increased its reach, the microwaves developed the ability to recover SyncE at their input and pass that frequency along to the far end. On the other side of the link, the remote receiver can recover that frequency and use it to transmit Ethernet onto the fiber, thereby passing the frequency along as SyncE. Figure 9-35 illustrates how that might work.

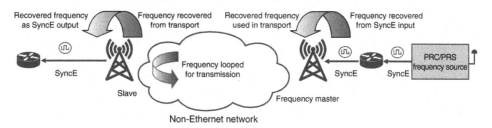

Figure 9-35 *Frequency Transport over a Non-Ethernet Transport*

Obviously, the best outcome is that the transport system can carry a reference frequency natively using some physical method. This tends to be a binary yes/no capability based on the equipment. But when looking at the transport of PTP messages over the transport, the situation is more nuanced because, of course, the transport can carry PTP, but how well can it do it?

Carrying Phase/Time over the Transport System

When the packet transport carries phase/time, the same issues arise as with the carriage of frequency. The issue of PDV is still a factor, but with the extra complication of asymmetry, which means a difference in message transit time between the forward and reverse direction. You have already seen that the asymmetry can also be caused by PDV and traffic load (see the section "Asymmetry" earlier in the chapter).

The answer to these issues must be applied to two areas: the network elements themselves and the transport between them. The earlier section "Clock Performance" deals with the first area, where the designer considers the performance of the (PTP-aware) network element and how it resists the accumulation of PDV and asymmetry. But what about the second area, the links in the transport system?

When network elements are linked together via some passive transport such as dark fiber, an Ethernet circuit running on that link has very little ability to inject any PDV or asymmetry into the timing signal. It is effectively a first-in, first-out pipe with no active components to delay or process the PTP messages. For a timing engineer, they are ideal pathways to transfer time signals, if they are available as an option.

But even seemingly simple Ethernet pluggable optics have increasing amounts of intelligence built into them, especially for higher interface speeds and those designed for long distances. There might be electronic components such as digital signal processors in the optical devices that exhibit dynamic and variable amounts of latency. Introducing support for extra capabilities like Forward Error Correction (FEC) and MACsec adds several points where extra variable delay can be introduced. For a detailed discussion on the issues raised for MACsec, see the section on it in Chapter 11.

The industry is working on methods to resolve these shortcomings, so this book will not name and shame any specific interfaces or optical devices because this area of work is proceeding very quickly. Some devices currently being deployed (for Ethernet) can contain significant buffering that introduces variable latency and therefore asymmetry. The size of the effect is limited, but it can add enough additional cTE and dTE to turn a Class C boundary clock into a Class A boundary clock just by using those optics. It is the general principle that is important—if the device contains active components, then part of the planning process is to learn what that device will do to the timing signal.

However, there are various transport technologies that have characteristics that make it difficult to accurately transport timing signals at all. This is because the methods used to communicate either are naturally asymmetric by design, have some form of time-based scheduling, or contain some dynamic buffering scheme. Any of these features makes those transport types poor choices to carry PTP messages as packet frames, or what is called *over-the-top*. This term refers to a situation where the PTP is carried just like every other IP or L2 frame, without any special handling beyond the usual QoS mechanisms.

The following is a list of some popular technologies that make over-the-top PTP difficult. They tend to be broadband access–type methods, which tend to be asymmetric because they are designed for consumer broadband, which has asymmetric traffic patterns (more download than upload). Because these systems have multiple generations, optional topologies and numerous versions, a detailed treatment would be overwhelming, so here are just the highlights:

- **Cable:** From the beginning, cable was inherently asymmetric, given that it was originally designed for one-way video distribution. Earlier versions of DOCSIS data services began using time to allocate cable modem slots for upstream transmission. Over time, cable systems are reducing this effect, and DOCSIS 4.0 will be another step in that direction.

- **DWDM optical:** Modern optical systems have levels of buffering built into them that suffer from non-static and asymmetric latency changes. The latest versions of these systems have introduced techniques to become PTP/timing aware. The techniques used include turning optical nodes inside the optical network into boundary clocks, running PTP over a separate link such as the optical supervisory channel (OSC), or somehow bypassing the problematic hardware.

- **Passive optical networking (PON):** There are numerous variations on PON, but many of them feature a natural asymmetry. For example, the popular GPON is broadcast in the downstream direction, but uses a TDMA technique in the upstream. This time-based scheduling causes aging of PTP packets that are waiting. Some newer technologies and later generations of PON are moving to a more symmetrical model or have adopted a native method to carry frequency, phase, and time.

- **DSL:** In a similar fashion to PON, DSL systems tend to be asymmetric because they are designed to support a market that generates asymmetric traffic patterns (with ADSL, asymmetry is in the name). Even if not the DSL itself, but the network nodes and switches serving the broadband market tend to have high levels of oversubscription. Like PON, there are better versions of DSL offering methods to alleviate the problem.

- **Microwave:** The case of frequency over microwave was covered in the previous section, but a similar issue occurs in microwave systems carrying PTP packets. The modulation of frames onto a radio channel can involve time-based scheduling and timeslots. There may be configuration options available to mitigate some of the worst effects, but you should always assume that microwave will need your attention. Thankfully, there are modern solutions to the microwave problem that allow these systems to successfully carry good-quality PTP and SyncE.

Each of these categories involves a very large number of details, and it is outside the scope of the book to go into each possible combination of technologies and how they all perform when carrying PTP messages. The point is that one of the most important tasks on a timing design is to discover all the forms of transport needed to carry both PTP messages and frequency signals and decide upon a remediation strategy for each of them. So, what would that remediation look like?

In the same way that transport systems can carry the frequency signal using their own native methods, phase/time can use a similar mechanism. Figure 9-36 illustrates a typical approach whereby the phase/time is recovered from the ingress port on the device, carried across the transport, and the PTP is re-created at the egress interface. In this diagram, the reference timing signal, coming from the right, is recovered at the input to the transport system via a PTP slave interface (for example). This phase and time signal is then used by the transport system to transport phase/time to the other components on the system using some other mechanism (which could even be some form of PTP). There is no carriage of over-the-top PTP traffic on the system; it is carried by some other mechanism native to the transport. At the remote end, the transport system recovers the phase/time from the native mechanism and uses that phase/time to re-create a PTP packet stream to the downstream devices.

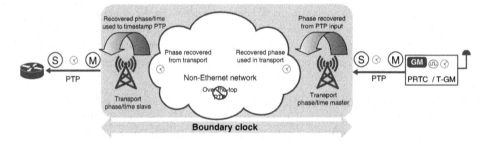

Figure 9-36   *Phase and Time Transport over a Non-Ethernet Transport*

To the PTP network, the link looks like an extended, distributed boundary clock with PTP extinguished at the ingress to the transport system and re-created on the egress side. In the same way that a network element can be a BC with a set of performance characteristics, so can a transport link. This allows the link to be budgeted and modeled for time error just like every other link in the timing chain. The limits for timing error allowed in timing distribution systems for telecoms are found in clause 7.1 of ITU-T G.8273.2.

Of course, phase/time can be injected into the transport system centrally and recovered where necessary; taking it from the data path isn't necessary. This is what many SDH/SONET systems did with frequency. The core nodes in these transport systems may be equipped with their own PRTC inputs to access phase/time and frequency sources for their own purposes.

The mechanism used to carry phase/time internally to the transport system varies based on the technology involved. It even varies based on the internal architecture of the system—for one example, see the following section "DOCSIS Cable and Remote PHY Devices" for details on timing inside the new packetized version of distribution for DOCSIS cable systems.

Non-Mobile Deployments

This chapter draws to a close by updating a few of the use cases mentioned in the earlier chapters and showing the deployments applied to those specific problems. Some other cases, such as circuit emulation, are covered earlier in the chapter in the discussion of frequency synchronization using packet-based methods.

The idea here is not to delve into the myriad details around every unique scenario, but to demonstrate how the knowledge gained up until this point can be applied to other cases. No matter what is the problem that requires solving, the solution will involve the same principles of frequency transfer, and phase/time alignment. It will entail using physical and packet methods in combination with sources of frequency, phase, and time such as atomic clocks and GNSS receivers.

DOCSIS Cable and Remote PHY Devices

Cable operators, under pressure for ever-higher bandwidth from their subscribers, are working on relieving bottlenecks on their hybrid fiber coax (HFC) networks by adopting a new architecture, one that replaces the analog fiber plant with packet-based transport. In this new design, the converged cable access platform (CCAP) core functions are split from the PHY (physical) function. The PHY function is moved further toward the edge of the network, and the two systems communicate with pseudowires over an IP/Ethernet network. This new network is called the converged interconnect network (CIN).

This evolution pushes the beginning of the cable portion of the network much closer to the subscriber and fills the gap in the middle with Ethernet-style devices. A new device called a Remote PHY Device (RPD) is used at the head of a cable run, and existing coax cable is used to complete the last mile. This allows the cable operator to offer much higher bandwidth services, on par with fiber services, but still retain the investment in its existing HFC plant. Figure 9-37 illustrates the building blocks of the new architecture.

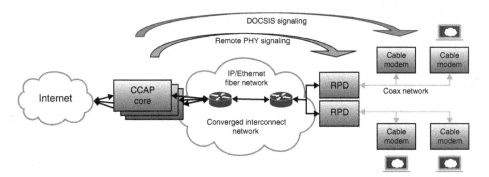

Figure 9-37 *Cable Network Using Remote PHY Architecture*

To enable data services, the cable industry uses a system called Data-Over-Cable Service Interface Specification (DOCSIS), which is terminated on a cable modem in the customer premises. DOCSIS has its own timing protocol known as the DOCSIS Time Protocol, or DTP. DTP is used to provide the needed frequency and phase synchronization to the cable modems to build a two-way data network.

Phase synchronization is needed to allow the cable modems to cooperate with other modems on the same cable (for instance, to use timeslots for uplink transmission). The requirement for phase in cable is currently somewhat relaxed, generally around a millisecond. However, as cable operators are now wishing to use their cable plants for backhauling 5G mobile traffic, the synchronization performance will need to equal that of mobile (see Chapters 10 and 11).

The basic problem is that DOCSIS previously was carried directly between the CCAP core and the CM, but now this has been replaced with the CIN packet network. As a result, the packet network needs to be able to pass frequency and phase synchronization to both the RPD and core devices. This enables the RPD to re-create the DTP on the coax to synchronize the cable modem and results in the CCAP functions being phase aligned with the CMs via the RPD.

Figure 9-38 illustrates how this new timing is applied to the remote PHY architecture. A PRTC with a T-GM providing PTP with (optional) SyncE is placed in this new CIN network. The recommended profile is the G.8275.2 telecom profile for PTS networks, using IP as a transport (many large operators deploy IPv6). Even with an unaware node or so, using PTP in this topology can very readily achieve one millisecond of phase alignment.

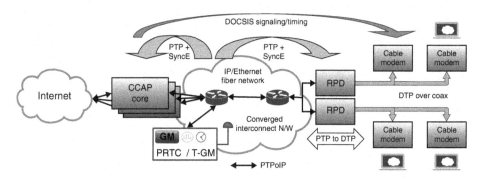

Figure 9-38 *Remote PHY Timing Architecture Using PTP and SyncE*

There are some variations on the deployment architecture, depending on which device originates the PTP and in which direction it flows. There is also increased interest in high-accuracy timing to support use cases that require timing to be delivered as a service at the microsecond level. This will involve a switch to using the G.8275.1 (PTPoE) profile with SyncE, mirroring the type of solution now being deployed for the mobile use cases (see Chapter 11).

For more information on timing for remote PHY, please see the reference to the CableLabs specifications at the end of the chapter, which are part of the DOCSIS 3.1 release. There are three references, *R-PHY* covering RPHY, *R-DTI* for timing the RPHY, and *SYNC* that covers the requirements on cable equipment for providing synchronization and timing as a service over the DOCSIS network (with particular focus on mobile backhaul).

Power Industry and Substation Automation

Chapter 2, "Usage of Synchronization and Timing," describes the uses of packet-based transport to carry phase/time and frequency for other industries, with a focus on the power industry. The power industry has had a long history of using accurate time to run its widely distributed infrastructure and monitor the distribution grid of alternating current across most of the globe.

These power networks typically utilize an architecture based on distributed sources of time. The primary sources in the substations are GNSS enabled and may contain many legacy timing signals, such as Inter-Range Instrumentation Group (IRIG) interfaces. Some requirements are unique to the power industry, and for this reason there are profiles of PTP specific to that industry. Chapter 7 includes information on the various types of profiles supporting the power industry, including the *Power Profile*.

The power industry seeks to improve its ability to manage its electrical distribution networks under all circumstances, including during widespread outages of GNSS systems (specifically GPS). The concern with vulnerability to GNSS outages is not unique to the power industry, nor is it particularly recent, and increasing political pressure is being applied to address it. Power industry members are addressing this concern through a combination of several approaches:

- Improving the resilience and accuracy of the GNSS receivers they deploy as PRTC as sources for their substation automation, phasor measurement units (PMU), and supervisory control and data acquisition (SCADA) systems around the grid.

- Using the transport network to carry timing signals from other remote GNSS + T-GM sites to improve the dispersion of their time sources over a wider area to survive localized outages.

- Looking at using a limited number of atomic sources to increase the holdover capability of their reference clocks. This allows them to increase the autonomy for the timing signal sources for their monitoring and control systems.

To implement this strategy, as covered in Chapter 2, utilities are working on projects to deploy PTP telecom profiles across the WAN to supply redundant timing signals in the case of localized outages of their timing sources. This solution is robust and is increasingly deployed in several other industries as well.

The unique issue to address is that when the PTP timing signal using the telecom profile reaches the substation, there needs to be a device that can handle the interworking between the telecom profile used for the WAN and the Power Profile inside the substation. In Power Profile terms, this is known as a boundary clock—one that can recover the clock over the WAN from the remote T-GM and provide a signal to the Power Profile transparent clocks inside the substation. Many primary sources already located in the substation can increasingly support being a slave to the telecom profile. This is exactly what the interworking function needs to do. See Figure 9-39 for an illustration.

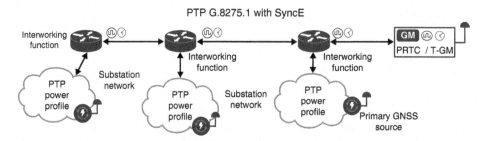

Figure 9-39 *Providing Resilient Backup Timing for Electricity Substations*

What is the requirement for phase/time accuracy in the power application? That is a difficult question to provide a single answer, because a lot depends on the specific application. The explanation from customers is that the closer the alignment the better, but certainly requiring accuracy to the level of microseconds is a common request. So, once again, it is important to understand the end-to-end budget.

A similar approach to providing some resilience to the local GNSS primary source is an increasingly common project under way in many industries, although the power industry is especially active. Every industry has several of their own special issues or requirements to deal with, but there is usually a vibrant vendor community who supports those needs and understands those trade-offs. What they are looking for is help from people like you to bring the expertise in carrying a timing signal over the wide area using packet-based techniques. The tricky bit is usually where the two parts come together.

Summary

This chapter covered a lot of ground, focused on real-world considerations when designing and deploying a timing synchronization network using PTP and SyncE/eSyncE. It starts out with the beginning of the timing chain, the sources of time and the physical signals and connectors used. Increasingly, the use of the time signal is done by an application with an embedded slave clock, so there is decreasing need for physical cabling at the edge of the network. As you will see in Chapter 12, testing and accuracy verification using BITS or 1PPS ports is the exception.

The chapter then covered several methods to carry frequency signals across the packet network, both using packet techniques (including PTP) and using physical methods in a packet network (SyncE). Some of the packet techniques are especially suitable for specialized use cases such as circuit emulation of TDM circuits. The next section covered the transport of phase/time using packet networks and several approaches to its deployment and design.

Then the discussion moved to timing performance in packet networks with a detailed breakdown of the metrics behind time error, asymmetry, and PDV. Then this knowledge was applied to clock performance, describing the performance of both reference sources and PTP clock types. This was reinforced with examples of how to use this information for budgeting an end-to-end network time error.

The chapter finished with a general examination of the performance issues that can arise when transporting timing through packet transport systems and what can be done to control and remediate those systems. The chapter concluded with two use cases from the power and cable industries, showing the different timing issues those industries are currently concerned with.

Now that you have an overview of all the aspects that must be considered in the design and deployment of a synchronization solution, it is time to take a very detailed look at one specific application. So, Chapter 10 will detail the requirements for timing for the mobile applications, and Chapter 11 will thoroughly examine the possible solutions to deliver upon those requirements. But again, these lessons apply more widely than just mobile, although the mobile case is a very well-documented example to examine.

References in This Chapter

Cable Television Laboratories (CableLabs)

"Remote DOCSIS Timing Interface Specification." *Data-Over-Cable Service Interface Specifications: DCA – MHAv2*, CM-SP-R-DTI-I08-200323, Version I08, 2020. https://www.cablelabs.com/specifications/CM-SP-R-DTI

"Remote PHY Specification." *Data-Over-Cable Service Interface Specifications: MHAv2*, CM-SP-R-PHY-I15-201207, Version I15, 2020. https://www.cablelabs.com/specifications/CM-SP-R-PHY

"Synchronization Techniques for DOCSIS Technology Specification." *Data-Over-Cable Service Interface Specifications: Mobile Applications*, CM-SP-SYNC-I01-200420, Version I01, 2020. https://www.cablelabs.com/specifications/CM-SP-SYNC

IEEE Standards Association

"IEEE Standard for a Precision Clock Synchronization Protocol for Networked Measurement and Control Systems." *IEEE Std 1588:2002*, 2002. https://standards.ieee.org/standard/1588-2002.html

"IEEE Standard for a Precision Clock Synchronization Protocol for Networked Measurement and Control Systems." *IEEE Std 1588:2008*, 2008. https://standards.ieee.org/standard/1588-2008.html

"IEEE Standard for Precision Clock Synchronization Protocol for Networked Measurement and Control Systems." *IEEE Std 1588:2019*, 2019. https://standards.ieee.org/standard/1588-2019.html

International Telecommunication Union Telecommunication Standardization Sector (ITU-T)

"G.703: Physical/electrical characteristics of hierarchical digital interfaces." *ITU-T Recommendation*, 2016. https://handle.itu.int/11.1002/1000/12788

"G.781: Synchronization layer functions for frequency synchronization based on the physical layer." *ITU-T Recommendation*, 2020. https://handle.itu.int/11.1002/1000/14240

"G.813: Timing characteristics of SDH equipment slave clocks (SEC)." *ITU-T Recommendation*, 2003, with Corrigenda 1 (2005) and Corrigenda 2 (2016). https://handle.itu.int/11.1002/1000/13084

"G.823: The control of jitter and wander within digital networks which are based on the 2048 kbit/s hierarchy." *ITU-T Recommendation*, 2000. https://www.itu.int/rec/T-REC-G.823-200003-I/en

"G.8260: Definitions and terminology for synchronization in packet networks." *ITU-T Recommendation*, 2020. https://handle.itu.int/11.1002/1000/14206

"G.8261: Timing and synchronization aspects in packet networks." *ITU-T Recommendation*, Amendment 2, 2020. http://handle.itu.int/11.1002/1000/14207

"G.8261.1: Packet delay variation network limits applicable to packet-based methods (Frequency synchronization)." *ITU-T Recommendation*, Amendment 1, 2014. https://handle.itu.int/11.1002/1000/12190

"G.8262: Timing characteristics of synchronous equipment slave clock." *ITU-T Recommendation*, Amendment 1, 2020. https://handle.itu.int/11.1002/1000/14208

"G.8262.1: Timing characteristics of enhanced synchronous equipment slave clock." *ITU-T Recommendation*, Amendment 1, 2019. https://handle.itu.int/11.1002/1000/14011

"G.8263: Timing characteristics of packet-based equipment clocks." *ITU-T Recommendation*, 2017. https://handle.itu.int/11.1002/1000/13320

"G.8265.1: Precision time protocol telecom profile for frequency synchronization." *ITU-T Recommendation*, Amendment 1, 2019. https://handle.itu.int/11.1002/1000/12193

"G.8271: Time and phase synchronization aspects of telecommunication networks." *ITU-T Recommendation*, 2020. https://handle.itu.int/11.1002/1000/14209

"G.8271.1: Network limits for packet time synchronization with full timing support from the network." *ITU-T Recommendation*, Amendment 1, 2020. https://handle.itu.int/11.1002/1000/14210

"G.8271.2: Network limits for time synchronization in packet networks with partial timing support from the network." *ITU-T Recommendation*, Amendment 2, 2018. https://handle.itu.int/11.1002/1000/13768

"G.8272: Timing characteristics of primary reference time clocks." *ITU-T Recommendation*, Amendment 1, 2020. https://handle.itu.int/11.1002/1000/13769

"G.8272.1: Timing characteristics of enhanced primary reference time clocks." *ITU-T Recommendation*, Amendment 2, 2016. https://handle.itu.int/11.1002/1000/13325

"G.8273.2: Timing characteristics of telecom boundary clocks and telecom time slave clocks for use with full timing support from the network." *ITU-T Recommendation*, 2020. https://handle.itu.int/11.1002/1000/14213

"G.8273.3: Timing characteristics of telecom transparent clocks for use with full timing support from the network." *ITU-T Recommendation*, 2020. https://handle.itu.int/11.1002/1000/13770

"G.8273.4: Timing characteristics of telecom boundary clocks and telecom time slave clocks for use with partial timing support from the network." *ITU-T Recommendation*, 2020. https://handle.itu.int/11.1002/1000/14214

"G.8275: Architecture and requirements for packet-based time and phase distribution." *ITU-T Recommendation*, 2020. https://handle.itu.int/11.1002/1000/14016

"G.8275.1: Precision time protocol telecom profile for phase/time synchronization with full timing support from the network." *ITU-T Recommendation*, 2020. https://handle.itu.int/11.1002/1000/14215

"G.8275.2: Precision time protocol telecom profile for time/phase synchronization with partial timing support from the network." *ITU-T Recommendation*, 2020. https://handle.itu.int/11.1002/1000/14216

Internet Engineering Task Force (IETF)

Mills, D. "Simple Network Time Protocol (SNTP) Version 4 for IPv4, IPv6 and OSI." *IETF*, RFC 4330, 2006. https://tools.ietf.org/html/rfc4330

Mills, D., J. Martin, J. Burbank, and W. Kasch. "Network Time Protocol Version 4: Protocol and Algorithms Specification." *IETF*, RFC 5905, 2010. https://tools.ietf.org/html/rfc5905

Schulzrinne, H., S. Casner, R. Frederick, and V. Jacobson. "RTP: A Transport Protocol for Real-Time Applications." *IETF*, RFC 3550, 2003. https://tools.ietf.org/html/rfc3550

Chapter 9 Acronyms Key

The following table expands the key acronyms used in this chapter.

Term	Value
1000BASE-T	gigabit Ethernet over copper wiring (802.3ab)
1 GE, 10 GE	1-gigabit Ethernet, 10-gigabit Ethernet
1PPS	1 pulse per second
3G	3rd generation (mobile telecommunications system)
4G	4th generation (mobile telecommunications system)
5G	5th generation (mobile telecommunications system)
AC	attachment circuit

Term	Value
ACR	adaptive clock recovery
ADSL	asymmetric digital subscriber line
APTS	assisted partial timing support
ARB	arbitrary (timescale)
BC	boundary clock
BITS	building integrated timing supply (SONET)
BNC	bayonet Neill-Concelman (connector)
CAT-5, CAT-6, etc.	category 5, category 6, etc. (cable)
CCAP	converged cable access platform
CEM	circuit emulation
CF	correctionField
CIN	converged interconnect network
CRC-4	Cyclic Redundancy Check 4 (SDH)
cTE	constant time error
dB	decibel
DCR	differential clock recovery
DIN	Deutsches Institut für Normung (German Institute for Standardization)
DOCSIS	Data-Over-Cable Service Interface Specification
DSL	digital subscriber line (a broadband technology)
dTE	dynamic time error
dTE_H	dynamic time error with high-pass filter
dTE_L	dynamic time error with low-pass filter
DTP	DOCSIS Time Protocol
DWDM	Dense Wavelength Division Multiplexing (optical technology)
EEC	Ethernet equipment clock
eEEC	enhanced Ethernet equipment clock
EMI	electromagnetic interference
E1	2-Mbps (2048-kbps) signal (SDH)
eNB	evolved Node B (E-UTRAN)—LTE base station
eNodeB	evolved Node B (E-UTRAN)—LTE base station

Term	Value
ePRTC	enhanced primary reference time clock
ePRTC-A	enhanced primary reference time clock—class A
ePRTC-B	enhanced primary reference time clock—class B
ESF	extended super frame
ESMC	Ethernet synchronization messaging channel
FDD	frequency division duplex
FEC	Forward Error Correction
FPGA	field-programmable gate array
FTS	full timing support (for PTP from the network)
GE	gigabit Ethernet
GM	grandmaster
GNSS	global navigation satellite system
GPON	gigabit passive optical network
GPS	Global Positioning System
HFC	hybrid fiber coax (cable system)
IEEE	Institute of Electrical and Electronics Engineers
IERS	International Earth Rotation and Reference Systems
IETF	Internet Engineering Task Force
IP	Internet protocol
IPv4, IPv6	Internet protocol version 4, Internet protocol version 6
IRIG	Inter-Range Instrumentation Group
ITU	International Telecommunication Union
ITU-T	ITU Telecommunication Standardization Sector
L2	layer 2 (of the OSI model)
L3	layer 3 (of the OSI model)
LPF	low-pass filter
LSF	low smoke and fume
MAC	media access control
MACsec	media access control security
max|TE|	maximum absolute time error
MCX	micro coaxial connector

Term	Value
MMCX	micro-miniature coaxial connector
MPLS	Multiprotocol Label Switching
MSP	mobile service provider/mobile operator
MTIE	maximum time interval error
NMEA	National Marine Electronics Association
NTP	Network Time Protocol
OCXO	oven-controlled crystal oscillator
OSC	optical supervisory channel
PABX	private automatic branch exchange
PDV	packet delay variation
PEC-M	packet-based equipment clock—master
PHY	PHYsical Layer (of OSI Reference Model)—an electronic device
pktSelected-2wayTE	packet-selected two-way time error
PLL	phase-locked loop
PMU	phasor measurement unit
PON	passive optical networking (optical broadband)
POP	point of presence
ppb	parts per billion
ppm	parts per million
PRC	primary reference clock (SDH)
PRS	primary reference source (SONET)
PRTC	primary reference time clock
PRTC-A	primary reference time clock—class A
PRTC-B	primary reference time clock—class B
PTP	precision time protocol
PTPoE	PTP over Ethernet
PTPoIP	PTP over Internet protocol (IP)
PTS	partial timing support
QL	quality level

Term	Value
QoS	quality of service
RFC	Request for Comments (IETF document)
RJ-45, RJ-48c	registered jack 45, 48c
RMS	root mean square
RPD	Remote PHY Device
RTP	real-time transport protocol
SCADA	supervisory control and data acquisition
SDH	synchronous digital hierarchy (optical transport technology)
SFP	small form-factor pluggable (transceiver)
SMA	SubMiniature version A (connector)
SMB	SubMiniature version B (connector)
SONET	synchronous optical network (optical transport technology)
SP	service provider (for telecommunications services)
SSM	synchronization status message
SSU	synchronization supply unit (SDH)
SyncE	synchronous Ethernet—a set of ITU-T standards
T1	1.544-Mbps signal (SONET)
TAI	Temps Atomique International
T-BC	Telecom Boundary Clock
T-BC-A	Telecom Boundary Clock—Assisted
T-BC-P	Telecom Boundary Clock—Partial support
TC	transparent clock
TDEV	time deviation
TDM	time-division multiplexing
TDMA	time-division multiple access
TE	time error
T-GM	Telecom Grandmaster
TNC	threaded Neill-Concelman (connector)
ToD	time of day
T-TC	Telecom Transparent Clock
T-TSC	Telecom Time Slave Clock

Term	Value
T-TSC-A	Telecom Time Slave Clock—Assisted
T-TSC-P	Telecom Time Slave Clock—Partial support
UBX	u-Blox
UTC	Coordinated Universal Time
UTP	unshielded twisted-pair
V.11	ITU-T specification for differential data communications
VLAN	virtual local-area network
WAN	wide-area network

Further Reading

Refer to the following recommended sources for further information about the topics covered in this chapter.

Internet Engineering Task Force (IETF)

Bryant, S. and P. Pate. "Pseudo Wire Emulation Edge-to-Edge (PWE3) Architecture." *IETF*, RFC 3985, 2005. https://tools.ietf.org/html/rfc3985

Malis, A., P. Pate, R. Cohen, and D. Zelig. "Synchronous Optical Network/ Synchronous Digital Hierarchy (SONET/SDH) Circuit Emulation over Packet (CEP)." *IETF*, RFC 4842, 2007. https://tools.ietf.org/html/rfc4842

Stein, Y., R. Shashoua, R. Insler, and M. Anavi. "Time Division Multiplexing over IP (TDMoIP)." *IETF*, RFC 5087, 2007. https://tools.ietf.org/html/rfc5087

Vainshtein, A. and Y. Stein. "Structure-Agnostic Time Division Multiplexing (TDM) over Packet (SAToP)." *IETF*, RFC 4553, 2006. https://tools.ietf.org/html/rfc4553

Vainshtein, A., I. Sasson, E. Metz, T. Frost, and P. Pate. "Structure-Aware Time Division Multiplexed (TDM) Circuit Emulation Service over Packet Switched Network (CESoPSN)." *IETF*, RFC 5086, 2007. https://tools.ietf.org/html/rfc5086

ITU-T. "GSTR-GNSS: Considerations on the Use of GNSS as a Primary Time Reference in Telecommunications." *ITU-T Technical Report*, 2020. http://handle.itu.int/11.1002/pub/815052de-en

Chapter 10

Mobile Timing Requirements

In this chapter, you will learn the following:

- **Evolution of cellular networks:** Covers the generations of cellular networks and how they have evolved from the first generation to the fifth generation.

- **Timing requirements for mobility networks:** Discusses the various modulation techniques and how they have evolved and been adopted in mobile networks. Also explains how synchronization requirements have evolved along with each change in modulation technique.

- **Timing requirements for LTE and LTE-A:** Briefly discusses modulation techniques and use cases implemented by Long Term Evolution. Several use cases are included to explain the synchronization requirements and illustrate the mobile performance issues if these requirements are not met. However, detailed examination of every technique and use case is beyond the scope of this book.

- **Evolution of the 5G architecture:** Outlines the evolution of the 5G network architecture, including the system architecture, definitions, and terminologies, as well as the requirements of the network. Discusses the overall architecture changes in 5G to help establish how the time synchronization architecture must evolve with it.

- **5G New Radio synchronization requirements:** Captures several major use cases for 5G mobile and outlines the time synchronization requirements evolving around those use cases.

The traditional telecom network has relied on accurate frequency distribution to optimize transmission and operate time-division multiplexing (TDM) transport connections. However, with evolved wireless networks, the distribution of accurate time and phase (as well as frequency) has become necessary for modern telecom-based services and the underlying infrastructure. In wireless services (for example, GSM, CDMA, WiMAX, LTE, or 5G), the air interface demands stringent synchronization, even though the end-user service (for example, mobile broadband Internet access) might not require synchronization.

Timing and synchronization are critical components of an efficient mobile network, and hence the transport network plays a very critical role in distributing and maintaining accurate time and synchronizing phase signals. If accurate time and phase distribution is not properly designed, implemented, and managed, mobile networks will suffer a dramatic (negative) effect on the efficiency, reliability, and capacity of the mobile services. Mobile subscribers will likely suffer dropped calls, interrupted data sessions, and a generally poor user experience.

The main characteristics of a well-implemented timing and synchronization network include the following:

- **Accuracy:** Network elements that are frequency synchronized (syntonized) to the nominal frequency (small tolerance allowed).

- **Time budget:** Phase error delivered across the network within the timing budget required both for the application use cases and time division duplex (TDD) radios.

- **Convergence:** Requires a short time interval to "lock on" and align frequency and phase to the sources of time.

- **Holdover time:** Provides a reasonable buffer period for the network to maintain time accuracy so that the radio units can maintain their frequency and phase within specification if they lose their connection to the reference clocks or sources of time.

- **Resilience:** The ability of the network to adapt to failures and network rearrangements—for example, an outage of the GPS signal.

- **Cost:** The design is cost-effective to build, monitor, and operate.

- **Adaptability and flexibility:** The architecture can be implemented across different topologies and transmission technologies.

It is very important to understand radio synchronization requirements, the use cases, and the trade-offs involved to architect a good synchronization network design. For that reason, this chapter discusses the evolution of mobility networks, radio access network architectures, and their synchronization requirements.

Evolution of Cellular Networks

The first generation of mobile network was launched by Nippon Telegraph and Telephone (NTT) in Tokyo in 1979. The version of mobile was analog in nature, implemented on simple frequency-division multiple access (FDMA) technology. In 1983, the USA launched its first 1G cellular networks, and Motorola produced one of the first mobile phones to be used across various states. Following the first generation, the mobile networks have been evolving to a new generation roughly every 10 years.

The second generation (2G) of cellular network was launched in Finland in 1991. The starting point for the 2G mobile technologies was the Global System for Mobile communications (GSM) and Interim Standard 95 (IS-95), also known as cdmaOne. Both were fully digital and supported voice and circuit-switched data communication at low rates up to 14.4 Kbps. GSM adopted a hybrid of FDMA and time-division multiple access (TDMA), while IS-95 used hybrid FDMA and code-division multiple access (CDMA) techniques.

2.5G was the first technology to introduce packet-switched data services in addition to circuit-switched voice and data services in the mobile networks. The main 2.5G standard, General Packet Radio Service (GPRS), introduced the packet-switched (PS) domain to the GSM core network. GPRS coding schemes adopted error correction and more than one timeslot to achieve higher data rates in a GSM-based radio access network (RAN). For the cdmaOne family, packet-switching was introduced by IS-95B, and was deployed by South Korea starting in 1999.

In 1998, the 3rd Generation Partnership Project (3GPP) was initiated to define a third-generation (3G) mobile system based on 2G GSM and to standardize network protocols across different vendors. In parallel, the 3rd Generation Partnership Project 2 (3GPP2) defined an alternate 3G network based on CDMA. 3G was first lunched by NTT Docomo in 2001.

The first 3G systems were as follows:

- Enhanced Data rate for GSM Evolution (EDGE), also widely referred as pre-3G or 2.75G, was a superset to GPRS and could function on any GPRS-based network. EDGE used 8PSK modulation techniques to achieve higher data rates; for example, a downlink rate up to 384 kbps.

- Universal Mobile Telecommunication System (UMTS) shared the same core network as GSM/GPRS or EDGE. However, the UMTS RAN architecture was completely different, as the radio used wideband code-division multiple access (WCDMA). UMTS could achieve the same download speed, 384 kbps, as EDGE.

- cdma2000 1xRTT (Single-Carrier Radio Transmission Technology), specified in IS-2000, achieved downlink rate up to 153.6 kbps using the quadrature phase-shift keying (QPSK) modulation technique.

1x Evolution-Data Optimized (1xEV-DO), which was defined in IS-856, and High Speed Packet Access (HSPA) are considered as part of the 3.5G network generation. 1xEV-DO used adaptive modulation and coding (AMC); it implemented Forward Error Correction (FEC) with higher code rates for better spectrum efficiency and an improved signal-to-noise ratio (SNR) to achieve a higher downlink rate of 2.4 Mbps. 1xEV-DO used only packet-switched data networks, unlike cdmaOne and cdma2000.

HSPA covers both High Speed Downlink Packet Access (HSDPA) and high-speed uplink packet access (HSUPA), which improved upon the performance of 3G. HSDPA, widely known as 3.5G, improved the UMTS downlink rates by using AMC and improved scheduling algorithms to achieve a peak rate of 14 Mbps. HSUPA, widely known as 3.75G, used similar techniques with a dedicated data channel to achieve uplink data rate up to 5 Mbps.

4G was first deployed in 2009 in Stockholm, Sweden, and Oslo, Norway, using the LTE standard. Originally, the 3GPP 4G system was referred to as System Architecture Evolution (SAE), with a core network known as the Evolved Packet Core (EPC), and only the RAN was called LTE. However, the LTE name became more popular, and the industry now uses it to refer to the whole system.

As Figure 10-1 illustrates, 4G LTE has a core and RAN network that are completely different from the UMTS architecture. The 4G LTE EPC is fully packet-switched and does not support circuit-switched domains; in LTE even voice is carried as data using voice over IP (VoIP).

The 4G LTE RAN uses a different orthogonal frequency-division multiple access (OFDMA) technique to achieve peak data rates of up to 300 Mbps. The LTE Advanced (LTE-A) that followed introduces techniques that allow multiple carriers to use ultra-wide bandwidth, up to 100 MHz, to achieve data rates up to 3000 Mbps.

Worldwide Interoperability for Microwave Access (WiMAX) was created by the WiMAX Forum and is based on the IEEE 802.16 set of standards from 2005. WiMAX was designed as an alternative to cable or digital subscriber line (DSL) to provide last-mile wireless broadband access.

In 2011, a mobile version of WiMAX (based on IEEE 802.16e-2005 and 802.16m-2011) became a candidate technology for 4G adoption in competition with the LTE-A standard. WiMAX was originally designed to provide data rates up to 30–40 Mbps, but with the release of the 802.16m standard in 2011, it was able to provide up to 1000 Mbps for fixed stations. Between LTE-A and WiMAX, LTE-A seemed to be the better approach and is widely preferred in 4G deployments over WiMAX.

The 5th generation (5G) mobile network is the next major phase of mobile telecommunications standards. The exponential growth of wireless data services driven by mobile Internet broadband and smart devices triggered the development of the 5G cellular network. 5G provides higher data rates, enhanced end-user quality of experience (QoE), reduced end-to-end latency, and lower energy consumption.

To satisfy the wide range of use cases, 5G requires access to spectrum across a range of frequency bands:

- **Low band:** Frequencies less than 1 GHz are best suited to provide wide coverage for mobile subscribers as well as support Internet of things (IoT) services. These bands are especially well-matched for high mobility, coverage over a wide area, and reception deep in structures and vegetation.

- **Mid-band:** Situated between 1–6 GHz, these frequencies provide coverage as well as capacity. The majority of 5G commercial networks are deploying in the 3.3–3.8 GHz range. Operators are also allowed to reuse traditional 4G bands such as 1.8 GHz, 2.3 GHz, and 2.6 GHz for 5G services. To address the long term increase in demand for bandwidth, the bands between 3–24 GHz are also allocated for 5G use and future expansion.

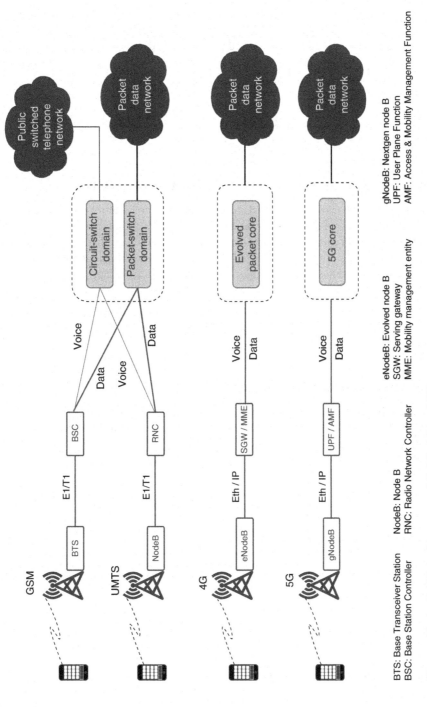

Figure 10-1   *RAN Architecture—2G, 3G, 4G, and 5G*

BTS: Base Transceiver Station   NodeB: Node B
BSC: Base Station Controller   RNC: Radio Network Controller

eNodeB: Evolved node B
SGW: Serving gateway
MME: Mobility management entity

gNodeB: Nextgen node B
UPF: User Plane Function
AMF: Access & Mobility Management Function

- **High band:** Currently 26 GHz, 28 GHz, and 40 GHz are being considered globally to support ultra-high-speed broadband use cases. To expand the capacity requirements, the millimeter wave (mmWave) bands have attracted significant interest for use in 5G networks. Although mmWave, from 30–300 GHz, serves only a small coverage area, that short range also allows efficient reuse of the spectrum and limits the inter-cell interference.

5G was first deployed in South Korea by SK Telecom in April 2019 using 3.5 GHz spectrum.

Timing Requirements for Mobility Networks

Traditionally, radio networks required strict frequency stability at the air interface to: minimize disturbance between adjacent radio base stations (BS); facilitate call handover; and fulfill regulatory requirements on interference. In addition to that, the backhaul transport methods (for example, E1/T1, SDH/SONET, or Ethernet) require tight frequency synchronization to allow accurate multiplexing and de-multiplexing.

To understand the impact of these requirements on the timing design, the following section covers the fundamentals of radio technology. The sections that follow cover each radio technology and describe the reason why timing becomes important for that technology.

Multi-Access and Full-Duplex Techniques

To distinguish and separate signals between different transmitters and receivers, 2G and 3G cellular networks adopted multi-access techniques such as FDMA, TDMA, and CDMA. Figure 10-2 illustrates each of the techniques, which are further described in the list that follows.

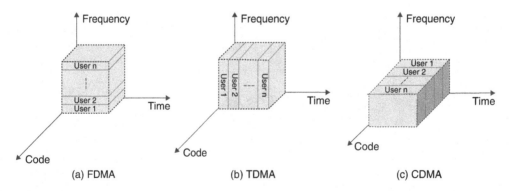

Figure 10-2 *Multi-Access Techniques*

■ FDMA is the process of dividing one frequency channel or bandwidth into multiple nonoverlapping subchannels and allocating a subchannel to each user with which the user's data is sent. The radio then modulates the data on a carrier wave at the subchannel frequency. Note that FDMA can also (rapidly) assign the user to a new frequency in the band, which is a way to reduce interference and improve service—a technique known as *frequency hopping*.

■ TDMA is a technique to divide a single frequency channel into multiple timeslots. These timeslots can be used by different transmitters; for example, user 1 uses timeslot 1 and user 2 uses timeslot 2. TDMA allows the entire channel bandwidth to be used by each user in turn during their given timeslot. Each radio modulates the data on a carrier wave at the frequency of the channel, but only during their assigned transmission timeslot.

■ CDMA works on a principle where *spread-spectrum* techniques are used to distinguish between different users sharing a frequency—each user has access to the entire bandwidth for the entire time duration. Many different signals with unique *spreading codes* can be modulated onto a single carrier to support multiple users. Each radio modulates the data on a carrier wave at the frequency of the channel but only using the assigned spreading code.

Now that the techniques for access to the radio resource have been covered, the next step is to see how the separation between transmission and reception is handled to enable simultaneous two-way communication, or *full duplex*.

As illustrated in Figure 10-3, radio systems implement two key techniques to enable full duplex: frequency division duplex (FDD) or time division duplex (TDD).

FDD uses two frequency bands at the same time: TDD uses same frequency for uplink and
f_1 for uplink and f_2 for downlink   downstream but different time slots

Figure 10-3 *FDD Versus TDD*

■ FDD allocates a channel in two different frequency ranges—one channel is used for each direction of transmission: uplink and downlink.

A guard band separates the two frequency bands to avoid any interference. For example, the upper half of a band may be used for transmission and the lower half of the band may be used for reception. In FDD, the guard band does not affect the throughput of the user because it is like a highway median strip—a separation of the "lanes" of traffic going in different directions.

- TDD uses a single frequency for both uplink and downlink and assigns different timeslots for transmission and reception. For example, the system may allow an individual user to transmit for 4 ms and then require them to receive for the next 4-ms timeslot.

  Guard times (or guard periods [GP]) are used to separate the transmit and receive timeslots and allow time for the radio to switch between transmit and receive modes (which may take several microseconds). Additionally, the guard time reflects the propagation time between the two radios. For example, if the user equipment (UE) transmitter is 10 km away, the signal will take 33 µs to get from handset to cell site, then the receiver at the cell site must allow 33 µs for the signal to arrive before it can use the same frequency to transmit.

  For short range, this is a trivial problem, but for longer range, using guard time in this way consumes more and more of the available time for signal transmission—reducing the effective throughput of the system. On the other hand, although using smaller guard times provide higher throughput, that only works in combination with strict synchronization between the handset and the cell site.

There are many advantages and disadvantages of FDD- and TDD-based implementations and, depending on the deployment requirements, mobile operators adopt one over the other. Table 10-1 captures a quick comparison of FDD versus TDD (UL is uplink, and DL is downlink).

In summary, mobile telecommunication systems have evolved to use various combinations of multiple access and duplexing techniques. Some examples:

- Advanced Mobile Phone Service (AMPS) was the first cellular system to assign a pair of frequencies to facilitate full-duplex operation. The AMPS system is an example of a pure FDD full-duplex FDMA system.

- GSM uses TDMA and FDD in combination. A given user is allocated both an uplink and downlink channel of 200 kHz (and so, FDD) separated by a wide guard band (in many countries, this gap is/was 45 MHz). The single channel in each direction is further divided into eight timeslots to support eight users using TDMA.

- In comparison, the UMTS, 4G, and 5G architectures allow both FDD and TDD duplexing techniques in their radios.

Impact of Synchronization in FDD and TDD Systems

Irrespective of the technology generation or mobile protocols utilized, all base stations must support a frequency accuracy of 50 parts per billion (ppb) at the radio air interface. To meet this requirement at the radio, the standards specify that a frequency with an accuracy of at least 16 ppb is available at the base station. This is implemented by providing a frequency signal of that accuracy, traceable to a primary reference clock (PRC) or primary reference source (PRS).

Table 10-1 *Comparison of FDD-Based Versus TDD-Based Communications*

Parameter	FDD	TDD	FDD or TDD
Spectral Efficiency	Due to separate frequency bands for UL and DL, spectrum efficiency is low.	TDD requires only one frequency channel for both UL and DL and hence is more efficient.	TDD
Latency	Because transmission and reception happens simultaneously, FDD is more suitable for low latency communication.	Because transmission and reception cannot happen at the same time, there is time delay in TDD, and latency is higher for TDD based communication.	FDD
Guard Band/ Guard Time	A guard band is used to separate the UL and DL bands.	UL and DL transmission is separated by guard time, which allows for base station switching time between transmitter and receiver. Because no data can be transmitted during this time, this part of the spectrum is sacrificed.	FDD
Traffic Asymmetry	Because UL and DL band size is predetermined, it is not possible to make dynamic changes in spectrum to match capacity demands.	Distribution of UL or DL bandwidth can be optimized by adjusting the timeslots to match traffic capacity with demand.	TDD
Distance	Can be transmitted over longer distances without any limitation of frame structure.	Longer GP must be introduced to accommodate the effect of multipath and propagation delay.	FDD
Throughput	For a given size of spectrum, FDD gives better peak throughput due to less overhead frame structure.	Due to more overhead in frame structure, to meet the same performance as FDD, TDD requires more radio sites for a given spectrum.	FDD
Time Synchronization	UL and DL are in separate bands; hence no need for phase alignment.	Because the radios are continuously switching between UL and DL within the same band, accurate phase alignment is mandatory.	FDD
Features Support	Because UL and DL transmission bands are different and predefined, features like multiple antenna schemes, beam formation, and beam alignment are not efficient for FDD.	Multiple input multiple output (MIMO), beam forming, beam alignment, and so on are easier to implement with TDD.	TDD

To understand the need for frequency accuracy, consider the example of the handover process between cell sites. In FDD-based systems, the handset, or user equipment (UE), acquires two frequency channels from the base station (BS): one for transmission and the other for reception. If the BS has an accurate source of frequency, then the radios will be very closely tuned to the allocated, expected frequency and the UE will be locked onto them.

When the UE moves from its current BS to a neighboring BS, as shown in Figure 10-4, it needs to acquire two new frequency channels from the neighboring BS for the call to seamlessly continue. The radio in the UE switches to the new channels and tries to continue the call. If the frequency offset between adjacent base stations is too large (more than 50 ppb), the UE radio will not be able to lock onto the frequency being used by the new BS, and the call will be dropped.

This is referred as *frequency error*—the difference between the actual frequency used by the BS and the frequency that the BS should be using. The limit in frequency error needs to be maintained across the BS to deliver a better user experience, such as successful handover processing, cell throughput, and so forth. To ensure that all the connected BSs are within the frequency synchronization limits, a common source is used for frequency and data clock generation.

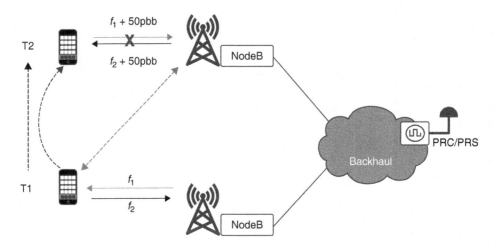

Figure 10-4   *Call Handover Scenario—FDD*

Additionally, the frequency of the carrier output of the transmitter must also be accurate to guarantee there is no overlap of signal that would cause interference between neighboring base stations. This interference and crosstalk between overlapping signals would impair the voice quality and generally degrade the user experience.

In TDD-based systems, the UE and BS must be not only frequency synchronized but phase synchronized as well. In TDD, to allow full-duplex operation, the uplink and downlink share the same frequency channel but utilize different timeslots. Accurate phase/time synchronization is critical for error-free communications between UE and BS, as both radios alternate between transmit and receive. If either end loses synchronization, their transmit timeslots overlap, and the two ends talk all over the top of one another.

As shown in Figure 10-5, accurate phase/time synchronization is also critical to reducing service outage at the point of handover from one BS to the adjoining BS. Furthermore, if neighboring base stations are not synchronized and use different uplink and downlink slots on the same channel, then interference occurs at the overlapping edges between cells.

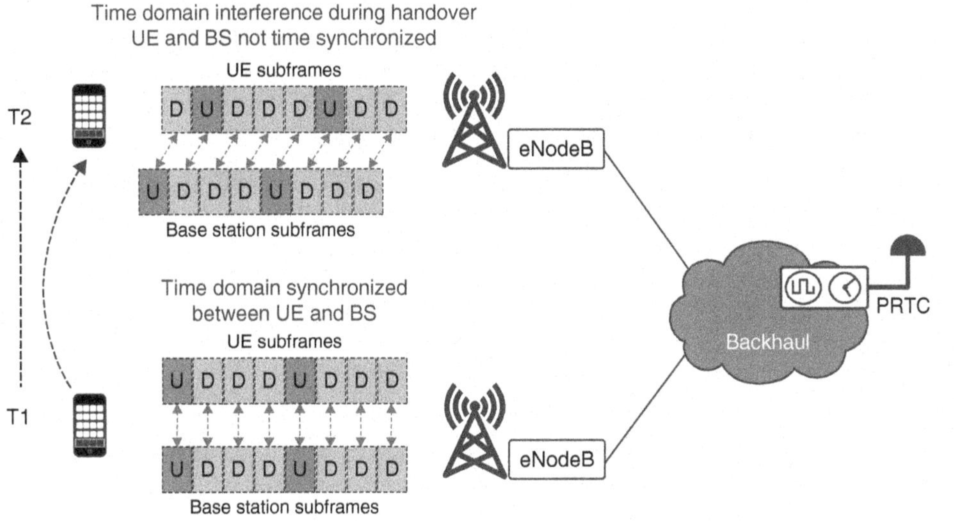

Figure 10-5 *Call Handover Scenario—TDD*

The preceding section mentions that TDD uses a guard time to prevent overlap between transmission and reception. The guard time should be sufficiently long to allow the signal from the UE to reach the base station before the base station shuts down its receiver and switches to transmit mode. Consequently, the guard time is decided based on two main factors:

- **Propagation delay:** This is proportional to the distance separating the UE and the BS. The BS informs the UE when to start the uplink transmission. The greater the distance between UE and base station, the longer the transmitter must wait before it receives a reply to its transmission, because the receiver must wait for all the data in that timeslot to arrive before beginning to transmit a reply.

- **Switching time:** The time required to switch between transmit and receive mode.

The network operator configures the guard time based on the cell size.

The preceding paragraphs confirm that every adjacent BS must be in strict phase/time synchronization with all its neighbors (for example, to allow smooth handover of any UE between them). For that reason, the standards organizations have agreed that TDD requires a phase alignment of less than 3 µs between base stations to function correctly.

This requirement is satisfied by synchronizing all the base stations to a common phase and time reference, so the operator makes sure every BS is aligned to within ±1.5 µs of a commonly agreed time reference. This common time standard is normally Coordinated Universal Time (UTC), so if every BS is within ±1.5 µs of UTC, then every BS will be within 3 µs of all its neighbors. This phase/time alignment to UTC is delivered either through the backhaul network or via timing radio signals received by a global navigation satellite system (GNSS) receiver.

The BS in turn continuously aligns the phase/time on the UE using the radio signals. If the UE is not able to align its time with the BS, it needs to reattach to the BS and regain its synchronization. These requirements for timing are explained with more detail in the following section "Timing Requirements for LTE and LTE-A."

The ITU-T G.823/G.824 recommendations for transport protocols (such as E1/T1 or SDH/SONET) specify an accuracy of 16-ppb frequency delivered by the backhaul network to meet the overall 50-ppb budget at the radio interface. Hence, in 2G or 3G FDD-based networks, T1/E1 links were widely used to provide frequency accuracy to the base stations. Table 10-2 summarizes the requirements for 2G/3G FDD- and TDD-based networks.

Table 10-2   *Synchronization Requirements for 2G and 3G Networks*

Application	Frequency Accuracy (Network)	Frequency Accuracy (Air Interface)	Phase Accuracy	References
GSM	16 ppb	50 ppb	—	3GPP TS25.411
				3GPP TS25.431
CDMA,	16 ppb	50 ppb	±3 µs	3GPP2 C.S0010-B
CDMA2000			±10 µs (for less than 8 hours during holdover)	3GPP2 C.S0010-C
UMTS-FDD	16 ppb	50 ppb	N.A.	3GPP TS25.104
UMTS-TDD	16 ppb	50 ppb	±2.5 µs	3GPP TS25.105
				3GPP TS25.402
TD-SCDMA	16 ppb	50 ppb	3 µs	3GPP TS25.123

Timing Requirements for LTE and LTE-A

Chapter 4, "Standards Development Organizations," outlines the role of 3GPP in the evolution in generations of mobile technologies. Technical specifications are still continuously evolving from the early days of 2G GSM, through to the 5G of the present day and onto future releases to meet new demands for services, performance, and features.

3GPP documents are structured as releases, whereby each release has a set of new features added into it. The first set of 3GPP releases were termed Phase 1 and Phase 2. However, later, the release was named based on the year of the anticipated release. For example, Release 96 was planned for approval in 1996 and released in Q1 1997. After Release 99 of the UMTS/WCDMA standard, 3GPP reverted to using specific release numbers. For example, Release 2000, which defines the UMTS all-IP core standard, was renumbered as Release 4.

LTE was first introduced in 3GPP Release 8; however, the work continued to add many use cases and enhancements to LTE specifications beyond Release 8. Release 10, also known as LTE-Advanced (LTE-A), specifies the more advanced features of LTE. And in Release 13, LTE got the new marketing name of LTE-Advanced Pro (LTE-A Pro). Figure 10-6 provides a quick view of some of the major use cases added by LTE, LTE-A, and LTE-A Pro.

This section focuses on covering some of these techniques and use cases in context to explain the change in time synchronization requirements for LTE and LTE-A. However, covering every use case and detail of these techniques and its implementation is beyond the scope of this book.

One of the key differences between 3G and 4G LTE is the adoption of orthogonal frequency division multiplexing (OFDM) as the modulation technology for LTE. As Figure 10-7 illustrates, OFDM divides each channel into multiple narrower sub-carriers. The spacing is chosen in such a way that sub-carriers are *orthogonal* to each other, which means the crosstalk is eliminated and inter-carrier guard bands are not required.

For downlink, LTE deploys orthogonal frequency division multiple access (OFDMA), which is a multiuser variant of OFDM. As shown in Figure 10-8, a subset of sub-carriers is assigned to different users so that multiple users can transfer data simultaneously.

For uplink, modified OFDMA has been used as part of single carrier frequency-division multiple access (SC-FDMA). The main objective of SC-FDMA is to use lower peak-to-average power ratio (PAPR) than the OFDMA technique. PAPR is the power level of the highest instantaneous power compared to the average power level. Having a large PAPR requires a linear transmit amplification circuit that can operate over a wide range of power, which is costly, inefficient, and not deployable for mobile phones or any form of battery-powered UE.

Portable UE requires a nearly constant power level when operating, and hence SC-FDMA uses a single carrier transmission schema to transmit symbols sequentially, as shown in Figure 10-9. The individual transmission is now shorter in time but wider in the frequency domain with higher data rate and lower PAPR.

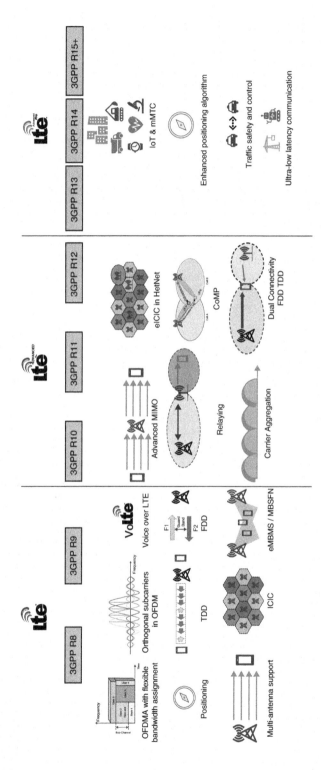

Figure 10-6 *LTE and Its Evolution*

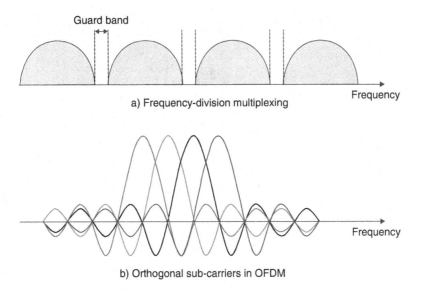

a) Frequency-division multiplexing

b) Orthogonal sub-carriers in OFDM

Figure 10-7 *FDM Versus OFDM*

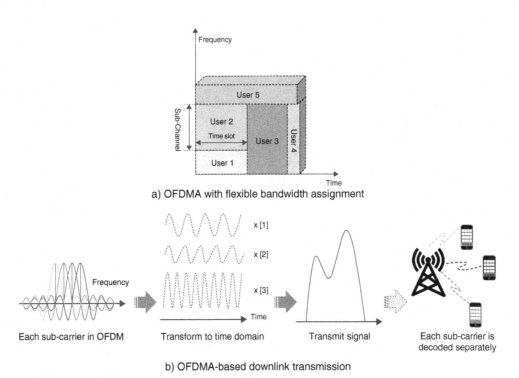

a) OFDMA with flexible bandwidth assignment

b) OFDMA-based downlink transmission

Figure 10-8 *OFDMA as a Multiuser Access Scheme for Downlink*

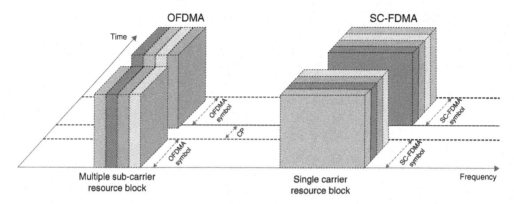

Figure 10-9   *OFDMA Versus SC-FDMA*

OFDM Synchronization

Both frequency accuracy and phase/time alignment are critical to ensure adequate performance of any OFDM-based system. OFDM and OFDMA techniques are extremely sensitive to the carrier signals being off frequency and having timing offset errors. Inaccurate compensation of frequency offset destroys orthogonality among sub-carriers and so produces inter-carrier interference (ICI) because it is the orthogonality that is keeping the sub-carriers separate.

To understand how the synchronization requirements arose, it is necessary to understand how time domains are maintained in the LTE frame structure as well as the duplex schema for uplink and downlink transmission.

OFDM Frame Structure

In OFDM transmission, several hundred sub-carriers are transmitted over the same radio link to the same receiver. Any corruption of the frequency-domain structure of the OFDM sub-carriers leads to the loss of inter-subcarrier orthogonality and results in interference between the sub-carriers (ICI).

OFDM systems can also suffer from what is called *time-dispersive channel propagation*, which simply means that any signal subject to multipath propagation will see the signal dispersed over time. When suffering multipath, some signals have a longer path than others and therefore take longer to propagate from transmitter to receiver. This means that the same information is received by the receiver at several different moments of time. These signals arriving over time can obviously interfere with each other, and this is known as inter-symbol interference (ISI).

To retain the properties and integrity of the OFDM signal and to prevent ISI and ICI due to time-dispersive channel propagation, a guard interval known as a *cyclic prefix* (CP) is allocated to each OFDM symbol. To better understand CP, consult the multipath scenario shown in the Figure 10-10.

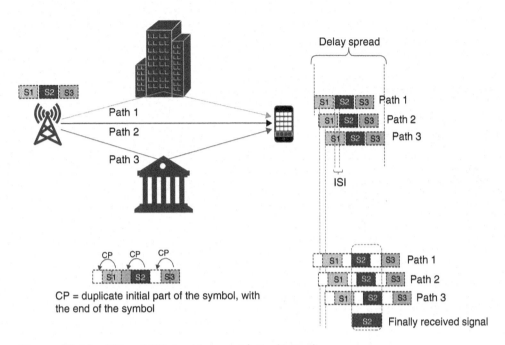

Figure 10-10 *ISI and ICI Avoidance with Cyclic Prefix*

The total delay between the first copy of the data received and the last copy of the data of the same signal is called the *delay spread*. If the length of the OFDM symbol that carries the useful data is smaller than the delay spread, there is interference between the symbols. But if the OFDM symbol size is increased, then the interference can be reduced. One technique to do this is to copy or duplicate the initial part of the symbol (the prefix) with the end of the symbol. Here, the CP length needs to be larger than the delay spread in the environment where the system is intended to operate.

In LTE, the synchronization of the BS (known as the evolved node B [eNB] in LTE terminology) and UE is achieved at the physical layer. As shown in Figure 10-11, the LTE radio frame has a length of 10 ms and each frame is divided into ten equal-sized subframes of 1 ms. Each subframe consists of two equal-sized slots of 0.5 ms each. With a 15-kHz sub-carrier spacing, the useful LTE symbol time to transmit data is 66.7 μs. However, the total symbol time is the useful symbol time plus the CP. Hence, each slot can have either normal CP that is 4.7 μs with seven useful symbols or extended CP that is 16.67 μs with six useful symbols to transmit.

The differing length of the CP is used to address different use cases. For example, the normal CP is used in the case of small-diameter cells in urban areas where multipath occurs over relatively small distances but with high data rates. The extended CP is used in rural areas with large-diameter cells using extended delay spreads but with lower data rates.

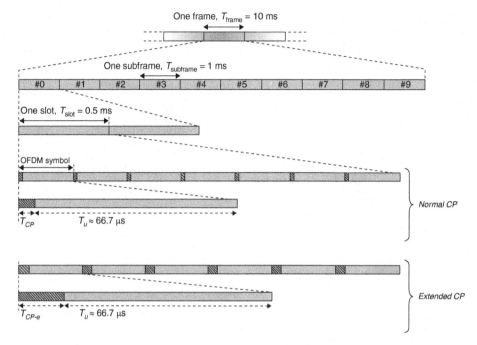

Figure 10-11 *OFDM Frame Structure*

Using the CP technique makes OFDM symbols immune to time dispersion and mitigates ISI or ICI issues. However, the selection of the appropriate CP design is sensitive to timing errors in the cell, and synchronization inaccuracy can move portions of the symbol data outside the CP coverage area and cause signal degradation.

LTE TDD Frame Structure

In the case of TDD, the UL and DL occur on the same carrier frequency but in separate time domains. In each LTE frame, some subframes are allocated for uplink, and some subframes are allocated for downlink. Switching between downlink and uplink is carried out using a special subframe. 3GPP defines different configuration sets; for example, one of the options is shown in Figure 10-12.

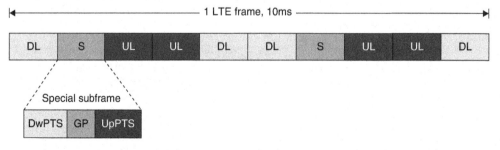

Figure 10-12 *Frame Structure of TDD Subframe*

The special subframe is split into three parts:

- Downlink pilot time slot (DwPTS)

- Guard period (GP)

- Uplink pilot time slot (UpPTS)

The DwPTS is part of the subframe and smaller in size, but used to carry downlink transmission data, whereas the UpPTS is very small and does not carry any transmission data. The GP is a duration where neither uplink nor downlink transmission occurs, and this period allows for the radio to switch between downlink and uplink.

The length of the GP depends on many factors. First, it should be sufficiently large so that the eNBs and the radio device get enough time to switch from downlink to uplink. Second, the GP must be long enough to ensure that there is no interference between uplink and downlink transmission.

To minimize interference, within a cluster of eNBs, the same subframe configuration is applied across all eNBs. Such a configuration ensures that all these eNBs are either transmitting or receiving at the same time. However, this simultaneous operation means that the eNBs must be accurately synchronized to avoid a UL transmission interfering with a DL transmission from an adjacent eNB. 3GPP (TS 36.133) has specified that the neighboring eNBs must be phase aligned to within 3 µs for smooth handover and to avoid any kind of interference between eNBs.

Cell Search and Synchronization

Cell search is a method for a device to acquire time and synchronization information (transmitted by every radio frame) when the UE initially attaches to the LTE system. When a device or UE attaches to the system, it synchronizes with the OFDM symbol, slot, subframe, half-frame, and radio frames using the synchronization signals.

Each eNB is differentiated by a unique identifier called the cell ID, which is not really unique and can be reused on multiple eNBs if they are placed far enough apart. Two synchronization signals are defined for this purpose:

- The primary synchronization signal (PSS)

- The secondary synchronization signal (SSS)

These two signals are used in combination to reduce the complexity of the cell search process.

The UE first decodes the PSS and SSS to determine the cell ID of the nearest eNB. The UE also acquires information on the carrier frequency, frame timing, CP length, and duplex mode. The UE can then request resources for uplink transmission and establish uplink synchronization with the new eNB. Because PSS and SSS signals are transmitted periodically, the UE can continually utilize them.

UE to BS Synchronization (TDD)

In the preceding sections, you saw that one of the factors influencing the length of the GP is the propagation delay between the UE and the eNB. Once the UE is connected, the eNB keeps estimating the propagation delay, called the timing advance offset, T_A. The offset needs to be calculated for each connected device to coordinate its transmission. A special subframe is used in TDD to adjust the GP to ensure a smooth switch between downlink and uplink according to the received T_A offset.

As the UE receives the value of T_A (T_A is in the range of 0–1282 for LTE) from the eNB, it is important that each device acquires the synchronization information before it starts data transmission. Each additional increment in the value of T_A (0.5208 µs) allows for approximately 156 meters extra round-trip distance between the UE and the eNB (0.5208 µs at 300,000 km/s is 156 m).

For 4G LTE, the value range of 0–1282 for T_A allows for a cell radius up to 100 km.

BS to BS Synchronization (TDD)

The time synchronization between adjacent eNBs is important to avoid cell interference. When a UE is moving from one eNB to another, the relative timing difference between subframe timing boundary becomes important. It is related to the time duration of the CP and the propagation delay between the cells. For smaller cell sizes, this is about 3 µs.

Consequently, in TDD deployment scenarios from small cells up to full-sized macro cells, the standards specify a mandatory phase alignment within 3 µs as the synchronization requirement. This phase synchronization accuracy is defined as applying at the eNB antenna connectors.

In the case of a small cell or home eNB (HeNB), there is a technique whereby it derives its timing from the signal transmitted by a neighboring synchronized eNB. This is referred to as *synchronization using network listening* and is shown in Figure 10-13. This has an advantage in those environments where obtaining a phase/time signal from another source might be difficult (such as a GNSS signal inside the home).

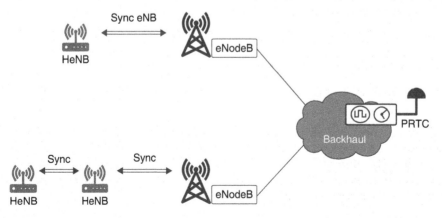

Figure 10-13 *Synchronization Using Network Listening*

The 3GPP 36.133 standard defines the synchronization requirements for wide area eNB and HeNB that are outlined in Table 10-3.

Table 10-3 *Time Synchronization Requirements for TDD Radios*

Wide Area eNB	Cell Radius	Requirement
Small cell	≤ 3 km	≤ 3 μs
Large cell	> 3 km	≤ 10 μs
Home eNB	**Propagation Distance**	**Requirement**
Small cell	≤ 500 m	≤ 3 μs
Large cell	> 500 m	≤ 1.33 + $T_{propagation}$ μs

$T_{propagation}$ is the propagation delay between the small cell and the eNB selected as the network listening synchronization source. If the small cell or HeNB obtains synchronization without using network listening, the small cell requirement applies.

If the coverage area of a cell overlaps with another cell of a different radius, then the cell phase synchronization accuracy corresponding to the larger of the two cell sizes applies to the overlapping cells.

Multi-Antenna Transmission

One of the fundamental technologies introduced with the adoption of OFDM or OFDMA is support for different multi-antenna transmission techniques. The performance of wireless communications is degraded by poor propagation—events such as signal fluctuations, interference, dispersion, fading, path loss, and so on. Multi-antenna systems are used to improve the efficiency and quality of the received signal.

Consider a radio where a data stream is transmitted from a single transmit antenna to a single receive antenna (known as single in, single out). This data stream may get lost or corrupted before the receiver can recover the data—due to any number of anomalies caused by poor conditions in the propagation of the signal. However, if the same data stream can be transmitted to multiple receive antennas over different paths, there is a better chance for the receiver to properly recover the signal.

The introduction of additional paths on a single frequency to improve the reliability of the overall system is known as *spatial* or *antenna diversity*. For a system with N_T transmitters and N_R receivers, the maximum number of diversity paths is $N_T \times N_R$— although the improvement diminishes with an increasing number of paths. Diversity can be introduced at either the receiver end (single input, multiple outputs), the transmitter end (multiple inputs, single output), or both ends (multiple input, multiple output—often called MIMO).

When this technique is used to divide the given data stream into multiple parallel sub-streams and transmit each on a different antenna, it is known as *spatial multiplexing*. With this technique, a high-rate signal is split into low-rate streams, and each stream is transmitted from different transmit antenna in the same frequency channel (see Figure 10-14). Receivers can separate these streams if received with sufficiently different spatial signatures. Note that with MIMO, the number of interfering signals also increases.

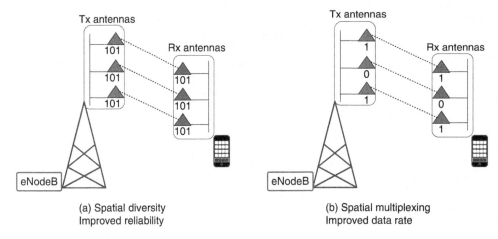

Figure 10-14   *MIMO: Spatial Diversity and Spatial Multiplexing*

In a MIMO system, it is necessary to discover the signal characteristics of each radio path between each combination of transmit and receive antennas. This requires an accurate phase/time alignment at each antenna to identify the required spatial alignment of the paths. 3GPP 36.104 and 3GPP 38.104 define the time alignment error (TAE) budget required for transmission diversity in LTE and 5G as shown in Table 10-4.

Table 10-4   *Time Synchronization Requirements for Transmission Diversity or MIMO*

E-UTRA (LTE)	Time Alignment Error
LTE MIMO or transmission (Tx) diversity transmissions, at each carrier frequency	Maximum 65 ns

Although this is a very low value for phase alignment, it is not a significant problem because it is a relative alignment—it applies only between the member of an antenna array connected to a single radio. Chapter 11, "5G Timing Solutions," provides more discussion on relative time errors and cluster-based timing.

Inter-cell Interference Coordination

At the network level, *interference* is caused when UEs in neighboring cells attempt to use the same resource at the same time. For example, as shown in Figure 10-15, at the edge of the cell, two UEs (UE$_A$ and UE$_B$) might experience interference because 1) both are using the same frequency (f_2) and high transmit power to communicate; and 2) they do not have any knowledge about the neighboring UE; and 3) the eNBs are independently scheduling their radio resources.

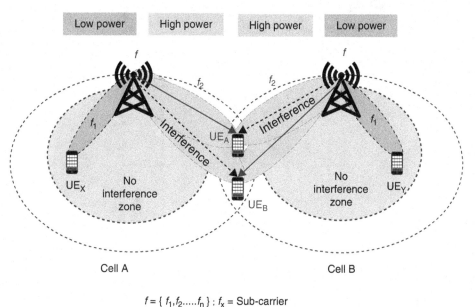

$f = \{ f_1, f_2 ..... f_n \}$; f_x = Sub-carrier

Figure 10-15   *Inter-cell Interference at the Cell Edge*

To mitigate such problems, 4G LTE and LTE-A adopted several technologies known as inter-cell interference coordination (ICIC) and enhanced inter-cell interference coordination (eICIC).

ICIC, defined in 3GPP Release 8, reduces inter-cell interference by using different frequency resources at the edge of the cell. eNBs that support this feature can generate interference information for each used frequency and exchange that information (frequency, transmit power, and so on) with neighboring eNBs over the X2 (inter-cell signaling) interface defined by 3GPP.

As shown in Figure 10-16, if the UEs (UE$_A$ and UE$_B$) at the edge of cell A are using the same frequency f_2, one of the cells, for example, cell B, will switch to a different frequency, f_3, to communicate with UE$_B$ at the edge of the cell while continuing to use f_1 or f_2 with low transmission power at the center of the cell to avoid interference.

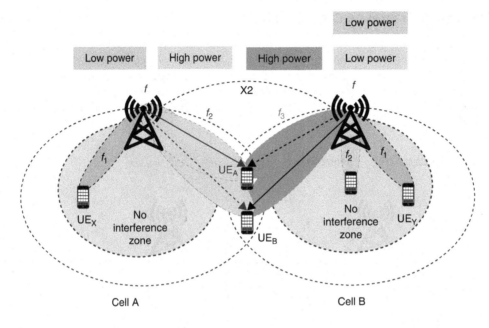

$f = \{ f_1, f_2.....f_n \} ; f_x =$ Sub-carrier

Figure 10-16   *Inter-cell Interference Coordination*

Enhanced Inter-cell Interference Coordination

Traditionally, cell sites were mostly homogenous, which means that each cell site was much the same as the rest (usually large, "macro" cells). So, *homogeneous networks* are ones where all the cell sites are of the same type and function. With more recent developments, there is a trend toward the *heterogeneous network* (sometimes called the *HetNet*), which has a mix of low-powered small cells and large macro cells.

The ICIC technique is used to manage the resource requirements, such as frequency (bandwidth) and power, at each eNB and the interactions between eNBs. It does this by allocating frequency resources to UEs using data channels, while other resource allocations (for example, power) are managed through the control channels.

Unlike data channels, which can use different frequency ranges, the control channel uses the entire carrier bandwidth. In homogeneous networks, this is not a big problem because there is not much difference in the transmit power values from adjacent eNBs, and hence no significant inter-channel interference is caused by the control channel.

In heterogenous deployments, the UE receives the same frequency signal from multiple cells, and it selects the signal from the cell with the highest power. However, selecting the cell that transmits at the highest power implies that the UE may often select the macro cell over the small cell.

But the macro cell signal will have a higher propagation delay and path loss to the attached UE compared to a nearby small cell. And this will obviously not be optimal for the network from an uplink coverage and capacity point of view (implementing small cells is pointless if every UE selects the macro cell instead).

Second, because in heterogenous networks the low-powered small cells are deployed along with macro cells in overlapping control channels, the ICIC technique is only partially effective because it cannot mitigate the control channel interference between the small cells and the macro cells.

eICIC, defined in 3GPP Release 10, is an advanced version of ICIC to support the heterogenous environment. Whereas ICIC uses different frequency ranges to control interference on the data channels, eICIC instructs the UEs at the edge of the cell to use different time ranges to avoid interference on the control channel. They do this by introducing the concept of the Almost Blank Subframe (ABS), which allows the UEs at the edge of the cell to use the same radio resources, but in different time ranges.

As shown in Figure 10-17, eICIC uses alternate timeslots for the overlapping cells, and for that to work there must be accurate phase/time synchronization between the macro eNB and small cells.

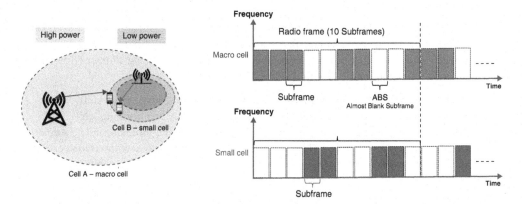

Figure 10-17 *eICIC with ABS*

In summary, ICIC handles interference using frequency domain partitioning, whereas eICIC provides its solution with time-domain partitioning. Although ICIC demands accurate frequency synchronization, eICIC in a heterogenous network needs tight and accurate phase alignment.

For eICIC to work, you must define the maximum absolute time alignment error between any pair of eNBs that are on the same frequency with overlapping coverage area, as outlined in Table 10-5.

Table 10-5   *Synchronization Requirements for ICIC and eICIC*

Application	Type of Coordination	Phase Accuracy	Reference
ICIC	Inter-cell interference coordination	—	—
eICIC	Enhanced inter-cell interference coordination	3 μs	3GPP TS 36.133

Coordinated Multipoint

While eICIC reduces the level of interference for the UEs at the edge of the cell, the throughput of the UE is still limited to the frequency and time partitioned resources available from a single eNB or BS. But what advantage could arise if the UE were able to operate with multiple eNBs at the same time?

Coordinated Multipoint (CoMP) transmission and reception, a technique defined in 3GPP Release 11, enables multiple eNBs to coordinate their transmission and reception to enhance the service quality for the UEs at the edge of the cell. CoMP improves coverage, cell-edge throughput, and system efficiency. 3GPP defines several types of CoMP as illustrated in Figure 10-18.

Coordinated scheduling (CS) and coordinated beamforming (CBF) are enhanced versions of ICIC where frequency resource partitioning happens dynamically, and changes are applied every time scheduling is performed. Coordinated scheduling allocates different frequency resources to UEs located at the edge of cells.

As shown on the left of Figure 10-18, UE_A is receiving frequency f_2 from eNB_A, while UE_B is receiving frequency f_3 from eNB_B. With coordinated beamforming, eNB_A and eNB_B can use the same frequency resource, f_2, as shown, and coordinate and schedule spatially separated resources to UEs at the cell edge, thereby avoiding interference. eNB_A could serve as master eNB for UE_A, and eNB_B could serve as the master eNB for UE_B.

Note   Use of the terms "master" and "slave" is ONLY in association with the official terminology used in industry specifications and standards, and in no way diminishes Pearson's commitment to promoting diversity, equity, and inclusion, and challenging, countering, and/or combating bias and stereotyping in the global population of the learners we serve.

The eNBs working together and cooperating with each other make up what is known as a *coordination set*. And these scheduling and beamforming decisions are communicated using channel site information (CSI) data shared between the eNBs over the X2 interface. These dynamic coordination techniques require the X2 connectivity between eNBs in the coordination set to be very low latency.

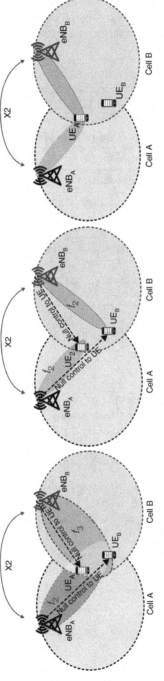

Figure 10-18 *Coordinated Scheduling, Beamforming, Joint Processing and Joint Transmission*

Joint processing (JP) is very similar to the CS. However, in joint processing, the user data is available to multiple eNBs, and a scheduling decision determines which eNB should handle transmission to the UE. With joint transmission (JT), multiple eNBs can send data to the edge UE at the same time and using the same frequency. This technique not only reduces interference but improves the received power at the UE and hence the signal quality and data throughput.

Because all the eNBs and/or small cells in the coordination set are sending and receiving radio waves to the UE at the same time, then accurate phase/time synchronization is mandatory for any CoMP scenario to work. Table 10-6 summarizes these requirements.

Table 10-6   *Time Synchronization Requirement for CoMP*

Application	Type of Coordination	Phase Accuracy	Reference
CoMP moderate	UL coordinated scheduling	≤ 3 μs	3GPP TS 36.104
	DL coordinated scheduling		3GPP TS 36.133
CoMP tight	DL coordinated beamforming	≤ 3 μs	3GPP TS 36.104
	DL noncoherent joint transmission		3GPP TS 36.133
	UL joint processing		
	UL selection combining		
	UL joint reception		

Carrier Aggregation

Carrier Aggregation (CA) increases the overall channel bandwidth by aggregating two or more carriers to transmit data in an aggregated spectrum. CA was introduced in 3GPP Release 10 and allowed up to five 20-MHz carriers to be aggregated—allowing for a total bandwidth up to 100 MHz. In Release 13, this was extended to support up to 32 carriers, allowing for the total bandwidth to rise to 640 MHz.

Note that the aggregated component carrier does not need to be contiguous in frequency domain (meaning each carrier can be from a different band in the available spectrum). This means that, as shown in Figure 10-19, different combinations of inter- and intra-band CA are possible, namely

- Intra-band CA
  - Using contiguous carrier components (frequencies adjacent to each other)
  - Using noncontiguous carrier components (nonadjacent frequencies)
- Inter-band CA using noncontiguous carrier components

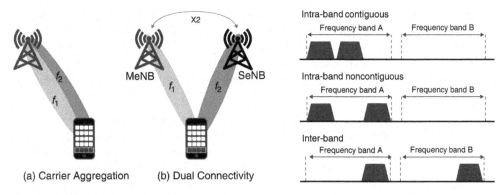

Figure 10-19 *Different Types of Carrier Aggregation*

When a pair of eNBs support CA, only a very small time interval is available to complete the scheduling for all the carrier components participating in sending the data frame. This means that a tightly coupled, low-latency connection is required between these eNBs to schedule and control their contribution. Additionally, they both need to be very closely aligned in time/phase because they are both cooperating to send related data to the UE at the same time.

Given these requirements, it is safe to assume that the participating antennas are either co-located or connected to their eNB with a direct link. Specifically, for intra-band contiguous CA, the synchronization requirement is so tight that one can assume that all the antennas in the CA scheme are installed at the same location. Carrier Aggregation synchronization requirements are outlined in Table 10-7.

Table 10-7 *Time Synchronization Requirement for Transmission Diversity and MIMO*

Application	Time Alignment Error	Reference
LTE intra-band noncontiguous CA with or without MIMO or Tx diversity	260 ns	TS 36.104
LTE inter-band CA with or without MIMO or Tx diversity	260 ns	
LTE intra-band contiguous CA with or without MIMO or Tx diversity	130 ns	
LTE MIMO or Tx diversity at each carrier frequency	65 ns	
Maximum transmission time difference in intra-band noncontiguous and inter-band CA	30.26 μs	

Dual Connectivity

3GPP Release 12 introduces the concept of Dual Connectivity (DC), which allows the use of two eNBs to communicate with the UE. Whereas CA allows the UE to simultaneously transmit and receive data on multiple carriers from a single eNB, DC extends it to allow the UE to transmit and receive data simultaneously from two cell groups or eNBs. These consist of a master eNB (MeNB) and a secondary eNB (SeNB), as shown in Figure 10-19. The MeNB and SeNB coordinate the scheduling between themselves using the X2 interface. This technique is widely used to improve overall performance when introducing small cells into heterogeneous networks.

However, when using DC, each eNB handles its own scheduling and maintains its own timing relations with the UE. Hence, the X2 interface latency and synchronization requirements are somewhat relaxed in contrast to the CA use cases. However, there can still be a requirement to coordinate between the two eNBs, depending on the type of operation.

Dual Connectivity operations are either synchronous or asynchronous. In the synchronous mode of operation, the maximum absolute timing mismatch between data transmitted by MeNB and SeNB is critical. For a given UE, the two eNBs must align a data transmission within a finite window offset from each other. You cannot have a situation where one eNB is sending current data and the other sending related data many seconds later. In an asynchronous mode, there is little or no relation between the two streams, so there are no strict limits tying the two separate transfers together. In technical terms we define this as follows:

- In synchronous operation, the maximum receive time difference (MRTD) and maximum transmit time difference (MTTD) from MeNB and SeNB should be within a certain threshold. These two metrics measure the relative difference for transmission and reception between each of the eNBs (primary and secondary).

- In asynchronous operation, the UE can perform the operation without a specific MRTD or MTTD.

The MRTD and MTTD synchronization requirement considers relative propagation delay, transmit time difference, and multipath delay propagation of radio signals from MeNB and SeNB antenna ports and requires phase accuracy as outline in Table 10-8.

Table 10-8 *Time Synchronization Requirement for LTE Dual Connectivity*

Application	MRTD/MTTD Error	Reference
LTE Synchronous Dual Connectivity	33 μs/32.47 μs	3GPP TS 36.133
LTE Asynchronous Dual Connectivity	NA	

Multimedia Broadcast Multicast Service (MBMS)

Another use case in 4G LTE that requires accurate phase/time synchronization between eNBs is the Multimedia Broadcast Multicast Service (MBMS). MBMS transmits the same content to multiple users located within a predefined MBMS service area, letting all the users subscribed to MBMS simultaneously receive the same multimedia content.

MBMS services are frequently used to provide (TV-like) broadcast multimedia services via the LTE network in, for example, venues such as sporting arenas.

Using the OFDM modulation, multiple cells can transmit truly identical and mutually time-aligned signals into the MBMS service area. In this case (as shown in Figure 10-20), the transmission received from multiple cells will appear as a single multipath transmission to the subscribed UE devices. This kind of transmission is referred to as an MBMS Single Frequency Network (MBSFN). Not only can an MBSFN area consist of multiple cells, but a single cell can also be part of multiple MBSFN areas.

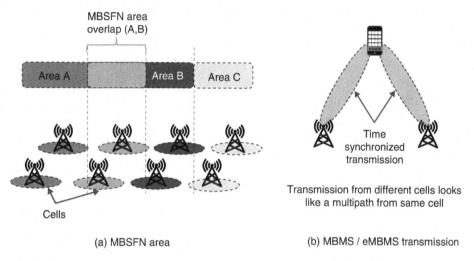

(a) MBSFN area   (b) MBMS / eMBMS transmission

Figure 10-20 *MBMS Area and Parallel Transmission of the Same Data Streams*

There is also a specification for evolved MBMS (eMBMS), which uses MBMS over LTE and is also referred to as *LTE Broadcast*. One of the major differences between MBMS and eMBMS is that eMBMS allows dynamic network resource allocation. For example, an operator can choose to dedicate its network capacity in a specific MBSFN area to broadcast a specific event, and once the event is over, it can reallocate those resources back to regular traffic.

With MBMS based on MBSFN, the timing alignment must be accurate to within few microseconds to ensure that the signals from multiple cells appear as signals from a single cell. Hence, all transmissions require a tight synchronization at the eNB or BS level; that accuracy requirement is outlined in Table 10-9.

Table 10-9 *Time Synchronization Requirement for MBSFN*

LTE MBSFN	Time Alignment Error	Reference
Cell phase synchronization accuracy at the BS antenna connectors	5 µs	TS 36.133

Positioning

Positioning refers to functionality in the RAN that allows it to determine the location of the UE. Most UEs now have an embedded GNSS receiver to provide tracking and positioning data and to enable location-based services. However, there are conditions where either the GNSS service is not available (indoors) or the UE itself does not have a GNSS receiver (such as small IOT sensors).

Hence, in 3GPP Release 9, positioning support is introduced into the RAN with the *LTE* Positioning Protocol (LPP). This technique can determine the location of the UE by taking measurements using special reference signals transmitted regularly from several cell sites.

The technique used to do this is known as the Observed Time Difference of Arrival (OTDOA). The UE measures the time difference between the arrival times of special reference signals sent from a minimum of three eNBs and reports these time differences to a specific device in the network.

The key entity in the mobile core network that handles positioning is the Evolved Serving Mobile Location Center (E-SMLC). The E-SMLC is responsible for providing assistance data and calculating the positions. The E-SMLC gets the LPP to request time-of-arrival data from the UE and supplies reference information to allow the UE to receive the signal. The E-SMLC uses the UE measurements and combines them with data on the location of eNBs to calculate the UE's position.

The eNBs participating in LPP must be time-synchronized very accurately and reliably to provide precise location accuracy. At the speed of light, each nanosecond of error in timing translates into about 0.3 m of error in position. Hence, for an accurate determination of position, the synchronization accuracy requirements are dependent on the location accuracy required.

For example, to achieve 40–60 m location accuracy, the maximum relative phase/time error needs to be less than around 200 ns between a minimum of three participating eNBs. Of course, both the geometry of the eNB around the UE and the number of participating signals are also factors that strongly influence the possible accuracy. The synchronization requirements for location accuracy or positioning are more stringent than those needed for communication.

No matter what the level of accuracy in location that is achievable, it is even harder to try to determine an accurate height (such as knowing on which level inside a building the UE is). Despite these stringent demands, the regulators are increasing the requirements for accuracy in location data, mainly to support the emergency services (these standards are known as E911 in North America and E112 in Europe).

Current time synchronization requirements are shown in Table 10-10, whereas more accurate and stringent location-based services are not yet defined and for further study.

Table 10-10 *Time Synchronization Requirements for Location-Based Services*

LTE	Time Alignment Error	Reference
Location-based services using OTDOA	200 ns (relative)	TR 37.857
(40 to 60 meter of location accuracy with minimum three base stations)		ITU-T G.8271 2020-03

Synchronization Requirements for LTE and LTE-A

The common names 4G LTE and LTE-A are overarching terms that cover a range of techniques, services, and features rather than any single technology. Not every feature will be deployed everywhere, and that leads to variations in any real-world deployments and different requirements for each case.

For example, small cells may be deployed to improve system throughput, increase cell density, and provide more user bandwidth at lower cost. Typically, they will only be deployed in appropriate locations, such as densely trafficked urban downtown areas. Whatever the feature or technology being adopted, without providing the type of synchronization required to support it, the desired objectives will not be achieved.

Table 10-11 summarizes the synchronization needs for several applications and use cases and indicates the impact of not meeting those requirements.

Table 10-11 *Synchronization Compliance for LTE Applications*

Application	Need for Compliance	Impact of Noncompliance
LTE-FDD	Call initiation	Call interference and dropped calls
	Timeslot alignment	Packet loss/collisions and spectral inefficiency
LTE-A MBSFN	Proper time alignment of video signal decoding from multiple BTS	Video broadcast interruption
LTE-A MIMO/ CoMP	Coordination of signals to/from multiple base stations	Poor signal quality at edge of cells, poor location-based service accuracy
LTE-A eICIC	Interference coordination	Spectral inefficiency and service degradation

For the basic LTE and LTE-A radio services, the frequency synchronization requirements remain the same as for the earlier generations of radio, namely 50 ppb at the air interface and 16 ppb at the BS or eNB interface. In addition to the frequency requirement, some other services have phase requirements because the operator applies specific technologies and techniques in the radio.

Table 10-12 summarizes the frequency and phase requirements for LTE.

Table 10-12 *Summary of Time Synchronization Requirements for LTE Radio Services*

Application	Frequency Accuracy	Phase Accuracy	Reference
LTE-FDD	±50 pbb	—	—
LTE-TDD Wide area BS	±50 pbb	3 µs for small cell (< 3 km cell radius) 10 µs for large cell (> 3 km cell radius)	3GPP TS 36.133
LTE-A/LTE-A Pro	±50 pbb (wide area) ±100 pbb (local area) ±250 pbb (Home eNB)	Wide area 3 µs for small cell (< 3 km cell radius) 10 µs for large cell (> 3 km cell radius) Home eNB 3 µs for small cell (< 500 m cell radius) ±1.33 µs + air propagation delay for large cell (> 500 m cell radius)	3GPP TS 36.133 3GPP TR 36.922
LTE MBSFN	±50 pbb	5 µs	3GPP TS 36.133
LTE-TDD to CDMA 1xRTT and HRPD handovers	±50 pbb	10 µs (timing deviation) ±10 µs (up to 8 hours during holdover)	3GPP TS 36.133

However, for some of the techniques covered in the previous sections, there are specific phase synchronization requirements with differing levels of strictness based on the specific technique being implemented. For those use cases, the phase accuracy requirements at the eNB interface are captured in Table 10-13.

Table 10-13 *Summary of Phase Synchronization Requirements for LTE-A Applications*

LTE-Advanced	Type of Coordination	Phase Accuracy	Reference
eICIC	Enhanced Inter-cell interference coordination	≤ 3 μs	3GPP TS 36.133
CoMP moderate	UL coordinated scheduling	≤ 3 μs	3GPP TS 36.104
	DL coordinated scheduling		3GPP TS 36.133
CoMP tight	DL coordinated beamforming	≤ 3 μs	3GPP TS 36.104
	DL noncoherent joint transmission		3GPP TS 36.133
	UL joint processing		
	UL selection combining		
	UL joint reception		
MIMO	Tx diversity transmission at each carrier frequency	65 ns	3GPP TS 36.104
Location accuracy	LTE positioning using OTDOA for 40–60 m with minimum three base stations	200 ns	3GPP TR 37.857 ITU-T G.8271 2020-03
Carrier Aggregation and MIMO	LTE intra-band noncontiguous CA with or without MIMO or Tx diversity	260 ns	3GPP TS 36.104
	LTE inter-band CA with or without MIMO or Tx diversity	260 ns	
	LTE intra-band contiguous CA with or without MIMO or Tx diversity	130 ns	
	LTE MIMO or Tx diversity at each carrier frequency	65 ns	
	Maximum Transmission Time Difference in intra-band noncontiguous and inter-band CA	30.26 μs	
LTE Dual Connectivity	Synchronous Dual Connectivity	33 μs MRTD 32.47 μs MTTD	3GPP TS 36.133
	Asynchronous Dual Connectivity	—	

Evolution of the 5G Architecture

As each generation of mobile network has evolved, the technology has also progressed and changed, which eventually leads to having to make changes in the architecture. For example, 3G/UMTS was a very successful implementation mostly based upon the WCDMA technology, while 4G/LTE was driven by the adoption of OFDMA. However, 5G is driven more by the use case than the technology itself, which needs a change in thinking about the architecture.

The primary use cases can be classified into three major categories:

- **eMBB (enhanced mobile broadband) and fixed wireless access (FWA):** eMBB services are focused on broadband Internet access for the mobile user and are generally categorized as human-centric communications. Having ever higher-speed and more-reliable connectivity while in motion has been the main driver for every successive mobile generation.

    The demand for more bandwidth with better coverage is increasing, which has resulted in a broad range of use cases with different challenges. These range from the use of dense hotspots to long-range, wide-area coverage and using both together as a combined system—hotspots are used to provide high data rates and high capacity in high traffic areas, while wide-area coverage solutions provide seamless mobility. Each one demands different performance and capacity requirements.

    High-speed access in densely trafficked areas (stadiums, offices, malls, and so on), broadband connectivity everywhere (rural, suburban, highway, and so on), and high-speed transport (trains, planes, and so on) are the key market categories of eMBB services.

- **mMTC (massive machine type communications):** This is purely a machine-centric communication segment, whereby a massive number of connected devices communicate over a very low-bandwidth connection. These connections are neither delay sensitive nor latency sensitive. It is likely that such machines or sensors are autonomous, deployed at difficult-to-reach locations and required to support a very long battery life.

    The applications that can be supported by mMTC are being dreamed up and deployed all the time, so the possible uses for this technology are limitless. Examples include smart metering, environmental (air and water quality) monitoring, fleet management, weather station reporting and forecasting, farming, smart cities (waste, traffic, and parking management), and just about everything else you can imagine. Any reason to support the collection of the measurements from a massive number of sensors will use mMTC services and technologies.

- **uRLLC (ultra-reliable and low-latency communication):** Sometimes also referred as critical machine-type communication (cMTC), these human- and machine-centric communication cases demand stringent low latency as well as high reliability and availability. The use cases for such requirements apply to industries such as aviation, healthcare, factory automation, smart grids, oil and gas, transportation, and many others.

LTE and its evolutionary developments are very mature and capable technologies; however, there are many emerging requirements that were not possible to address with LTE. Hence, 3GPP initiated a development of new radio access technology known as New Radio (NR). NR for 5G is equivalent to LTE for 4G or UMTS to describe 3G technology.

3GPP Release 15 defines the first phase of the 5G NR standard, which reuses many of the underlying structures and features of LTE, focusing on the tight integration of 4G LTE and 5G radio access technology (see Figure 10-21). However, being a totally new radio technology, NR is not required to maintain full backward compatibility but to address a different set of technical solutions.

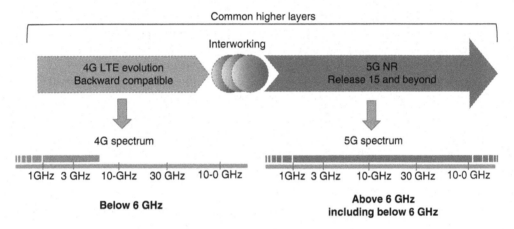

Figure 10-21 *Evolution of 4G LTE Along with 5G NR*

5G is defined in at least two phases, with phase 1 specified in Release 15 and phase 2 starting from Release 16. With Release 15, approved in mid-2018, 5G NR defines a set of specifications that delivers features like superior throughput, low latency, extreme reliability, device density, spectrum flexibility, heterogeneity, and energy efficiency to support existing and new sets of use cases.

Two deployment options are defined for 5G in Release 15:

- The *non-standalone* (NSA) architecture, where the 5G RAN and its New Radio interface is used in conjunction with the existing LTE Core Network, thus making the NR technology available without network replacement.

- The *standalone* (SA) architecture, where the NR is connected to the 5G Core network, which will allow the full set of 5G phase 1 (Release 15) services to be supported.

As part of phase 2, Release 16 and Release 17, new feature extensions will be covered to address forward-looking use cases and deployment scenarios across eMBB, mMTC, and uRLLC. More details on this topic are in the upcoming section "5G System Architecture."

At the time of writing, many operators are deploying 5G and NR networks in NSA mode to coexist with their existing LTE/LTE-A networks as a temporary step toward a full

5G deployment. In this configuration, only 4G services are supported, but this allows the features offered by the 5G New Radio (such as lower latency) to be available without a massive migration.

5G is an evolved architecture, and to understand the synchronization requirements for 5G, it is necessary to understand the basis of its architecture and requirements defined around the system design. The following sections discuss some of these aspects, after which the details of the 5G synchronization requirements are presented.

5G Spectrum

Access to the correct type of spectrum is one of the most important factors needed to ensure success for 5G. Lower frequency bands (for example, under 1 GHz) have good coverage but lower capacity to carry data. Meanwhile, higher frequency bands (from 24 GHz to 100 GHz) offer very high capacity but with limited coverage and poor penetration into structures. These extremely high-frequency bands have only small, millimeter-scale wavelengths, and so are known as mmWave.

There is a compromise between these two cases, using middle-frequency bands (between 1–6 GHz), that offers a mix of both coverage and bandwidth. To support the many and varied use cases, 5G requires a range of spectrum across lower, middle, and higher bands to deliver the desired coverage in combination with the high throughput.

5G targets new bands of spectrum, both above and below 6 GHz. Currently, there are two primary categories of frequency, known as FR1 and FR2, defined for 5G NR technology:

- **Frequency range 1 (FR1):** 410–7125 MHz

- **Frequency range 2 (FR2):** 24,250–52,600 MHz

Some bands of these frequency ranges cover licensed spectrum as well as unlicensed or even *lightly licensed* sections of the total spectrum. Lightly licensed includes parts of the spectrum, such as Citizens Broadband Radio Service (CBRS), that are underutilized and can be allocated dynamically to network operators on demand. In the United States, this auction of 3.5-GHz CBRS spectrum concluded in mid-2020, with Verizon and Dish spending the most on spectrum.

An earlier U.S. auction of mmWave bands (37, 39, and 47 GHz) had attracted many successful bidders, with Verizon, AT&T, and T-Mobile taking the lion's share. At the time of writing, the spectrum auction for what is called C-band (3.7–3.98 GHz) has just been announced, with Verizon, AT&T, and T-Mobile spending about 96% of the total auction bids. Other auctions of mid-band spectrum are to follow.

Spectrum licensing is a very complex and rapidly evolving landscape. A small number of examples from the recent past in the U.S. market gives a taste of the process. Whatever the market, it is critical for operators to secure a combination of bands to meet both coverage and capacity requirements across their heterogenous deployments, both within traditionally licensed bands and in combination with other models of spectrum licensing.

5G Frame Structure—Scalable OFDM Numerology

4G LTE supports carrier bandwidth up to 20 MHz with "fixed" 15-kHz spacing between sub-carriers, and these values were fixed for LTE. However, to support diverse spectrum bands and various heterogeneous deployment models, 5G introduces flexible spacing for sub-carriers known as scalable numerology or multi-numerology. An OFDM *numerology set* is a set of values consisting of the number of sub-carriers, the sub-carrier spacing, slot duration, and CP time duration values used in a specific deployment.

The preceding section on 4G shows that the CP is utilized to avoid ISI and ICI in OFDM radio deployments. OFDM defines spectrum efficiency by the number of sub-carriers that can be carried in a specific frequency range; the more sub-carriers, the more data that the device can transmit or receive. With the sub-6-GHz bands, a narrow sub-carrier spacing (for example, 15 kHz) is used with longer OFDM symbols, which allows a larger CP to be used. With a larger CP, there is less ISI, and the signal has more tolerance to fading.

When using high-frequency bands such as mmWave, however, the carrier frequency gets increasingly sensitive to any frequency drift and phase misalignment. A wider sub-carrier spacing (for example, 120 kHz) can be used to reduce the interference and phase issues, but this wider sub-carrier spacing reduces the available space for the CP.

Hence, with NR, the sub-carrier spacing is no longer fixed but is adjusted to different values for different frequency bands. There is a numerology factor, μ, that is used to scale the size of the sub-carrier spacing based on the frequency band (it uses $2^{\mu} \times 15$ kHz). As shown in Table 10-14, as the numerology, or value of μ, increases, the number of slots in a subframe increases, which increases the number of symbols sent in a given time, resulting in more bandwidth.

Table 10-14 *5G Numerology Structures and Corresponding Maximum Bandwidth*

μ	Frequency Range	Sub-carrier Spacing (kHz)	Cyclic Prefix/ Guard Space (µs)	OFDM Symbol Duration (µs)	OFDM Symbol with CP (µs)	Max BW (MHz)
0	FR-1	15	4.69	66.67	71.35	50
1		30	2.34	33.33	35.68	100
2		60	1.17 | 4.17	16.67	17.84	100
2	FR-2	60	1.17 | 4.17	16.67	17.84	200
3		120	0.59	8.33	8.91	400

CP length is based on the path delay information, and the BS and UE determine this information with synchronization signals (using the PSS and SSS frames). Hence, based on the path propagation delay details from each UE, 5G NR defines a CP for each transmission stream.

An operator can choose to use different numerology values simultaneously in each cell to support different 5G use cases—services can utilize different OFDM numerologies, multiplexed on the same frequency channel. The caveat is that mixing different numerologies on a carrier can cause interference with sub-carriers of another numerology. Synchronization can eliminate the major source of the interference by coordinating the frame timing. Therefore, the synchronization accuracy requirements are even more stringent with 5G frame structure and timeslot duration than they were with previous generations of radio.

5G System Architecture

In parallel to its work on aspects of the NR radio, 3GPP is working on the overall system architecture, including the 5G Core network (5GCN) and the 5G RAN. These components have the following responsibilities:

- The 5GCN is responsible for control functions like authentication and access, charging, end-to-end connection setup, mobility management, policy control, and many similar functions in the control plane of the mobile network. It is also responsible for the User Plane Function (UPF), although the core maintains a clear split between the user and control plane functions.

- The 5G RAN is responsible for all radio-related functionality, including scheduling, radio-resource handling, transmission, coding, multi-antenna schemes, and so on.

The following section details the basic features of the 5GCN. A detailed treatment of the architecture is beyond the scope of this book because it is unnecessary information for anybody wanting to deploy a timing solution for 5G mobile.

5GCN

The principles behind the 5G system architecture are based on the fundamental vision that it must co-exist with the existing LTE-A network and then evolve over time. When an operator chooses how to deploy their radio networks, they select technology for the mobility bands and capacity bands, which could result in them having different core networks. The 5GCN architecture needs to take account of this.

The 5GCN architecture captures two main deployment options, as illustrated in Figure 10-22:

- **Non-standalone (NSA) Architecture:** This architecture allows early adopters of the 5G RAN to utilize the existing LTE radios and the 4G EPC network. This option supports only 4G services, but characteristics of 5G NR can be leveraged (for example, low-latency applications).

  In brief, NSA allows dual connectivity between the 4G and 5G access networks and is best applied when the goal is to improve eMBB services. The deployment of NR can provide capacity enhancements to the existing 4G LTE network without major network investments.

- **Standalone (SA) Architecture:** In this architecture, the NR-capable device connects to a 5GCN. This allows the full set of 5G services, so it can be considered a fully mature 5G NR deployment.

Various NSA and SA deployment options are outlined as part of the 3GPP TR 38.8xx and TS 38.xxx series of recommendations. Figure 10-22 illustrates several of the possible options and shows the flow of control and user plane traffic in both the SA and NSA cases. In a later update, 3GPP dropped the option 6 and option 8 specifications, as it did not consider these 5G options to be viable.

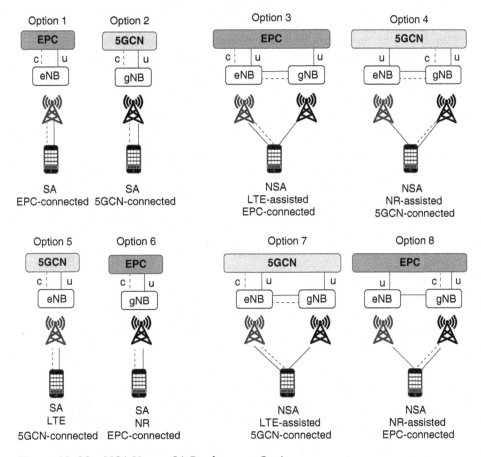

Figure 10-22 *NSA Versus SA Deployment Options*

The 5GCN builds upon the 4G EPC with three new areas of enhancement:

- **Service-based architecture (SBA):** The 5GCN is built on the principle that the 5G system will support a wide range of services that demand different characteristics and performance. Hence, the architecture is kept modular, customizable, and flexible to adopt future requirements and scale.

  For example, SBA uses API-based interfaces that allow services to be dynamically discovered, network entities to be added, and network services to be added without impacting any aspects of the system. The 5GCN is designed to use new technologies such as Network Function Virtualization (NFV) and Software-Defined Networks (SDN) to make it more agile and flexible to deploy.

■ **Network slicing:** One of the most critical aspects of 5G is to support multiple service types on a common physical network infrastructure. A *network slice* can be defined as a logical virtual network instance that meets an individual performance need for an identified user, service, or customer using the common network infrastructure. Resources can be dedicated to a single slice or shared between slices without compromising their associated requirements of data flow isolation, quality of service, bandwidth, latency, jitter, independent billing and charging, and so forth.

■ **Control and user plane separation (CUPS):** 3GPP defined CUPS in Release 14 as a foundation technology to address time-critical (low-latency) use cases. To deliver latency-sensitive applications successfully, it is necessary to bring the application functions to deliver the service much closer to the end user.

The CUPS architecture allows operators to flexibly place these functions wherever needed to support time-sensitive applications without affecting the functionality of the core network. The control plane functions can still be deployed in the centralized location, while the user plane functions (delivering the service) can be placed closer to the end user. With 5G adopting an SBA, it is simpler to achieve a cloud-native distributed architecture with CUPS.

Radio Access Network Evolution

To understand the 5G deployment requirements, you must start with a high-level understanding of the evolution of the RAN architecture and how the 5G RAN architecture is fundamentally designed to adopt not only CUPS, but the disaggregation and virtualization of the base station.

Note It is not helpful that there are different sets of nomenclature used when describing the different generations of the 3GPP standards, but unfortunately, the situation for the RAN is much worse. Depending on which standards documents you consult and what standards organization the RAN architecture is based upon, each component of the RAN can have one of three or four different names. To reduce the likely confusion that would result from using different terms depending on how a topic is discussed, this book tends heavily toward the 4G/5G 3GPP and O-RAN terminology, even if some organizations use completely different terms.

Mobile networks are continuously evolving to support much higher frequency bands because that allows much higher data rates. On the downside, the propagation properties of the radio waves suffer in the higher frequency bands; signals are easily absorbed/attenuated, so the coverage area falls away very quickly with increasing frequency. This means that for the operators to retain good coverage with higher frequency bands, many additional BS sites must be built—a process called *densification*.

Operators adopt either of the following two approaches to increase the coverage—the decision of which to use may depend on several factors such as population density:

- Densification by continuing to increase the number of macro sites.

- Densification with a layered approach where many low-powered small cells are deployed under the coverage areas of the macro layer. These sites tend to be small, indoor or under cover, and located in high-traffic areas.

In both cases, densification causes interference on both the current channel and adjacent channels, resulting in deterioration in the signal quality and hence adversely affecting the user experience.

To support heterogenous deployments and minimize issues with densification, the standard *distributed* eNB-based architecture evolved to support a *centralized* eNB architecture. The distributed architecture is known as the distributed RAN (DRAN), while the centralized version is called the centralized RAN (CRAN). Figure 10-23 shows the transition from the (traditional) DRAN to the CRAN architecture, with a comparison between the two architectures.

In conventional 4G LTE, the Radio Unit (RU) connects to the Baseband Unit (BBU) or eNB for baseband processing using the Common Public Radio Interface (CPRI) or the Open Base Station Architecture Initiative (OBSAI) interface. In this RAN architecture, all the components are situated out at the remote cell site with the baseband processing, radio functions, and antennas all co-located there.

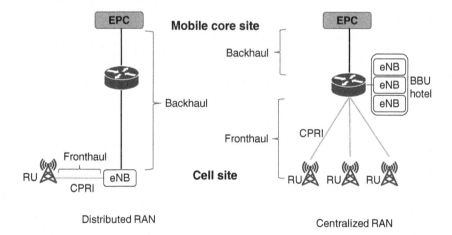

Figure 10-23 *Distributed RAN Versus Centralized RAN Architecture in LTE*

In the DRAN architecture, the CPRI network is local to the cell site and the BBU/eNB connects directly to the mobile core over the backhaul network, as illustrated in Figure 10-23.

However, using dark fiber or passive optical networks, CPRI can be carried over an extended distance (tens of km), which allows the baseband functions of a group of cell sites to be pooled into what is known as a *baseband hotel* (BBH) while the radio functions (RF) remain at the remote cell site. In this CRAN architecture, the CPRI interface is extended up to the centralized BBH location, and this extended CPRI network is referred to as the *fronthaul network.*

CPRI, being a synchronous protocol, requires a frequency synchronization signal. But the RAN also has phase/time synchronization requirements to support the advanced radio techniques covered in the preceding sections. The advantage to using CPRI is that it natively supports transfer of synchronization over its interface, so CPRI carries the necessary phase/time information. This means that the timing signal needs only to be delivered to the BBU, and CPRI looks after the distribution of that signal to the RU.

So, it is important for the network engineer to understand that the phase/time signal needs to be delivered only to the BBU/eNB, wherever it is located. From that point onward, CPRI looks after the synchronization between the BBU/eNB and the RU. The different timing points for phase alignment are shown in Figure 10-24.

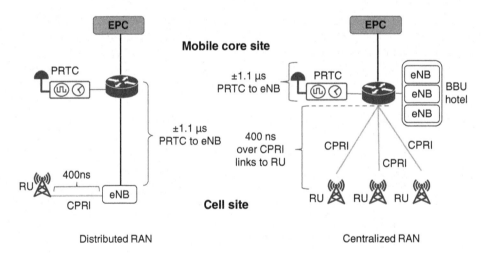

Figure 10-24   *Time Synchronization Architecture with CPRI-Based Fronthaul*

As shown in Figure 10-24, the network phase requirement of ±1.1 μs is to be delivered to the BBU/eNB. The other 400 ns of budget to meet the overall radio requirement of ±1.5 μs is consumed in the radio equipment and the CPRI network.

By centralizing the processing, the CRAN architecture can better utilize advanced techniques like eICIC and CoMP to improve the efficiencies and quality of the available radio resources. This reduction of the latency between the BBUs/eNBs allows closer radio coordination, supports improved spectrum utilization, and reduces cost by optimizing the amount of equipment and improving operational efficiency.

CPRI over the Packet Network

CPRI can be thought of as digitized radio signals because it represents the raw output after all the *baseband* radio processing has taken place (the same in the receive direction, CPRI is carrying raw digitized radio signals before radio processing in the BBU).

There are three important characteristics of CPRI traffic that need to be explained because they impact directly on 5G:

- It is a TDM-like transmission, meaning that it is always communicating and that it must transmit at specific time intervals. It can be thought of as being like an E1/T1 circuit, and like E1/T1, CPRI circuits need frequency synchronization.

- CPRI is defined by a consortium consisting of a small number of RAN vendors. Although the specification is public, the actual implementation of the upper layers and the contents of the messages are mostly proprietary. This means that there is no opportunity for multi-vendor interoperability in the traditional fronthaul even when every device is using CPRI.

- As it basically represents the raw form of the radio signals, CPRI is quite an inefficient way to transmit high-data-rate traffic. One analogy is that it is like sending photos of an email printout rather than the much smaller ASCII (or Unicode) data making up the message. With the increasing focus of 5G on much higher data rates to support eMBB, CPRI is showing its limitations and the amount of bandwidth it consumes is increasing more quickly than the increase in the user data traffic—it has simply reached the end of its scalability.

Another shortcoming is that because CPRI was traditionally designed to connect the RU with the BBU, it is fundamentally a point-to-point, TDM-style synchronous protocol. For this reason, there is very little flexibility in CPRI to build complex networks and adopt efficiencies like multiplexing (aggregating lower-bandwidth signals together into a single higher speed link). Other systems, such as those based on packet technology, do this very easily.

In summary, the CPRI protocol is not efficient enough to support 5G bandwidth demands or flexible enough to implement a scalable RAN architecture. So, the CPRI approach has to either evolve or be replaced.

Two evolutions are underway that resolve these shortcomings with CPRI in the 5G RAN:

- The separation of functions inside the RAN is being changed so that not as much data needs to be carried over the fronthaul network (see the upcoming section "RAN Functional Splits").

- CPRI is being migrated from a TDM-based to a packet-based protocol (such as Ethernet) to gain efficiencies and allow statistical multiplexing and increased flexibility. The participants in the CPRI consortium have standardized this evolution and named it enhanced CPRI (eCPRI).

On the other hand, many industry participants and standards development organizations are proposing other (packet-based or optical) technologies to replace CPRI, which will have the additional advantage that it is open and interoperable between vendors. One of those is the O-RAN Alliance, but there are others (see Chapter 4).

Beyond bandwidth inefficiency, CPRI also has a very strict delay budget. In practice, this need for a low-latency interconnect requires that for CRAN every RU component must be within a certain radius from the BBU. Even moving away from CPRI to a packet-based system like eCPRI, this requirement will still be there—there are still latency requirements between the BBU/DU and the RU in the 5G packet-based RAN.

5G RAN

The 5G RAN architecture includes measures to ensure backward compatibility, enhanced spectrum efficiency, greater energy efficiency, and reduced cost of deployment. Further, to support the evolution toward the 5GCN architecture, the 5G RAN is designed to support both SA and NSA deployments.

Figure 10-25 shows how elements of both an LTE and a 5G RAN architecture work together with either a 4G EPC or a 5GCN. Again, notice that the terminology is quite different between the two architectures for both the control plane (c) and user plane (u) connections.

In the LTE architecture, the eNB connects to the mobility management entity (MME) to terminate the control plane and to the serving gateway (SGW) for the user plane. Similarly, in the 5G NR architecture, the 5G version of the eNB (the gNB) terminates the control plane interface at the Access and Mobility Management Function (AMF) and user plane at the User Plane Function (UPF).

In the LTE assisted architecture, the base station consists of the following:

- The eNB is a node that supports LTE UE handsets and devices and connects the LTE user plane and control plane protocols to the EPC.

- The en-gNB is an NR gNB that connects over the 4G RAN to the EPC.

In the 5G RAN, the base station has these possible modes:

- The gNB is a node that supports NR UE handsets and devices and connects the NR user plane and control plane protocols to the 5GCN.

- An ng-eNB, or next generation eNB, is a node that supports LTE devices and connects the LTE user plane and control plane protocols to the 5GCN.

Like the LTE eNB, the 5G gNB is responsible for all radio-related functions: for example, radio resource management, admission control, connection establishment, security functions, routing the user-plane and control-plane traffic to the 5GCN, QoS flow management, and so on.

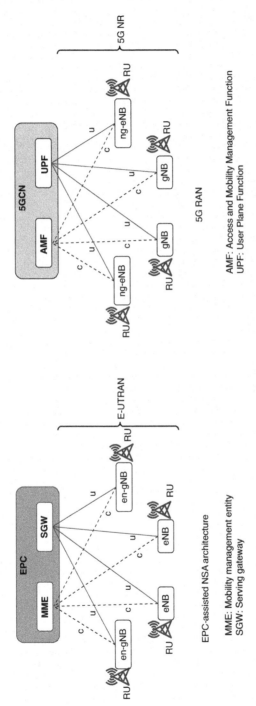

EPC-assisted NSA architecture

MME: Mobility management entity
SGW: Serving gateway

AMF: Access and Mobility Management Function
UPF: User Plane Function

Figure 10-25   *5G RAN Architecture*

In LTE, the eNB-to-eNB connection is handled by the X2 interface and is used for scheduling and coordination. Similarly, in 5G the gNBs coordinate with each other over an Xn interface (similar to the X2 interface in LTE) to support mobility, radio resource coordination, Dual Connectivity, and so on.

The 5G RAN is designed based on the principle of disaggregating the RAN and transport functions and seeks to unlock the benefits of centralization and virtualization techniques in its deployment. For this reason, it is very important to understand that the 5G RAN defines gNB as a logical function and not a single physical implementation.

The 5G RAN supports a functional split between the levels of radio processing to allow the concentration of a portion of the RAN processing into a centralized architecture. This gives operators the option to split the RAN function into two parts, the Central Unit (gNB CU) and Distributed Unit (gNB DU), with an F1 interface defined between them. See Figure 10-26 for an illustration of this architecture.

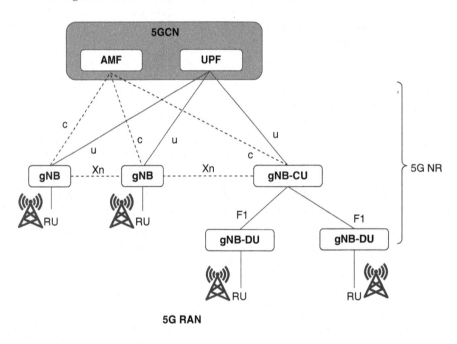

AMF: Access and Mobility Management Function
UPF: User Plane Function

Figure 10-26   *5GCN and RAN Components*

In theory, this F1 is an open interface that introduces the possibility of interconnection between a CU and a DU from different manufacturers. This split of the CU and DU means that these functions can be hosted in different locations, as shown in Figure 10-27. The transport network that connects DU and CU over the F1 interface is called the *midhaul network*.

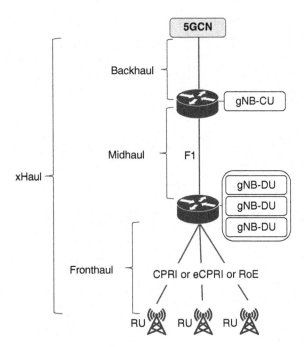

Figure 10-27   *5G Disaggregated Cloud RAN*

Having the gNB split into separate functions for the DU and CU brings flexibility in hardware and software implementation, scale, and cost. Adding in a definition for an open architecture and open interfaces allows for multi-vendor deployments and interoperability in the 5G RAN. Distributing the processing allows closer coordination, better load management, and higher spectrum utilization by selectively offloading low-latency functions closer to the user.

The following section discusses the gNB functional splits that allows this to happen.

RAN Functional Splits

With CPRI, the disadvantages of continuing with that form of functional split is clear, especially in those cases where massive bandwidth is a requirement (such as the eMBB use cases). However, what if the fronthaul network did not carry such low-level radio traffic, but was transporting data in an interim, semi-processed form?

For this reason, the radio processing has been divided into eight distinct steps, which are possible points where the centralized functions can be separated from the distributed functions. Option 8, as used by CPRI, is the lowest point on this pyramid. This means that just about all the processing to generate the radio signal is done at the centralized DU and only the very lowest-level RF functions are decentralized out to the cell site. But that means CPRI needs to carry a lot of data between the two sites.

Figure 10-28 provides an overview of the different possible split options for a disaggregated 5G RAN architecture. In both upstream and downstream directions, the radio signals go through a series of signal processing blocks; and the interfaces between these blocks are potential split options for 4G as well as 5G RAN. The functions on the left up to the point of the split are centralized, and the functions to the right of the split are decentralized to the radio site. CPRI option 8 is on the far right.

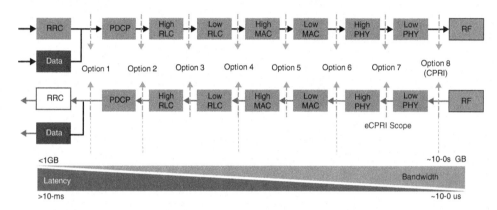

Figure 10-28   *Functional Split Options*

Options with the lower numbers (1, 2) are known as *high-level splits*, whereas those with large numbers (7, 8) are known as *low-level splits*. When RAN engineers talk of low-level splits, they mean the split is lower down the processing stack, but the option number is higher. Conversely, a high-level split has a low option number. With each split option, the latency requirements change, and the amount of bandwidth required to carry a specific amount of user traffic also changes.

Selecting a low-level split option (toward the right) increases the bandwidth (and thus the inefficiency) and tightens the latency requirements. But the high level of centralization with these low-level splits allows much tighter coordination between radios at the BBU/DU, improving the quality and spectrum efficiency using methods seen in the preceding sections.

With high-level split options (toward the left), the bandwidth requirement scales more in line with user traffic and allows statistical multiplexing advantages. Being able to deploy high-layer splits (for example, option 2) relaxes the latency and bandwidth requirements. Therefore, option 2 is a commonly chosen split level for those use cases that require high bandwidth.

Controlling multiple RUs with a centralized DU deployment allows load-balancing between different RUs, real-time interference management, CoMP with higher spectrum gains, seamless mobility, and resource optimization.

Because there are different functional split options, the operator needs to evaluate which of the valid splits are suitable for their deployment requirements and at which locations. For more information, 3GPP TR 38.801 captures the functional split requirements and details the advantages and disadvantages of each.

5G Transport Architecture

The evolution in the 5G RAN toward disaggregation and using functional splits segments the architecture into three main parts: the Central Unit (CU), the Distributed Unit (DU), and the (remote) Radio Unit (RU). Depending on the split of the functions, the transport network needs to support different characteristics in the areas of bandwidth, jitter, latency, and time synchronization.

Depending on location of the DU function, there are multiple deployment scenarios for 5G NR as part of a CRAN or DRAN architecture. As already described in the preceding section, the interface between the RU and the DU is known as the fronthaul network, the interface between the DU and the CU (the F1 interface) is called the midhaul network, and the interface between the CU and 5GCN core is the backhaul network. Figure 10-29 illustrates the basic four scenarios:

- **Scenario 1: RU and BBU/eNB/gNB (DU + CU) at the cell site location:** This is an implementation of a fully distributed DRAN architecture and is a very common deployment for 4G LTE. In this case, the cell site is connected to the core network over a backhaul network. These deployments are very common, especially for macro and rural sites.

- **Scenario 2: Centralized BBU (DU + CU) with RU radio functions at the cell site:** In this classic CRAN architecture, the RU connects to the BBU at a centralized BBH over a fronthaul network (normally using CPRI). This deployment model is already quite common for 4G LTE deployments in some parts of the world (for example, East Asia) and especially for denser urban areas with higher traffic loads.

- **Scenario 3: The RU and DU are integrated at the cell site but the CU is centralized:** In this CRAN architecture, the RU and DU are co-located and integrated at the cell site and any fronthaul network is local to the site, but the higher-layer functions of the CU are centralized and connected to the DU/RU over a midhaul network.

- **Scenario 4: The RU, DU, and CU are all in different locations:** This is a truly disaggregated CRAN architecture where the RU, the DU, and the CU functions are split over fronthaul, midhaul, and backhaul networks.

With the introduction of functional splits, there needs to be much more flexibility concerning the types of data being carried through the fronthaul (and midhaul) networks. This change in focus away from a single-use system like CPRI is allowing operators to build packet-based fronthaul networks, using technology such as Ethernet. For this to succeed, CPRI needed to be redefined to use packet-based transport instead of its fixed TDM-like transport.

Currently, there are two main standards that defines the packetization of CPRI:

- Radio over Ethernet (RoE) defines encapsulation and mapping of radio protocols over Ethernet frames. RoE is defined by the IEEE 1914.1 and 1914.3 standards.

- eCPRI provides an Ethernet- or IP-based fronthaul protocol definition between the RU and DU with support for deployment of functional splits. Note the terminology: eCPRI refers to DU functions as the eCPRI Radio Equipment Control (eREC) and refers to the RU functions as eCPRI Radio Equipment (eRE).

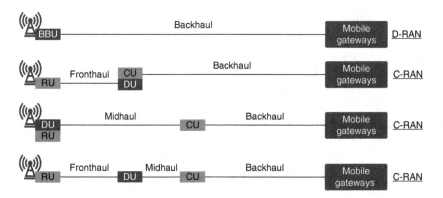

Figure 10-29 *5G NR CRAN and DRAN Architecture*

Table 10-15 outlines the difference between the CPRI, eCPRI, and RoE protocols.

Table 10-15   *CPRI Versus eCPRI Versus RoE*

CPRI	eCPRI	RoE
TDM-based design, synchronous protocol	Packet-based asynchronous protocol	Ethernet-based asynchronous protocol
Constant data rates, no packet delay variation	Variable data rates, sensitive to packet data variation	Variable data rates, sensitive to packet data variation
-	Not compatible with CPRI	CPRI compatible modes: CPRI structure agnostic and CPRI structure aware
Heavy header, high bandwidth	Bandwidth optimized, statistical multiplexing benefits	Bandwidth optimized, statistical multiplexing
Point-to-point protocol	Point-to-point and point-to-multipoint and ring topology support	Point-to-point and point-to-multipoint and ring topology support
Proprietary in nature	Vendor-defined fields for proprietary definition	Open standards-based definition
No L3 definition, no support for Ethernet or IP	Ethernet- or UDP/IP-based protocol	Ethernet-based protocol

For more details, please refer to CPRI specification 7.0, eCPRI specification 2.0, eCPRI Transport Network specification 1.2 and IEEE 1914.3 specification documents.

Both eCPRI and RoE also recognize many of the elements in IEEE 802.1CM-2018, "Time-Sensitive Networking for Fronthaul" (see Chapter 4). As it says in the introduction to 802.1CM:

> This standard defines profiles that select features, options, configurations, defaults, protocols and procedures of bridges, stations, and LANs that are necessary to build networks that are capable of transporting fronthaul streams, which are time-sensitive.

Both eCPRI and the 802.1CM standard include important recommendations for the Ethernet-based fronthaul network—including QoS, latency, jitter, and frame loss ratio (FLR) requirements between the RU and the DU. Additionally, a packet-based fronthaul network must be designed to keep the worst-case network delay below the end-to-end latency requirements. Please refer to the eCPRI Transport Network specification 1.2, eCPRI specification v2.0, and IEEE 802.1CM standards for more details on CPRI and eCPRI.

In addition to eCPRI and RoE, the O-RAN Alliance is also working on fronthaul specifications. The O-RAN Alliance uses the eCPRI framework and defines specifications for an open packet–based fronthaul that supports a multi-vendor environment for 5G RAN. These O-RAN fronthaul specifications support both 5G NR and LTE RAN. They define control, user, and synchronization plane specifications as well as detailed signal formats and messages for multi-vendor RAN operation over the management plane. Please refer to ORAN specifications for more details, especially the fronthaul Control, User and Synchronization Plane Specification (currently release 5.0).

The most important thing to understand is that because CPRI has changed into a packet-based implementation (either eCPRI, O-RAN, or something else), the formerly synchronous protocol between the BBU and the radio (or between the DU and RU) has disappeared and been replaced by a nondeterministic packet stream. The result is that the synchronization requirement is no longer being "handled for you" by CPRI. The fronthaul network must now carry the phase/time and frequency synchronization all the way down to the RU.

In general, the requirement for synchronization applies only to the DU and the RU (they must be synchronized with each other—to a level that depends on the implementation) and *not* to the CU (beyond normal operational needs such as timestamps in log files). However, the synchronization requirements demanded by the advanced radio techniques (see the following section "5G New Radio Synchronization Requirements") must be delivered all the way to the RU because the requirements are specified as being delivered to the antenna (rather than the BBU when using CPRI).

And for that, the network engineer needs to rely on the familiar tools of PTP and SyncE (see Chapter 11), not only in the backhaul network, but now also in the midhaul and fronthaul networks.

Network Slicing

The combination of flexible radio technology splits, the categorization of different use cases, and the CUPS architecture allows the 5G RAN to adopt what is known as *network slicing*. Network slicing allows an operator to offer different categories of services over the shared physical network infrastructure. As shown in Figure 10-30, with network slicing, operators can deploy applications at different locations depending on their latency, jitter, and synchronization requirements and adopt the necessary RAN split to support the bandwidth.

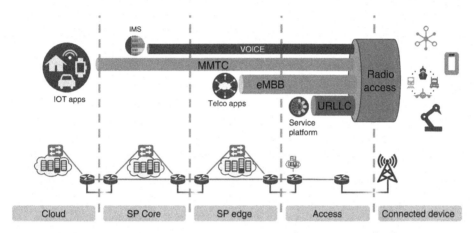

Figure 10-30   *Network Slicing*

Network slices can be a dedicated resource or a shared resource or can even span across multiple parts of the network or across multiple operators. The idea is that the operators will deploy a *slice type* that satisfies the requirements of multiple traffic flows that align to the same characteristics.

A good way to think about network slicing is that it is a form of QoS that ensures that the required conditions and resources are available and are guaranteed end-to-end for the use of any application that needs to use it. The full timing support profile, G.8275.1, is recommended with network slicing to avoid complexity and maintain accurate timing across all categories of services. However, various deployment use cases are evolving with 5G slicing and are currently still open for further study.

5G New Radio Synchronization Requirements

The preceding sections discussed synchronization use cases and the requirements for 4G LTE and LTE-A, and 5G NR adopts many of the same use cases. But 5G uses more flexible spectrum and deploys it across a wider range of heterogenous cell sites. In addition to the basic needs of LTE, 5G also adds deployment scenarios and radio techniques that demand stringent synchronization.

For LTE, the TS 36.104 and TS 36.133 3GPP technical standards represent the key documents that describe base station transmission and reception requirements. For 5G, the equivalent documents are TS 38.104 and TS 38.133. For numerous use cases, these documents define requirements for the maximum TAE that is allowed at the air interface.

The RAN standards development organizations, such as eCPRI, IEEE 802.1CM, and O-RAN, reflect these requirements and categorize them into four unique categories, A+, A, B, and C, as summarized in Table 10-16. Note that although some values are not specified to fall into a particular category, they can still have a timing requirement and are identified as "Not specified in 3GPP."

Table 10-16   *Timing Accuracy Categories*

3GPP Defined 5G NR Feature	RAN	
	LTE	**NR**
MIMO or Tx diversity transmission	Category A+	Category A+
Intra-band contiguous Carrier Aggregation	Category A	BS Type 1: Category B
		BS Type 2: Category A
Intra-band noncontiguous Carrier Aggregation	Category B	Category C
Inter-band Carrier Aggregation	Category B	Category C
TDD	Category C	Category C
Dual Connectivity	Category C	Category C
CoMP	Not specified in 3GPP	Not ready in 3GPP
Supplementary uplink	Not applicable for LTE	Not ready in 3GPP
In-band spectrum sharing	Not ready in 3GPP	Not ready in 3GPP
Positioning	Not specified in 3GPP	Not ready in 3GPP
MBSFN	Not specified in 3GPP	Not ready in 3GPP

These requirements, as illustrated in Figure 10-31, are further presented as relative and absolute time error (TE) between the various components of the architecture. The absolute TE is defined against a commonly agreed reference time (UTC) traceable to a PRTC—the previously mentioned ±1.1 μs at the interface to the radio. Alternatively, relative TE is the absolute value of the maximum allowable TE between any two RU devices.

Figure 10-31 is used in the following discussion to explain the timing requirements outlined in Table 10-17.

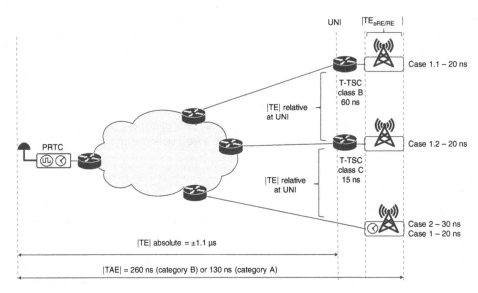

Figure 10-31 *Time Accuracy Definition*

The user-network interface (UNI) represents the handoff point between the transport network and the radio device (the last interface before it connects to the radio). The term $|TE_{RE}|$ (from eCPRI) represents the maximum absolute TE in the radio equipment (RU)—see Figure 10-31. Finally, the TAE represents the maximum allowable time error between any two antenna ports (which always also includes the 3 μs requirement from 3GPP).

Table 10-17 covers the TE requirements for these different categories of timing.

Here are the definitions for the different deployment cases:

- **Case 1:** The Telecom Time Slave Clock (T-TSC) is integrated inside the radio equipment, meaning that the final PTP hop ends inside the radio. Case 1 includes two subcases:

  - **Case 1.1:** The performance of an integrated T-TSC is assumed to be as per ITU-T G.8273.2 Class B specification, including budget of cTE and dTE, which is 60 ns.

  - **Case 1.2:** An integrated T-TSC is assumed to perform at total maximum absolute TE of 15 ns; could be considered as per ITU-T G.8273.2 Class C T-TSC; and as defined in eCPRI Transport Network V1.2.

- **Case 2:** The T-TSC is not integrated inside the eRE or RE but is deployed in a device (like a cell-site router) sitting in front of the radio. The timing signals are delivered to the RE via a physical interface such as the 1 pulse per second (1PPS).

Table 10-17 *Timing Accuracy Requirements*

Category	Time Error Requirements at UNI |TE|			3GPP TAE Requirements at Antenna Ports
	Case 1		Case 2	
	Case 1.1	Case 1.2		
A+	—	—	20 ns relative	65 ns
A	—	60 ns relative	70 ns relative	130 ns
B	100 ns relative	190 ns relative	200 ns relative	260 ns
C	1100 ns absolute		1100 ns absolute	3 µs

One important point to note is that for the Category A+ requirement covering the MIMO and Tx diversity needs, the TAE number is internal to the equipment specification and does not impose a requirement on the transport network. The requirement is still there, but it is between two components "inside the box" and therefore is not a concern for the network engineer.

The limit for the absolute time error internal to the RE, or $|TE_{RE}|$, depends on the case and the category, as shown in Table 10-18.

Table 10-18 *Absolute Value of the Internal Time Error, $|TE_{eRE/RE}|$, for the Radio Equipment*

	A+	A	B	C
Case 1.1 and 1.2	N/A	20 ns	20 ns	20 ns
Case 2	22.5 ns	30 ns	30 ns	30 ns

Be careful to understand the difference in implementation between the relative and absolute timing requirements. When a cell site is required to be within 3 µs of any other radio, this is implemented by ensuring that every radio is within ±1.5 µs of a common reference time (UTC). This is an absolute requirement because every radio is measured against a fixed, external yardstick.

However, when there is a relative requirement between two radios, then it simply means that radio 1 and radio 2, cooperating to deliver a service to their connected UEs, need to agree with each other to within the relative time limit. It does not mean that a radio on one side of the country needs to be aligned within that same relative time limit with every other radio in the network or some external measure such as UTC.

As these two radios are close to each other, both of their timing signals probably come from a common master source that is relatively close by—as shown in Figure 10-32. Mostly, these devices are connected to the same segment of the fronthaul or midhaul network, so getting them within a relative time budget of 260 ns or even 130 ns of each other is much easier to achieve than trying to synchronize the whole network to this required level of accuracy. There is further discussion on this topic in the section "Relative Versus Absolute Timing" in Chapter 11.

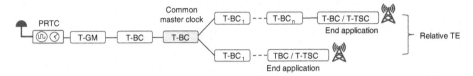

Figure 10-32   *Relative Time Budget with Respect to Common T-BC*

Remember that the relative time budget is an important part of timing the CRAN, but the absolute limit still applies. The nodes need to be simultaneously within their corresponding limits for both absolute and relative time.

This section covered the details of the timing requirements in the 5G NR architecture and showed that the complexity really arises when discussing timing for the evolution of the 5G disaggregated CRAN. Timing the DRAN is much more straightforward for the network engineer, but it is clear that most attention on delivering accurate time will need to be paid in the evolved RAN.

Relative Time Budget Analysis

As shown in Figure 10-31 and Figure 10-32, to maintain the relative time error budget between two radios at the antenna interface, the transport design needs to ensure that the maximum absolute time error (max|TE|) from the common master clock to the end radio interface is within the defined limits. As a rule of thumb, the network time budget from the common master to the radios should be less than or equal to half of the relative requirement to ensure that the relative time error between two radios is achieved.

The maximum absolute relative time error (max|TE_R|) at the UNI between two radios that are sourcing their time from a common master clock can be calculated for category A and B deployment use cases (see Table 10-17) by using the following equations.

- Category A

  Case 1.2 and case 2 are both relevant for category A deployments. The maximum relative TE can be determined per these equations:

  Case 1.2: max|TE_R| = 130 ns − 2 × |$TE_{eRE/RE}$| − 2 × |$TE_{T\text{-}TSC}$| = 60 ns

  Case 2: max|TE_R| = 130 ns − 2 × |$TE_{eRE/RE}$| = 70 ns

  To calculate the maximum relative TE, max|TE_R|, for case 1.2, |$TE_{eRE/RE}$| at the input of the eRE/RE is 20 ns (as per Table 10-18) and the defined |$TE_{T\text{-}TSC}$| for case 1.2 is 15 ns.

  In case 2, the T-TSC is integrated into the radio equipment and |$TE_{eRE/RE}$| at the input of the eRE/RE is 30ns (as per Table 10-18).

Remember, these values of 60 ns and 70 are the maximum relative TE that is allowed between the two paths to the radio at each respective UNI. It defines the network budget for TE allowed between the UNI for each radio and the common master clock.

■ Category B

Case 1.1, case 1.2, and case 2 are relevant for category B deployments. The maximum relative TE for category B deployment can be determined as per these equations:

Case 1.1: $\max|TE_R| = 260\ ns - 2 \times |TE_{eRE/RE}| - 2 \times |TE_{T\text{-}TSC}| = 100\ ns$

Case 1.2: $\max|TE_R| = 260\ ns - 2 \times |TE_{eRE/RE}| - 2 \times |TE_{T\text{-}TSC}| = 190\ ns$

Case 2: $\max|TE_R| = 260\ ns - 2 \times |TE_{eRE/RE}| = 200 ns$

To calculate the maximum relative TE, $\max|TE_R|$ for case 1.1, $|TE_{eRE/RE}|$ at the input of the eRE/RE is $|TE_{eRE/RE}|$ is 20 ns (as per Table 10-18), the $|TE_{T\text{-}TSC}|$ for case 1.1 is 60 ns as per ITU-T G.8273.2 Class B T-TSC that includes the budget of cTE and dTE values.

To calculate the maximum relative TE, $\max|TE_R|$ for case 1.2, $|TE_{eRE/RE}|$ at the input of the eRE/RE is 20 ns (as per Table 10-18) and $|TE_{T\text{-}TSC}|$ for case 1.2 is 15 ns.

In case 2, the T-TSC is integrated into the radio equipment and $|TE_{eRE/RE}|$ at the input of the eRE/RE is 30ns (as per Table 10-18).

To help achieve these stringent requirements on the network between a common master clock and the end application, the ITU-T has also defined a new Class C clock performance level for T-BC, T-TC, and T-TSC clocks in the G.8273.2 recommendation. This is mainly to reduce the 60-ns TE that results from the previous Class B performance in the T-TSC down to 15 ns.

In addition, the ITU-T introduced a new level of performance for the PRTC in the ITU-T G.8272 standard, and introduced the enhanced PRTC in ITU-T G.8272.1. SyncE performance was also enhanced in the ITU-T G.8262.1 recommendation to ensure that the network clock distribution has an improved physical layer frequency signal to assist the PTP timing signal distribution. These changed were made principally to improve the timing performance in the fronthaul network.

Once again, do not forget that in spite of all the discussion about relative time error, there is still the absolute time requirement, which remains at ±1.1 μs from the PRTC to the handover point to the radio equipment—and depicted as Category C in Table 10-17.

Network Time Error Budget Analysis

With the functional split in the RAN and then deploying the BBU as virtualized DU and CU functions, the timing synchronization design aspects for 5G xHaul architecture require extra attention and care. For 5G services, knowing when and how to apply the relative timing budget requirements (alongside the absolute one) within the network architecture is important to ensure the successful deployment of the services.

Within the virtualized 5G RAN architecture, the requirements are broadly categoried as follows:

- Time alignment error limits between the PRTC/T-GM and the radio air interface

- Time alignment error limits between the PRTC/T-GM and the UNI of the virtualized DU and CU function

As a best practice, it is recommended to have a common PRTC/T-GM for the virtualized DU and the RUs associated with that DU. As depicted in Table 10-17, the absolute time error budget between the PRTC and the RU antenna remains as ±1.5 µs; similarly, the time error requirement between the PRTC and the virtualized DU also must remain within the ±1.5 µs budget, keeping the network budget as ±1.1 µs. However, operators could have different deployment scenarios with virtualized DU architecture, as described in the sections that follow.

Option 1

The RUs are directly connected to the DU, and the DU acts as a synchronization master to all connected RUs. As shown in Figure 10-33, this could have several options:

- The DU has a PRTC/T-GM directly connected to it.

- The DU is receiving clock from the network PRTC/T-GM using PTP and SyncE.

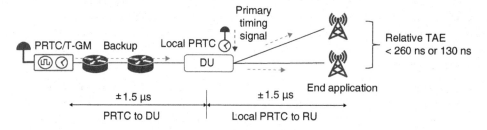

a) PRTC/T-GM is directly connected to DU

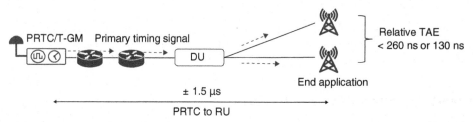

b) DU is receiving clock over the network

Figure 10-33 *DU as Synchronization Master and Distributing Clock to Connected RUs*

In this option, the DU needs to support the SyncE or eSyncE clock definition according to ITU-T G.8262 or G.8262.1. In suboption (b), the DU needs to comply with the G.8275.1 PTP telecom profile with a minimum ITU-T G.8273.2 Class B performance as a T-BC clock.

Option 2

The DU and RU are connected over multiple network hops, as shown in Figure 10-34:

- A centralized PRTC/T-GM distributes clock to the DU and RU as a primary reference

- The PRTC/T-GM is deployed on the network that connects the DU and RU. The centralized PRTC/T-GM is used as the backup timing source.

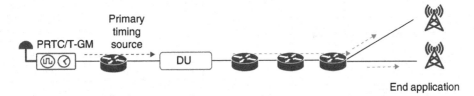

a) PRTC/T-GM is deployed in the backhaul network

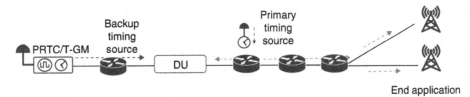

b) PRTC/T-GM is deployed in the network connecting DU and RU

Figure 10-34 *Network PRTC Distributes Clock to DU and RUs Connected over Multiple Network Hops*

In this option, the nodes distributing the timing signal to the DU and RUs need to comply with the G.8275.1 PTP telecom profile with a minimum ITU-T G.8273.2 Class B performance as T-BC clocks. These network nodes also need to support the SyncE or eSyncE clock definition according to ITU-T G.8262 or G.8262.1. However, the DU or RU is not mandated to use the SyncE frequency source to achieve its own frequency synchronization and instead may use only PTP to achieve frequency as well as phase synchronization.

As shown in suboption (b), if the network topology has a backup timing source, the DU and RUs can use the backup timing source to maintain longer holdover during a failure of the primary timing source. Features like APTS or BMCA are useful in such deployments.

Option 3

The PRTC is directly connected to the RU as shown in Figure 10-35, which has two possibilities:

- The DU and RU receive clock from the same PRTC/T-GM.

- The DU and RU receive clock from different PRTC/T-GMs.

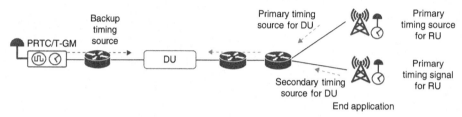

a) PRTC/T-GM is deployed at RU, distributing clock to DU

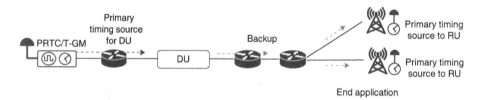

b) PRTC/T-GM is deployed at RU, DU receives clock from network PRTC/T-GM

Figure 10-35 *Network PRTC Distributes Clock to DU and RUs Connected over Multiple Network Hops*

In this option, the nodes distributing the timing signal from the RUs to the DU need to comply with the G.8275.1 PTP telecom profile with a minimum ITU-T G.8273.2 Class B performance as T-BC clocks. These network nodes also need to support SyncE or eSyncE clock definition according to ITU-T G.8262 or G.8262.1. However, the DU or RU is not mandated to use the SyncE frequency source to achieve its own frequency synchronization and instead may use only PTP to achieve frequency as well as phase synchronization.

Network Time Error Budget Analysis: Summary

In the three preceding scenarios, when the local PRTC/T-GM is deployed at the DU or RU, the local PRTC/T-GM becomes the primary source of the timing signal. A network PRTC/T-GM located in the backhaul network can be used as a backup timing source to maintain timing alignment during any local PRTC failure.

Because the CU does not require a stringent timing requirement, there is no specific timing error limit defined for the CU. However, the CU requires a normal level of system alignment with UTC for purely operational reasons. One need would be to allow the network engineers to perform fault isolation by correlating timestamped entries in error logs from all systems in the network. Another use is that the CU may require UTC time for reasons related to cryptography, such as certificate validation.

For further reading, ITU-T G.8271.1 defines a reference architecture and network model with network limits made up of T-BC and T-TSC clocks with Class B or Class C performance from G.8273.2. These devices are used to carry the timing signal to the RAN across the backhaul network or between the DU and RU in the fronthaul network.

Synchronizing the Virtualized DU

The virtualized DU (vDU) is designed to be deployed on x86-based commercial off-the-shelf servers. As shown in Figure 10-36, the server accepts the ability to plug network interface cards (NIC) into the PCIe slots with timing capabilities built into the card. These (Ethernet) interfaces can be used for midhaul or fronthaul connectivity.

Additional ASIC-based hardware acceleration cards are also used so that the intensive baseband processing can be offloaded from the primary CPU complex. Other PCIe slots are used to provide additional capabilities using peripheral or auxiliary cards.

NIC cards with timing support, or *time sync NICs*, are essential to maintain the timing accuracy on the DU server platform because the x86 cannot do it unassisted. It is the time sync NIC that provides the hardware to support the T-GM, T-BC and T-TSC clocks for the PTP telecom profiles at high accuracy.

An important point to understand—for high-accuracy timing—is that the x86 platform itself is not (and should not be) the PTP clock in the path of the timing between the DU and RUs but will normally recover its timing from the PTP hardware clock (PHC) on the NIC card. The x86 uses a Linux package, *ptp4l*, to accurately synchronize the time on the NIC card with the remote T-GM. The ptp4l process handles PTP traffic on the NIC port, updates the NIC PHC, and tracks the synchronization status.

The x86 can also run the *phc2sys* process to update the system real-time clock from the PHC in the NIC card (the PHC is aligned to a remote grandmaster using PTP). The NIC card is the source and distributor of accurate PTP time for the x86 server and other devices in the network. The x86 server does not have the hardware capability to align its system real-time clock with the PHC on the NIC card to a high degree of accuracy.

For this reason, in the case where an x86 server has multiple NIC cards, the server cannot distribute time accurately to the multiple NIC cards from the system real-time clock. Each NIC card has its own PHC, and the server cannot accurately (at the level required for mobile) transfer time between the cards. The NIC cards must be daisy chained together externally and some mechanism used to distribute the timing signal between each NIC card in the system.

Figure 10-37 illustrates the typical deployment scenario using a virtualized DU. Once the DU server gets the accurate time and phase signal from a connected PRTC, the NIC card can be configured as a T-GM to distribute the clock to other connected servers in daisy chain configuration (or other NIC cards in the same server).

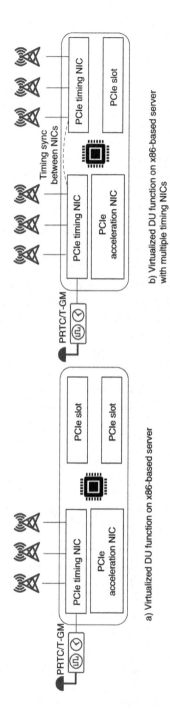

a) Virtualized DU function on x86-based server

b) Virtualized DU function on x86-based server with multiple timing NICs

Figure 10-36   *Virtualized DU Server Architecture*

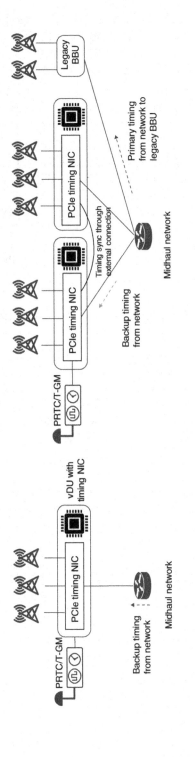

Figure 10-37 *vDU Deployment at Cell Site Location*

The (multiple) interfaces on the time sync NIC are used to connect to midhaul network elements, fronthaul routers, and RU devices. If any legacy BBU or eNB is also deployed at the site, it is possible to use timing from the DU or take timing from another network source such as a router/switch in the midhaul or backhaul network.

Because the CRAN architecture using the vDU is heavily dependent on the time sync NIC card inside the server, the reliability of the solution is reduced using this approach—especially when the single NIC is used as a source of PTP and daisy chained to other vDUs in the site.

Similarly, this approach affects the timing error and scalability of the timing source at the site. The daisy chaining adds hops between the vDU and the associated RUs, adding to the time error. And since many of these NIC cards have a very limited number of interface ports, using a daisy chain takes two ports from each card (they typically only have three or four ports). This limitation also makes it very hard to scale the timing network without introducing excessive hops and NIC cards. For this reason, a network-based fronthaul timing distribution is highly recommended when more than one vDU server is deployed at the site.

Maximum Received Time Difference Versus Time Alignment Error

The synchronization requirements in the 3GPP specifications are expressed either in terms of MRTD or TAE. For DC or CA use cases, excessive MRTD between the transmissions from the cooperating base stations to the UE can cause problems, because some of the information will arrive too early and/or the other will be too late for the UE to use. For example, according to the 3GPP TS 28.133, inter-band CA in FR1 is allowed a budget of up to 33 μs MRTD. Higher frequencies (FR2) with shorter range have a lower budget. See the earlier section "Dual Connectivity" for more details.

Any calculation of MRTD also needs to consider the propagation delay of approximately 3.3 μs per km. Therefore, a difference of 9.5 km in distance between each cooperating base station and the UE accounts for around 31.5 μs MRTD. This loss of 31.5 μs of budget through the propagation delay means that the two radios must be phase synchronized to within 1.5 μs to allow MRTD to work. This is a tighter requirement than the general 3 μs allowed between two neighboring TDD radios.

The intra-band CA use case between LTE and NR base stations, also known as EN-DC (E-ULTRA NR Dual Connectivity), is defined for co-located deployments. In co-located deployment, there is no need to consider the propagation differences between the UE and the cooperating base stations because it is the same for both radios.

Table 10-19 summarizes the detailed phase and frequency requirements for 5G deployments for FR1 and FR2 frequency ranges when using CA or DC.

Note that the values of ±1.5 μs, 260 ns, 130 ns, and 65 ns align with the requirements outlined in the RAN specifications from organizations like CPRI, eCPRI, IEEE 8021.CM, and O-RAN.

Table 10-19 *Time Synchronization Requirements for Various 5G Use Cases*

Application	TAE or MRTD (Air Interface)	Specification
5G NR TDD	±1.5 μs TAE	3GPP TS 38.133
Inter-band synchronous EN-DC (LTE & 5G NR Dual Connectivity)	33 μs MRTD	3GPP TS 38.133
NR inter-band CA	33 μs MRTD (FR1)	3GPP TS 38.133
	8 μs MRTD (FR2)	Clause 7.6.4
	24 μs MRTD (FR1/2)	
NR intra-band synchronous EN-DC (TDD-TDD or FDD-FDD)	3 μs MRTD	3GPP TS 38.133
		Clause 7.6.3
NR inter-band CA; with or without MIMO or Tx diversity	3 μs TAE	3GPP TS 38.104
		Clauses 9.6.3.2, 9.6.3.3
NR intra-band noncontiguous CA; with or without MIMO or Tx diversity	3 μs TAE (FR1)	3GPP TS 38.104
	260 ns TAE (FR2)	Clauses 9.6.3.2, 9.6.3.3
NR intra-band contiguous CA; with or without MIMO or Tx diversity	260 ns TAE (FR1)	3GPP TS 38.104
	130 ns TAE (FR2)	Clauses 9.6.3.2, 9.6.3.3
NR MIMO or Tx diversity transmission at each carrier frequency	65 ns TAE	3GPP TS 38.104
		Clauses 9.6.3.2, 9.6.3.3

Summary

The different generations of mobile network have always required strict frequency synchronization, and several very effective solutions have been available to address that need. Lately, network engineers see that phase/time synchronization has become mandatory for some of the more advanced technologies and techniques in the RAN networks.

However, some of these features or services were often not business critical, so operators always had the option to delay the rollout of new features when they saw that phase/time synchronization was needed to make it work. What has changed now is that 5G has moved from FDD- to TDD-based radio systems, so deferring deployment of phase and time synchronization solutions is no longer an option.

With 5G, the absolute time error limit must be met across the whole mobile network. Within that absolute number, there are other, much smaller relative error limits that must be met between cooperating radio equipment in the fronthaul and midhaul. Some of these limits apply at the UNI—the output of the timing distribution network—or at the output of the radio antenna (TAE).

The problem has been clearly defined, and the requirement has been clearly stated, so the next step is to define a timing distribution methodology to meet those challenges. Chapter 11 discusses the available technologies and techniques for the distribution of timing and how to apply them to the 5G deployment case.

References in This Chapter

3GPP

"Evolved Universal Terrestrial Radio Access (E-UTRA); Base Station (BS) radio transmission and reception." *3GPP*, 36.104, Release 8. https://portal.3gpp.org/desktopmodules/Specifications/SpecificationDetails.aspx?specificationId=2412

"Evolved Universal Terrestrial Radio Access (E-UTRA); Measurement Requirements." *3GPP*, 36.801, Release 8, 2007. https://portal.3gpp.org/desktopmodules/Specifications/SpecificationDetails.aspx?specificationId=2481

"Evolved Universal Terrestrial Radio Access (E-UTRA); Requirements for support of radio resource management." *3GPP*, 36.133, Release 8. https://portal.3gpp.org/desktopmodules/Specifications/SpecificationDetails.aspx?specificationId=2420

"NR; Base Station (BS) radio transmission and reception." *3GPP*, 38.104, Release 15, 2017. https://portal.3gpp.org/desktopmodules/Specifications/SpecificationDetails.aspx?specificationId=3202

"Study on indoor positioning enhancements for UTRA and LTE." *3GPP*, 37.857, Release 13, 2016. https://portal.3gpp.org/desktopmodules/Specifications/SpecificationDetails.aspx?specificationId=2629

"Study on new radio access technology Physical layer aspects." *3GPP*, 38.802, Release 14, 2017. https://portal.3gpp.org/desktopmodules/Specifications/SpecificationDetails.aspx?specificationId=3066

"Study on new radio access technology: Radio access architecture and interfaces." *3GPP*, 38.801, Release 14, 2017. https://portal.3gpp.org/desktopmodules/Specifications/SpecificationDetails.aspx?specificationId=3056

Common Public Radio Interface (CPRI)

"Common Public Radio Interface: eCPRI Interface Specification." *CPRI*, eCPRI Specification V2.0, 2019. http://www.cpri.info/downloads/eCPRI_v_2.0_2019_05_10c.pdf

"Common Public Radio Interface: Requirements for the eCPRI Transport Network." *CPRI*, eCPRI Transport Network V1.2, 2018. http://www.cpri.info/downloads/Requirements_for_the_eCPRI_Transport_Network_V1_2_2018_06_25.pdf

Federal Communications Commission. "Wireless E911 Location Accuracy Requirements." Federal Register, 2020. https://www.federalregister.gov/documents/2020/01/16/2019-28483/wireless-e911-location-accuracy-requirements

IEEE Standards Association

"IEEE Standard for Local and metropolitan area networks – Time-Sensitive Networking for Fronthaul." *IEEE Std 802.1CM-2018*, 2018. https://standards. ieee.org/standard/802_1CM-2018.html

"IEEE Standard for Local and metropolitan area networks – Time-Sensitive Networking for Fronthaul – Amendment 1: Enhancements to Fronthaul Profiles to Support New Fronthaul Interface, Synchronization, and Synchronization Standards." *IEEE 802.1CMde-2020*, 2020. https://standards.ieee.org/ standard/802_1CMde-2020.html

International Telecommunication Union Telecommunication Standardization Sector (ITU-T).

"G.823, The control of jitter and wander within digital networks which are based on the 2048 kbit/s hierarchy." *ITU-T Recommendation*, 2000. https://www.itu.int/ rec/T-REC-G.823/en

"G.824, The control of jitter and wander within digital networks which are based on the 1544 kbit/s hierarchy." *ITU-T Recommendation*, 2000. https://www.itu.int/ rec/T-REC-G.824/en

"G.8262, Timing characteristics of synchronous equipment slave clock." *ITU-T Recommendation*, Amend 1, 2020. http://handle.itu.int/11.1002/1000/14208

"G.8262.1, Timing characteristics of enhanced synchronous equipment slave clock." *ITU-T Recommendation*, Amend 1, 2019. https://handle.itu.int/11.1002/ 1000/14011

"G.8271, Time and phase synchronization aspects of telecommunication networks." *ITU-T Recommendation*, 2020. https://handle.itu.int/11.1002/1000/14209

"G.8271.1, Network limits for time synchronization in packet networks with full timing support from the network." *ITU-T Recommendation*, Amend 1, 2020. https://handle.itu.int/11.1002/1000/14527

"G.8273.2, Timing characteristics of telecom boundary clocks and telecom time slave clocks for use with full timing support from the network." *ITU-T Recommendation*, 2020. https://handle.itu.int/11.1002/1000/14507

"G.8272, Timing characteristics of primary reference time clocks." *ITU-T Recommendation*, Amend 1, 2020. http://handle.itu.int/11.1002/1000/14211

"G.8272.1, Timing characteristics of enhanced primary reference time clocks." *ITU-T Recommendation*, Amend 2, 2019. http://handle.itu.int/11.1002/1000/14014

"G.8275.1, Precision time protocol telecom profile for phase/time synchronization with full timing support from the network" *ITU-T Recommendation*, Amend 1, 2020. http://handle.itu.int/11.1002/1000/14543

O-RAN Alliance. "O-RAN Fronthaul Control, User and Synchronization Plane Specification 5.0." *O-RAN*, O-RAN.WG4.CUS.0-v05.00, 2020. https://www.o-ran. org/specifications

Chapter 10 Acronyms Key

The following table expands the key acronyms used in this chapter.

Term	Value
1G	1st generation (mobile telecommunications system)
1PPS	1 pulse per second
1xEV-DO	1x Evolution-Data Optimized
1xRTT	Single-Carrier Radio Transmission Technology
2G	2nd generation (mobile telecommunications system)
3G	3rd generation (mobile telecommunications system)
3GPP	3rd Generation Partnership Project (SDO)
3GPP2	3rd Generation Partnership Project 2 (CDMA2000)
4G	4th generation (mobile telecommunications system)
5G	5th generation (mobile telecommunications system)
5GCN	5G Core network
ABS	Almost Blank Subframe
AMC	adaptive modulation and coding
AMF	Access and Mobility Management Function
AMPS	Advanced Mobile Phone Service
API	application programming interface
ASCII	American Standard Code for Information Interchange
BBH	baseband hotel
BBU	Baseband Unit
BS	base station
CA	Carrier Aggregation
CBF	coordinated beamforming
CBRS	Citizens Broadband Radio Service
CDMA	code-division multiple access
cMTC	critical machine-type communication
CoMP	Coordinated Multipoint
CP	cyclic prefix
CPRI	Common Public Radio Interface
CRAN	centralized radio access network

Term	Value
CS	coordinated scheduling
CSI	channel site information
CU	Central Unit (3GPP)/Control Unit (O-RAN) (5G radio access network)
CUPS	control and user plane separation
DC	Dual Connectivity
DL	downlink
DRAN	distributed radio access network
DSL	digital subscriber line (a broadband technology)
DU	Distributed Unit (5G radio access network)
DwPTS	downlink pilot time slot
E1	2-Mbps (2048-kbps) signal (SDH)
eCPRI	enhanced Common Public Radio Interface
EDGE	Enhanced Data rate for GSM Evolution
eICIC	enhanced inter-cell interference coordination
eMBB	enhanced mobile broadband
eMBMS	evolved Multimedia Broadcast Multicast Service
eNB	evolved Node B (E-UTRAN)—LTE base station
EN-DC	E-UTRA (LTE) NR Dual Connectivity
eNodeB	evolved Node B (E-UTRAN)—LTE base station
EPC	Evolved Packet Core
eRE	eCPRI Radio Equipment (CPRI)
eREC	eCPRI Radio Equipment Control (CPRI)
E-SMLC	Evolved Serving Mobile Location Center
E-UTRA	Evolved Universal Terrestrial Radio Access (LTE)
EV-DO	Evolution-Data Optimized
FDD	frequency division duplex
FDMA	frequency-division multiple access
FEC	Forward Error Correction
FLR	frame loss ratio
FR1	frequency range 1
FR2	frequency range 2

Term	Value
FWA	fixed wireless access
gNB	next (5th) generation node B
gNB-CU	gNodeB Central Unit
gNB-DU	gNodeB Distributed Unit
GNSS	global navigation satellite system
GP	guard period
GPRS	General Packet Radio Service
GPS	Global Positioning System
GSM	Global System for Mobile communications
HeNB	home eNB
HetNet	heterogeneous network
HRPD	High-Rate Packet Data
HSDPA	High Speed Downlink Packet Access
HSPA	High Speed Packet Access
HSUPA	High Speed Uplink Packet Access
ICI	inter-carrier interference
ICIC	inter-cell interference coordination
IoT	Internet of things
IS	Interim Standard
ISI	inter-symbol interference
JP	joint processing
JT	joint transmission
LPP	LTE Positioning Protocol
LTE	Long Term Evolution (mobile communications standard)
LTE-A	Long Term Evolution—Advanced (mobile communications standard)
LTE-A Pro	Long Term Evolution—Advanced Pro (mobile communications standard)
max|TE|	maximum absolute time error
max|TE$_R$|	maximum absolute relative time error
MBMS	Multimedia Broadcast Multicast Service
MBSFN	MBMS Single Frequency Network
MeNB	master eNodeB

Term	Value
MIMO	multiple input multiple output
mMTC	massive machine-type communication
mmWave	millimeter wave (radio frequency bands)
MRTD	maximum receive time difference
MTTD	maximum transmit time difference
NIC	network interface card
NFV	Network Function Virtualization
ng-eNB	next generation evolved node B
NR	New Radio
NSA	non-standalone
NTT	Nippon Telegraph and Telecom (Japan)
OBSAI	Open Base Station Architecture Initiative
OFDM	orthogonal frequency-division multiplexing
OFDMA	orthogonal frequency-division multiple access
OTDOA	Observed Time Difference of Arrival
PAPR	peak-to-average power ratio
PHC	PTP hardware clock
ppb	parts per billion
PRC	primary reference clock (SDH)
PRS	primary reference source (SONET)
PS	packet switched
PSS	primary synchronization signal
QoE	quality of experience
QoS	quality of service
QPSK	quadrature phase-shift keying
RAN	radio access network
RE	Radio Equipment (CPRI)
REC	Radio Equipment Control (CPRI)
RF	radio frequency/function
RoE	radio over Ethernet
SA	standalone
SAE	System Architecture Evolution

Term	Value
SBA	service-based architecture
SC-FDMA	single carrier frequency-division multiple access
SDH	synchronous digital hierarchy (optical transport technology)
SDN	Software-Defined Network
SeNB	secondary eNodeB
SGW	serving gateway
SNR	signal-to-noise ratio
SONET	synchronous optical network (optical transport technology)
SSS	secondary synchronization signal
T1	1.544-Mbps signal (SONET)
T_A	timing advance
TAE	time alignment error
TDD	time division duplex
TDM	time-division multiplexing
TDMA	time-division multiple access
TD-SCDMA	time-division synchronous code-division multiple access
TE	time error
$|TE_{RE}|$	absolute time error in radio equipment
T-TSC	Telecom Time Slave Clock
Tx	transmission
UE	user equipment
UL	uplink
UMTS	Universal Mobile Telecommunications System
UNI	user-network interface (from the MEF)
UPF	User Plane Function
uRLLC	ultra-reliable and low-latency communications
UTC	Coordinated Universal Time
UpPTS	uplink pilot time slot
vDU	virtualized Distributed Unit
VoIP/VOIP	voice over Internet protocol (IP)
WCDMA	wideband code-division multiple access
WiMAX	Worldwide Interoperability for Microwave Access

Further Reading

Refer to the following recommended sources for further information about the topics covered in this chapter.

3GPP

"Base Station (BS) radio transmission and reception (TDD)." *3GPP*, 25.105, Release 1999. https://portal.3gpp.org/desktopmodules/Specifications/ SpecificationDetails.aspx?specificationId=1155

"Evolved Universal Terrestrial Radio Access (E-UTRA); Physical Channels and Modulation." *3GPP*, Release 8. https://portal.3gpp.org/desktopmodules/ Specifications/SpecificationDetails.aspx?specificationId=2425

"NG-RAN; F1 general aspects and principles." *3GPP*, 38.470, Release 16, 2020. https://portal.3gpp.org/desktopmodules/Specifications/SpecificationDetails. aspx?specificationId=3257

"NR; NR and NG-RAN Overall description; Stage-2." *3GPP*, 38.300, Release 16 (16.2.0), 2020. https://portal.3gpp.org/desktopmodules/Specifications/ SpecificationDetails.aspx?specificationId=3191

"Release 15." *3GPP*, 2019. https://www.3gpp.org/release-15

"Study on Central Unit (CU) – Distributed Unit (DU) lower layer split for NR." *3GPP*, 38.816, Release 15, 2017. https://portal.3gpp.org/desktopmodules/Specifications/ SpecificationDetails.aspx?specificationId=3364

"Synchronisation in UTRAN Stage 2." *3GPP*, 25.402, Release 1999. https:// portal.3gpp.org/desktopmodules/Specifications/SpecificationDetails. aspx?specificationId=1185

"System Architecture for the 5G System (5GS)." *3GPP*, 23.501, Release 15, 2016. https://portal.3gpp.org/desktopmodules/Specifications/SpecificationDetails. aspx?specificationId=3144

"UTRAN overall description." *3GPP*, 25.401, Release 1999. https://portal.3gpp.org/ desktopmodules/Specifications/SpecificationDetails.aspx?specificationId=1184

Common Public Radio Interface: http://www.cpri.info

Dahlman, E., Parkvall, S. and J. Skold

4G: LTE-Advanced Pro and The Road to 5G, 3rd Edition. Academic Press, 2016.

5G: 5G NR: The Next Generation Wireless Access Technology, 1st Edition. Academic Press, 2018.

Global System for Mobile Communications Association (GSMA): https://www.gsma. com/

NGMA: https://www.ngmn.org/

O-RAN Alliance Specifications: https://www.o-ran.org/specifications

Chapter 11

5G Timing Solutions

In this chapter, you will learn the following:

- **Deployment considerations for mobile timing:** Covers the different factors to consider when building a solution for 5G timing. These factors reflect the choices that a 5G operator has made about the form of network that they are building.

- **Frequency-only deployments:** Covers those cases where phase synchronization is not a requirement. This is unlikely to include most 5G networks, but many LTE and LTE-A networks may fall into this category, and many times will be coexisting alongside the new 5G elements.

- **Frequency, phase, and time deployment options:** Details the different solutions to apply to each situation, based on the constraints that the operator must deal with.

- **Midhaul and fronthaul timing:** Expands on previous chapters discussing the fronthaul evolution and explains the various topologies underlying modern radio access network (RAN) designs.

- **Timing security and MACsec:** Follows preceding discussions about precision time protocol (PTP) security to deal with other aspects of security such as using IPsec and MACsec for transport security.

It is time to bring together all the threads running through the preceding chapters. The early chapters presented the underlying principles and features of timing and synchronization. Then, the middle chapters included the particulars of the appropriate tools to carry synchronization across a network. Then, later chapters looked at how to design a synchronization network and started to introduce the details of applying these tools to assist the rollout of 5G mobile. Chapter 10, "Mobile Timing Requirements," then covered many of the requirements of synchronization for the 5G backhaul and fronthaul networks.

This chapter now goes through the numerous approaches to deploying a synchronization solution to support all the different 5G network use cases. This takes into account factors like what is the appropriate tool for each situation, what are the pros/cons behind different network topologies, and what is the best way to solve the problem in each circumstance.

Deployment Considerations for Mobile Timing

The most important thing to recognize before designing a timing solution is the scope of the problem, and that is determined by the sort of mobile network that the operator is building. Another large factor is the constraint about what sort of transport options are available for the backhaul (and fronthaul) network. For these reasons, this section covers the major considerations when choosing the best solution for each type of network.

If the reader has an active role in the rollout, it is important to consult with the appropriate people on the details of deployment to correctly define the problem. For example, 5G has introduced a lot of flexibility and complexity in the fronthaul, so the network timing engineer must cooperate closely with the RAN team. Missing out on correctly understanding a key requirement can invalidate the total design and be a very expensive mistake for an operator to correct.

If the authors, having talked about timing with many, many operators all around the world, could give one simple piece of advice, it would be to be sure you understand the problem in some detail. The sections that follow include numerous general principles about deploying timing, but the engineer deploying timing must understand their individual situation thoroughly.

Never stop asking for clarification, requesting more details, and questioning anything you are told.

Network Topology

Building a robust timing solution typically includes a distributed network of global navigation satellite systems (GNSS) combined with the transmission of timing signals over the transport network. Given that assumption, the following two very important factors greatly influence the design:

- **The topology:** The features of the nodes in the network and how they are distributed. The features would be those characteristics that influence how well the nodes process and forward timing signals such as PTP.

- **The transport:** Characteristics of the underlying transport network between the nodes, which relates to how accurately and faithfully it carries timing signals.

The topology is important in two fundamental ways:

- The shape of the network (for example, whether it is a ring or a hub-and-spoke layout)

- The size or length of the network (number of nodes in a ring, or the length of a chain)

The basic issue is to determine how the timing signal can be distributed (with redundancy) to the end application and meet the network timing budget with the lowest cost. If the GNSS receivers are located at the radio site, then the characteristics of the transport and network are not important. But the more the source of timing is centralized, the more important the network design becomes to carry the timing signal.

Although the design must consider how the nodes are geographically dispersed, distance itself is not a huge factor because PTP compensates for distance. Carrying timing over very long distances might, however, require features on the transport system that affect the timing design (for example, requiring a long-reach optical transport method that introduces asymmetry). In this book the transport system refers to the technical method used to carry the frequency signal and PTP messages from node to node.

It follows then that the design must always consider the technology of the underlying transport if it is to be used to carry a timing signal. The simplest transport to deal with is essentially dark fiber—pulses of light go into one end of the fiber-optic cable and they appear out the other end, with no additional processing.

As a rule, however, the more complex the transport, the more likely it is to introduce impairment to the accurate transport of time. When the transport has active components, or intelligent devices in the link between nodes, the engineer must assume that those components are likely to introduce either asymmetry, packet delay variation (PDV), or both. To transport timing with PTP, both these characteristics are a problem. Although asymmetry can be tolerated when transporting only frequency with PTP, it very severely impacts phase accuracy.

Vendors and standards organizations are updating many of the transport technologies to support and/or improve the accurate transport of phase and frequency. The mechanisms to achieve this vary widely depending on the type of transport being considered. In summary, if one of these problematic systems lies in the path of a timing signal, it needs to contain design features that make it what is known as *timing aware*.

Additionally, if frequency is being carried using a physical technique such as synchronous Ethernet (SyncE) or enhanced SyncE (eSyncE), then you must ensure that the end-to-end network can faithfully carry that frequency (even if the network is designed with different transport types besides Ethernet). Again, the method to achieve this will depend on the technology of the transport because it normally requires using some native method to carry frequency (a microwave link is a common example).

At one end of the link, the device must be able to recover the SyncE frequency, convert it into some native method to carry frequency, recover that frequency at the far end, and apply that same frequency signal to the Ethernet port facing the next network element (NE). Figure 11-1 illustrates the interworking with different transport types to carry a timing signal such as frequency.

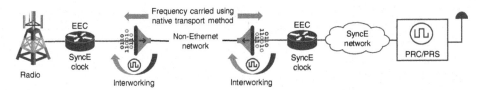

Figure 11-1 *Carrying Frequency Across Different Transport Types*

Much the same as the transport links, the NEs carrying the packet-based timing signal can fall into the category of being either timing-aware or timing-unaware:

- The timing-unaware device simply switches packets or messages like any other device, and has no mechanisms to deal with these packets in a way that could be considered timing aware. It has no ability to process time-sensitive packets beyond the normal mechanisms such as quality of service (QoS).

- The timing-aware device recognizes packets (such as PTP) that are transporting a timing signal and treats them specially. For a PTP-aware NE, this would include such features as being a boundary clock (BC) or a transparent clock (TC) for the profile of PTP being used.

A packet network that contains *only* NE nodes that are PTP-aware connected by a transport that consists only of timing-aware devices is known as having *full timing support* and is often referred to as FTS. Packet networks that do not offer that level of support to transport timing are referred to having *partial timing support* (PTS), meaning that only a subset of the nodes understand and process PTP. Other elements in the network treat PTP messages just like any other layer 2 (L2) or layer 3 (L3) packet.

Recall from Chapter 8, "ITU-T Timing Recommendations," that the ITU-T has defined network limits for phase/time synchronization based on both these use cases: G.8271.1 for the FTS case and G.8271.2 for PTS. There is also a recommended PTP telecom profile to support each of those use cases: G.8275.1, PTP over Ethernet (PTPoE) for FTS; and G.8275.2, PTP over IP (PTPoIP) for PTS.

Use Cases and Technology

The most fundamental question to answer before designing a timing solution for a mobile network is, "What sort of mobile network is this design for?" The answer to that question should help the operator and the timing engineer decide upon the form of timing that is needed and the degree of accuracy it must deliver.

The generation of mobile technology, the services enabled, the radio type, and the architecture of the RAN will determine whether only frequency is needed or whether the combined frequency, phase, and time synchronization is required. Chapter 10 covered the various flavors of LTE, LTE-A, and 5G networks and the type of timing they need in more detail.

To summarize the basics, here are a few general rules (they all require frequency):

- All time division duplex (TDD) radios require phase/time.

- 5G radio or New Radio (NR), because they are TDD, requires phase/time.

- LTE (4G) networks that use frequency division duplex (FDD) radios normally only require frequency.

- LTE-A networks might require phase/time depending on the services and radio methods being enabled. Features such as Multimedia Broadcast Multicast Service (MBMS), enhanced inter-cell interference coordination (eICIC), Coordinated Multipoint (CoMP), and Carrier Aggregation (CA) normally require phase/time.

- Sophisticated antenna techniques such as multiple input, multiple output (MIMO) require phase alignment, although the requirement to be phase-aligned might apply only to the antenna array connected to a single radio.

  Note that MIMO normally does not apply to low-frequency bands because the antenna array ends up being very large (low frequency means long wavelength means large antenna arrays).

- Any radio transmission scenario that mentions a single frequency network (SFN) requires phase synchronization. Note that broadcasting also uses this technology, so digital radio and television transmitters normally require phase synchronization.

- Deployments of small cells generally requires phase, especially when interference coordination is part of the integration of small cells into the network (see the next section, "Small Cells Versus Macro Cells").

- Many techniques to do with determining the position/location of the user equipment (UE) require phase alignment, but especially when using Observed Time Difference of Arrival (OTDOA).

Please see Chapter 10 for detailed coverage on timing requirements for mobile technologies.

At least for the early deployments of 5G, there are three generally recognized use cases:

- eMBB (enhanced mobile broadband) and fixed wireless access (FWA), the focus of 3GPP Release 15 (Rel15). FWA is where a 5G network is used to provide communications services to a fixed location, like home broadband.

- uRLLC (ultra-reliable and low-latency communications), the focus of Release 16 (Rel16).

- mMTC (massive machine-type communications) and industrial Internet of things (IoT), the focus of Release 17 (Rel17).

The use case normally does not determine the timing requirements for the 5G network being built; however, you should understand that the use case might require the deployment of a technology that can change the timing requirements.

As an example, the deployment of mobile infrastructure to support very high-bandwidth eMBB might entail the rollout of very high-frequency millimeter wave (mmWave) radios. The coordination of those bands with lower-frequency mobility bands might require phase synchronization between the cooperating radio transceivers (and the mmWave radios are likely to be TDD anyway).

Small Cells Versus Macro Cells

An important aspect to consider is the type of cell site that is being deployed. Although it might be more correct to look at the range and capability of different cell site types on a spectrum, this section divides them into two quite broad categories, *macro cells* and *small cells*. Whichever way you want to define them, the important thing is to understand how the category of cell site influences the timing solution—a timing solution that might work for one type might not be suitable for another.

Generally, these different types of radio sites are deployed to meet local conditions and requirements and include the capability to hand off and coordinate between them. They are deployed alongside each other and in overlapping coverage areas in what is called a *heterogeneous network*, or HetNet. The HetNet approach is used to provide a mosaic of coverage, with complex interoperation between the radio site and even with other access technologies such as Wi-Fi.

The degree of coordination between these sites can vary on a spectrum from quite loose on one end to very tight on the other. In the tighter coordination, technologies such as CoMP are used to service the subscriber device using the radios working in concert. Other deployments might work more independently of each other, but still require ICIC and eICIC to stop the overlapping radios from degrading each other's signal. Chapter 10 has a lot of information on these techniques.

For some additional reading, MEF 22.3 also gives a good overview about radio coordination; see sections 5.2.1 and Appendix D, both named "Radio Coordination."

A macro cell is what you might see when driving along a major road, and it would feature some or all the following characteristics:

- A large tower or radio antenna with good sky view (for GNSS antennas).

- Deployed widely in rural areas, but less so in dense urban areas. Macro cells tend to have high output, long range, and be spaced relatively widely apart. They are designed more to provide mobility and coverage rather than high-bandwidth broadband, so they tend to transmit in the lower bands.

- Are older installations originally built for previous generations of mobile.

- Connect to the backhaul network via specialized transport technology, such as microwave. The remote location means that fiber connections might not be available and accurate timing may not be available without remediation and upgrades to the deployed equipment.

- Feature expensive radios with good oscillators (even rubidium) for holdover.

- Cell site equipment is generally secure and inaccessible to the public. Most times the equipment to run the site is locked in a shelter with an enclosed fence and monitored by cameras.

- May already feature GPS receivers. In some parts of the world (such as in North America), the macro cell might already have GPS receivers built in, for example, to support previous radio standards such as code-division multiple access (CDMA).

- Substantial footprint of equipment. There will be infrastructure at the site to support power, space, environmental controls, and so on (although this can be constrained in more remote locations, for example, where mains power is not available).

Meanwhile, the small cell has the following characteristics (engineers might consider microcells, picocells, or even femtocells as small cells):

- Relatively low-cost devices deployed in urban canyons, indoors, or under cover. Therefore, either there is no option to deploy GNSS or it is available only at considerable cost (to buy access to a good sky view, rent space for the GNSS antennas, and gain permission to run the antenna cables back to the indoor radios).

- Deployed mainly in areas of high population density and sheltered locations with poor coverage (so, not in rural areas). Small cells tend to have lower output, have shorter range, and be deployed in very large numbers in high-traffic areas. They are designed more for coverage inside buildings or increased bandwidth than to support high mobility. They could also use higher frequency bands and even mmWave.

- Frequently newer deployments in shared locations and in areas that cannot be well secured by the operator. It could be easy for unauthorized personnel to gain access to the equipment, or at least access the cables connecting it to the backhaul network.

- Good connection options to the backhaul network. The site would normally be well served by fiber or *Carrier Ethernet* (CE) solutions, allowing for simple deployment of a timing over transport solution.

- Poor oscillators with almost no satisfactory holdover performance. Their timing performance is normally poor when deprived of a reference signal. For example, they might have no support for SyncE at all, and once they lose their PTP timing signal, they just switch off their transmitters rather than attempt holdover.

- Coordinate their radio signals with macro cells that could be providing overlapping coverage. Imagine a large macro cell situated beside a major highway rest stop, but the restaurant, restrooms, and surrounding rest areas are covered by small cells.

- Lacking substantial environmental control and supporting infrastructure (being indoors). Alternatively, they can be deployed in outdoor cabinets or in limited spaces (such as street poles). This means that they can suffer from conditions such as extreme temperature variations.

Figure 11-2 shows the difference between a macro cell on the left (it shows only the top of the mast) and a small cell on the right (outside a building in Bangalore). The macro cell has three clear sectors (120 degrees each) and the small cell has a GNSS antenna attached to it.

Figure 11-2   *Macro Cell Versus Small Cell (Sources: left, © Justin Smith / Wikimedia Commons, CC-BY-SA-2.5; right, © Rohanmkth / Wikimedia Commons, CC-BY-SA-4.0)*

The correct approach to providing timing for these different categories should become clear when reading their main features. These two broad categories need to be handled differently, certainly from the following aspects:

- Ability to deploy a GNSS-based solution

- Level of holdover performance (oscillator quality)

- Resilience to failure and need for backup or redundancy

- Transport options toward the backhaul network (and to neighboring sites)

A large macro cell (in North America, anyway) might already have GPS equipment with a good stable oscillator, so providing any form of timing over the transport or some redundancy scheme might be unwarranted. It already has a source of frequency, phase, and time with good holdover performance, and additional effort might not be required.

On the other hand, the small cell has none of those advantages. If small cells are all located indoors, that would make it very difficult to provide GNSS receivers for every small cell in that facility. For something like a campus or shopping center, for example, it might make more sense to deploy a GNSS receiver with a PTP grandmaster (GM) in a rooftop location and carry phase/time to every small cell over the local LAN cable infrastructure by using PTP.

For resilience, losing the function of a single small cell in a shopping center might have limited effect. Frequently, there is overlapping coverage from other cells, and an outside macro cell can also cover any dead spots. For operators, it might not be worth trying to provide expensive redundancy for what are relatively cheap devices if enough of them are deployed, and it is the same for timing. Of course, using a single centralized GNSS receiver and GM with PTP can take the whole site off the air (except for what an overlapping macro cell could service) should it ever fail.

On the other hand, a macro cell on a mountaintop going offline (by losing its timing signal) might take down every subscriber in the valley below, so redundancy is a more important consideration. An advantage to these high locations is that GNSS signals are generally coming from well above the horizon, so a shielded antenna that blocks signals from below the horizon greatly reduces the risk of jamming. Any possibility of short-term jamming can be handled by robust holdover performance.

For holdover, the small cell is probably going to be quite poor in holdover performance, so that it needs assistance from the transport network to maintain an accurate timing signal. It makes sense to provide resilience should the local GNSS receiver fail or get jammed. This requires putting another receiver at an alternative location and bringing a backup source of timing into the local campus from there. The alternative site should be out of range of a local jammer, at least 2–3 km away for urban sites or 5–10 km in more suburban or rural settings.

Transport capabilities are also an important consideration. If the transport network is not timing-aware, then placing a GNSS receiver without a form of backup is the best worst option. For the example mountaintop deployment, it is very unlikely that such a site would be well-served by fiber links, and it might be reachable only by microwave. Providing a timing solution over the transport might require modification of that microwave equipment. As previously discussed, for many locations, that might be neither necessary nor cost efficient.

Redundancy

In the section on small cells and macro cells, one aspect to consider in the design of a timing solution is redundancy and resilience. Given the deployment location and type, sometimes redundancy can be improved without spending a lot of money. For example, many small cells might be connected to the backhaul network via only a single network link. It makes little sense to provide extensive redundancy in the timing solution when the link to the device is a single point of failure. Deciding on the level of redundancy required is a cost/benefit trade-off for the operator.

For the point of view of redundancy for a timing solution, the following is a list of the aspects to consider:

- Failure of the GNSS source of time signals, both local (such as a local jamming event) and systemic (the entire constellation has some problem).

- Failure in the transport carrying a timing signal from a remote primary reference time clock (PRTC) and GM (assuming the use of SyncE and PTP).

- Redundancy of PRTC and GM sources, spaced widely around the network but not too far away from the cell sites that need to consume the signals.

- Increasing the duration of holdover; for example, by using a physical source of frequency as a method to enhance holdover (see the section "Holdover" later in the chapter for more information).

- Designing the network to ensure redundancy in the SyncE signals.

- Deploying several extremely stable reference clocks (such as atomic clocks) to allow autonomous operation of the timing solution (meaning it can run without an external source).

- Methods that improve the ability of a PTP-unaware network to accurately carry timing signals during a GNSS outage in the local area. Figure 11-3 shows how a device at the end of the network, such as a cell-site router, can have a local primary source of time with a PTP slave port recovering clock from a remote T-GM.

Note   Use of the terms "master" and "slave" is ONLY in association with the official terminology used in industry specifications and standards, and in no way diminishes Pearson's commitment to promoting diversity, equity, and inclusion, and challenging, countering and/or combating bias and stereotyping in the global population of the learners we serve.

One commonly proposed method is to also utilize the assisted partial timing support (APTS) technique as outlined in ITU-T G.8273.4, where the connection to the remote T-GM is across an unaware network.

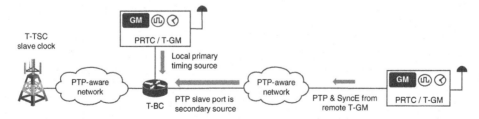

Figure 11-3 *GNSS Primary Source with Remote PTP Backup*

The following sections address each of these considerations in turn.

GNSS Redundancy

Currently, there are only limited alternatives available to replace GNSS systems as an accurate source of frequency, phase, and time at a remote location or for an isolated piece of equipment. See Chapter 3, "Synchronization and Timing Concepts," for much more information on the features and capabilities of the various GNSS systems.

This lack of an alternative worries experts who are concerned about the vulnerability of GNSS systems. That risk can arise from bad actors denying the signal or even normal events such as disruptive space weather conditions or even human error. There are several organizations lobbying for alternative methods of timing distribution beyond the current space-based radio systems, one being the *Resilient Navigation and Timing Foundation* (https://rntfnd.org/).

One obvious way to provide backup would be to build another, much more powerful, terrestrial radio signal for positioning, navigation, and timing (PNT). The U.S. Congress has initiatives in this area, although it does not seem to be moving with any great urgency. For further information, please consult the *National Timing Resilience and Security Act* of 2018 and the U.S. President's Executive Order 13905, "Strengthening National Resilience Through Responsible Use of Positioning, Navigation, and Timing Services."

This section is not meant to be a comprehensive guide to every method available to harden a GNSS-based timing system, but here are some of the options available that give the operator a better chance of surviving an event:

- Use hardened and resilient receivers, with better anti-jamming and anti-spoofing capabilities. These mechanisms are not foolproof, but they help. Previous historical events have shown that many older models of receiver are quite fragile.

- Incorporate improved antenna technology, an area that does not receive nearly enough consideration. Even basic shielding and a more directional antenna can help mitigate the signal from ground level jamming.

  For example, there is a class of antenna known as controlled reception pattern antennas (CRPA), which use adaptive antenna techniques to accept signals from the direction of the satellites and ignore the jamming signal. These devices are still high-end units with primarily military applications, but antenna design is a very important factor for the resilience of a GNSS receiver.

- Implement multiband, multi-constellation receivers and use any available augmentation signals (see Chapter 3). For example, Galileo had an extended outage during July 2019, and even GPS had an "anomaly" on January 25–26, 2016 due to an error in a data upload. Having a receiver that can survive outages or instability in a whole system is a clear advantage.

  Happily, most receivers produced in recent years have multi-constellation capability, so be sure to activate the feature and make sure that the antenna systems and radio filters allow reception of the desired signal bands.

■ Use a remote source of frequency, phase, and time, such as PTP with SyncE. This will help the cell site survive a local GNSS outage. If the transport connected to the cell site has PTP-unaware elements in it, then using APTS is an obvious solution (see the section "Assisted Partial Timing Support with G.8275.2" later in the chapter).

Engineers understand that a major outage of GNSS-based systems is not a matter of "if" but a matter of "when." For example, one risk that is increasing up until a predicted maximum in 2025–2026 is the worsening space weather resulting from the next peak in the solar cycles. Whatever the cause, at some point there will be a widespread outage.

PTP and SyncE Redundancy

It is obvious that any timing solution should not rely on a single point of failure, and this certainly applies to the reference sources of frequency, phase, and time. One common oversight that many engineers have is to focus more on the resilience of the PTP signal and not take full advantage of what SyncE can do to build a robust and accurate timing solution.

It is vital to have multiple sources of PTP available within a limited number of hops away (how many hops depends on the timing budget and the accuracy of the intervening boundary clocks). This does not mean that putting one PRTC+GM in the capital city and another in the second largest city of a country is a good design. For a large, complex network, there might be tens or even hundreds of these devices deployed further out in the network in the *aggregation* or *pre-aggregation* layers.

These PRTC+GM devices can be standalone devices from a vendor dedicated to producing them, or they can be embedded in a network router or other piece of equipment. Sometimes they are no larger than small form-factor pluggable (SFP) that can be plugged into a router's Ethernet port to provide SyncE and PTP to the router.

The best performance for carrying timing signals across a transport network comes from combining PTP with a physical frequency signal such as that from SyncE. SyncE has one other great advantage, which is to provide an extended holdover should the PTP signal disappear (see the following section "Holdover"). For this reason, the SyncE flow throughout the network should be designed to provide the maximum level of redundancy.

In the case where a small cell is deployed in a location that is served with only a single network link, not much can be done because the single backhaul link is the single point of failure. In fact, many small cells might not be able to use a SyncE signal at all and might only function with a valid PTP feed, so redundancy of a source of PTP is more important in this case.

If a larger macro cell is connected by multiple network links, however, then it is normally no extra cost to run SyncE on each of those links, or at least on enough of them to survive multiple link failures. Normally, having SyncE is just a matter of configuring it to be enabled, so it is a very good idea to have it on and not need it rather than need it and not configure it.

Holdover

The first thing to understand about holdover is just how hard it is to do for long periods with everyday equipment at reasonable cost. So, it is important to do some expectation setting. The authors frequently answer detailed questionnaires on timing requirements from mobile operators. These questionnaires are part of a process of selecting the network equipment that an operator wants to use in its 5G-ready backhaul and fronthaul network (see the section "Writing a Request for Proposal (RFP)" in Chapter 12, "Operating and Verifying Timing Solutions").

It is not unusual to see a holdover requirement for a cell-site router (CSR) that must be able to hold the 1.1-μs phase accuracy for TDD radios for wildly unrealistic periods. Many times, operators purchasing new equipment require that CSRs should support a holdover of 24–72 hours. Some others ask for multiweek holdover or even up to several months! This is possible, but only in combination with (very expensive) atomic clocks. The reality is that an operator should only expect to keep that level of phase alignment for several hours in an everyday network element.

As the name suggests, the CSR is a device sitting in every cell site, and, obviously, because of the large numbers of cell sites, cost is a major factor in deciding on the win-ning vendor. But the technical specification might specify that this multiweek holdover is a mandatory requirement—meaning that the device will be denied any chance of selection unless it complies. And yet, the purchaser expects the CSR with all its Ethernet ports, components, and features to cost less than 20% of the cost of a device that can meet this holdover requirement.

The ideal situation would be to put an enhanced PRTC (GNSS receiver plus cesium clock) at every cell site. That would deliver amazing accuracy and very long-term holdover capa-bility. But that would cost something like \$50K–\$100K per site and OPEX of around \$5–7K per year. A simple and totally robust timing solution is simple to deploy if one has unlimited funds.

However, for a normal NE or router with a Stratum 3E oscillator, you should expect a standalone 1.1 μs phase holdover time of around 4–8 hours, not days or weeks. Of course, in combination with SyncE (assisted holdover), this time improves dramatically. The way to deliver robust, long-term holdover at reasonable cost is in the design of the network solution, not relying on gold-plated capabilities of the nodes in it.

Note that there is both frequency holdover and phase holdover, although they are related because both holdover types rely upon the signal from the same oscillator. Frequency holdover is usually quite good, but it is phase holdover that quickly moves out of speci-fication and causes issues. The reason that phase moves more quickly than frequency is that a small offset in frequency continues to accumulate phase error although the fre-quency might not wander any further. For example, frequency could be very stable with an offset of 10 ppb, which is still very accurate, but it means phase is accumulating 10 ns of phase error every second.

There are two forms of holdover, which are normally discussed separately. *Short-term holdover* (*transient response*) is due to transients or temporary rearrangements in the network. This is a normal situation due to everyday conditions in a network, links bouncing, node failover, fiber cuts, power outages, and so on. This occurs regularly in a timing network and should be of little concern because timing can disappear for a short while with no impact on the radio network.

The second form is *long-term holdover*, which is normally when the chain of traceability from the source of frequency, phase, and time to the end radio application in the radio site is broken. The long-term holdover measures the ability of a cell site to survive as an autonomous timing clock without those reference signals (this is what most people mean when they talk holdover).

ITU-T specifications for clock performance (such as ITU-T G.8273.2) normally include separate requirements for transient response and long-term holdover, both with and without physical frequency (SyncE) support and for both constant and varying temperature. For example, G.8273.2 (10/2020) Table 7-9, titled "Performance allowance during loss of PTP input (MTIE) for T-BC/T-TSC classes A and B with constant temperature," indicates that a clock can increase its MTIE by 25.25 ns per second and still pass. Think how badly misaligned the phase would be after several weeks of that level of holdover performance.

Third-Party Circuits

An important consideration when building an accurate timing distribution network is the characteristics of the backhaul and fronthaul network. Depending on the situation in individual markets, to connect their cell sites to the mobile infrastructure, the mobile service provider (MSP) may either substantially own its own network or rely upon other service providers.

These third-party organizations offer services such as Metro Ethernet or Carrier Ethernet circuits to the MSP, so this book will refer to them as Carrier Ethernet (CE) service providers (SP). They are also referred to by numerous other names, including *alternate access vendors* (AAV) or third-party circuit providers. Figure 11-4 shows the basic service offered by a CE SP.

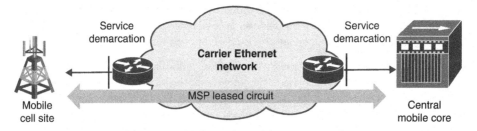

Figure 11-4   *Carrier Ethernet Service Provider*

Many of the services offered by the CE SP are defined by the *MEF Forum* (MEF), which is an industry association of more than several hundred member companies. MEF has released *Implementation Agreements* (IA) that detail the transport services needed to support the backhaul needs of mobile networks. The appropriate documents to consult are known as 22.3 and the amendment to it as 22.3.1. For more details on MEF, see Chapter 4, "Standards Development Organizations."

The first problem with using CE SP circuits is that any PTP- (and SyncE-) aware nodes in the third-party network can only run a single PTP clock instance belonging to a single clocking domain. Any significant CE SP is not likely to allow its own CE nodes to run as clocks in the timing domain of an external MSP. Even if the CE SP did allow this, there is no possibility for the CE SP to be able to host a second MSP because its own nodes are already clocks belonging to the timing domain of the first MSP.

One option is that the CE SP configures its network nodes as transparent clocks (TC) because a TC can be used to update the correctionField in PTP messages belonging to multiple domains. SyncE has no equivalent capability, and unless the hardware used to build the CE network changes fundamentally, this restriction is not likely to be lifted soon. There are some innovative suggestions as to how to facilitate a network in multiple timing domains, but doing so is not a realistic option with any currently available technology. Note that MEF has marked the transport of G.8275.1 across a CE network with TC support as a use case "for further study" (see MEF 22.3: 13.4, "Performance of time synchronization architecture").

For MSP, the only option is to ask the CE SP to carry its PTP traffic transparently. This could involve using PTP over IP (PTPoIP) or some other technique such as encapsulating the PTP (L2 or L3) in a VPN, tunnel, or pseudowire. No matter which technique is used, the PTP will be transported without any underlying support from the network (and no SyncE).

This introduces all sorts of issues with the network performance and accuracy of the timing signal. For this to work, the MSP must be able to specify and order a service (or circuit) that it can rely on to carry PTP accurately but without on-path hardware assistance. This means being able to order circuits from the CE SP that have very limited PDV and almost no asymmetry. Delivering and operating such circuits under varying network traffic conditions is very difficult.

The other problem is, how can the MSP monitor the timing and be assured that the phase is arriving accurately when it appears at the cell site? The MSP can use APTS to be able to correct for asymmetry in the CE network, but that has its own trade-offs (see the section "Assisted Partial Timing Support with G.8275.2" later in the chapter). For both local phase/time assurance monitoring and APTS deployments, a local GNSS receiver would be needed at the cell site and the CE SP would be merely delivering a backup source of time via the PTP.

On the other hand, the CE SP must guarantee that accurate phase/time can be recovered by its customers when the PTP the CE SP is carrying for them is receiving no on-path assistance. To be successful, the CE SP must design and deliver a service that has only a few hundred nanoseconds of asymmetry and extremely low PDV. The CE SP must also be able to monitor the timing performance of this network and demonstrate compliance to its service level agreements (SLA) for timing performance. This requires careful network engineering. MEF has defined some of these service characteristics in MEF 22.3 section 12.3.1, "Performance for Synchronization Traffic Class," and section 13, "Synchronization."

See the MEF website (https://www.mef.net/) for its documentation on providing time-aware backhaul services.

Basically, the best option from a timing point of view is being able to lease dark fiber, but this can be difficult to procure in every location where it is needed by the MSP. The other alternative is for the CE SP to offer to deliver *time as a service* (TAAS).

Time as a Service

The earlier section "Third-Party Circuits" shows that there are basically three options available to carry timing where the MSP does not own or control their own backhaul network. The options for the CE SP are

- Carry the MSP (packet) timing signal transparently using some tunneling technology without any attempt to provide on-path hardware assistance at the intervening nodes in the CE network.

- Carry the MSP timing signal using a TC at every step in the CE network to update the correctionField of the transit packets. For this to work, the PTP messages must be visible to the underlying hardware (the tunnel labels and tags might make it difficult for the TC to recognize PTP messages).

- Provision/sell some timing transparent circuit/service—such as dark fiber—that has no active components between the two ends of the fiber.

There is a fourth option available, and that is where the CE SP delivers an accurate (but independent) PTP message stream with a stable SyncE at any location where the MSP needs it. The major difference is that the timing signal comes from inside the CE network and is not part of the MSP timing domain. For this reason, it is frequently called time as a service.

Figure 11-5 illustrates how this works.

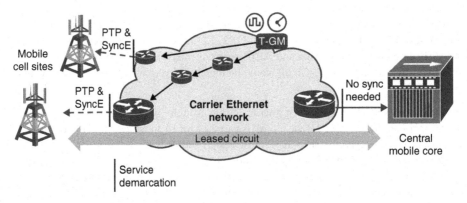

Figure 11-5 *Carrier Ethernet SP Offering Time as a Service*

This is similar in concept to having a standalone GNSS receiver at the cell site. If the frequency is within specification (traceable to PRC/PRS) and the phase is well within 1.1 μs of UTC, then it does not matter where the timing signal comes from. Of course, the MSP must ensure that its timing domain is closely aligned to UTC for this scenario to work (the MSP could not run an autonomous timing source without it being calibrated or aligned to UTC).

So, a CE SP can build its network into one that supports accurate timing and use it to offer MSP and other customers a traceable timing signal as a service. If that supplied (for example) 200–300 ns offset from UTC and an accurate SyncE, most operators would leap at it. Many would happily pass their timing distribution problems onto somebody else.

Of course, the attractiveness of such a service comes down to cost, but the authors are of the opinion that it would gain a lot of interest and are aware of several large SPs that are currently trialing or planning such a network.

Frequency-Only Deployments

The solution options for deploying frequency timing are covered in Chapter 9, "PTP Deployment Considerations," for packet and in Chapter 6, "Physical Frequency Synchronization," for non-packet (such as SyncE). This section focuses more on the deployments typically selected by most mobile operators. See the relevant chapters for other use cases such as circuit emulation.

A source of accurate frequency has always been a required element of radio installations, going back almost a full century. The thing that has changed is the method of achieving it. For MSPs, the options were typically provided by the PDH/SDH/SONET backhaul network; however, for 3G and later network generations, those options began to disappear. The following sections examine what sort of solutions the mobile operators have deployed for frequency synchronization.

Solution Options

In summary, here is a list of the possible options to deploying a timing solution to support accurate and stable frequency synchronization:

- GNSS receiver inserted everywhere that a frequency signal is needed and feeding a physical TDM circuit (such as E1/T1 SSU/BITS), or a 2-MHz/10-MHz signal over a short-run cable.

- A standalone atomic clock acting as a PRC/PRS wherever frequency is needed.

- TDM E1/T1 or SSU/BITS network link to bring frequency from a remote PRC/PRS source over a frequency distribution network.

- Synchronous Ethernet (SyncE) frequency from a remote PRC/PRS source.

- Packet-based methods and protocols such as PTP or NTP, or some other packet technique such as adaptive clock recovery (ACR) or differential clock recovery (DCR). See Chapter 9 for more details of these solutions.

- Some radio technique; this is no longer common, although ground-based radio technology (for example, eLORAN) might make a comeback as alternatives to GNSS receivers.

First, deploying atomic clocks (as a PRC/PRS) wherever frequency is needed is possible, but it is prohibitively expensive unless it is needed at only a few locations. The normal way would be to use these clocks as sources of frequency placed at strategic sites and then transport frequency via the network. Traditionally, that was how SSU/BITS networks were designed.

The vast majority of MSPs used atomic clocks to provide a frequency signal and then built a standalone SSU/BITS network to carry that signal to their main facilities, central offices, and points of presence (POP). To get that frequency from the local POP out to the mobile cell site, the TDM-based backhaul link from the nearest POP carried it via line-clocking along with the data and voice circuits.

As TDM was replaced (eventually) by Ethernet, SyncE became widely adopted, but only where it was possible, given that technical constraints exist. For instance, for sites that are difficult to reach, the only transport available might have been microwave, which originally offered only limited support for SyncE. For these and similar reasons, most MSPs settled on one of the following options:

- Install a GNSS receiver at the cell site giving frequency with a short-run cable (such as 10-MHz signal over coax cable)

- Maintain a legacy TDM network link (at least one) to the cell site to carry only frequency (it was no longer required for voice and data, as that was using Ethernet)

- Bring a SyncE timing signal over the backhaul network where technically possible

- Transport frequency with a packet-based method, almost exclusively using the G.8265.1 PTPoIP telecom profile for frequency (at least one major mobile equipment vendor used NTP instead)

To understand the trade-offs when using deployments based on GNSS, refer to the upcoming section "GNSS Everywhere" because the trade-offs are identical to using them for phase/time use cases.

Using the frequency recovered from a TDM backhaul circuit is what many SPs already have a long history with. But many are now getting rid of it, as it means maintaining another separate network (such as SSU/BITS) solely for the purpose of carrying frequency. Not only that, but the equipment is very expensive and, in many cases, becoming impossible to maintain as vendors drop support for TDM and therefore the cost to maintain these links soars. Another factor was that cell site routers increasingly began to drop support for TDM connectivity. Therefore, many SPs are actively decommissioning or have already decommissioned TDM connectivity.

The current situation is that SyncE is the recommended solution because most equipment in the backhaul network supports it natively, and many transport options increasingly support it. With the improved performance of the enhanced eSyncE now becoming available, nothing can match it for frequency stability.

The last alternative is to use a packet-based transport for frequency, which normally means PTP. This solution is recommended only where SyncE is not available as an option. NTP did have some role to play here, but operators regarded it as proprietary and only one vendor widely adopted it. It is no longer being actively proposed. And for PTP, it really makes sense to use the ITU-T G.8265.1 profile because it is designed to interoperate with SDH/SONET and SyncE (although you could recover frequency with G.8275.1 and G.8275.2).

The next question to address is, what conditions would cause an engineer to choose SyncE rather than PTP or vice-versa?

G.8265.1 Packet-Based Versus SyncE

At several points, this book examines the strengths and trade-offs between using PTP or SyncE to carry frequency. In summary, the main points for the MSP to consider in the decision include the following:

- Using G.8265.1 is easy because there is no on-path support needed (it is not even allowed). This means that an implementation only requires a PTP GM next to a PRC/PRS and a PTP slave clock where the frequency is required with an IP network connecting them together.

- The G.8265.1 telecom profile carries PTP with IPv4 (IPv6 optional), so transporting the PTP and configuring the paths between the slave and master is straightforward.

- The main downside of the PTP approach is that the engineer needs to minimize any PDV in the network for the slave to be able to accurately recover frequency. This means (as a minimum) the configuration of high-priority QoS on the intervening nodes carrying the PTP. See Chapter 9 for more information on the effect of PDV on end-to-end timing performance.

- SyncE is much quicker to align, more stable (not influenced by traffic load and PDV), and easier to manage and run.

- SyncE deployment does require the engineer to design loop-free timing paths and manually configure the interfaces to complete the path.

- Besides support in each intervening node, SyncE also requires every type of transport system to carry frequency accurately and transparently (examples would be microwave or optical).

As a rule, the recommendation is to run SyncE wherever you can and PTP wherever you must. Because SyncE and G.8265.1 are designed to easily interoperate, a mixed deployment is quite straightforward. This allows relatively seamless conversion between TDM, SyncE, and PTP.

Frequency, Phase, and Time Deployment Options

As previously pointed out, all 5G networks, and an increasing number of LTE-A networks, are requiring traceable phase and time signals in addition to the existing need for frequency. The following section of the book covers the various alternatives to delivering accurate frequency, phase, and time to the cell site, with recommendations and best practices.

This section concentrates on the main ±1.1 µs phase alignment to UTC at the point of demarcation to the radio device from the backhaul network. The extra requirements needed in midhaul and fronthaul networks that result from the most recent innovation in the RAN are covered later in the chapter in the section "Midhaul and Fronthaul Timing." You need to understand the details of the backhaul case before looking to extend this knowledge to the RAN.

The major options considered by most MSPs include the following:

- Install a GNSS receiver as a PRTC everywhere that synchronization is required.

- Distribute timing signals from a remote PRTC over the backhaul network with full timing support from the network in combination with SyncE.

- Distribute timing signals from a remote PRTC over the backhaul network with partial timing support from the network (SyncE optional).

- Use a combination of the preceding two methods with interworking between them.

The GNSS-only (or GNSS-mostly) case is covered in the upcoming section "GNSS Everywhere," although some of the benefits and drawbacks of using only a GNSS solution are covered in multiple places throughout the book, most recently in the section "Redundancy" earlier in this chapter.

Network Topology/Architecture for Phase (G.8271.1 Versus G.8271.2)

There are two basic network topologies available when building a timing distribution network, and each of the cases is covered by an ITU-T recommendation:

- **G.8271.1:** The MSP owns its own transport infrastructure and has the possibility to build and support a fully aware PTP and SyncE network.

- **G.8271.2:** The MSP is leasing CE and other third-party circuits and for that reason (or some other constraint) is not able to build a fully aware timing network.

The first topology, using PTP and SyncE carried over an FTS network, is the best option because it delivers best-in-class performance and is being widely deployed and supported by the timing industry. The second case, using PTP over a PTS network, is the second-best option and requires careful engineering to be able to meet the timing performance requirements. See Chapter 9 for the details of these reasons.

The first case requires the use of the ITU-T G.8275.1 telecom profile (PTPoE) in combination with SyncE. Of course, every node between the GM and the final slave device must be timing-aware for it to be an FTS network. This means that every active network element must be either a Telecom Boundary Clock (T-BC) or a Telecom Transparent Clock (T-TC).

The second topology is implemented with the ITU-T G.8275.2 telecom profile (PTPoIP), with SyncE being an optional assistance to carry the frequency. It is very important that every node in the PTS network that can support being a T-BC or a T-TC is configured as such. The more unaware nodes in the chain between the master and the slave, the more difficult it will be to get accurate clock out to the cell site.

There is substantial variation in the performance characteristics between the G.8275.1 and G.8275.2 models of PTP transport. This results from the major difference in the two profiles: one being at layer 2 and hop-by-hop, and the other being at layer 3 (IP) that can transit network nodes that will not process PTP. Table 11-1 describes the main trade-offs between the two approaches.

Table 11-1 *Fully Aware Network Versus Partially Aware Network*

Feature	Fully Aware Network	Partially Aware Network
Network model	G.8271.1	G.8271.2
Telecom profile	G.8275.1 PTPoE	G.8275.2 PTPoIP
IP routing	Not applicable	Problems with rings, asymmetry
Transit traffic	Not allowed	Problems with PDV, asymmetry
Performance	Best	Variable, nondeterministic, and dependent on traffic load and type
Configuration model	On the physical port	Attached to an L3 device
PTP over bundles	No issue	Solutions available for T-BC connected to another T-BC; otherwise a problem
Asymmetry	Reduced	Can have high dynamic asymmetry
PDV/jitter	Timestamping on wire and the lack of transit nodes limits its buildup	Uncontrolled amounts of PDV/jitter introduced at unaware nodes, which accumulates the further along the chain

These differences will be explained in the following sections of this chapter.

GNSS Everywhere

At the first glance, the simplest solution is to place a GNSS receiver at every location that needs frequency, phase, and time. This is a good solution and has been deployed in some parts of the world starting from several decades ago (such as North America CDMA networks that needed phase alignment). There are many vendors supplying products for this purpose. There are even pluggable SFP devices that can provide PTP and SyncE to an Ethernet port for a local source of PTP and SyncE on the router.

However, this solution has some considerations that need to be addressed:

- **GNSS solution only suitable for some sites:** GNSS is a good solution for sites, typically macro sites, where there is a large tower with a good view of the sky and space to support a straightforward installation. Unless the tower is owned by the MSP, however, the cost of renting a spot on the tower to accommodate the antenna can be significant.

  This approach is less popular with current 5G rollouts, because 5G needs a lot more radios to be deployed (referred to as *densification*). These new radios are increasingly sited indoors or in urban canyons where a sky view is impossible or severely limited (see the earlier section "Small Cells Versus Macro Cells").

  In these urban environments, besides not having a good sky view, obtaining permission for cable access to a rooftop antenna can be a problem (and an expensive one). So, a solution is needed for those environments. One solution is to put one or two GNSS receivers somewhere in the facility and run PTP and SyncE locally around the site (such as around the building, stadium, or campus).

  There are also GNSS antenna-like devices that combine the antenna with a GNSS receiver and a PTP T-GM. These are located at the roof site and are connected to copper Ethernet cables so that they can use Power over Ethernet as a power source. This device allows the timing signal to be carried as PTP and SyncE over Ethernet twisted pair to the radio equipment. This can be much easier to deploy than having to carry GNSS RF signals over unwieldy and expensive coaxial cable from the antenna to the receiver inside the building.

- **Resilience in the case of local (or system) GNSS outage:** Multi-constellation devices can help survive a systemic issue with one of the systems but may not help with a local outage, such as jamming (discussed in the following bullet point). Note also that there is a (small) possibility that space weather events could disrupt all space-based signals.

  One answer to these vulnerabilities is to deploy the GNSS receiver with an atomic clock holdover—which is very expensive. The other approach is to have multiple backup PRTC and T-GM sites located outside the area potentially disturbed by any local event. Time is then transported with PTP and SyncE between all the facilities that need a synchronization source but have lost their local GNSS signal.

- **Susceptibility to jamming, either intentional or incidental:** This is becoming a real problem, although (good) receivers are aware that jamming and signal denial events are underway and can protect themselves. Having multi-constellation receivers might not help, because these jamming devices tend to block all GNSS signals, not just the L1 signal from GPS (for example).

The main remedy is to have multiple backup PRTC and T-GM sites with sufficient geographical dispersion so that at least one GNSS source is outside the range of the jammer. Luckily, most jamming events cover a limited area, but it is possible, particularly with a state actor, to suffer a severe denial of signal over a very wide area. In that case, the only remedy might be to call the air force. Some countries have developed ground-based alternatives (such as eLORAN) to rely on in case of interference with space-based signals (see the section "Resilience" in Chapter 3).

- **Spoofing:** A bigger threat, increasing all the time, is the ability to "spoof" the GNSS signals to mislead the receivers into receiving the incorrect time. The receivers might not always be able to detect these attacks, especially from sophisticated actors. Multiband and multi-constellation devices can help mitigate this threat for now because this makes spoofing much harder (although an attacker could attempt to spoof one GNSS signal while jamming the other constellations and bands). As mentioned previously, there are options such as antenna technologies becoming available to help filter out false signals—which helps in very high-threat environments.

    GNSS systems are also moving to authentication mechanisms to ensure the signal is valid. One example is the Open Service Navigation Message Authentication (OSNMA) developed by Galileo. It makes use of the Timed Efficient Stream Loss-tolerant Authentication (TESLA) protocol for authentication. There are already receivers available that support this feature and, at the time of writing, the signals are already available for testing.

- **Geopolitical risk:** Some operators do not want to rely on devices that have "geopolitical risk," whether real or imagined; however, in some geographies, bad actors (including nation states) are always a real and substantial threat to a clean GNSS signal.

GNSS-based systems are wonderful tools, and their use as sources of time is very valuable, but their drawbacks must be understood and mitigated. The operator needs to design and prepare the network for that day when the space signal is either not available or becomes untrustworthy.

Full On-Path Support with G.8275.1 and SyncE

The first network topology to consider is the case where the network provides full on-path support at the protocol level. For this case, the ITU-T G.8275.1 telecom profile combined with SyncE is the best-in-class method to carry frequency, phase, and time across the transport. Most operators globally are using this approach wherever it is possible for them to do so.

Figure 11-6 illustrates what the network looks like in such a deployment. The timing signal is sourced from a PRTC (typically a GNSS receiver) either as a standalone device or embedded in some aggregation router. These signals are used to generate a SyncE signal for the frequency, and the T-GM function generates a PTP message stream for the phase and time.

Figure 11-6 *Transport of PTP and SyncE Across an FTS Network*

Every network element in the path between the T-GM and radio equipment is configured as a T-BC operating in *hybrid mode* with SyncE for frequency (it can also be a T-TC). Typically, the radio equipment has the last PTP clock in the chain embedded inside it and functions as a Telecom Time Slave Clock (T-TSC).

For a successful deployment, the network design must address the following points:

- Every node in the path is required to support PTP and SyncE, meaning that each node needs to be a T-BC or a T-TC (up to a maximum of eight T-TC clocks). This PTP support cannot be just some random PTP profile, but specifically the PTP G.8275.1 telecom profile, together with SyncE carrying frequency in hybrid mode. See Chapter 12 for details on specifying the requirements for network devices.

- The level of performance of any PTP-aware device must meet the performance specifications for boundary/slave clocks in G.8273.2 and/or transparent clocks in G.8273.3. Similarly, the SyncE clock must meet the specifications for either G.8262 or G.8262.1 Ethernet clocks.

  Obviously, the better the performance of the individual T-BC or T-TC, the more accurate and stable the timing signal will be at the destination and the more hops it can be carried without excessive degradation.

- The transport system is required to support PTP "awareness" from end to end. Any device in the transport system that processes packets in any meaningful way needs mechanisms to limit the timing noise and phase error that it introduces.

- The transport system is required to support transparent carriage of SyncE. Any transport device between network elements must be able to faithfully carry the frequency without introducing excessive jitter and wander. It must also honor and faithfully replicate the Ethernet synchronization messaging channel (ESMC) quality level values from SyncE.

- The network must introduce only a very small amount of asymmetry and PDV. With full on-path support and PTP-aware transport, these factors are well controlled, but for the PTS approach, the network designer is responsible.

The main benefit to the FTS approach is the last bullet point. When a timing solution is designed according to these principles, the resulting performance of the timing is deterministic and largely predictable. It is for these reasons that G.8275.1 and SyncE carried over an FTS network is the recommended approach.

But what happens then the FTS case cannot be supported by the network and the transport system? For that case, the other option is to use the mechanism designed for the PTS case.

Partial Timing Support with G.8275.2 and GMs at the Edge

The second network topology to consider is the case where the network provides either zero or only partial on-path support. For this case, the ITU-T G.8275.2 telecom profile is deployed as the method to carry frequency, phase, and time across the transport. The use of SyncE is optional but recommended if possible. As a rule, operators are deploying this approach only in limited circumstances (where they have no other options), and in many cases, only in combination with APTS.

Figure 11-7 illustrates what the network looks like in such a deployment. The timing signal is sourced from a PRTC (typically a GNSS receiver) either as a standalone device or embedded in some aggregation router. These signals are supplied to a T-GM function that generates the PTP message stream to carry phase and time. Optionally, SyncE is used to carry a frequency signal, but if SyncE is not available, then the frequency is also carried by the PTP.

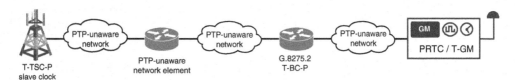

Figure 11-7   *Transport of PTP and (Optional) SyncE Across a PTS Network*

If possible, network elements in the path between the T-GM and radio equipment should be configured either as a Telecom Boundary Clock with partial support (T-BC-P) (it can also be a Telecom Transparent Clock with partial support [T-TC-P]). Typically, the radio equipment has the last PTP clock in the chain embedded inside it and functions as a Telecom Time Slave Clock with partial support (T-TSC-P). The more nodes that can be configured as T-BC-P or T-TC-P, the easier it will be to make the solution work and the better the result will be.

Because of the nondeterministic performance of the timing signal across such an unaware network, the GM needs to be quite close to the slave clock to reduce the amount of degradation of the timing signal. For this reason, some vendors refer to these distributed PRTC+T-GM as *Edge Grandmasters* because they are located toward the (outer) *edge* of the backhaul network.

Figure 11-8 shows a typical deployment topology for the PTS use case in combination with a T-GM deployed toward the network edge. The key to success is to ensure that there are very few hops between the "Edge GM" and the slave clock. If there are any significant forms of asymmetry in the network, then this rapidly becomes a problem.

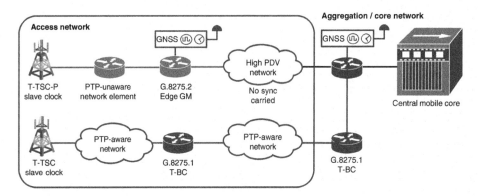

Figure 11-8 *Grandmaster Deployment at Network Edge*

For a successful deployment in unaware networks, the design must address the following points:

■ Reduce PDV to the absolute minimum. Exactly how much PDV can be tolerated is very hard to quantify.

■ Reduce asymmetry to the absolute minimum. How much is too much depends on the overall timing budget, but every microsecond of asymmetry introduces 0.5 μs of phase offset.

Reducing PDV for PTP messages passing through an unaware node is not easy. One obvious answer to limit the effects of PDV is to configure the transit node as a PTP-aware clock (TC or BC), if possible. Note that PDV accumulates for every step along the unaware timing chain, and the most effective way to control it is to use a T-BC. That is because it terminates the old PTP stream from the master and regenerates a new stream toward the next slave, effectively resetting the PDV to zero.

Consider the following points about PDV in an unaware transit node:

■ Any queuing within the node contributes to PDV. This can be partially avoided by designing a set of QoS policies that ensure the PTP traffic is treated with the highest priority.

■ The type of traffic and mix of frame lengths can contribute to PDV. In a converged network, the small PTP messages can be multiplexed in with enterprise or other types of data streams. Non-PTP traffic may consist of packets with a very large maximum transmission unit (MTU) size, what are known as *jumbo frames*. Even when PTP messages are in a high-priority queue, a jumbo frame that has already started to transmit will cause the following PTP message to have to wait.

One method to reduce this effect is to have the device support a time-sensitive networking (TSN) technique known as *frame preemption* (see the section "IEEE TSN" in Chapter 4). Frame preemption allows the transmission of a non-express packet to be interrupted and an express packet to be forwarded before recommencing the transmission of the long frame. The other end of the link does the reassembly of the preempted frame (note that this feature requires hardware support at both ends of the link).

Of course, as interface speeds increase, the time taken to send a jumbo frame decreases, and so the time waiting for transmission of those frames decreases. Despite that, this technique is still considered useful for very tight timing requirements in the fronthaul—if interface speeds are low (less than, say, 25–50 Gbps).

Remember that this applies only for cases with transit PTP traffic because a properly designed boundary clock will not suffer this problem (the timestamping is not done until the frame is ready to transmit). Similarly, for a T-TC, any waiting time for the transit PTP packet will be reflected in the value written to the correctionField, effectively nullifying the delay.

For details of frame preemption, see IEEE 802.1Q-2018 (which absorbed 802.1Qbu-2016) and 802.1CM-2018, "Time-Sensitive Networking for Fronthaul," and the 802.1CMde-2020 amendment.

■ The type and design of the network element itself generates PDV. Some devices just suffer from high PDV, because reducing it was not a design goal, or the engineers designed the device with some feature that causes it (such as oversubscription). Remember that the main issue is not the time taken to switch the PTP message but rather the variability in this time.

Reducing asymmetry for PTP messages passing through an unaware node is also not easy. As with PDV, the straightforward answer to limit asymmetry is to configure the transit node as a PTP-aware clock (TC or BC) wherever possible.

Consider the following points about asymmetry in an unaware transit node:

■ Queuing in an unaware node can cause asymmetry, especially when the amount and length of queuing varies based on the direction that the data is travelling (which is usually the case). If the PTP messages between master and slave are delayed more in one direction than the other, then that is asymmetry.

■ Routing can cause asymmetry because routing protocols can calculate a different path for traffic in the upstream direction than that for the downstream traffic. This is especially the case in rings, where a slave clock can have two paths to the same GM, but with very different characteristics (the number of hops, for instance). Swapping between those two paths can also play havoc with the recovered timing signal.

■ Traffic patterns and behavior can also cause asymmetry. If the PTP messages encounter a heavily congested traffic stream in one direction, and yet flow relatively unaffected in the other direction, that is also asymmetry. Because the traffic flows are constantly changing, this form of asymmetry is dynamic and cannot be compensated for.

- The type and design of the network element itself can cause asymmetry. Some devices suffer from asymmetry internally because reducing it was not a design goal. This arises in devices that feature design choices such as oversubscription, whereby the device (usually for cost reasons) has internal bottlenecks in the packet processing path. For example, some nodes have asymmetry in the path in the direction of the user-network interface (UNI) to network-network interface (NNI) compared to the NNI-UNI direction.

- Using PTP over bundles or aggregated links can also introduce asymmetry. This technology of forming multiple physical interfaces into one virtual link is formally known as a Link Aggregation Group (LAG) and specified in IEEE 802.1AX-2020 (formerly IEEE 802.3ad).

  When a PTP message is queued to a LAG interface, the driver must select one physical link from the group down which to forward that message. That decision may be made based on some algorithm, such as one based on hashing several fields from the IP header. Each end of the link makes that decision somewhat autonomously and therefore the link chosen by each side of the link can differ. This difference in link selection in opposite directions can also cause asymmetry.

When two PTP BCs are connected to each other, then it is possible to ensure that the PTP implementation can be enhanced to enable the engineer to configure the LAG to use a specific physical link to transmit PTP messages.

The LAG standard also includes the Link Aggregation Control Protocol (LACP) that is used to control a LAG link. It is used to communicate and negotiate parameters between the two ends of the link. When two T-BCs are connected to each other, the PTP implementation can be enhanced to use LACP to negotiate the link selection.

On the other hand, with PTP-unaware nodes, the NE at each end of the link has no idea how to treat transit PTP messages, and so it just treats them according to its configured link selection algorithm. Figure 11-9 illustrates this problem.

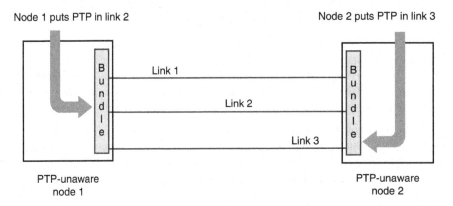

Figure 11-9 *Asymmetry in LAG Bundles in PTP-Unaware Nodes*

Note that there are other forms of asymmetry that could arise from factors external to the network elements and instead arise in the links between them. This can be as simple as a mismatch in length between the transmit and receive fibers, where somebody might have patched one direction of the path with a 1-m patch cable and the other with a 10-m cable.

It can also arise where a fiber may have been cut and the repairer added a length (loop) of spare fiber when splicing it back together. Using bidirectional optics (where upstream and downstream are carried over a single fiber on separate laser frequencies) is one technique that mitigates this problem, although it introduces another.

The use of different lambdas (frequencies) in the optical signal in a fiber cable causes asymmetry when the upstream and downstream signals use different lambdas. These different frequencies travel at different speeds through the fiber and introduce a measurable asymmetry (multiple factors influence the size of this effect—it may be quite small, especially for short cable runs). This is an issue with bidirectional optics, although it is possible to model and correct it to a large extent, which is why bidirectional links are used to carry highly accurate time signals such as White Rabbit (discussed in Chapter 7, "Precision Time Protocol").

However, unlike dynamic asymmetry that is induced by changing traffic patterns, these forms of asymmetry are static and therefore can be measured and compensated for. See the sections "Asymmetry, "Correction and Calibration," and "Carrying Phase/Time over the Transport System" in Chapter 9.

Multi-Profile with Profile Interworking

The major approaches (using GNSS, FTS, and PTS) to delivering frequency, phase, and time are dealt with in the preceding sections, but many situations in living, breathing production networks are not so uniform and do not always conform to a single approach. In some circumstances, there will be a difficult situation to overcome that requires a combination of techniques or a compromise outside the "one size fits all" solution.

This is the case when the characteristics of the network or underlying transport require the engineer to combine the approaches from the FTS and PTS use cases. In those network locations where it is possible, G.8275.1 with SyncE is deployed using a PTP-aware network and transport, but G.8275.2 is used where the network no longer offers the level of PTP awareness that is needed.

One common situation that arises is where the radio equipment may embed some device, such as a "dumb" layer 2 switch that understands no PTP or SyncE. Because that device is embedded into a standalone piece of equipment, a cell site router might be required to turn the G.8275.1 signal it is receiving into G.8275.2 (PTPoIP) to "hop over" the PTP-unaware L2 switch. Another case might be that the PTP slave clock in the radio itself supports only G.8275.2 and not G.8275.1.

Figure 11-10 illustrates what a typical deployment of this nature looks like. The timing signals may be carried from the PRTC+T-GM using the "best in class" G.8275.1 with

SyncE and terminate at the last T-BC before the PTP-unaware segment begins. That T-BC performs a *profile interworking* function that terminates the G.8275.1 message flow and initiates a G.8275.2 flow to cross the PTP-unaware network.

This capability is also referred to as *multi-profile support* because it features the ability to configure a T-BC with a different profile on each PTP port.

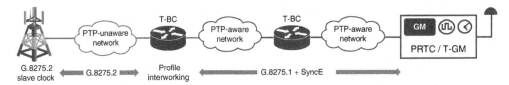

Figure 11-10   *Mixed Telecom Profile Deployment with Profile Interworking*

Profile interworking is not as simple as it sounds, although it is somewhat easier between the G.8275.1 and G.8275.2 profiles because they have a lot of commonality in the way they work and the PTP features they support. The section "Profiles" in Chapter 7 covers the ways that different PTP profiles vary and explains how complex it can be to do interworking between other combinations of PTP profiles. For example, there are different fields that must be translated as they are received on one slave port and transmitted out on another master port in a different profile. This is particularly the case for the fields in the Announce message, which require translation of values that depend on the profile. The most obvious field that needs translation is the domain number. Since the range of domain values for G.8275.1 is between 24–43, and the range of domain values for G.8275.2 is 44–63, then a common value for domain number cannot be used across the end-to-end network.

The engineer must configure these value translations on the device performing the interworking function, at least for domain number. Other translation values can be assumed automatically (such as values like priority1 that are fixed in G.8275.1 and G.8275.2) or allowed to remain unchanged by default. This approach works fairly readily when interworking between these two profiles but may be much more complex for other profile pairs.

Note   The defaulting of many parameters does not work for other combinations of profile interworking, which might need different values of fields like clockClass, clockAccuracy, and offsetScaledLogVariance. A good example is the difference in clockClass values between G.8275.1 and the G.8265.1 telecom profile for frequency. Sometimes a straight 1:1 translation in values is simply not possible.

Assisted Partial Timing Support with G.8275.2

If the G.8275.2 PTPoIP profile causes rapid accumulation of TE and PDV when running over an unaware network that lacks support from BC and TC clocks, what is its purpose?

If you cannot rely upon the profile to avoid asymmetry and carry time accurately, then why have it as an option?

It is already clear that one possibility is to deploy the profile only in a very narrow range of circumstances, such as only over well-designed networks and using a very small number of intervening nodes between a remote edge GM and the slave. In fact, there is a note in the G.8275.2 recommendation that states just that. However, this profile is perfectly adequate as a timing solution for those cases where the phase alignment accuracy is more relaxed. One example is for timing the DOCSIS cable system using the remote PHY architecture.

The use case that led to the development of the G.8275.2 profile was where the MSP did not own its own backhaul network and needed to rely on third-party SPs to provide circuits to interconnect the MSP's mobile infrastructure. These operators typically relied upon GNSS at every cell site to supply phase and frequency synchronization but were becoming concerned about the lack of a backup in case of a local GNSS outage. The solution that the ITU-T came up with was APTS.

The APTS design takes advantage of the fact that the cell site already houses GNSS equipment. Figure 11-11 illustrates the concept. During normal operations, the GNSS is the primary source of phase/time and frequency for the radio equipment. The APTS clock, in this case a slave clock called the Telecom Time Slave Clock with assisted partial support (T-TSC-A), has a slave port that is receiving a timing signal from a remote T-GM across an unaware third-party network.

If the GNSS signal becomes unavailable and the primary source has lost its traceability to a reference, this slave port will be selected as a backup by the best master clock algorithm (BMCA).

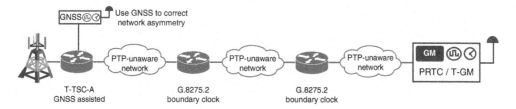

Figure 11-11   *Assisted Partial Timing Support*

The extra component that makes this work is that during normal operations, while the GNSS is the primary source, the slave clock also recovers the timing signal from the remote T-GM and measures any difference in the recovered clock against the accurate GNSS time signal. Any offset between the two time signals is most likely to come from asymmetry in the third-party network links between the slave and the T-GM. This measurement is taken constantly and used to determine a correction value for the PTP slave port.

If the primary GNSS source ever disappears, and the slave port is selected as a backup source, then the last measured value of the asymmetry is used to correct the timing signal from the remote PTP T-GM. It is the use of the local GNSS source (or any other source of timing signal) to *assist* the PTP slave to recover accurate time that gives this solution its name.

In theory, this is a nice solution; however, you need to understand a couple of things when considering this design. Remember from Chapter 9 that there are two basic forms of asymmetry, static and dynamic. If the asymmetry being measured by the T-TSC-A (or Telecom Boundary Clock with assisted partial support [T-BC-A]) is mostly static in nature (for example, differing fiber lengths), then measuring the value and using it to compensate the recovered PTP time will deliver quite accurate results.

On the other hand, if the asymmetry is largely dynamic, and prone to mostly random and dramatic movements in value, then this method has limited use. This is because the last recorded value (even with some filtering to reduce the jitter) will not remain valid for the next measurement period. Fairly reasonable and common behavior such as rapidly changing traffic patterns would lead to this form of dynamic noise.

The second factor to consider is that a complex backhaul network is hardly a static setting because it is constantly changing and adapting to changes in the environment. Equipment heats up, links go down, routing protocols remap preferred paths, people dig up cables, and so on. The point to make is that the last-known-good value is only good for as long as the current situation remains stable.

Once there is a significant change in the path between the T-TSC-A and the T-GM, the measured asymmetry becomes stale and could quickly become highly erroneous. And with IP routing looking after all those details at a lower layer, the T-TSC-A slave clock will not know anything happened (unless the transient resulted in a very sudden and dramatic phase shift). And even if the slave does detect a very sudden shift in phase, what is it to do about it? Should it ignore that shift as erroneous or follow the phase data it is getting from the T-GM? Trying to second-guess the data received from an authoritative timing signal is a dangerous assumption.

From the discussion of these points, it is clear that the APTS approach is a valid one, but that it has some limitations that you need to understand before deploying and relying on it. Although it would never deliver the degree of accuracy and certainty that the FTS approach delivers, in some cases, operators might only have limited options and APTS will certainly help.

Midhaul and Fronthaul Timing

The goal in the preceding sections is to outline the possible trade-offs in designing a timing solution for the backhaul—delivering the ±1.1-µs phase alignment at the handoff point to the radio equipment and/or ±1.5-µs phase alignment at the radio antenna. But as described in Chapter 10, with the evolution of the RAN, there are additional (relative) timing requirements between components of that RAN.

For that reason, the engineer needs to consider the extension of the backhaul timing design into the midhaul and fronthaul networks and take these additional requirements into account. The first consideration is the topology and type of RAN being deployed.

As outlined in Chapter 10, there are two basic topologies of 4G/5G RAN, the centralized RAN (CRAN) and the distributed RAN (DRAN), defined by the location of the RAN components.

Chapter 10 also covered the fact that there are basically three components to the modern 5G network: the Central Unit (CU), the Distributed Unit (DU), and the Radio Unit (RU). Figure 11-12 (taken from Chapter 10) illustrates the combination of topologies that these units support with fronthaul (FH), midhaul (MH), and backhaul (BH) networks.

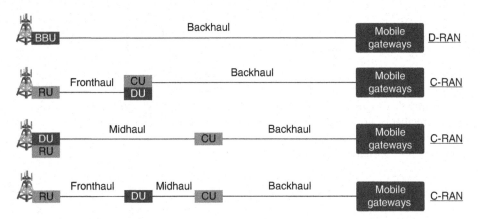

Figure 11-12 *5G RAN Topologies*

The first two topologies apply to 4G and 5G networks (where the Baseband Unit [BBU] is a combined DU/CU from 5G), and the second two topologies apply to 5G. See the sections "5G RAN" and "Radio Access Network Evolution" in Chapter 10 for more details on 4G and 5G RAN evolution.

There are three main methods related to time distribution (time and phase) from a timing source to slave clocks at a RAN base station (BS) site:

- Using a GNSS receiver at RAN sites where time/phase is required. In previous generations of radio, a GNSS receiver would use short-run cables to give physical timing signals (to carry phase, time, and frequency) to the radio, but current generations of equipment use PTP and SyncE for this purpose.

- Using a PTP-unaware network with a packet-based method based on PTP (G.8275.2) carried "over the top" from the MSP core to the RAN BS.

- Using a combination of Ethernet physical layer (SyncE) for links combined with a packet-based method based on PTP (G.8275.1) carried on an FTS network.

Technology Splits

Chapter 10 provides extensive coverage of the technology splits for RAN deployments. This section is simply intended to show how the different technology splits influence the timing design. The basic principle for RAN timing is that normally both the DU and RU need synchronization, but the CU does not (only time of day to simplify operations). Of course, the RAN and radio team from the MSP can clarify the details of what their implementation really requires.

The higher levels of the radio stack tend to be housed in the CU, the middle layers in the DU, and the lower layers in the RU. The point of demarcation between the various levels is defined by the split option and numbered from 1 (high in the stack) to 8 (low in the stack). The CU/DU demarcation is normally based on an option 2 split, and the two functions are interconnected by what is called the F1 interface, carried over the MH network (third and fourth cases depicted in Figure 11-12). The CU side of that split requires no phase synchronization, but the DU side does.

For this reason, it makes sense to introduce the timing signals at the site of the DU because this is the first component in the RAN chain that requires it. You can place a PRTC and T-GM higher in the network and carry the signals through the BH network, but the number of hops to the DU quickly consumes the phase alignment budget. The downside is that this means you must deploy a larger number of PRTC and T-GM devices when they are distributed like this.

The next factor to consider is the technology split between the DU and RU and what the connection is between them. Additionally, the location where the DU and the RU are situated matters—whether they are co-located or distributed over a FH network (third and fourth cases depicted in Figure 11-12).

In summary, this decision comes down to whether the architecture of the RAN is centralized or distributed, or, in other words, whether the goal is a DRAN or a CRAN; and for a CRAN, where the various components are located.

Centralized RAN and Distributed RAN

Recall from Chapter 10 that there are numerous positive benefits to building a CRAN over building a DRAN. These benefits materialize (or become cost effective) only when the CRAN is applied to cell sites in dense metro areas—the DRAN is still the dominant architecture outside those areas. But the selected approach influences the timing design quite markedly; the lesson to take away here is that if timing the DRAN is difficult, timing the CRAN can be even more so.

Note   Although the selected architecture of the RAN determines the timing requirements, there are also other consequences of the decision that limit the layout of the various RAN components. The major consideration is latency, and the selected RAN type and the technology split can have dramatic effects on the latency requirements. Some of these latency budgets are very small (say, 100 µs), and that imposes severe geographical constraints on distance between these elements. A centralized RAN is still distributed out close to the radio units for this reason—it is only partially centralized. Latency is also why a CRAN can be a problem to deploy in more remote locations because the distances between elements are too large.

For many types of small cell and devices based on mmWave, the DU and RU are commonly co-located, and there is no FH network. These can have a centralized CU and so are connected to the MH network (see the third case in Figure 11-12). These tend to be deployed at very high-traffic locations in dense urban settings.

Wherever the location of the CU, the timing is only required at the cell site and the requirement for ±1.1 μs absolute phase alignment always applies. There may also be additional, tighter requirements to achieve inter-cell radio coordination (see the upcoming section "Relative Versus Absolute Timing").

The design is similar for many macro cells, especially those in more remote and rural locations; everything is at the cell site, and there is only a connection to the BH network (the first case from Figure 11-12). As in the small cell case, the timing is only required at the cell site and the same ±1.1-μs phase requirement always applies.

It is commonly the mobility or mid-band sites that tend to centralize their DU or BBU equipment into a shared facility remote from the cell site, and these are deployed mainly in high-traffic, metro areas (the first and fourth cases depicted in Figure 11-12). This CRAN architecture connects the DU/BBU to the RU over the FH network.

In 4G deployments, the FH network was based on the lowest level option 8 split that used transport methods such as the Common Public Radio Interface (CPRI). CPRI is a TDM-like synchronous protocol, so the same ±1.1-μs phase was delivered to the DU/BBU, and CPRI took care of the frequency and phase out to the RU at the cell site. For phase timing of mobile networks, this is one of the easiest deployments to build a timing solution for.

For 5G, CPRI now shows its limitations (mainly excessive bandwidth needs in the FH because of the option 8 split), so there is now a migration to transform the CRAN to use an FH network utilizing packet-based transport.

Packetized Radio Access Network

With the option 8 (CPRI) split, the timing signal only needs to be made available at the DU/BBU, with CPRI doing the job for the last few kilometers (limited to about 20 km for latency reasons). But with the increase in bandwidth coming with 5G, the option 8 split is being replaced with a technology split higher up the stack, most commonly a version of option 7. This increases efficiency, but it means CPRI is no longer appropriate, so the fronthaul network is adopting a packet-based method such as Ethernet.

The problem with adopting a packet-based network is that the FH network is replacing CPRI (that natively supports transfer of synchronization) with a nondeterministic transport (Ethernet). Because there is now a "timing gap" between the DU and the RU, the engineer needs to extend the timing solution all the way to the RU. That introduces some additional complexity for the timing solution designer that was not previously visible. As you saw in Chapter 10, significant amounts of coordination are required between the various antennas connected to different RUs or antennas connected to RUs on adjacent DUs. All that coordination requires accurate phase synchronization, and that synchronization is now not being handled or transported by CPRI.

This responsibility now falls on the shoulders of the timing solution. So, the new packet-based FH network needs accurate phase/time synchronization and must meet the new relative timing requirements between the components of the FH network.

Relative Versus Absolute Timing

Up until now, the only concern has been the requirement to deliver the "traditional" ±1.1 µs phase alignment to the radio equipment, either at the combined DU/RU site in the DRAN or at the DU/BBU site in the CRAN. Currently, there is a new range of requirements from some of the radio coordination techniques and new services defined in the later releases of the 3GPP specifications. Tables 10-13 and 10-19 in Chapter 10 summarize these requirements for both LTE-A and 5G New Radio (NR), respectively.

However, these new requirements are *relative* requirements (as opposed to *absolute*). Absolute timing requirements, such as the ±1.1/1.5 µs for TDD radios, means that every piece of radio equipment must be aligned to within ±1.5 µs of some absolute time standard, which has been agreed is UTC. By this method, every cell site on the network will be within 3 µs of any of its neighbors, and the whole network will be within ±1.5 µs of UTC.

A relative requirement is only between two or more pieces of equipment, such as an RU and a DU, and says very little about the phase alignment compared to an RU on the other side of the country. So, when two RUs are coordinating with each other to deliver a combined data stream to a UE visible to both, these devices have to be closely aligned in phase relative to each other. This might be a value of ±130 ns or even ±65 ns, which sound like very, very small numbers.

Note that the relative requirement does not override the absolute ±1.5-µs phase alignment to UTC. Both constraints must apply simultaneously; the RU must be within the ±1.5 µs for the TDD radio and it must also be within the required relative alignment to the other RU it is coordinating with.

However, recognize that these relative requirements are between devices that are close together (less than 20 km or so) and, in many cases, either directly connected with each other or are only one or two hops apart. It is not like the timing solution must deliver around ±130 ns of accuracy across the whole network or the whole country. Figure 11-13 illustrates relative versus absolute timing within pieces of equipment clustered closely together.

As previously mentioned, the absolute requirement (requirement one) still applies, but the figure shows the additional relative requirement (requirement two) within the cluster of devices coordinating with each other (sites 9, 7b, and 8b). In this topology, the timing signal is coming from a PRTC/T-GM located seven to nine hops away from the edge devices. Given what you have read about the accuracy of time degrading after transiting numerous boundary clocks, delivering requirement two over that many hops might seem to be very challenging. However, note that the sites numbered 9, 7b, and 8b all source their timing signals via a common source, node 5. Since that common reference source is only two to four hops away, the error introduced between the PRTC/T-GM and node 5 can be discounted because it is common to the three devices in this cluster. The relative requirement is not quite so onerous as it sounds, but it is still difficult.

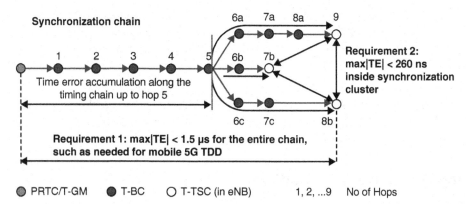

Figure 11-13 contains:

Synchronization chain

Requirement 2: max|TE| < 260 ns inside synchronization cluster

Time error accumulation along the timing chain up to hop 5

Requirement 1: max|TE| < 1.5 µs for the entire chain, such as needed for mobile 5G TDD

⬤ PRTC/T-GM   ⬤ T-BC   ○ T-TSC (in eNB)   1, 2, ...9   No of Hops

Figure 11-13 *Relative and Absolute Timing—Based on Figure VII.1 of ITU-T G.8271.1*

Given the difficultly with transporting highly accurate phase/time across many hops, could this use case support a "GNSS everywhere" style of solution, where both the RU and DU are synchronized using their own GNSS receivers? The performance for a PRTC Class A device is that it should be accurate to within 100 ns or better when verified against the applicable primary time standard (UTC). This is basically the level of accuracy guaranteed by the GPS system with a basic receiver.

That means that the time recovered by a GPS receiver at the RU site could be up to 200 ns off when compared to the signal recovered from another receiver at the DU site (one could be +100, and the other be –100). The offset is unlikely to be that large, since both receivers are quite close to each other and, most likely, are tracking the same satellites under very similar atmospheric conditions. However, it is theoretically possible that they can diverge by up to 200 ns. According to vendors of GNSS chipsets, a more realistic number might be 30–40% of that, but there is no guarantee.

This means that the requirement for a phase alignment of ±65 ns or ±130 ns may not be possible with the GNSS at each site, simply because the Class A PRTC is not accurate enough. In that case, it may be better to use PTP than separate GNSS receivers. It is for this reason that the ITU-T defined a new class of PRTC, known as the Class B PRTC, that offers an accuracy of 40 ns rather than 100 ns. To achieve that, the vendors are adopting techniques to improve the recovered accuracy from the satellite systems, including dual-band and multi-constellation receivers.

For larger networks, an increasing number of hops will introduce more time error than separate GNSS receivers, and using a PTP-aware FTS design may not meet the very tight CRAN packet FH timing requirements. Depending on the topology and budget, the better solution might be GNSS (where there are more hops in the FH) or PTP (where there are fewer). The exact dividing line will depend on the level of T-BC performance of the nodes and the timing error in the transport system. It is this exact scenario that led to the adoption of the Class C accuracy requirement for the T-BC in G.8273.2.

Note that throughout this chapter, the term SyncE is used to refer to both the SyncE (G.8262) and eSyncE (G.8262.1) recommendations for physical transport of frequency in a packet network. The use of SyncE is widely accepted and the adoption of eSyncE is gaining momentum; within a couple of years, eSyncE will be the default option. Currently, the requirement to use eSyncE over SyncE is most important in the FH networks in combination with Class C T-BC clocks.

For further information on T-BC performance, see the section on G.8273.2 in Chapter 8. See the Chapter 3 section "Sources of Frequency, Phase, and Time" for further details on the PRTC. Details of the performance of the PRTC and the ePRTC are covered in G.8272 and G.8272.1, respectively.

Timing Security and MACsec

The Chapter 7 section "PTP Security" discusses the pros and cons of implementing security protections for the PTP-based timing distribution. It also covers the security features newly adopted in the IEEE 1588-2019 edition of PTP. See 1588-2019 Section 16.14, "PTP integrated security mechanism," and the informative Annex P from the same document.

However, Chapter 7 also explains that there is little use to encrypting PTP messages because the data in the messages is well known and even encrypted PTP messages can be readily detected by their behavior and size. It also points out that a delay attack is a much simpler and effective method to corrupt a PTP timing signal than any complex technical attack.

But there is theory and there is reality and then there is politics. In previous generations of mobile networks, the totality of the equipment to run the network was largely under the control of the MSP. Most of it was housed in large, protected facilities, while the cell site was a locked container at the bottom of a tower surrounded by a barbed wire fence, observed by cameras and protected with alarms. Links between the two were either some cable buried underground or microwave dishes on the top of towers and buildings.

With the increasing densification of the 5G network and the adoption of many more small cell–type radio devices in a wide variety of facilities, the MSP can no longer guarantee the physical security of portions of the network. This means it is becoming easier for third parties to get access to the links feeding into the equipment and even to the device itself, especially in the FH and MH.

For this reason, most MSPs are looking at encrypting portions of their RAN to safeguard against interception or alteration of user or control plane data. The fact that almost all the user data is already encrypted at a higher layer or by the application does not change the political and reputational cost of suffering a breach. Similarly, regulators in some countries are mandating encryption of segments of the transport network to safeguard the data integrity of subscribers.

PTP Security

The current versions of the ITU-T telecom profiles have not adopted any of the security mechanisms from the IEEE 1588-2008 version of the PTP standard. With the publication of the new security Annex P in the 1588-2019 release, one would expect that some security features will be incorporated into future revisions of the telecom profiles. For that reason, some of the basic elements of implementing a more secure version of the transportation of timing are outlined in the following section, "Integrity Verification from IEEE 1588-2019."

In the new edition of the 1588 standard, a multipronged approach to security has been developed and is described in the informative Annex P. All the prongs are optional for PTP and not required in a standard implementation (and are not yet adopted by the telecom profiles). Several prongs can be applied in parallel; they are not mutually exclusive. The four "prongs" of the approach in Annex P are as follows:

- **Prong A:** Authentication and integrity verification

- **Prong B:** The use of transport security

- **Prong C:** Architectural guidance

- **Prong D:** Monitoring and management mechanisms

The following sections expand on these concepts and examine how they can be applied to future deployments. The focus is on one item: the encryption of the transport links with MACsec.

Prong A: Authentication Field and Security Processing

This prong includes methods to enhance the security of the timing signal using a field to carry a checksum that is used by the receiver:

- Section 16.14 of IEEE 1588-2019 describes a type, length, value field (TLV) that can be appended to messages. This TLV includes an Integrity Check Value (ICV) or *hash* to allow validation of the source of the message and confirm that the message has not been altered during transmission. None of the standard data fields in the PTP message is encrypted, only this added check value allows the receiver to guarantee that it has received the data without having been altered.

- Generation and validation of the ICV requires a shared key at sender and receiver. This method for generation and sharing of keys is currently outside the scope of the standard, but it requires the building of a robust and secure infrastructure for key generation and distribution.

Prong B: Transport Encryption and Security

This prong includes methods to enhance the security of the timing signal by encrypting the PTP messages in transit. There are two possible methods for doing this:

- **IPsec:** *IP security* provides a security mechanism to protect data at the network layer (layer 3), which is the IP layer. That makes it possible to use IPsec for those profiles of PTP that use IP for their transport (namely G.8265.1 and G.8275.2). It is not applicable for layer 2 profiles of PTP (such as G.8275.1).

IPsec is implemented as end-to-end encryption, which makes using a T-TC in the path almost impossible because it cannot read the PTP message and update the correctionField. Adding a T-BC to the path is also difficult because IPsec requires a lot of hardware resources to run at the speed needed to allow full encryption at line rate (such as multiple 10GE or 25GE links on a T-BC).

■ **MACsec:** *Media Access Control security* provides a security mechanism to protect data at the link or MAC layer (layer 2) or the Ethernet layer. This means that MACsec can be used for PTPoE profiles such as G.8275.1 as well as for all the PTPoIP profiles.

MACsec is normally implemented in the hardware of the interface, and any reasonable implementation offers the ability to run at the line rate of the interface (although MACsec adds extra bytes, slightly reducing the throughput). Of course, the hardware to do this adds to the cost of deploying the network elements, as the MSP must install only equipment that supports MACsec.

Figure 11-14 illustrates the basic differences between IPsec and MACsec implementations, and how MACsec is a "hop-by-hop" approach, whereas IPsec is an "end-to-end" approach because the intervening nodes never see the data in the clear. On the other hand, MACsec encrypts the data as it goes onto a link and decrypts it at the other end, so the data is in the clear at transit nodes. Figure 11-14 does not intend to suggest that both approaches should be implemented together.

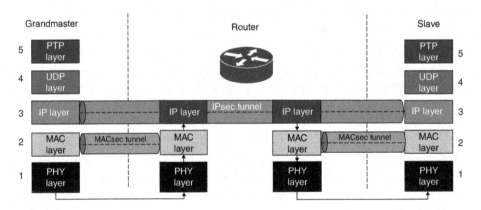

Figure 11-14 *IPsec Versus MACsec*

Having the data in the clear is not regarded as the best security, but for packet-based timing, it is an advantage because it allows T-BC and T-TC to offer on-path support as the timing signal transits the node.

The downside to encrypting the transport of PTP messages is that the encryption and decryption processes of messages can introduce variable delay that could affect time accuracy. More details about that are provided in the upcoming section "MACsec."

Prong C: Architectural Guidance

This prong includes architectural approaches to increase the reliability of the timing signal by implementation and design. The possible methods include the following:

- Redundant grandmasters, in case of failure or compromise of the primary source.

- Redundant links, in case of failure or the link being compromised.

- Redundant time sources and paths (for example, by using multiple, independent sources of time). This form of deployment is not that common in the case of mobile timing; normally, operators only deploy redundancy of GMs and paths/links.

This book has dealt with many of these issues under separate topics.

Prong D: Monitoring and Management

This prong includes monitoring and management methods for ensuring the security, accuracy, and integrity of the timing solution. There are several parameters described in Annex J of IEEE 1588-2019 that can be used for performance monitoring. By observing those parameters and analyzing them for anomalies, such as unexpected values or unforeseen variations, alerts can be generated and acted upon.

Chapter 12 provides more on the topic of operational deployments and timing verification.

Integrity Verification from IEEE 1588-2019

The ICV (from prong A) is calculated by applying a cryptographic hash function to ensure that unauthorized modifications of a message can be detected by the receiver. The AUTHENTICATION_TLV that carries this value is appended to the PTP message that it is validating. The ICV is calculated according to a security algorithm selected and shared during the negotiation of the security parameters by the key distribution mechanism. Figure 11-15 shows the format of the TLV.

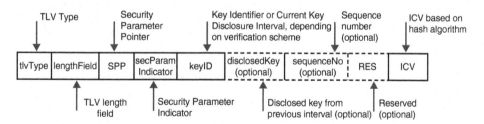

Figure 11-15   *The AUTHENTICATION TLV*

The implementors of the ICV mechanism are required to support at least the HMAC-SHA256-128 algorithm, as defined in Federal Information Processing Standards (FIPS) PUB 198-1, which uses the SHA-256 hash combined with a keyed-hash message authentication code (HMAC) shortened to 128 bits of output. HMAC codes are also used in IPsec implementations as well as the Transport Layer Security (TLS) encryption package that is the basis of web browser, email, and other network encryption schemes.

An HMAC is a type of message authentication hash or checksum that combines a cryptographic hash function and a secret cryptographic key. By doing this, the HMAC can be used to validate the integrity of the data as well as ensure the authenticity of the source. Because the hash includes a secret key (previously shared between sender and receiver by a key exchange), the ICV cannot be recalculated by an eavesdropper and the data cannot be modified without detection by the receiver. Of course, the key must be kept secret from any eavesdropper, or they could recalculate a valid ICV after changing the data in the message.

There are two processing schemes proposed to use an ICV for integrity validation. Their names and main features are as follows:

- **Immediate security processing:** All intermediate and end nodes have shared the current key in advance, and the PTP data can be immediately validated through standard cryptographic techniques.

- **Delayed security processing:** The key used to generate the ICV is distributed only after it is no longer used (and cannot be used to generate new ICV fields). This allows the clock to validate the already-received PTP data at some point in the future once the key has been replaced. This method requires the receiver to store the messages until the required keys have been shared, and then the stored PTP messages can be verified.

Of course, not having the current key in the delayed security processing mode makes it impossible to modify the correctionField at intermediate transparent clocks without invalidating the ICV. One approach to circumvent that problem is to assume the updatable fields (such as correctionField) are zero during the calculation of the hash, which means those fields can be changed without invalidating the ICV—which seems to defeat the point of the whole scheme.

As illustrated in Figure 11-15, it is also possible to combine both approaches on the same messages, so that the PTP message could contain two TLVs, one for immediate processing and one for delayed processing. The TLV also includes a sequence number to prevent replay attacks (whereby previously transmitted messages are re-sent later to mislead the slave clock with stale timestamps).

There are two basic key distribution methods defined in IEEE 1588-2019 depending on the processing model:

- **GDOI:** *Group Domain of Interpretation*, RFC 6407, is the scheme applied for group key distribution to support immediate security processing for PTP ICV authentication.

- **TESLA:** *Timed Efficient Stream Loss-tolerant Authentication*, RFC 4082, is the scheme applied when using delayed security processing. It is a very clever arrangement that allows immediate validation of the source of the message but delays the ability to validate the content until the key from the previous key interval has expired.

Further discussion of these mechanisms and fields is beyond the scope of this book, not least because they have not yet been adopted by the main profiles. This is background information to give you an understanding of what might make its way into the popular profiles. For further information on the ICV and AUTHENTICATION TLV, please see Section 16.14 of the IEEE 1588-2019 standard.

MACsec

For reasons outlined previously, an increasing number of MSPs are looking to adopt transport-level security to protect the information within their networks, not least because elements of the network might be more accessible to nonauthorized person-nel than previously. This is not just a problem with MSPs, but also cable operators, because increasingly they are adopting common packet-based transport standards such as Ethernet. Previously, only the very specialized hacker had access to a device to eaves-drop on cable protocols or Asynchronous Transfer Mode (ATM) cells, but now Ethernet devices are ubiquitous.

The major backbone pipelines of the Internet, such as high-speed interconnection between data centers and cloud services, are increasingly encrypted, and the technology most often deployed to protect any data "in motion" is MACsec. The MACsec protocol was originally defined by IEEE 802.1AE in 2006 and is now specified in the latest ver-sion, IEEE 802.1AE-2018. Because it is used to protect the very high-speed data rates needed to connect data centers and cloud services, MACsec supports close to the full line rate of the interface.

On transmission, MACsec takes the existing Ethernet frame as input and turns it into a MACsec frame, adding a small amount of data in the MACsec security tag and the vali-dation field. Figure 11-16 shows the formation of the MACsec frame, where SA and DA are the source and destination (MAC) address of the Ethernet frame and FCS is the frame check sequence.

The MACsec tag is inserted before the Ethertype and an ICV is inserted before the FCS. MACsec supports both using just the integrity checking alone (against the ICV) or combining it with encryption of the payload.

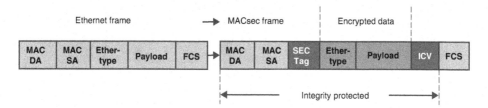

Figure 11-16 *MACsec Frame Format*

On reception of a MACsec frame, the network element verifies the ICV field and decrypts the MACsec frame (if encrypted). When the NE determines the egress direction, if that path is configured to apply MACsec, it (optionally) encrypts the mes-sage, calculates the ICV, and forwards the message, with the MACsec tag, down the link to the next hop.

The MACsec security tag contains a small amount of information, the most important of which is the packet number. This packet number is used to protect against replay attacks. It used to be a 32-bit value, but with very high-speed links, there is now a 64-bit extended packet number that can be used to reduce the frequency of rollover when using the smaller value. It also contains an Ethertype (88:E5) to indicate that this frame is now a MACsec frame.

So, MSPs are looking at deploying MACsec in those sites that might be vulnerable to interception, and this raises a few challenges regarding the transport of PTP timing signals. When PTP messages are secured with MACsec, there are time error effects introduced by the implementation:

- Constant time error (cTE), due to the difference in time taken to encrypt and decrypt PTP messages at a single node

- cTE, due to asymmetry based on the choice of the point at which the message is timestamped on reception versus transmission

- cTE, due to the difference in time to encrypt a message at one end of the link with the time needed to decrypt it at the other end

- dTE, due to the difference in time to encrypt and/or decrypt successive messages

It is possible to mitigate some of these effects with two-step clocks, but a lot depends on the capabilities of the physical electronic part that is doing the encryption (commonly the PHY). Figure 11-17 and Figure 11-18 detail the issues with timestamping when combined with MACsec on both one-step and two-step clocks, respectively.

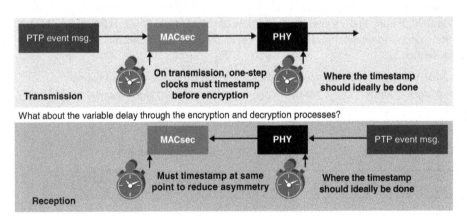

Figure 11-17 *PTP Timestamping Combined with MACsec on One-Step Clocks*

Problems with the implementation in the one-step clock can be somewhat mitigated by using a two-step clock, whereby the T1 timestamp is returned in the Follow_Up message that is sent immediately after the Sync message. This saves the PHY from having to write an accurate time estimate for the time of the transmission before the encryption has taken place.

Figure 11-18 illustrates how the process works with a two-step clock.

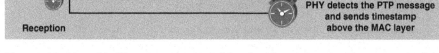

Figure 11-18 *PTP Timestamping Combined with MACsec on Two-Step Clocks*

In the two-step clock, the PHY must be able to determine which encrypted packets are PTP messages that need timestamping and share that information with the component that is generating the Follow_Up message. Similarly, on reception, the PTP must try to determine which packets are PTP messages and forward the reception timestamp to the PTP servo to match to the incoming message.

The issues with MACsec combined with PTP can be summarized as follows:

- The timestamp must be inserted into a packet *before* it goes through the MACsec block, after which the packet cannot be modified. Therefore, for one-step clocks, the use of PHY-based timestamping does not work. It is possible to try to predict the time that encryption will take and timestamp the PTP message before encryption with the predicted (future) time of transmission. Some vendors claim good results doing this.

- Two-step clocks *might* be able to help mitigate this problem and still use PHY time-stamping, but if the packet is encrypted, the PHY will not be able to recognize the outgoing Sync as a PTP message, so other means of tagging the PTP messages are required inside the network element.

- Even if only integrity checking (no encryption) is being used and the PHY is MACsec aware, the timestamp still must be read back and applied to the Follow_Up message, complicating the implementation.

- On reception, the PHY will not recognize an encrypted packet as being PTP, so it would have to timestamp every packet and then match up the timestamps with the correct packets after MACsec processing. Some hardware implementations do this already (store the timestamp until the packet has been identified as PTP), so it is not impossible, but it does complicate the hardware design.

■ A PHY might be able to recognize a MACsec packet that only uses integrity checking (no encryption) if it is MACsec aware.

What does this all mean? It means that using PTP over a MACsec-enabled link will not generate the same accuracy as a link that is not using MACsec. So, the current choice is either one of the following:

■ Have a secure link

■ Have an accurate timing transfer

The PHY vendors have been working to resolve this issue for a couple of years, and components are now becoming available to the industry that allow both features simultaneously at an acceptable level of timing accuracy.

The lesson to take away is that you should not expect, unless you have confirmed otherwise, that putting PTP over a link enabled with MACsec will deliver accurate phase/time. There is not much point in buying a top-quality T-BC that performs at the level of Class C but then lose 200 ns of time accuracy in the MACsec implementation.

Summary

This chapter outlined the factors to consider when deploying a timing solution to support 4G/LTE, LTE-A, and 5G mobile networks. The basic aspects to consider are the type of mobile technology being used, the topology and size of the network (including who provides the circuits), and the type and reach of the cell site. These elements will determine whether the radio system requires frequency synchronization only or phase/time/frequency synchronization and will indicate the best approach to constructing the timing solution.

The design of the timing solution will likely incorporate a combination of GNSS and atomic clocks as a source of time, together with a transport network carrying the timing signal using packet and/or physical timing signals. The details of the design in any one segment of the network requires knowledge of the local factors such as the type and characteristics of the transport system and the ability of the network and nodes to support timing-aware features—such as whether the network elements support being a PTP T-BC.

The evolution of the RAN is a special case, especially where the operator has adopted the centralized CRAN architecture and has introduced technology splits in the radio processing stack—connected by an FH (and possibly MH) network. This introduces a need for much higher accuracy phase alignment, although within clusters of cell sites coordinating with each other—known as relative phase alignment.

One vital consideration is the ability of the timing solution to survive any possible events that could lead to disruption of the timing signal, either from space or within the transport network. This involves the use of devices with outstanding holdover performance, autonomous sources of time (such as atomic clocks), and clever network design to provide redundancy in the timing signal.

The final aspect to consider is the security of the timing signal, both in the PTP message stream itself and within the transport links. This is covered in prongs A and B of Annex P of IEEE 1588-2019 and includes the ability to use MACsec in the links without sacrificing phase/time accuracy.

Chapter 12 covers the final topics when implementing a timing solution, which includes hardware design, procurement, acceptance testing, assurance, and monitoring as well as troubleshooting and field testing.

References in This Chapter

3GGP. "Evolved Universal Terrestrial Radio Access (E-UTRA); Base Station (BS) radio transmission and reception." *3GPP*, 36.104, Release 16 (16.5.0), 2020. https://www.3gpp.org/DynaReport/36104.htm

IEEE Standards Association

"IEEE Standard for a Precision Clock Synchronization Protocol for Networked Measurement and Control Systems." *IEEE Std 1588-2002*, 2002. https://standards.ieee.org/standard/1588-2002.html

"IEEE Standard for a Precision Clock Synchronization Protocol for Networked Measurement and Control Systems." *IEEE Std 1588-2008*, 2008. https://standards.ieee.org/standard/1588-2008.html

"IEEE Standard for a Precision Clock Synchronization Protocol for Networked Measurement and Control Systems." *IEEE Std 1588-2019*, 2019. https://standards.ieee.org/standard/1588-2019.html

"IEEE Standard for Local and metropolitan area networks-Media Access Control (MAC) Security." *IEEE Std 802.1AE-2018*, 2018. https://standards.ieee.org/standard/802_1AE-2018.html

"IEEE Standard for Local and metropolitan area networks — Time-Sensitive Networking for Fronthaul." *IEEE Std 802.1CM-2018*, 2018. https://standards.ieee.org/standard/802_1CM-2018.html

"IEEE Standard for Local and metropolitan area networks — Time-Sensitive Networking for Fronthaul – Amendment 1: Enhancements to Fronthaul Profiles to Support New Fronthaul Interface, Synchronization, and Synchronization Standards." *IEEE Std 802.1CMde-2020*, 2020. https://standards.ieee.org/standard/802_1CMde-2020.html

International Telecommunication Union Telecommunication Standardization Sector (ITU-T)

"G.8273.2: Timing characteristics of telecom boundary clocks and telecom time slave clocks." *ITU-T Recommendation*, 2020. https://handle.itu.int/11.1002/1000/14507

"Q13/15 – Network synchronization and time distribution performance." *ITU-T Study Groups*, Study Period 2017-2020. https://www.itu.int/en/ITU-T/studygroups/2017-2020/15/Pages/q13.aspx

Internet Engineering Task Force (IETF)

Arnold, D., and H. Gerstung. "Enterprise Profile for the Precision Time Protocol with Mixed Multicast and Unicast Messages." *IETF*, draft-ietf-tictoc-ptp-enterprise-profile-18, 2020. https://tools.ietf.org/html/draft-ietf-tictoc-ptp-enterprise-profile-18

Mills, D. "Network Time Protocol (NTP)." *IETF*, RFC 958, 1985. https://tools.ietf.org/html/rfc958

Mills, D. "Network Time Protocol Version 3: Specification, Implementation and Analysis." *IETF*, RFC 1305, 1992. https://tools.ietf.org/html/rfc1305

Mills, D. "Simple Network Time Protocol (SNTP) Version 4 for IPv4, IPv6 and OSI." *IETF*, RFC 4330, 2006. https://tools.ietf.org/html/rfc4330

Mills, D., J. Martin, J. Burbank, and W. Kasch. "Network Time Protocol Version 4: Protocol and Algorithms Specification." *IETF*, RFC 5905, 2010. https://tools.ietf.org/html/rfc5905

Perrig, A., D. Song, R. Canetti, J. Tyger, and B. Briscoe. "Timed Efficient Stream Loss-Tolerant Authentication (TESLA): Multicast Source Authentication Transform Introduction." *IETF*, RFC 4082, 2005. https://tools.ietf.org/html/rfc4082

Weis, B., C. Rowles, and T. Hardjono. "The Group Domain of Interpretation." *IETF*, RFC 6407, 2011. https://tools.ietf.org/html/rfc6407

MEF

"Amendment to MEF 22.3: Transport Services for Mobile Networks." *MEF Amendment*, 22.3.1, 2020. https://www.mef.net/wp-content/uploads/2020/04/MEF-22-3-1.pdf

"Transport Services for Mobile Networks." *MEF Implementation Agreement*, 22.3, 2018. https://www.mef.net/wp-content/uploads/2018/01/MEF-22-3.pdf

Mills, D. *Computer Network Time Synchronization: The Network Time Protocol on Earth and in Space*. CRC Press, Second Edition, 2011.

Mills, D. et al. "Network Time Synchronization Research Project." *University of Delaware*, 2012. https://www.eecis.udel.edu/~mills/ntp.html

National Institute of Standards and Technology (NIST)

"Secure Hash Standard (SHS)." *NIST*, FIPS PUB 180-4, 2015. http://dx.doi.org/10.6028/NIST.FIPS.180-4

"The Keyed-Hash Message Authentication Code (HMAC)." *NIST*, FIPS PUB 198-1, 2008. https://doi.org/10.6028/NIST.FIPS.198-1

Rohanmkth. "Small Cell by Samsung.jpg." Wikimedia Creative Commons, Attribution-Share Alike 4.0 International license (https://creativecommons.org/licenses/by-sa/4.0/deed.en). https://commons.wikimedia.org/wiki/File:Small_Cell_by_Samsung.jpg

Smith, J. "Cell-Tower.jpg." Wikimedia Creative Commons, Attribution-Share Alike 2.5 license (https://creativecommons.org/licenses/by-sa/2.5). https://commons. wikimedia.org/wiki/File:Cell-Tower.jpg

SMPTE. "SMPTE Profile for Use of IEEE-1588 Precision Time Protocol in Professional Broadcast Applications." *SMPTE ST 2059-2:2015*, 2015. https://www.smpte.org/

U.S. Department of Homeland Security. "Report on Positioning, Navigation, and Timing (PNT) Backup and Complementary Capabilities to the Global Positioning System (GPS)." *Cybersecurity and Infrastructure Security Agency*, 2020. https://www.cisa.gov/sites/default/files/publications/report-on-pnt-backup-complementary-capabilities-to-gps_508.pdf

Chapter 11 Acronyms Key

The following table expands the key acronyms used in this chapter.

Term	Value
3GPP	3rd Generation Partnership Project (SDO)
3G	3rd generation (mobile telecommunications system)
4G	4th generation (mobile telecommunications system)
5G	5th generation (mobile telecommunications system)
AAV	alternate access vendors
ACR	adaptive clock recovery
APTS	assisted partial timing support
ATM	Asynchronous Transfer Mode
BBU	Baseband Unit
BC	boundary clock
BH	backhaul (network)
BITS	building integrated timing supply (SONET)
BMCA	best master clock algorithm
BS	base station
CA	Carrier Aggregation
CDMA	code-division multiple access
CE	Carrier Ethernet
CoMP	Coordinated Multipoint
CPRI	Common Public Radio Interface

Term	Value
CRAN	centralized radio access network
CRPA	controlled reception pattern antenna
CSR	cell-site router
CU	Central Unit (3GPP)/Control Unit (O-RAN) (5G radio access network)
DA	destination address
DCR	differential clock recovery
DOCSIS	Data-Over-Cable Service Interface Specification
DRAN	distributed radio access network
DU	Distributed Unit (5G radio access network)
E1	2-Mbps (2048 kbps) signal (SDH)
eICIC	enhanced inter-cell interference coordination
eLORAN	enhanced Long Range Navigation
eMBB	enhanced mobile broadband
ESMC	Ethernet synchronization messaging channel
eSyncE	enhanced synchronous Ethernet
FCS	frame check sequence
FDD	frequency division duplex
FH	fronthaul (network)
FIPS	Federal Information Processing Standard
FTS	full timing support (for PTP from the network)
FWA	fixed wireless access
GDOI	Group Domain of Interpretation
GM	grandmaster
GNSS	global navigation satellite system
GPS	Global Positioning System
HetNet	heterogeneous network
HMAC	keyed-hash message authentication codes
IA	Implementation Agreement (MEF Forum)
ICIC	inter-cell interference coordination
ICV	Integrity Check Value
IEEE	Institute of Electrical and Electronics Engineers

Term	Value
IoT	Internet of things
IP	Internet protocol
IPv4, IPv6	Internet protocol version 4, Internet protocol version 6
IPsec	IP security
ITU	International Telecommunication Union
ITU-T	ITU Telecommunication Standardization Sector
L2	layer 2 (of the OSI model)
L3	layer 3 (of the OSI model)
LACP	Link Aggregation Control Protocol
LAG	Link Aggregation Group
LAN	local-area network
LTE	Long Term Evolution (mobile communications standard)
LTE-A	Long Term Evolution—Advanced (mobile communications standard)
MAC	media access control
MACsec	media access control security
MBMS	Multimedia Broadcast Multicast Service
MEF	MEF Forum (formerly Metro Ethernet Forum)
MH	midhaul (network)
MIMO	multiple input multiple output
mMTC	massive machine-type communications
mmWave	millimeter wave (radio frequency bands)
MSP	mobile service provider/mobile operator
MTU	maximum transmission unit
NE	network element
NNI	network-network interface
NR	New Radio
NTP	Network Time Protocol
OSNMA	Open Service Navigation Message Authentication (Galileo)
OTDOA	Observed Time Difference of Arrival
PDV	packet delay variation
PHY	PHYsical Layer (of OSI Reference Model)—an electronic device

Term	Value
PNT	positioning, navigation, and timing
POP	point of presence
PRC	primary reference clock (SDH)
PRS	primary reference source (SONET)
PRTC	primary reference time clock
PTP	precision time protocol
PTPoE	PTP over Ethernet
PTPoIP	PTP over Internet protocol
PTS	partial timing support
QoS	quality of service
R15/Rel15	Release 15 (of the 3GPP standards)
R16/Rel16	Release 16 (of the 3GPP standards)
R17/Rel17	Release 17 (of the 3GPP standards)
RAN	radio access network
RF	radio frequency/function
RFP	request for proposal
RU	Radio Unit (5G radio access network)
SA	source address
SFN	single frequency network
SFP	small form-factor pluggable (transceiver)
SLA	service level agreement
SP	service provider (for telecommunications services)
SSU	synchronization supply unit (SDH)
SyncE	synchronous Ethernet—a set of ITU-T standards
T1	1.544-Mbps signal (SONET)
TAAS	time as a service
T-BC	Telecom Boundary Clock
T-BC-A	Telecom Boundary Clock—Assisted
T-BC-P	Telecom Boundary Clock—Partial support
TC	transparent clock
TDD	time division duplex

Term	Value
TDM	time-division multiplexing
TE	time error
TESLA	Timed Efficient Stream Loss-tolerant Authentication
T-GM	Telecom Grandmaster
TLS	Transport Layer Security
TLV	type, length, value
TSN	time-sensitive networking
T-TC	Telecom Transparent Clock
T-TC-P	Telecom Transparent Clock—Partial support
T-TSC	Telecom Time Slave Clock
T-TSC-A	Telecom Time Slave Clock—Assisted
T-TSC-P	Telecom Time Slave Clock—Partial support
UE	user equipment
UNI	user-network interface (from the MEF)
uRLLC	ultra-reliable and low-latency communications
UTC	Coordinated Universal Time

Further Reading

Refer to the following recommended sources for further information about the topics covered in this chapter.

ITU: https://www.itu.int

Resilient Navigation and Timing Foundation: https://rntfnd.org/

MEF, Carrier Ethernet for 5G: https://www.mef.net/service-standards/additional-mef-domains/5g/

Operating and Verifying Timing Solutions

In this chapter, you will learn the following:

- **Hardware and software solution requirements:** Describes the components that make up a PTP-aware device and the principles underlying the design of an accurate clock.

- **Writing a request for proposal:** Gives hints and tricks about writing an effective RFP to help properly select network elements for a network-based timing solution.

- **Testing timing:** Introduces the equipment, methods, and test cases involved when testing both the behavior and performance of either a transport network or a standalone clock.

- **Automation and assurance:** Covers aspects of configuring and managing the timing network using automation tools. It begins with the Simple Network Management Protocol (SNMP) before moving on to more modern approaches with NETCONF/ YANG.

- **Troubleshooting and field testing:** Covers aspects such as common problems that engineers encounter, troubleshooting, clock accuracy monitoring, field testing, and extensive information on global navigation satellite system (GNSS) receiver performance.

The chapters up to this point have covered many of the theoretical aspects of deploying a timing solution, the requirements and design for a mobile deployment, and the ITU-T recommendations supporting all aspects of building a timing solution. This chapter answers questions about how to design, build, acquire, certify, and run a timing solution.

The order of topics in this chapter is meant to closely shadow how an operator might start with deployment of a real timing solution with practical lessons learned from development, rollout, and operation. The idea is to start with an understanding of what you are buying when you are looking to buy physical devices. The first section addresses the design of timing-aware equipment, because you must understand the internal hardware trade-offs to understand what you are buying. Much like buying a car, you should know the basics of what goes into selecting an automatic or manual transmission before signing up. This section shows you the difference between a well-designed implementation and one likely to disappoint.

The next section gives some practical advice that helps when selecting equipment through some type of procurement (buying) process. The idea is to find the best equipment to do the job at the minimum possible cost. At the end of this process, you end up with a short list of possible vendors.

The next step then is to take a sample of equipment from each of the vendors and put it in a testing lab; therefore, the third section covers the testing of timing in a lot of detail. This step is valuable for staff learning, equipment certification, and building confidence in the components selected for the solution.

Following this is a section on automation and assurance. Once the equipment has been ordered, the design is being fine-tuned by network engineers, and now the network operations staff has to figure out how it is going to be rolled out and incorporated into their network. This involves not just configuration but also being aware of the solution performance as well as monitoring the network and verifying every radio is receiving an accurate timing signal.

The final section describes potential deployment headaches and some of the day-to-day operational issues that might arise. It includes sections on devices that monitor the network time (probes) as well as field testing remote devices to isolate any faults. There is also a section on understanding the finer points of GNSS receivers and troubleshooting.

Hardware and Software Solution Requirements

Requirements for timing transfer are evolving toward improved precision and accuracy. And that means the timing distribution networks must be designed carefully so that the timing signal meets the overall synchronization requirements after crossing numerous network nodes and under a wide range of network conditions.

Even when the networks are designed properly to carry timing information, special design considerations are required for the equipment transporting the clock synchronization. Engineering this equipment to provide synchronization services starts with selecting the right hardware components and designing to minimize any systematic loss of accuracy.

This section covers what goes into the construction of a PTP-aware node and how these components are combined to meet the performance goals of the node. There is more to making a good PTP clock beyond just gluing a few off-the-shelf parts together. It requires the mating of good hardware to feature-rich software and connecting it all together into a system that minimizes time error (TE) and maximizes the desired performance.

Making a good PTP clock also requires a good track record from the component vendors making the individual timing components, skill from the software engineers controlling it all, and, most of all, experience of the hardware designers who decide how to put it all together. It is this combination of components, skills, and experience that turns loose pieces of steel, quartz, and silicon into a very precise clock. The quality of the implementation makes a huge impact, so the reputation and past performance of the clock vendor matters. Given these factors, it is very much worth looking at the details of what is involved in a well-designed PTP clock, and how the major pieces fit together.

Frequency and phase synchronization each require their own separate engineering when it comes to designing equipment to support them. The sections that follow discuss the main hardware and software components as well as the reference architectures for each.

Synchronous Equipment Timing Source

Before discussing the hardware and software architecture of synchronous equipment, it is important to understand the core hardware component of the clock, an element called the synchronous equipment timing source (SETS) device. This device provides the necessary controls to manage timing references, clock recovery, and clock signal generation for synchronous equipment.

The performance requirements of SETS devices are defined in ITU-T G.8262 for SyncE Ethernet equipment clocks (EEC), G.8262.1 for enhanced Ethernet equipment clocks (eEEC), and ITU-T G.813 for SDH or synchronous equipment clocks (SEC). Figure 12-1 illustrates the main components and interconnection of these components of a SETS device.

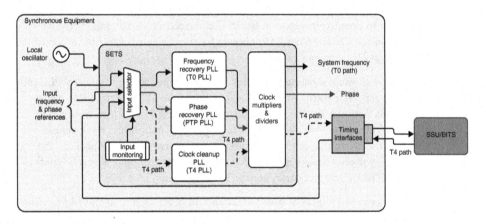

Figure 12-1 *SETS Block Diagram*

The list that follows explains the functionality of each component:

Note Use of the terms "master" and "slave" is ONLY in association with the official terminology used in industry specifications and standards, and in no way diminishes Pearson's commitment to promoting diversity, equity, and inclusion, and challenging, countering and/or combating bias and stereotyping in the global population of the learners we serve.

■ **System phase-locked loops (PLL):** Instead of equipment having only a single system PLL, more recent and better equipment designs have multiple system PLLs. A good and versatile SETS device in synchronous equipment is usually equipped with three PLLs, as shown in Figure 12-1:

   ■ **Frequency recovery PLL:** This PLL, sometimes also referred as the T0 PLL (as labeled in Figure 12-1), is used to recover the frequency reference signal for the whole system. This PLL recovers frequency from the selected frequency reference and outputs the recovered frequency to be used as the system frequency. The output frequency is fed to different ports (such as SyncE-enabled ports) and potentially to other components of the system requiring the recovered frequency.

   ■ **Phase recovery PLL:** The SETS device also hosts a separate PLL to recover phase independently alongside the frequency recovery PLL. The reference source for phase recovery could be a phase reference signal input to the equipment, such as a 1 pulse per second (1PPS) signal from an embedded GNSS receiver port or a dedicated 1PPS port on the equipment. Refer to the section "1PPS" in Chapter 3, "Synchronization and Timing Concepts," for details on 1PPS.

   In the absence of these reference sources, this PLL is used by the PTP slave software to recover phase from PTP packets. Having a separate PLL for phase, different from frequency, keeps the phase and frequency domains separate and hence helps implement a good *hybrid mode* function on a network element (NE). With separate PLLs, the NE is also able to simultaneously output separate frequency and phase to downstream clocks. Recall that the term *hybrid mode* is used to refer to phase recovery from PTP, whereas *frequency synchronization* comes from a physical frequency source.

   Figure 12-1 shows a separate frequency reference input to the frequency PLL and the phase reference to the phase PLL. However, some SETS devices also have a connection between the frequency PLL and the phase PLL to feed (or receive) a recovered frequency or phase signal from one PLL to another. For example, although the phase PLL might facilitate phase recovery from PTP packets, the phase PLL will also require a frequency reference (for example, a SyncE input) to drive both phase and frequency for the whole system.

■ **T4 PLL:** Some SETS devices host yet another PLL, referred to as the T4 PLL. The primary function of this PLL is to pass the input timing reference to a dedicated output timing port to be monitored externally or to facilitate the clock cleanup by external synchronization supply unit/building integrated timing supply (SSU/BITS) equipment (refer to the section "BITS and SSU," in Chapter 6, "Physical Frequency Synchronization," for more details on clock cleanup).

   This PLL usually locks to an external frequency source separate from the T0 PLL and outputs the T4 frequency on a single, specific port. This is sometimes also referred as the *T4 path*, as shown in Figure 12-1. Note that for the clock cleanup case, the T4 path enables the selected input frequency reference to pass to external SSU/BITS equipment. And the SSU/BITS equipment can then feed the cleaned-up clock back to the equipment via the same SSU/BITS port. The synchronous equipment can use this cleaned-up clock as input reference for the *T0 path*, which is then used as input to the system frequency. Be careful not to generate a timing loop while using this mode.

■ **Input selector:** Because the equipment can have multiple frequency sources (explained further in subsequent sections), the SETS device usually contains an input selector that selects one from the many possible frequency sources and switches that signal to the frequency recovery PLL. The SETS device also provides the capability to monitor all the available reference sources to assist the frequency selection process to select the best input frequency source.

■ **Clock multipliers and dividers:** Clock multipliers and dividers are used to change a clock signal by a multiple of its frequency. With these multipliers and dividers, a SETS device can generate variations of the locked reference clock as outputs, which can be fed to different components of the equipment (requiring different frequencies to operate). For example, a 1-gigabit Ethernet interface requires 125 MHz and, at the same time, the 10-MHz front panel port requires the frequency to be 10 MHz.

Thus, a SETS device can take multiple frequency and phase inputs, select frequency and phase reference sources from these inputs, and then output multiple frequency and phase signals that are locked to a configured reference source. It also provides the capability to monitor clock quality and clock signal fail conditions, which allows the software to take appropriate actions for selecting the next best reference source.

Frequency Synchronization Clock

As discussed at different points in earlier chapters, there can be many frequency sources for network equipment (for redundancy and resiliency purposes). These frequency sources can primarily be categorized into two different categories:

■ **External timing interfaces:** Recall from Chapter 6 that a system can have dedicated timing interfaces. These interfaces are used only for frequency synchronization purposes and not for data transfer. The BITS and 10-MHz ports (from a 10-MHz interface on a primary reference clock/primary reference source or a GNSS receiver) are examples of dedicated timing interfaces.

■ **Line timing interfaces:** In addition to carrying network data, these interfaces carry timing information as well. In recent times, Ethernet interfaces supporting synchronous Ethernet (SyncE) are the most prevalent interfaces carrying frequency along with data.

Engineers expect that network equipment being used for clock synchronization is able to support both interface types. They also expect that there should be minimum possible restrictions on the types of ports that can be used for clock synchronization. For example, one expects all the Ethernet ports on an NE are SyncE capable, and the equipment allows configuration to recover frequency from any of these ports.

Figure 12-2 illustrates a high-level hardware design for the frequency synchronization for such equipment. It shows SyncE-capable Ethernet ports (on the front panel); external timing interfaces, such as the SSU/BITS and 10-MHz interfaces; and an embedded GNSS receiver with a connector for an external antenna.

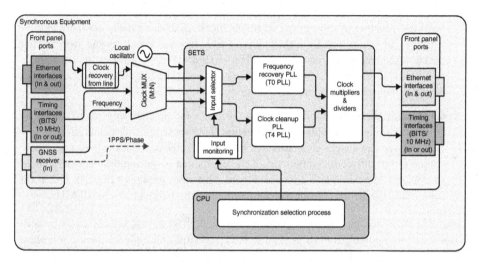

Figure 12-2 *SETS Architecture for Frequency Synchronization in Synchronous Equipment*

Note that as shown in the figure, some of these interfaces, such as SSU/BITS and 10 MHz, can be configured as *input* (receive frequency input) or *output* (output the system frequency) interfaces. Of course, the interfaces from the GNSS receiver can only act as inputs to the NE. Note also that Figure 12-2 shows different front panel ports on the left and right sides, but they are the same set of ports and are shown separately for better clarity. The ability for any one port to perform input or output should be flexible and controlled by configuration. Because this section talks about frequency synchronization only, the phase recovery PLL has been omitted in the figure for convenience and simplicity.

When looking at Figure 12-2, the following points list the prominent features of the hardware and software architecture needed for frequency synchronization:

- Although the frequency recovery PLL (sitting inside the SETS device) accepts a single frequency source to lock to, the equipment can accept numerous frequency sources. These sources are usually multiplexed in a hardware device—such as a field-programmable gate array (FPGA)—to feed only a few of these sources to SETS.

  This module is shown as the *Clock MUX (multiplexer)* in Figure 12-2. Consider an NE having tens of Ethernet interfaces (all SyncE capable), while the SETS device supports only a few (in the range of five to seven) input references. In such cases, the FPGA-based clock multiplexer acts as an initial frequency source selector, switching the selected input to the input selector based on the software configuration.

■ The SETS device has an *input selector* to choose the selected reference source for the system. It also has the capability to continuously monitor the input references, shown as *Input monitoring* in Figure 12-2. The input monitoring module in the SETS device updates the software on the state of the clock quality.

■ The frequency transfer using physical methods is usually accompanied by synchronization status messages (SSM) bits or Ethernet synchronization messaging channel (ESMC) protocol data units (PDU). These messages carry quality level (QL) information describing the quality of the frequency source being carried over the physical layer.

The selection of the best frequency source for the equipment is done according to the QL values, the physical link status, and the configuration on the equipment. This selection is also referred to as the *synchronization selection process* and is defined in ITU-T G.781 (Annex A). This synchronization selection process is usually implemented in a software module running on the control plane CPU, as shown in Figure 12-2.

■ Changes in QL values (or link status) triggers the synchronization selection process. This process involves selecting the best frequency source available on the equipment following the rearrangement. To achieve this, the SETS device monitors all the available frequency sources and, when necessary, triggers the software module to execute the synchronization selection process.

If the clock quality of the selected frequency source degrades (or the physical link goes down), the SETS device passes that information to the software module. The software then starts a new synchronization selection process triggered by this event.

■ Figure 12-2 shows the SETS device connected to the local oscillator, which is used by the PLLs in the SETS device for locking to any inputs and generating any output synchronized frequencies. Refer to Chapter 6 for a survey of the different types of oscillators and their impact on accuracy, stability, and precision of the clock recovery. Recommended practice dictates using a good stratum 3E oven-controlled crystal oscillator (OCXO) (or temperature-compensated crystal oscillator [TCXO] as the next-best option) as an oscillator when building good synchronous equipment.

Time- and Phase-Aware Clock

With phase synchronization now a key requirement for mobile networks, the hardware and software designs of NEs have evolved to be aware of time and phase. Although time and phase can be carried via PTP packets alone, the network timing engineer might encounter different situations where just packet-based time/phase transfer might not be good enough. Chapter 11, "5G Timing Solutions," covers many alternate deployment topologies.

To support multiple different network topologies of mobile networks, NEs are expected to support a variety of ways to accept time and phase from external sources. These sources can include a GNSS receiver (providing both time and phase) and/or dedicated ports for time of day (ToD) and 1PPS input.

If the system contains an embedded GNSS receiver, an antenna port on the front panel can be directly connected to an external antenna. The receiver will provide both ToD and 1PPS signals (in addition to the 10-MHz frequency signal) for the NE to generate PTP and SyncE. These signals are almost always internal to the device, so it is very difficult to monitor the signal between the receiver and the SETS device.

The NE is also expected to support dedicated 1PPS, 10-MHz, and ToD ports to receive respective signals from an external clocking and timing source. For measurement reasons, the NE should also support using these ports as output as well as input. Figure 12-3 shows the hardware components and signal paths of a time-/phase-aware NE.

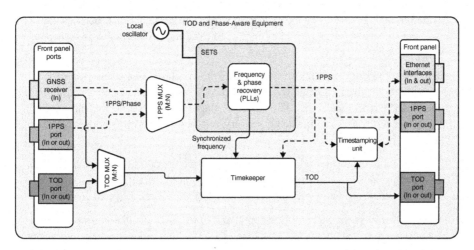

Figure 12-3   *SETS Architecture for Phase and Time Synchronization in Synchronous Equipment*

Because there can be multiple ToD and 1PPS inputs (from different front panel ports of the NE), a hardware-based multiplexer (usually implemented in an FPGA) is used to select the input from among the configured ports. The 1PPS multiplexer design has its own complexity because the expectation is that the multiplexer itself should not delay the 1PPS signal while passing it to the phase PLL in the SETS device (any delay results in phase error).

The PLL residing inside the SETS device recovers the phase from the 1PPS input. As illustrated in Figure 12-3, the SETS device uses the recovered phase to generate a 1PPS signal for internal use, and the same 1PPS signal can also be routed to the external 1PPS port for output. This allows a quick method for an engineer to check the phase offset of a device by use of external phase measuring equipment (see the section "Field Testing" toward the end of this chapter).

The output 1PPS signal from the SETS device is also carried to other hardware components on the system. The main components that need this 1PPS signal are the *timekeeper* and the *timestamping unit* (TSU), which uses the 1PPS signal to maintain an accurate time of day for the system. Maintaining and distributing this accurate time of day around the NE requires careful hardware design.

In the absence of PTP, the time of day is periodically received from external sources (such as once every second) by a combination of the external 1PPS and ToD signals. However, during this period between updates, the time of day needs to roll forward with the correct frequency for it to maintain good accuracy. To achieve this, the time of day is maintained in a piece of hardware (usually implemented in the FPGA, although some SETS devices also have the capability to maintain current time) that needs to be fed with the synchronized frequency for it to increment time accurately. Figure 12-3 represents this function as the *timekeeper* block.

Like 1PPS, the value of the time of day also needs to be distributed to other components of the NE. A good example is loading a representation of the time of day value from the *timekeeper* into the TSU, which is then used for timestamping of PTP packets (normally this mechanism uses the TAI timescale).

Designing a PTP-Aware Clock

So far in this chapter, the sections separately introduced a hardware and software architecture for frequency, phase, and time synchronization. Now, it is time to combine these separate elements to complete an architecture for an NE with full timing support.

Full timing support is usually implemented as an NE that supports PTP hybrid mode along with front panel ports to receive ToD and 1PPS from external sources. Because it requires both physical methods and PTP packets, the NE is expected to incorporate all the possible methods of clock synchronization in its design.

Figure 12-4 shows the hardware and software architecture for an NE with full timing support. Notice how the different components discussed separately in earlier sections come together for the design of an NE with full timing support. This section focuses on the design required to support the complete clock functionality, with particular focus on the PTP components.

PTP is designed to provide time transfer on a standard Ethernet network with a synchronization accuracy in the range of 10s of nanoseconds. This is possible only by leveraging special hardware components to assist PTP operations and accompanying them with well-built software algorithms. Remember that to achieve the most accurate synchronization, system designers need not only to choose the hardware components correctly but also carefully design such equipment to minimize any systematic loss of accuracy. It is the same when designing networks for synchronization services—the equipment participating in such networks needs to be carefully engineered to carry timing services.

The sections that follow discuss the different aspects of designing a good PTP-aware NE.

Hardware-Assisted Timestamping

The precision time protocol uses the timestamps carried in PTP event messages to recover phase (and optionally frequency if frequency distribution is not via the physical layer).

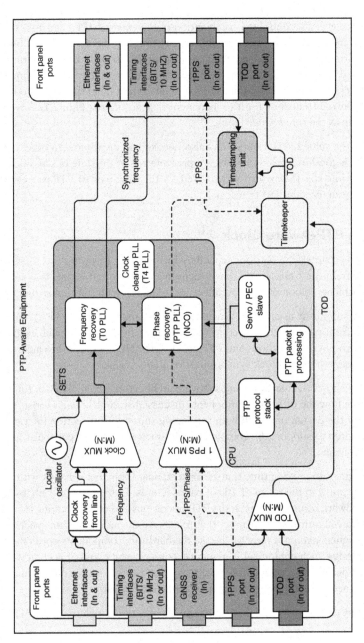

Figure 12-4 *SETS Architecture for Full Timing-Aware Synchronous Equipment Design*

To achieve the highest precision, timestamping of the PTP messages must be applied just before the packet's departure or immediately upon the packet's arrival. Performing the timestamping as close to the wire as possible is to make sure that the transit time between master and slave clock nodes is minimized.

Hardware-based timestamping is the key mechanism needed to achieve nanosecond levels of accuracy. The best implementations of PTP can be achieved only when utilizing a hardware element to assist in timestamping the PTP event packets. Furthermore, those timestamps must be both accurate and use a high-resolution clock with a fine granularity.

Hardware timestamping engines are required to achieve the best accuracy (in the range of nanoseconds or better) and they are not impacted by software-induced latencies. On the other hand, software-based timestamping is susceptible to unpredictable interrupts and variable operating system latencies, and the final accuracy of such implementations will be very low (typically in the millisecond range).

As shown in Figure 12-4, this hardware element is referred to as the TSU and is placed between the Ethernet media access control (MAC) and the Ethernet PHY transceiver to precisely mark the arrival or departure time of PTP packets. The accuracy (or inaccuracy) of the TSU, however small it is, can be directly observed on the eventual phase accuracy. The closer the timestamp for each individual PTP message is to the real ToD, the more accurate the PTP implementation will be.

The resolution (or granularity) of the TSU also affects how accurately the packets can be timestamped. If the minimum value that the TSU can represent is limited to (say) 1 microsecond, then the timestamp cannot represent the precision of the ToD, even if it has a completely accurate ToD available. Any jitter in the ToD transfer or the timestamps will affect the solution when the slave clock uses the PTP formula to calculate the offset from master. Recall that the formula for a slave clock implementation is

$$\text{offset from master} = (t_2 - t_1 + t_3 - t_4) / 2$$

So, for example, if the resolution of a timestamp from the TSU is 1 microsecond, this means that the TSU can add jitter of up to 1 microsecond to the timestamps. With such a low resolution, even a perfect PTP implementation and ideal transport (with zero packet delay variation [PDV] and no link asymmetry), the resultant accuracy cannot be better than 500 ns.

Note According to Institute of Electrical and Electronics Engineers (IEEE) standard 1588-2019 (clause 7.3.4.2), the timestamp shall be the time at which the event message timestamp point passes the reference plane marking the boundary between the PTP node and the network.

Interestingly, this clause from 1588-2019 implicitly indicates that the timestamping boundary for any NE must be at precisely the same point for both arrival and departure of the messages. As shown in Figure 12-5, the points $t_{ingress}$ and t_{egress} would be the correct and symmetric locations on the reference plane to record the timestamps.

So, even for the case of timestamping with hardware assistance, if the timestamp points of the received and transmitted PTP packets are not aligned, it results in asymmetry that will degrade the accuracy of the recovered phase. This is illustrated in Figure 12-5, where the asymmetric timestamping points are marked as $t_{ingress\text{-}skewed}$ and $t_{egress\text{-}skewed}$.

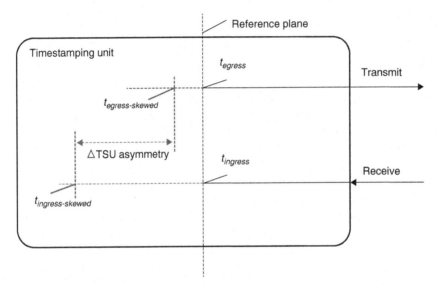

Figure 12-5 *Reference Plane for Timestamping Unit*

Finally, at the minimum TSU resolution, it should be possible to identify each individual packet on a link by a timestamp unique to that packet. This requires that the timestamp resolution must be less than the time it takes to transmit a minimum-size Ethernet frame of 64 bytes. Of course, the time taken to transmit that 64-byte (or 512-bit) packet depends on the link speed.

Including 8 bytes of preamble and 12 bytes of minimum inter-frame gap, the smallest frame becomes 84 bytes (or 672 bits) on the wire. It takes 67.2 ns to transmit 672 bits on a 10-Gbps link, 16.8 ns on a 40-Gbps link, and 6.72 ns on a 100-Gbps link. So, for a TSU supporting 100-Gbps interfaces, a minimum timestamp resolution of 6 ns is required.

Software: Control Plane, Servo, and Packet Processing

Designing and implementing a PTP-aware NE requires extensive software support as well as good hardware. These software modules need to be based on a flexible architecture so they can be adapted to constantly evolving recommendations from standards development organizations (such as ITU-T and IEEE).

From a software point of view, three key elements are required by a PTP-aware NE. The first element is the PTP *protocol stack* that usually runs on the host processor as a software module. The list that follows covers some features of the PTP protocol stack functionality (the list is not complete) to provide an idea of the functionality implemented in a PTP protocol stack. It must be able to

- Support configuration of different PTP clock node types (such as ordinary clock (OC), boundary clock (BC), and so on).

- Set up the PTP packet generator and receiver based on the node's PTP configuration and indicate the values of various message fields that are carried to the PTP peer.

- Track the status of PTP peers to manage the communications with them. On a slave node, this would include the implementation of the correct best master clock algorithm (BMCA) and similar functions.

A system designer could choose to design an implementation based on a separate dedicated processor for PTP functions. This would avoid problems occurring if the host processor is heavily loaded on a PTP master node—for example, it might become unable to communicate with all its slave nodes at the configured PTP message rate. On a slave node, CPU overload could cause nondeterministic response time, which becomes a problem when PTP packets are not received and processed in a timely manner. If this happens, the PLL will not be adjusted often enough, and that will decrease the accuracy of the recovered network clock.

The second element required by a PTP-aware NE is the PTP *packet processing engine*. As the name suggests, the primary aim of this engine is to generate and receive PTP packets either at the configured or negotiated packet rates.

The third key element of a PTP-aware NE is the PTP *servo algorithm*, commonly referred to simply as the *servo*. The goal of the servo algorithm is to synchronize the phase and frequency of the slave clock with that of the master clock. Thus, a high-quality servo is an essential component for high-accuracy synchronization.

Recall from Chapter 5, "Clocks, Time Error, and Noise," that a PLL is an electronic device that generates an output clock signal that is synchronized to an input clock signal. For frequency synchronization, a PLL uses a *phase comparator*, which compares the phase of the input signal versus the phase of the output signal and generates a voltage-based phase difference. This voltage is fed to a *voltage-controlled oscillator* (VCO) to adjust the output signal. Using this process, the PLL implements a continuous loop of adjusting the output frequency based on the input signal.

The servo performs a similar loop in software for the case of PTP and phase—the PLL (a hardware component) recovering frequency is emulated by the servo software algorithm to recover phase. To achieve this, the servo algorithm is assisted by another special hardware element known as a software-controlled PLL. This PLL is also known as a *numerically-controlled oscillator* (NCO) because the output signal from this PLL can be adjusted numerically (via software). This is just like adjusting a clock with numbers, such as increasing the frequency by some specific number (in parts per million) or adjusting the phase by a certain degree.

Figure 12-6 illustrates this loop in a simplified manner. The main functionality of the servo algorithm is to detect the difference between the phase information (received from the master) and the phase of its own clock. Depending on the difference, it adjusts the NCO to narrow the alignment of its clock to the master clock. The phase information received from the master is in the form of timestamps and the phase difference is calculated by the slave clock with the formula that calculates the offset from master.

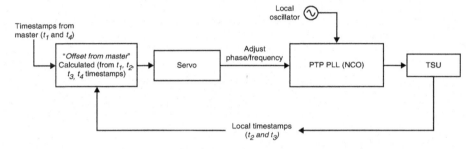

Figure 12-6   *PTP Servo Loop*

Note that Figure 12-6 is only for illustration purposes, and a lot more goes into the servo algorithm than shown here. In fact, improving the implementations of servo algorithms is an area of active research. For example, Figure 12-6 does not mention the filtering capabilities (like the loop filter in a VCO-based PLL) and other basic efficiency features. However, a few of the factors that can impact the servo algorithm accuracy (and hence the eventual node accuracy) are as follows:

- Imperfect compensation of the propagation delay (for example, due to propagation path asymmetries inside the network element)

- Timestamping jitter at both the transmitter and receiver ends, such as results from the low resolution of the TSU or other inaccuracies

- The quality and performance of the local oscillator being used on the NE

An efficient servo algorithm becomes a key factor of the PTP-aware NE design. Of course, it must be implemented using good hardware components and accurate timestamping. But a good servo cannot fix poor hardware, and neither can good hardware help a bad servo.

This section was intended to give you an idea of how the capabilities of a PTP-aware and SyncE-capable clock are implemented. You should see that much of the ability to provide nodes that can accurately transport time is not just a matter of lines of code. It takes numerous physical components that must meet their own design criteria, and they all cost significant money.

Now that you understand what you should request when you go out to the market to provide NEs for your timing network, the next section discusses the procurement of equipment and the request for proposal process.

Writing a Request for Proposal

This section covers the process behind soliciting, selecting, and procuring the equipment needed to build a timing network. The mechanism that most businesses and operators utilize is some form of the *request for proposal* (RFP), whereby the buyer puts together a description of the current situation and desired outcome, and vendors are invited to propose a comprehensive solution. The response to the RFP created by the vendor covers both commercial and technical aspects, with the technical part covering system performance and features.

Normally, as part of the solicitation, the buyer supplies a detailed list of requirements posed as a series of questions asking the vendors' level of support for numerous features. The details of these RFP requirements range widely from a couple of sheets of paper to many thousands of detailed questions. The timing subset of an RFP requirements list can also vary from a few general questions to many hundreds, digging into every possible aspect of the proposed solution.

Normally, these requirements are then classified as "mandatory," "optional," or "informational" as an indication of how much value the buyer places on that specific feature or behavior. The vendor then indicates its support for that requirement by providing an answer such as "Compliant," "Partially compliant," "Future roadmap," or "Not compliant." The conditions of some RFPs state that even a single "Not compliant" answer to a mandatory requirement eliminates the whole proposal without any future consideration!

It should be obvious from the preceding section that timing is not just a software solution. The characteristics of a clock, especially performance, are mostly driven by the hardware. The selection of components such as the oscillator, PLL, and PHY is critically important. Equally, the timestamping accuracy, signal asymmetry, media type, environmental controls, and overall system design are all very critical. Then, the combined performance of each device in the network directly maps into the final performance at the end of the chain.

For this reason, engineers writing an RFP should be mindful of the fact that many timing features, and certainly those that improve the performance, costs money. This is especially true when these timing features require specialized hardware (at non-trivial cost) to meet that requirement. A good example is holdover performance (see the sections on holdover in Chapter 9, "PTP Deployment Considerations," and Chapter 11). Vendors could build a PTP-aware network element that could stay aligned to 1 μs of holdover for several weeks, but it would be so expensive that nobody would buy it (yes, that level of holdover is frequently requested in RFP requirements).

So, the first piece of advice is to seriously think about the features that your solution requires and do not make a requirement mandatory unless it really is a deal breaker. For example, there are arguably better (and certainly cheaper) ways to deliver resiliency and redundancy in a timing design than to ask for atomic clock–level performance in thousands of low-end network elements.

On the other hand, vendors also understand that operators also define RFPs for brownfield deployments; hence, sometimes the requirements need to be defined in very fine detail to ensure that the requirements align or are compatible with the envisioned deployment scenarios. That is totally understandable, but just appreciate that timing characteristics are frequently hardware dependent, and poorly defined RFPs end up targeting more expensive equipment than might be required.

The second piece of advice is to allow the vendors some degree of flexibility in designing a solution and allow them to propose innovative solutions to the problem. A better approach might be to define the desired outcome rather than be too specific and lock the vendor into a traditional solution.

In summary, here are some tips to help get a result from your next timing RFP that might better reflect your needs and budget:

- Try to limit the list of features to what you reasonably need. Perhaps there are some "nice to have" features that might solve a problem that could perhaps arise in the future in a small subset of cases. That is completely understood, but it would be better not to have those features as a mandatory requirement for each class of equipment.

- Try to differentiate the features based on the type of machine being proposed and the location of the equipment in the network. There are networking, routing, and redundancy features that are standard for very large core routers that are uncalled for in a cell site router. The same is true for timing features, so it might help to tailor the RFP questions somewhat for the role and position of the network clock.

- There are standards defining almost every aspect of the timing solutions: from network topologies to clock behavior and performance. It saves a lot of work to specify, for example, "must meet Class B noise generation for a Telecom Boundary Clock (T-BC) according to G.8273.2" rather than to try to define clock performance in your own words. Be sure to include optional components of standards if they are important to your solution.

- Similarly, be more specific and avoid outlining features in an overly broad manner, such as "must support 1588," which is so vague and nonspecific that it is almost meaningless. It is an easy requirement to comply with, but it will not help you select a better device for a network timing solution.

- If you do not understand timing that well, then seek input from knowledgeable people who can help you design a better list of specifications for your next procurement. RFPs commonly have requirements like "must support Class C oscillator," which is impossible to comply with—since there is no such thing.

  The people answering the RFP questions will politely try to answer the question, but it will not help a selection process because it is basically a flawed question. On the other hand, hiring a consultant with 500 cookie-cutter questions in a "one size fits all" boilerplate does not help discern the best product for each role.

- Clearly understand the difference between clock requirements and network requirements. Asking for requirements such as "cell site router must support 5G" or "aggregation device must support G.8271.1" is not that useful because they are solution or network requirements. The clock could "support" those standards and solutions but still be a dreadful PTP and SyncE clock.

- There is a similar issue with standards for timing that specify architecture and concept. Some of them are very high level and you do not learn anything by asking the question. In fact, some standards are so conceptual that it is almost impossible to be noncompliant. The answers to those questions do not help you procure a better product.

- Do not ask the same questions repeatedly using slightly different terminology. One of the authors recently worked on a (very extensive) RFP that asked one question, in slightly different form, about a dozen times. When evaluating a vendor response, lack of support for this one (somewhat minor) feature would destroy the compliance score for that vendor's product.

- This can happen as well when the RFP is poorly divided into technologies. For example, the SyncE questions are in the timing section and repeated in the Ethernet or some other sections. This leads to possible inconsistency because the timing specialist did not see the timing question in the section on layer 2 (L2) Ethernet, and the L2 engineer did not understand the timing question properly.

- Do not expect support for the version of the ITU-T specification published last week to be in the router that will be delivered at the end of next month. If you need compliance to a new specification, then the better approach is to allow "Roadmap" as an answer with a committed delivery date, and do not require the answer to be "Mandatory, must comply."

- If you are going to ask about particular performance metrics, then unless you have a specific need or requirement, align them with standardized performance tests. For example, asking for the cTE, dTE_L, filter bandwidth, or input tolerance allows an apples-to-apples comparison between responses from different vendors.

  Alternatively, a question without specific test conditions allows the vendors to assume test circumstances that give them leeway to present their best result and prevent meaningful comparisons.

Hopefully these few tips will help you improve your next timing RFP and get the product that is as closely aligned to your needs as possible. Just as it is in the buyer's interest not to get the wrong product, it is in the vendors' interest not to sell you the wrong product for the job. That just causes pain for both parties.

Generally, the RFP process is a good one, but so is the request for information (RFI). Many times, when branching out into a new area of technology, the operator will send out an RFI to a wide range of vendors to get a feeling for the state of the industry. This is a good way to understand in what direction the mainstream is heading.

Another very helpful activity to understand timing in the various industries is to attend conferences dedicated to timing. In the telecommunications space, there are two conferences of note, one in Europe and one in North America:

■ The International Timing & Synchronization Forum (ITSF) is a four-day conference that switches locations each year among various interesting European cities. ITSF has a large and active steering group and always attracts a lively and enthusiastic set of delegates. Although focused primarily on telecoms, ITSF also features papers concerning timing in the finance, broadcast, automotive, smart grid, Internet of things (IoT), datacenter, and transport industries. It is usually held in the first week of November every year. See http://www.telecom-sync.com for details.

■ The Workshop on Synchronization and Timing Systems (WSTS) is a three-day conference sponsored by the National Institute of Standards and Technology (NIST) and the Alliance for Telecommunications Industry Solutions (ATIS), held somewhere in North America (up until recently, it was held annually in San Jose). The topic areas are much the same as those for ITSF. Although many of the timing experts attend both, the WSTS tends to have more of an American flavor. See https://wsts.atis.org/ for details.

WSTS is usually held between March and May every year, with the schedule alternating to line up with the ITU-T Q13/15 interim meetings held in North America. This allows ITU-T delegates to attend and present at the conference as part of a single business trip. Of course, with the COVID situation at the time of writing, that is less of a concern, and conferences are presumed to remain virtual for the next while.

So, now that you have written the RFP, selected a shortlist of possible vendors, and perhaps attended a few conferences, the next step is to run a test event, a *proof of concept* (POC) to validate the offered equipment. This launches the whole area of testing timing, which is quite a specialized area of expertise, so this chapter uses several of the following sections to cover it comprehensively.

Testing Timing

One of the first times that many engineers, newly introduced to timing, realize the difficulty of the whole topic of synchronization is when they first go into the lab and conduct testing. There is a dizzying number of testing options available, and you need a fair deal of background knowledge before starting in the lab. Well, to rephrase that, you *should* have substantial background knowledge before starting, but the experience of the authors suggests many engineers do not have sufficient knowledge to conduct testing. This leads to a high probability for unsuccessful tests, or at the very least, frustration and even public embarrassment.

This lack of background knowledge is the main reason that this chapter includes a separate section on testing timing. There actually is a copious amount of educational material on the topic available if you know where to find it (several pointers are provided in the "Further Reading" section at the end of the chapter). The problem is that many publicly available educational resources on the topic assume that you know what you are doing before you start testing.

The authors frequently deal with customers and prospective customers who have gotten themselves into difficulties by misapplying a test case to the wrong topology or judging the result using the wrong specification or metric. Most times, a quick phone call will uncover the mistake, and the tester can move on with their equipment certification.

The authors have seen numerous major test efforts where the results indicate that the equipment under test has failed one or more test cases. In many of those circumstances, either the test cases were wrongly executed or the result was compared against the wrong pass/fail criteria. For example, there are two broad categories of testing: end-to-end network testing and standalone clock testing. The topology, test cases, and expected result are different for each case.

So, this section will lead you through the fundaments of timing testing and point out the common mistakes that are very easy to make. First, Figure 12-7 should remind you of the most relevant ITU-T recommendations that provide a basis for conformance testing.

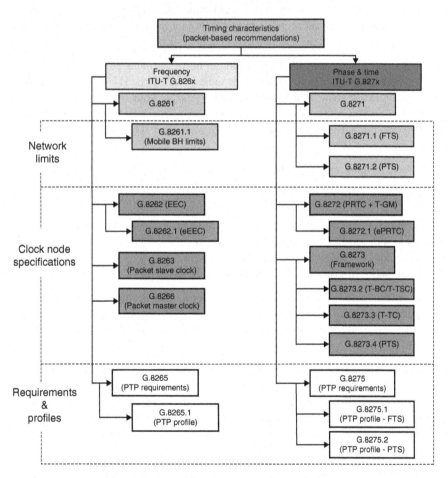

Figure 12-7 *ITU-T Recommendations*

Just about all the test cases and performance requirements are covered by one or more of these documents. Several other recommendations might come into the picture if your testing involves time-division multiplexing (TDM) type circuits, but the subset identified in Figure 12-7 should satisfy almost all situations. In the category of end-to-end network limits, a chain of devices under test (DUT) is tested to ascertain what happened to the timing signal after crossing multiple hops in the network. For clock node specifications, the DUT is tested standalone with specialized test equipment.

Overall Approach

The first thing to decide before starting the testing is what you need to test given the resources available. That is because there are possibly hundreds of different combinations of tests and scenarios that could entail being in the test lab for possibly months. Some of the test cases take days to execute (100,000 seconds or longer) and some cases require a settle time of at least 24 hours (or longer) before starting the test (holdover is one example). So, deciding what you want to prove or certify is an important consideration, as well as aligning the test cases with the equipment and time available.

The next important point is to realize that testing of network timing and the certification of clock and network performance are largely standardized (by the ITU-T). Yes, you can "roll your own" and test some scenarios that are important to your situation if you feel it is warranted. But remember that when you develop your own test case or method, you cannot claim the equipment is failing because it does not meet some limit in the ITU-T recommendations.

A good example of this is testing a T-BC against G.8273.2 noise generation limits—for example, to determine constant time error (cTE), but doing it while introducing impairment on the PTP input messages (such as adding PDV to the PTP messages sent to the DUT).

When certifying whether a clock is Class A, B, C, or D, the test method mandates that the output signal of the clock is measured while it receives "ideal inputs." To claim that the equipment does not meet the claimed class of performance is meaningless if the input signals were somehow impaired or degraded during the test.

But the fact that the tests are standardized is a very good thing because it allows engineers to make comparisons between vendors and equipment. This is an important feature of being able to select the right network equipment for the job (see the preceding section on writing an RFP).

Most of what is called "timing testing," both in this book and in the timing community, does not generally include functionality testing. Yes, it is a given that the network equipment should be able to function correctly according to the IEEE 1588 specification and the specific profiles. Some operators might like to confirm specific cases that concern them, such as the correct functioning of the leap-second logic, for example.

Some years ago, the IEEE did have a program called the IEEE Conformity Assessment Program (ICAP) that validated individual PTP implementations against 1588-2008 and whatever profile it claimed to support. Testing companies offered a service whereby they could test an implementation of PTP against the standards, using a very detailed test plan. ICAP turned out not to be commercially sustainable for the telecom profiles (only G.8265.1 was completed). The program is still active for the power profile, and you can even download a conformance test suite for it.

But the testing discussed here involves testing the timing signals and clock performance rather than specific behavior or even protocol compliance. Yes, if an interoperability issue arises, then packet traces and field validation are important tools to resolve those problems. But the topic here is mostly about performance, and therefore the testing requires dedicated equipment that can generate and measure time and frequency signals.

There are two fundamentally different approaches to testing:

- Standalone testing of a single network element against the performance recommendations, what the ITU-T refers to as *equipment limits*. This is normally a single clock connected back to back to a test device: input a time signal to the DUT and watch how it behaves or measure what appears at the output. A good example of this is testing a T-BC or Telecom Time Slave Clock (T-TSC) against the G.8273.2 performance recommendation.

- Network testing to determine the performance of a timing signal at the end of a chain of timing distribution. This form of testing tests the budgeting calculations and allowances in time error (see Chapter 9) to confirm that the synchronization signals meet the requirements when arriving at the final application.

Figure 12-8 illustrates these two approaches (for figures in this section, the M and S in circles refer to PTP ports in master or slave mode).

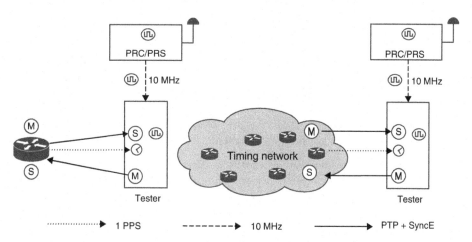

Figure 12-8 *Standalone Equipment Testing (left) Versus Network Testing (right)*

Each approach is underwritten by a group of ITU-T recommendations, whether it describes a network or a clock. Therefore, all the tests relating to the performance on a single node come from a clock recommendation (see Chapter 8, "ITU-T Timing

Recommendations," for more information). On the other hand, the network testing comes from ITU-T recommendations on network limits.

In the standalone clock test, the DUT is connected back to back to a timing tester. The tester generates timing signals as input to the DUT and compares the output to the generated input. However, in the network test case, there is a chain of network elements involved. To test network performance at the end of the timing distribution chain, the tester might be used to measure the output time signal carried from a remote primary reference time clock (PRTC) and Telecom Grandmaster (T-GM). The test unit can also be used as both input and output if the start and end of the timing chain are co-located.

Within each of these test types, the test equipment measures two signals:

- **Frequency testing:** Measures the interaction between clocks and a reference frequency signal. This means testing for aspects of frequency performance such as wander. Much of this form of testing has been inherited from the TDM technologies and standards. SyncE adds an extra dimension, but they have a lot in common.

  Because this testing involves frequency, the test equipment is connected via an interface carrying frequency. If the DUT has a BITS port, then it can be measured using it or via some legacy interface such as E1/T1. Testers that can perform these tests for TDM interfaces have been available for decades and are still operated by people using circuit emulation. Nowadays, Ethernet interfaces and SyncE inputs/outputs are more commonly used for testing.

- **Phase/time testing:** Measures what happens to the phase in interactions with the clock. This often involves measuring TE and calculating various statistics derived from those measurements. Therefore, the test equipment should include software to do those calculations, either as an embedded function or via some external application.

  Because this testing involves phase, the test equipment is connected via an interface that can signal phase. This is either a 1PPS signal (to measure output from the DUT) or PTP over an Ethernet interface. By using an Ethernet interface, it might be possible to capture phase/time (PTP) and frequency information (SyncE) at the same time. Testing the output of the ToD signal is possible on some test equipment, but almost no operators test it, unless they are interested in interoperability of the ToD feed.

The only other type of testing that might arise in some cases is the testing for PDV and packet selection. Not many operators perform this class of tests, although it can become relevant when some clock will not align over an unaware network that displays a high level of PDV. When mobile operators were first deploying G.8265.1 (frequency PTP), this type of testing was sometimes used because that profile does not allow on-path support—every node between master and slave is unaware. This chapter will not discuss this type of packet testing any further.

These are the main categories of testing that are available, and the testing obviously requires specialized equipment to perform. So, the next section covers the type of equipment that is used for accurately testing clocks and network scenarios.

Testing Equipment

Numerous companies specialize in designing and producing timing equipment, but few of them specialize in timing test equipment. There is one company that is very much the *de facto* standard for this class of equipment: Calnex Solutions, from Edinburgh, Scotland. If you go into any lab equipped to do timing testing, there's about a 90% chance that it will have Calnex equipment on site. There are other companies that make equipment to test timing, but the authors have little or no experience with the products from other companies.

For the testing of timing, Calnex currently offers three types of product:

- **Paragon-X**: Probably the most widely adopted product, it has been the mainstay of timing testing for a long while. It allows testing on 1-GE and 10-GE interfaces.

- **Sentinel**: A field-testing unit, which gives the engineer a portable device that can be taken to remote locations and used to take frequency and phase measurement. See the section "Field Testing" later in this chapter.

- **Paragon-neo**: An improved product from the Paragon-X, which can test up to 100-GE interfaces and measures to higher accuracy (Class C noise generation or eSyncE wander, for example).

Figure 12-9 shows a Paragon-X with the two sets of 10-GE (XFP and small form-factor pluggable-plus [SFP+]) ports on the top row, and the two sets of 1-GE (RJ-45 and SFP) ports below them.

Figure 12-9   *Calnex Paragon-X*

These ports are combination ports, allowing a number of possible connection options, but in total there are only two ports that can be active simultaneously: ports 1 and 2. The ports can be used either to measure or synthesize time signals depending on the test case being executed. There are also physical ports on the rear for 1PPS and 10-MHz reference signals (see following section, "Reference Signals and Calibration") as well as measurement ports for TDM signals like E1 and 2048 MHz.

One important consideration is to ensure that the equipment is rated to perform the accuracy of testing you require. Ensuring this likely requires some understanding of the specifications of the equipment you propose to use, and if this is not clear, then you need to consult with the vendor. Too often the authors see people try to test scenarios at a level of accuracy totally beyond the capability of the test measurement equipment.

With 5G increasing the interface speeds, many devices in the radio access network now carry timing over 25-GE and 100-GE interfaces. As the ITU-T increases the accuracy requirements in later versions of its recommendations, the vendors need to produce equipment that is capable of accurately testing it. Figure 12-10 shows one such device from Calnex, a Paragon-neo that allows connection of a larger choice of interface options, such as 25-GE (SFP28), 40-GE (QSFP+), and 100-GE (QSFP28 and CFP4) optics.

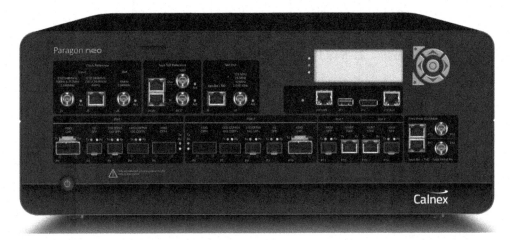

Figure 12-10 *Calnex Paragon-neo*

Other types of equipment might be used in some testing situations, and they will be mentioned where appropriate. A good example would be a standalone frequency or time interval counter or devices used for TDM testing, such as a bit error rate tester (BERT) or frequency tester. But you can do pretty much everything you would need with one of the previously mentioned dedicated timing testers. Note that some vendors for test equipment enable different features with optional licenses, so be sure to purchase the capability that you need.

One more point is that Calnex Solutions produces a wide range of very educational material on timing, PTP, and testing. Anybody who wants to get serious with testing timing is strongly recommended to refer to the Calnex technical library (a link is provided in the "Further Reading" section at the end of this chapter).

Reference Signals and Calibration

Previously, Figure 12-8 showed the tester connected to a PRC/PRS with a 10-MHz cable. This is used as a frequency reference for the oscillator of the device to ensure that the measurements are accurate and that the synthesized frequency signals are within specification. For example, it is impossible to accurately measure the wander on a SyncE signal without an accurate reference to compare it to. As a minimum, a tester needs to have an input reference, which is normally a 10-MHz reference signal from a PRC/PRS or PRTC or alternatively can be a BITS input (where BITS in a building is taking input reference from a PRC/PRS or PRTC).

For back to back phase testing, the Calnex generates its own time signals to send to the DUT and compares the return values to what it sent, so for that mode of testing, it does not need a 1PPS phase reference (see left-hand side of Figure 12-11). But if the Calnex is being used to test the phase at the final hop of a timing distribution, then it needs to have a 1PPS phase reference to compare the phase of the DUT with (see right-hand side of Figure 12-11).

Figure 12-11 shows a tester with the relevant input reference signals for the different cases.

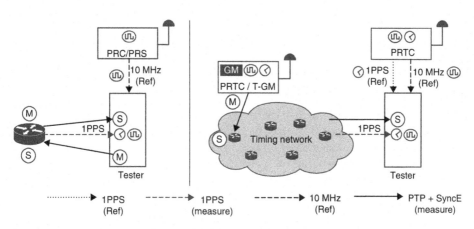

Figure 12-11 *Timing Tester with Reference Signals*

In the case shown on the right side of Figure 12-11, the tester is measuring the phase 1PPS or PTP signal coming from the router and comparing that with the reference phase from a PRTC. This is a test of whether the last clock is putting out a signal within the range of the budgeted phase error required by the application. For example, if the application requirement is that the phase is within ±1.1 μs of Coordinated Universal Time (UTC), then the tester should be able to measure the value coming out of the last clock in the chain to confirm it.

In that same topology (on the right), the tester can also measure the frequency wander that has accumulated in the transit across the network from the PRC/PRS. This requires only a 10-MHz reference signal on the tester. On the DUT, the tester measures either a SyncE signal or a E1/T1/2048-kHz signal from a BITS or TDM traffic port. This test signal is then compared to the wander requirement from the application (and the relevant ITU-T recommendation).

Because a reference signal is needed to measure phase alignment and frequency wander, some testers include a GNSS receiver inside the device. This is especially the case with field testers because they typically do not have any reference source available at a remote location. Engineers use these devices by first taking the tester outside and allowing the GNSS to align an accurate built-in oscillator. Then the engineers take the tester back inside to test the equipment, and the oscillator goes into holdover when it loses the GNSS signals. If the test device spends only a reasonably short interval in holdover, then it can accurately measure the phase (and frequency).

Note   When testing with a topology using two different GNSS receivers, where one is connected to the input of the test device(s) and the other is connected to (or integrated with) the tester, there is an additional error between the two receivers that must be considered. This is discussed further in Chapter 11 when considering GNSS timing sources for the fronthaul network.

One important detail that is often overlooked when trying to accurately measure phase (it does not matter for frequency) is cable length compensation. When the tester is measuring a PTP signal, it is not trying to recover the clock by solving the PTP equations and estimating the mean path delay. The tester just reads the timestamps in the event messages and applies the correctionField (CF).

Therefore, the time taken for the PTP signal to cross the cable between the DUT and the tester needs to be compensated for to get an accurate reading of the phase. For a 10-m fiber cable, this could be a significant error, in the order of 50 ns. Therefore, testers allow for the length of the Ethernet cable to be entered so that the phase can be adjusted to compensate. Similar logic applies to the 1PPS cable, both from the DUT and from the reference PRTC (about 5.1 ns per meter).

Another important point in understanding the source of TE also arises in the right-hand use case from Figure 12-11. Remember that according to G.8272, the accuracy of a Class A PRTC (PRTC-A) combined with a T-GM "should be accurate to within 100 ns or better" when verified against UTC. But the time tester is also connected to a PRTC-A to get a reference 1PPS phase signal, and that signal is also allowed to be up to 100 ns off from UTC. And that is with a good installation and accurate compensation of the GNSS antenna cable (see Chapter 11 on GNSS installation).

So, the issue is that the tester might be getting a 1PPS reference signal that differs by up to 200 ns from the signal being output by the T-GM! Of course, if the two receivers are seeing the same satellites from the same GNSS constellation and experiencing the same propagation conditions for the GNSS signal, then it is highly likely that the error would be less than this. But theoretically, an error of that significance could arise, so it is futile to try to measure the phase offset to the last 10 ns or so for this particular case.

If getting a much better measure is important, then the tester should take its 1PPS reference from the same PRTC to compensate for this error. The difference in phase alignment between the 1PPS output of the PRTC and the PTP output of the T-GM integrated inside the same PRTC should be only a few nanoseconds.

Testing Metrics

Table 12-1 is a summary of the metrics that are possible from different types of testing. Please consult Chapter 5 (for physical frequency) and Chapter 9 (for packet methods) for details on these metrics.

Table 12-1 *Metrics According to Test Types and Topology*

Topology	Sync.	Test Type	Metrics	Reference
Network	Frequency	Wander Generation	TIE, MRTIE, MTIE, TDEV	G.8261, G.8261.1, G.823, G.824
Network	Frequency	Wander Tolerance	TIE, MTIE, TDEV	G.8261, G.8261.1, G.812, G.813, G.823, G.824
Network	Frequency	Jitter Output	Peak-to-peak amplitude (UIpp)	G.8261, G.8261.1, G.823, G.824
Network	Frequency	Jitter Tolerance	UIpp	G.8261, G.8261.1, G.812, G.813, G.823, G.824
Network	Phase	Phase Alignment	max|TE|, 2wayTE, cTE, dTE (MTIE, TDEV)	G.8271.1, G.8271.2
Network	Phase	Phase Tolerance	max|TE_L|, MTIE, TDEV	G.8271.1, G.8271.2
Node SyncE	Frequency	Accuracy, hold-in, pull-in, pull-out	Parts per million (ppm)	G.8262, G.8262.1
Node SyncE	Frequency	Wander Generation	TIE, MTIE, TDEV	G.8262, G.8262.1, G.81x
Node SyncE	Frequency	Jitter Generation	UIpp	G.8262, G.8262.1
Node SyncE	Frequency	Noise Transfer	dB phase gain, TIE, TDEV	G.8262, G.8262.1
Node SyncE	Frequency	Transient Response	TIE, MTIE	G.8262, G.8262.1
Node SyncE	Frequency	Holdover	TIE, ns/s, ns/s^2	G.8262, G.8262.1
Node PTP	Phase	Noise Generation	2wayTE, cTE, dTE, MTIE, TDEV	G.8273.2
Node 1PPS	Phase	Noise Generation	TE, cTE, dTE, MTIE, TDEV	G.8273.2
Node	Phase	Noise Transfer	dB phase gain	G.8273.2, G.8262, G.8262.1
Node	Phase	Holdover	TE, dTE	G.8273.2

Note that there are additional (stricter) requirements for better-quality frequency clocks, such as wander generation for SSU and PRC clocks, so you need to consult the appropriate recommendation from the G.81x range. In G.823, there are also different limits for jitter and wander (output and tolerance) depending on the clock type and whether they are network or synchronization interfaces.

Testing PRTC, PRC, and T-GM Clocks

The testing of a PRTC and T-GM or a PRC/PRS frequency source is a somewhat different scenario, and testing this class of equipment is not that common. That is because it is specialized and requires measuring equipment not commonly found in most network labs. For reference, you can find more information on testing these clocks in Appendixes A.2 and B.2 of ITU-T G.8273.

The problem is that the performance of a PRTC based upon a GNSS receiver reflects not just the characteristics of the device, but also the behavior of the GNSS. Therefore, the vendor can only state what the equipment is capable of, not what it will deliver in a specific installation. For a quick check to ensure the equipment is functioning okay (or the antenna cable length compensation value seems reasonable), it would be fine to use one receiver to check another.

You could also compare the performance against a highly accurate and calibrated *reference receiver*, but the limitations of such an approach are clear: the reference receiver needs to have much better performance than the DUT, which might not be possible. These pieces of equipment must be well calibrated and are rarely seen outside facilities devoted to time measurement. There are various techniques to accurately compare PRTCs remote from each other using GNSS receivers, but these techniques are applied mainly in specialized timing laboratories.

An equivalent method is to measure the output of the PRTC against a time standard (atomic clock), given that this equipment is available to the engineer. Of course, the accuracy of that time standard is critical to the relevance of the test measurements. Most operators choose not to undertake this form of testing because it involves a considerable level of specialized equipment.

To overcome the fact that the GNSS is part of the measured performance, an external device is needed to generate standardized signals that the receiver can be measured against. Figure 12-12 illustrates one approach by introducing a piece of equipment known as a GNSS simulator (the most common variety is predominantly aimed at the GPS receiver market, but others exist).

In Figure 12-12, the GPS simulator accepts a 10-MHz reference input from a PRC/PRS for much the same reasons as a timer tester needs the signal. The GPS simulator produces GPS radio signals (most commonly for the L1 band), and the radio frequency (RF) output of the simulator is connected to the antenna port of the GPS receiver.

The simulator is configured to accept a position (latitude, longitude, and height) and a time (or it can take that information from another GNSS device). It then impersonates the combined signals of a fleet of GPS satellites and make the receiver think that it is at that location starting at that time. The simulator will continue to (accurately) roll the time forward (guided by the 10-MHz reference input), and the GPS receiver will recover the artificially produced time and phase.

The simulator also has a 1PPS output that is in phase with the time being generated (of course, the relative accuracy of these two signals needs validation). Besides the normal 10-MHz reference connected to the tester, the 1PPS out from the simulator is connected to the timing tester as a reference phase signal. Then the phase output by the GNSS receiver can be measured against the phase from the GPS simulator.

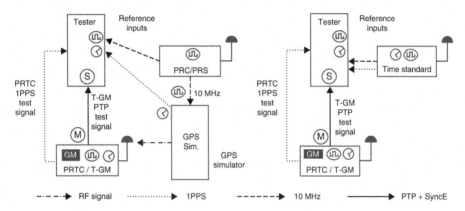

Figure 12-12 *Testing a PRTC and T-GM*

One piece of advice if you are interested in testing any of the highly accurate enhanced clocks (like the ePRTC and ePRC): because these devices have amazing amounts of stability over very long intervals, be prepared for extremely long test runs to verify their performance.

Phase Accuracy

For phase accuracy, the tester receives either or both of two test signals, PTP from the T-GM (or it could be 1PPS as well) and/or 1PPS from the PRTC (if separate devices). The tester compares the phase alignment between the signal from the GPS simulator and the output of the PRTC or the T-GM (or both).

For the PTP output, the appropriate metric on the tester is the two-way time error (2wayTE; see Chapter 9 for details). Compare the 2wayTE for the PTP messages, or the 1PPS max|TE| against that class of PRTC (PRTC-A, PRTC-B, ePRTC-A, or ePRTC-B). This measurement tells the engineer if the PRTC and/or the T-GM is within the required level of accuracy (100 ns for PRTC-A, 40 ns for PRTC-B, and 30 ns for the enhanced PRTC Classes A and B, ePRTC-A and ePRTC-B). This is a quick test that will validate that the recovery of phase in the receiver is aligned with the input signal. This confirms that the receiver will accurately source phase from a GPS signal.

Phase Wander and Jitter: Dynamic TE

G.8272 and G.8272.1 specify maximum time interval error (MTIE) and time deviation (TDEV) masks for wander generation, which, for MTIE, reaches the ceiling of 100 ns for PRTC-A, 40 ns for PRTC-B, and 30 ns for ePRTC. The MTIE is equivalent to the peak-to-peak value of the dTE. There are three basic ways to measure it:

- Measure the time interval error (TIE) for the 1PPS output at one sample per second and without any low-pass filtering. Calculate the MTIE and TDEV from that and compare those series to the G.8272 masks (or G.8272.1 for ePRTC clocks).

- Measure the TIE for the 2048-kHz, 2048-kbps, and 1544-kbps outputs passed through a 10-Hz low-pass measurement filter, with a maximum sampling time of 1/30 second. Calculate the MTIE and TDEV and compare them to the masks for G.8272 (or G.8272.1 for ePRTC clocks).

- Measure the 2wayTE for the PTP messages. These measurements are then passed through a moving-average low-pass filter of at least 100 consecutive TE samples. The sample will be based on the PTP packet rate, which is 16 per second for G.8275.1. Calculate the MTIE and TDEV and compare them to the masks for G.8272 (or G.8272.1 for ePRTC clocks).

The jitter requirements for frequency input and outputs are defined in the PRC specification of G.811 or the ePRC G.811.1 specification.

Holdover Testing

There are no official holdover performance requirements for PRTC-A and PRTC-B, although the ePRTC recommendations do have a holdover requirement (ePRTC clocks require connection to an atomic frequency standard to meet their expected levels of performance).

As an example, the requirement for holdover of an ePRTC-A is that it must be accurate to UTC within a margin increasing linearly from 30 ns to 100 ns over a 14-day period (after running it for 30 days in normal operation before starting the test). The requirements for ePRTC-B are for further study but will obviously be tighter than that. Owing to the equipment needed and the time needed to test to these levels, it is only of interest to timing experts, and further discussion is outside the scope of this book.

Frequency Testing

Frequency testing is very closely related to the original TDM testing from decades past. This book will not cover much in the way of frequency testing for TDM interfaces, as the market is well served for literature on this topic. Just be aware that these tests are aligned with standards that have been in force for many years.

Frequency measurement can be performed on any frequency signal transmitted by a piece of equipment. It could be a 10-MHz or 2048-kHz signal (or some other frequency), a BITS output such as E1/T1, or a TDM signal on a traffic interface. There are many (legacy) test devices that can be used to test these forms of signals, do the statistical analysis, and then check the results against the appropriate ITU-T mask—such as G.811, G.811.1, G.812, and G.813.

For frequency testing of a PRTC or PRC/PRS, the relevant recommendation is G.811 (see Chapter 8 for details on the ITU-T recommendations), and the enhanced PRC is in G.811.1. The testing of frequency for reference clocks is relatively simple because they are output-only devices, so conducting tests like those for noise tolerance and transfer makes no sense.

Frequency testing includes the following classes of tests:

- Frequency accuracy. For PRC/PRS/PRTC clocks this is 1 part in 10^{11} over one week from G.811; for ePRC/ePRTC clocks this is 1 part in 10^{12} over one week from G.811.1; and for other clocks this is 4.6 ppm in free-running state.

- Noise generation (jitter and wander) with MTIE and TDEV masks.

- Pull-in, hold-in, and pull-out (see Chapter 5; not used for PRC/PRS/PRTC clocks).

- Noise tolerance (jitter and wander; not used for PRC/PRS/PRTC clocks).

- Noise transfer (not for PRC/PRS/PRTC clocks).

- Transient response and holdover.

Again, most operators do not perform these types of tests unless they are primarily interested in the packet emulation of TDM circuits. This book focuses more on SyncE and eSyncE testing for frequency.

Functional and Other

Most operators are not overly interested in performing a lot of functional tests; however, there are some basic functions that could be relevant and interesting. For example, timing testers often have an option to perform PTP field validation. This ensures that the packets are sent, and protocol fields are filled, according to the standards.

Another example might be to validate some of the values in the Announce message, such as the values in the clock accuracy fields (particularly clockClass) when in normal operation versus when in holdover. Similar test cases to validate the ESMC behavior of the clock based on the reception and transmission of a range of SyncE QL values is also interesting.

Testing End-to-End Networks

The first major segment of a testing regimen is to test the timing distribution test network against the network limits from the ITU-T. There are three cases to consider:

- Testing for frequency wander when it has been carried with SyncE or eSyncE over the timing network

■ Testing for frequency wander when it has been carried over the timing distribution network using PTP

■ Testing for phase alignment when it has been carried over the timing distribution network using PTP

The following three sections deal with each case in turn.

Testing Networks for Frequency Wander (SyncE)

The goal of building a timing distribution network is to ensure that the timing signal arrives at the destination at the end of the timing chain with the quality needed by the end application. Therefore, once a proposed design has been accepted, the engineer needs to verify the calculations and assumptions underpinning it before deploying it into production. Similarly, once the new design is handed over to operations, situations may arise where the production network needs to be tested. Refer to the section "Field Testing" later in this chapter for details on how to achieve this.

For frequency synchronization, depending on the application, external information might quickly give an indication that the frequency is wandering out of range. A good example would be for circuit emulation, where a frequency mismatch would quickly lead to errors (such as slips) on the TDM attachment circuits or underruns or overruns in the dejitter buffers. Standard methods of monitoring circuit and interface statistics would quickly detect these anomalies.

For many other cases, or for pre-deployment compliance testing, the engineer might like to gain confidence in the quality of the delivered frequency through some tangible metrics. Figure 12-13 shows the basic topologies for testing using two different topologies with an appropriate test device.

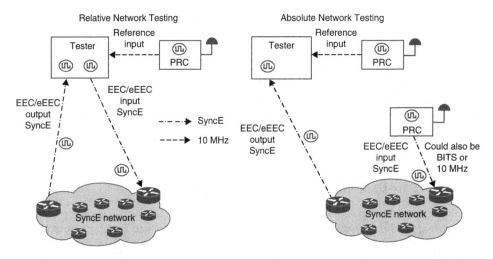

Figure 12-13 *Testing Frequency Wander in a Network*

The left-hand case, relative network testing, compares an output signal at the end of the network chain against the input signal produced by the tester, but with a reference frequency (PRC) as the ultimate source. The right-hand case, absolute network testing, measures the frequency signal at the end of a network chain after having been supplied by a separate source at the beginning of the chain.

Absolute network testing reflects a more realistic case because the end of the chain is normally located away from the source of frequency, such as would be the case when field testing. Remember that these constraints on the testing arise because of considerations around the maximum length of the cables. Reference inputs (such as 10 MHz) can run only over very short cable lengths, so the relative locations of the tester and the PRC need to be considered. If they are not co-located, then an additional PRC is needed.

Most commonly, an operator conducting these tests is concerned with frequency wander in a network topology, unless there is some specific reason to be concerned about some other parameter (such as jitter) that is causing issues with the final application.

Figure 12-14 shows the MTIE graph of the SyncE wander after crossing a (small) network of routers connected to an ePRTC with 10-GE links. Because the customer network uses option 1, and this is a network test, the appropriate mask to use when comparing the result is the G.8261 EEC Option 1 wander limit.

This capture comes from the Calnex Analysis Tool (CAT) post-processing a SyncE capture taken on a Calnex Sentinel field tester. The fact that the calculated MTIE stays well below the mask (on a logarithmic scale) means this is a very good result, which is as expected from a SyncE network.

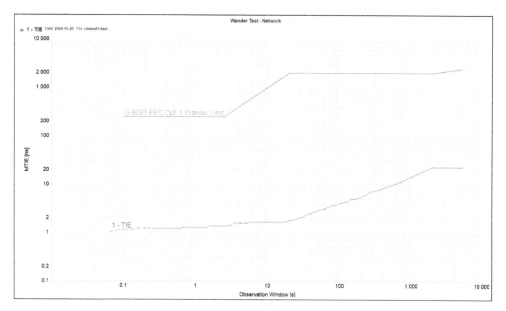

Figure 12-14   *Example of a Network SyncE Wander Test*

This is the simplest but most practical use of testing frequency—trying to determine what the signal is doing at the end of the chain and whether it will meet your needs. That will answer questions such as whether it is good enough to synchronize the emulated TDM circuit, or provide a stable source for a radio transmitter.

This test case makes sense where the clocks and transport system support SyncE. But if that support is not available, then the frequency must be carried using a packet-based approach. A test regime for that scenario is outlined in the following section.

Testing Networks for Frequency Wander (Packet)

There is one test case that is mentioned without too much detail, mainly because it is becoming increasingly less mainstream—the recovery of frequency using PTP rather than SyncE. Many mobile operators have been running the G.8265.1 telecom profile for frequency for quite some years, mainly to replace the TDM links that they used to syntonize their 2G and 3G radios. In networks that did not support SyncE, this form of PTP is a possible alternative.

The principal enemy of this solution is PDV, and because G.8265.1 does not allow any on-path support, the control of PDV through the intervening network elements is a very critical aspect to measure, understand, and control. Several models of timing test equipment, including those you saw previously, allow for testing and characterization of PDV if needed.

One interesting feature is the ability to test for a packet floor in the arriving PTP message stream (see "Packet-Based Frequency Distribution Performance" in Chapter 9 for a discussion of packet floors). The tester can see how many packets pass the criteria for packet selection to be used to calculate the timing solution.

This metric is known as the floor packet percent (FPP) and is defined as the percentage of packets that fall into the given fixed cluster range starting at the observed floor delay (the theory is covered in G.8260). A network limit for it is defined in G.8261.1 (basically 1% of packets should fall into the cluster around the floor delay). This is a handy test suite when you are having problems with PDV in a network.

The second thing commonly tested is much the same as covered in the preceding section, and that is the accuracy of the frequency recovered by the slave clock. Again, the tester device can measure a frequency output signal from the slave clock (as a E1/T1/2048-kHz signal) and measure the wander compared to a reference 10-MHz signal. This is compared to the packet-based equipment clock—slave—frequency (PEC-S-F) MTIE mask from G.8261.1 for a pass/fail determination.

One other approach would be where the test device directly negotiates a PTP packet stream with the remote G.8265.1 GM and measures the frequency wander based on timestamps in the arriving PTP messages. This approach tests the network for its ability to carry a packet frequency signal instead of testing the wander in the frequency recovered by a slave clock.

So, now that this test confirms that the frequency signal is accurate and stable, the next task is to test the phase alignment.

Testing Networks for Phase Alignment (Packet)

Like the case with frequency, the goal of building a phase/time distribution network is to ensure that the timing signal arrives with the quality required. For applications that require phase synchronization, maintaining that phase alignment is especially critical because there are numerous factors that work against it. So, thorough testing before deployment is crucial, both on the individual clocks themselves (see following sections) and on the end-to-end network.

End-to-end network phase/time testing is especially important to validate the lessons learned and assumptions made about the underlying packet transport. Additional tests that many operators like to perform include holdover testing when the PTP and SyncE reference signals to the last node are lost. Another is what happens to the phase alignment when there is a rearrangement and reconvergence somewhere upstream in the network and a clock must select a new grandmaster (GM).

Because this is a phase alignment test, it needs a phase signal, either PTP or 1PPS, from the last network node so that the tester can measure the phase as recovered by that device. The left-hand side of Figure 12-15 shows the case of a relative test, as opposed to the absolute test case on the right-hand side. Just as in the frequency case, the test engineer must consider the cable lengths because the reference signals can only use short-run cables, and that determines the relative location of the test device and the reference sources.

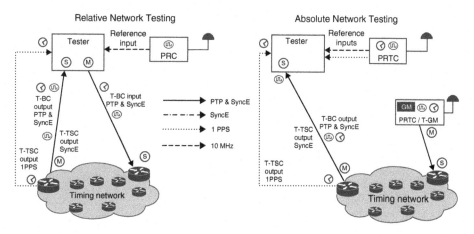

Figure 12-15 *Testing Phase Alignment in a Network*

Relative means that the phase is generated synthetically on the tester and given to the head of the timing chain and the phase from the last node is returned to the tester and compared to the input. The tester in this case is measuring the relative phase offset between the incoming and outgoing signals; there is no connection to UTC.

The absolute case is different because the phase that is input at the head of the timing chain is sourced from an independent source, such as a PRTC and T-GM combination with traceability to UTC. In this case, the tester must measure the output of the last node in the chain against a trusted source of UTC time.

Therefore, the tester needs a 1PPS reference from a UTC time source that it can compare against the phase received from the network. It could be the same PRTC as the one giving input at the head of the chain, but if the tester is remote from that device, then it must use a different PRTC as its source. That is why this case is called absolute: it is determining real-world phase/time. (The pedantic might say it is still a relative test against UTC, but many regard UTC as "absolute" time.)

As always, the tester also needs a frequency standard to condition its oscillator.

From the PTP test signal, the tester measures the 2wayTE (see Chapter 9 for details), and for a 1PPS input test signal, the tester measures TE. The engineer can then determine a pass/fail against any desired limit for max|TE|, such as the ±1.1/1.5-µs phase alignment limit for the 5G mobile case.

Figure 12-16 shows the MTIE graph of the phase TE after crossing a (small) network of routers connected to an ePRTC with 10-GE links. Because this is a network test, the appropriate threshold for pass/fail is the 1.1-µs 2wayTE network limit from G.8271.1 (the mobile requirement). As in the preceding case with SyncE, this capture comes from the CAT reporting tool and the Sentinel tester.

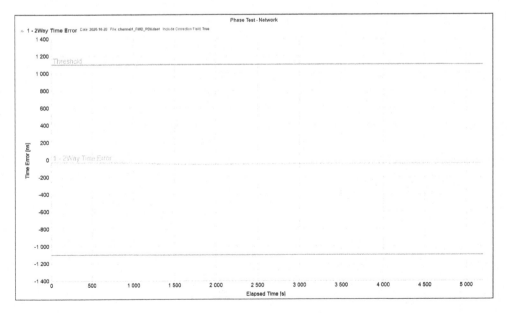

Figure 12-16 *Example of a Network Phase Time Error Test*

The max|TE| value of the measured TE is 61.5 ns with a peak-to-peak variation of 40 ns. An additional metric that is interesting is the combined cTE from the chain of T-BC clocks. For example, if there is a chain of five Class B T-BC nodes with 20 ns of cTE each, then the cTE could be calculated to ensure the cTE is less than 100 ns (5 times the G.8273.2 limit of 20 ns). In the case in Figure 12-16, the cTE (averaged over 1000 s) varied between –52 ns and –31 ns after having passed through four (Class B) T-BCs.

It is probably no surprise that this is one of the most commonly executed test cases, mainly because all engineers want to know how accurately their design carries time. Many engineers also like to know how much margin they have between the measured phase error and what their budget is.

Figure 12-17 shows the 1PPS phase alignment of the previous network when the last router before the tester is placed into holdover. Because this is a network test, once again the appropriate threshold limit is the 1.1-µs network limit from G.8271.1. As in the preceding examples, this capture comes from the CAT reporting tool and the Sentinel tester (this time measuring 1PPS).

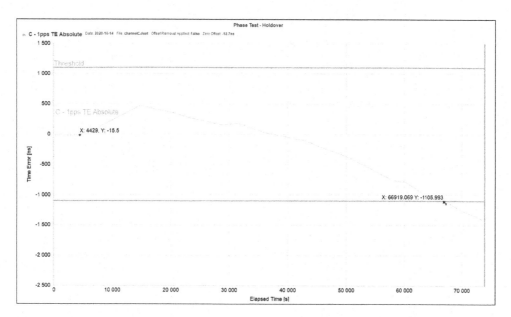

Figure 12-17 *Example of a Network Phase Holdover Test*

In this case, the PTP and SyncE signals were removed from the test node after 4429 s into the test, and the TE crossed the limit at about 66,900 s (just over 17 hours later). Perhaps it is slightly lucky, but this is a very good result for a (very recent model of) network element with a stratum 3E OCXO running in holdover (it is only testing the holdover capabilities of the last T-BC router in the network because the break is immediately upstream of the DUT, and so it is the isolated node).

Many operators like to perform holdover tests because they would like to know how the network and the network clocks will perform when faced with outages in the timing signal from their references.

One additional special case that needs mentioning is the testing of the assisted partial timing support (APTS) network topology, especially in the case of holdover (failure of the local reference clock). When the DUT is connected to its primary source, it is much like a T-GM or a T-BC with zero hops to the GM. When it loses that primary source of time, it is much like a T-BC on a partial timing support (PTS) network. The only assistance it has is the compensation mechanism to correct for the measured static asymmetry. Operators considering this option obviously want to test how well this approach works on a network that very closely matches their likely deployment.

Now that you have seen the basic cases of testing nodes end to end in a network, it is time to consider the performance of the clocks as standalone entities and verify the equipment limits.

Testing Frequency Clocks in a Packet Network

Until this point in the chapter, the testing has been about the performance of the timing signal in the end-to-end scenario—what the timing signal looks like after emerging from the last link in the chain. But the end-to-end performance is very much reliant on the performance of the nodes that make up that network, or what the ITU-T calls *equipment limits*.

The main clock of interest in these packet-based deployments is the SyncE (or eSyncE) clock, so the following section covers the testing of the metrics and behavior for those clocks.

Testing EEC and eEEC SyncE Clocks

The good thing about testing SyncE clocks (EEC and eEEC) is that it is somewhat less confusing than the case with TDM-type interfaces, and there are not as many recommendations to follow. Fortunately, most operators prefer SyncE as a frequency transport rather than some alternative based on TDM, and the testing is relatively straightforward.

The benefit of SyncE is that it is a broadly supported technology, is widely deployed, and performs very well. There might be a few unexpected results with early implementations of the newer eSyncE (as with all new technology), but testing SyncE almost never shows unanticipated results. One additional point when testing eSyncE is that frequency testers that are even a couple of years old may not have the accuracy to precisely measure the new generations of eEEC clocks.

Figure 12-18 illustrates the method to test the performance of SyncE clocks.

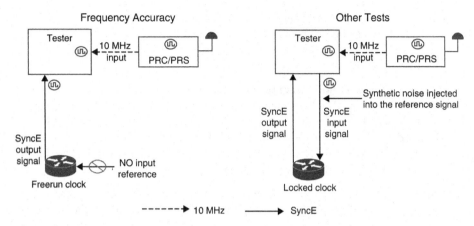

Figure 12-18   *Testing SyncE and eSyncE Equipment Clocks*

The testing for SyncE, much like testing the TDM technologies, also reflects the differences between SDH and SONET. For the North American markets, the SyncE for any network equipment should be configured as an option 2 EEC clock, and for other markets, it is usually option 1 (ITU-T). Some test cases differ in method and results depending on the option of SyncE being used. It makes sense that an individual operator will only test the option they intend to deploy. There is little point in performing option 1 tests when rolling out option 2 SyncE in North America.

The following represents the performance metrics of a SyncE clock from G.8262 (EEC) and G.8262.1 (eEEC):

- **Frequency accuracy:** Confirms that when running free, the clock's oscillator (driving the SyncE test signal) is accurate to within 4.6 ppm of the nominal frequency (of the Ethernet signal). The tester measures the output of the standalone DUT (without any reference inputs) to calculate the frequency offset and confirm it is within specification. The length of time for the test is not specified, but for eEEC clocks, very long periods have been proposed.

- **Pull-in, hold-in, and pull-out range:** The tester injects a frequency offset, and the engineer monitors the clock behavior for compliance. For pull-in, the large offset is applied to the input and then reduced until the DUT locks. For hold-in, the clock is locked, and then an offset of ±4.6 ppm is applied to the input while testing the DUT remains locked. For pull-out, the DUT is locked to the input, and an increasing frequency offset is applied until it unlocks.

- **Wander noise generation with the DUT in locked (and non-locked) state:** In locked state, the tester injects a wander-free reference and measures the output wander as TIE and compares the MTIE and TDEV to the masks in G.8262 (EEC) and G.8262.1 (eEEC). There are tests for constant and varying temperature and option 1 versus option 2.

- **Jitter noise generation with the DUT in locked state:** The test measures the peak UI over 60 seconds and compares it against the G.8262 recommendations. Because eEEC clocks must be compatible with EEC clocks, they should not exceed the G.8262 jitter generation limits either.

- **Tolerance to wander noise from the input (reference) signal:** The tester applies wander to the input signal and the engineer monitors the clock behavior for compliance. The amount of wander applied to the input is defined by MTIE and TDEV masks in G.8262 or G.8262.1 or other forms (sinusoidal) can be generated by the tester. The behavior of the clock can also be monitored through QL values generated by the DUT as well as TIE measurements against the nominal frequency.

  There is a difference in the G.8262.1 wander limits for clocks in networks consisting of only eEEC clocks versus those mainly consisting of EEC clocks. In the latter case, a higher amount of wander will accumulate, so the eEEC clock should be able to tolerate the higher wander from an EEC network.

- **Tolerance to jitter noise from the input (reference) signal:** The tester applies jitter to the input signal and the engineer monitors the clock behavior for compliance. The amount and characteristics of the jitter can be configured on a tester. Compliance, like all forms of tolerance testing, means that the clock did not trigger errors, switch references, or go into holdover.

  Because eEEC clocks must be compatible with EEC clocks, they should also meet the G.8262 jitter tolerance limits. Of course, a timing chain based solely on eEEC clocks will result in lower jitter accumulation due to the lower maximum bandwidth of 3 Hz and lower noise generation of the eEEC.

- **Characteristics of the transfer of noise through the clock:** This relates mainly to the bandwidth of the low-pass filter (for option 1 clocks; for option 2 there is a pass/fail TDEV mask in G.8262). The tester generates a series of input wander at different frequencies, both outside and inside the range of the passband. For option 1, the passband is 1–10 Hz, and for the eEEC, it is 1–3 Hz.

  For option 1 and eEEC, the tester measures the behavior of wander (phase gain less than 0.2 dB inside the passband) across the range of the frequencies. For option 2, the TDEV (from the measured TIE) is compared to the G.8262 Wander Transfer mask.

- **Transient response:** Testing phase error movements during a short-term reference switch (less than 15 s) as well as other longer-term scenarios. The input signal from the tester to the DUT is cut (or signaled to be invalid with ESMC), and the TIE of the output signal of the DUT is measured. The result of the TIE is compared to the G.8262 mask for short-term transient for option 1, G.8262.1 mask for eEEC, and the MTIE to the mask for option 2.

  For details of the methodology to test SyncE transients, there is some helpful information in Appendix II of G.8273.

■ **Holdover, or long-term transient:** Measures the wander (TIE) after a reference input signal is removed. For the eEEC, there is a mask for holdover in G.8262.1 and in G.8262 for both option 1 and 2 of the EEC clock. Of course, holdover tests should be initiated only following a period that allows the clock to settle down after setup. For frequency tests, that should be at least 15 minutes.

Of all these tests, the most popular performance value to test is the wander noise generation. Additionally, some operators like to confirm what the response is of the SyncE signal during an event like a switchover in the control plane processors. But performing every test in this list is not a trivial amount of effort, especially when considering the long run times and the number of possible platforms and different interfaces involved.

The only functional area that operators like to check is the behavior of ESMC to confirm that the QL transmission mechanism is behaving correctly. Especially popular is confirming that a router transmits the correct QL levels during a switchover between redundant processors. With the increasing adoption of the enhanced ESMC TLVs to support eSyncE, there might arise a few new scenarios of interest, particularly to confirm the correct interworking in both EEC and eEEC clocks.

Like the other test categories, the tester supports the ability to automatically set up the test parameters and methods and measure the results. It also comes with statistical software to calculate device metrics such as phase gain, MTIE, and TDEV and perform any necessary filtering. It also has all the pass/fail masks built-in to quickly determine the result. What it will not do is stop the unwary engineer from using the wrong mask to confirm pass/fail for the test set results.

So, now that you are happy with the frequency performance of your EEC/eEEC clock, it is time to determine how it performs as a time/phase clock using PTP.

Testing Standalone PTP Clocks

The frequency performance of the SyncE/eSyncE clocks is treated separately in the previous section, but the other facet to confirm is the performance of the network clocks as a PTP clock. In summary, that means testing the performance and behavior of all forms of BC, TC, and slave clock. Three foundation ITU-T recommendations guide the test engineer here: G.8273.2, G.8273.3, and G.8273.4.

The following sections will take each group of clocks in turn and explain the test cases that can be applied to each clock type. Each clock type will include tests for noise generation, tolerance to input noise, noise transfer through the clock, short-term transient response, and holdover (long-term transient).

Testing T-BC and T-TSC to G.8273.2

When operators are adopting network-based timing, most of the testing interest is split between T-BC/T-TSC performance and the phase error alignment after crossing the network. G.8273.2 contains the most important criteria for selecting PTP-aware network elements, so adherence to that recommendation is the prime area to investigate.

For reference, there is more information on testing these clocks in Appendixes A.3, A.5, B.3, and B.5 of ITU-T G.8273.

The idea is to test the performance of T-BC and T-TSC clocks using G.8275.1 in combination with SyncE (or eSyncE) in a full timing support (FTS) network. As this is the PTP method used most often for phase/time transport, this section starts with G.8273.2 compliance and covers it more thoroughly than others.

Figure 12-19 illustrates the method to test the performance of T-BC boundary clocks and T-TSC slave clocks.

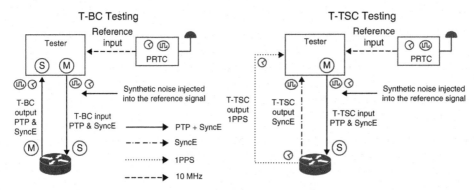

Figure 12-19   *Testing T-BC and T-TSC Clocks*

In all cases, the tester has a master port that gives the DUT a timing signal consisting of G.8275.1 PTP messages plus SyncE (or eSyncE). In some tests, the signal is an *ideal* input signal, meaning that no intentional impairment is applied to the signal. For other tests, the signal can be altered (for example, to test noise tolerance or transfer). As this is a back to back test, a 1PPS reference is not really needed because the tester generates its own phase and then reads it back from the DUT to determine a relative result.

The method of testing of the two clocks (T-BC versus T-TSC) is quite similar except for the following basic differences:

■ There is no PTP master port on a T-TSC (by definition); therefore, it cannot send the tester a PTP output signal. 1PPS is the only option available to measure phase on a T-TSC, whereas it is optional on the T-BC. The T-BC is the most important (and numerous) component of the timing delivery network, and the function of the T-BCs is to transport PTP, so it makes sense to measure their output time signal using PTP.

■ On a T-BC, the frequency is usually measured on the same port as PTP. If you can configure SyncE output on the T-TSC, you can measure frequency signals like any other EEC. Otherwise, some TDM interface such as a BITS port could be used to measure frequency (on both the T-BC and T-TSC).

- No PTP output limits are defined for the T-TSC because it has no role in passing on PTP timing signals. For this reason, it makes little sense for most operators to test a network element as a T-TSC because the end application is not hosted in the T-TSC router. (In the real network, the T-TSC would not be in a router.)

In the real network, the application device (for example, a radio transmitter) would contain the final T-TSC in the chain, and all the network elements would be configured as T-BCs. One drawback is that the radio unit (containing the T-TSC) might not be equipped with a 1PPS output port to measure its phase. In cases of general-purpose compute servers, getting a 1PPS signal output is difficult (it is available only on a NIC in the server). For radios, another option might be to measure the radio signals output from the transmitter to determine the phase alignment.

To not make this section overly complex with repetitions of the same message on the T-TSC compared to the T-BC, these differences are mentioned here at the start. However, understand that these differences apply throughout the rest of the following sections where they discuss methods for T-BC and T-TSC testing together. Remember also that before these test measurements (particularly holdover) start, some time is needed for the DUT to lock to the signal, and additional time is needed for the clock to settle.

Many operators no longer need a 1PPS signal to synchronize a piece of equipment, so there is little point in measuring it, and PTP is preferred. The only scenario in which 1PPS testing might make sense is when operators want to use it as a *measurement port*. This is where a mobile engineer with a field tester goes to the remote device, plugs a special tester into the 1PPS port, and immediately gets a readout of the phase alignment (see the section "Troubleshooting and Field Testing" later in the chapter).

Of course, to validate this approach, the operator should ensure that the 1PPS is an accurate representation of the phase on the device. For that reason, it makes sense to confirm that the 1PPS signal is closely aligned with the phase carried by the PTP signal. Most testers allow both the PTP and 1PPS TE from the DUT to be measured simultaneously.

One additional method of testing these sorts of clocks is to use a tester that is capable of passively monitoring the PTP signals. For this topology, the testers receive a copy of the PTP messages and perform measurements on them. This might be better suited to a scenario where the DUT does not support the ability to have measurement ports or where the clock is distributed across multiple devices (such as a microwave system acting as a Telecom Transparent Clock [T-TC]). Further information on testing a T-BC made up of two separate devices is provided later in the chapter in the section "Testing (Cascaded) Media Converters."

There is some detail behind each of the proposed tests, but clear reference information is available on each in the recommendations—in this case, mostly in G.8273.2 and the network requirements from G.8271.1. Additionally, there are references at the end of the chapter to numerous application notes, written by Calnex Solutions, that cover most testing scenarios.

The following represents a synopsis of the performance tests of a T-BC/T-TSC clock from G.8273.2:

- **Noise generation:** The tester applies an ideal signal and measures the phase error on the signal returning from the DUT (via PTP for T-BC and 1PPS for T-TSC). So, the tester generates time signals (PTP with SyncE), sends them out a master port, receives the return signal on a slave port, and measures the amount of TE introduced by the time signal passing through the DUT.

  The tester measures the 2wayTE (see Chapter 9) and confirms a pass/fail against the max|TE| T-BC clock limit configured on the tester. The appropriate limit would be 100 ns for Class A, 70 ns for Class B, and 30 ns for Class C. The value for Class D is specified only after filtering by a low-pass filter (LPF) (0.1 Hz) and allows a max|TE_L| of 5 ns.

  The tester calculates the cTE (averaged over 1000 s) and the MTIE and TDEV for the (filtered) dTE. Each of these calculated values is used to confirm compliance to the class (A, B, C, or D) for noise generation. For cTE, Class A is ±50 ns, Class B is ±20 ns, and Class C is ±10 ns (Class D is not yet specified). The dTE (after filtering) is compared to the relevant MTIE and TDEV mask for each class of performance (see Table 12-2 for details of the limits).

  As previously mentioned, if the DUT has a 1PPS output, you should confirm the result is also within specification and shaped the same as the result from the PTP. The peak-to-peak TE on the 1PPS should be close to that of the PTP 2wayTE. That ensures that the 1PPS can be used as an accurate measurement port after it is placed in a remote location.

  Remember that if you configure any impairment or do not have SyncE running, you cannot compare any result to the standardized values to determine the class of clock. Testing compliance to the recommendation requires following the rules on input signals. See section 7.1 of G.8273.2, which relates to noise generation, for detailed information on the various classes of T-BC/T-TSC.

- **Tolerance to phase/time and frequency noise from the input (reference) signals:** The tester applies a noise pattern to both the PTP message flow and the SyncE frequency signal while the engineer monitors the clock behavior. A pass result is when the DUT maintains its lock to the reference signal without switchover or moving into holdover.

  The PTP noise is a profile of PDV that is defined in G.8271.1 (it is an MTIE mask) and applied in either direction. This is effectively the same as the dTE network limit that would apply at the end of a chain of clocks. This makes sense, because the last clock in the chain, which receives the most dTE of all nodes, needs to be able to tolerate the maximum amount of dTE that the chain is permitted to accumulate.

  Simultaneously with the PTP noise, a frequency wander is applied to the SyncE input (see the noise tolerance test from the preceding section on the EEC/eEEC). According to the latest version of G.8273.2, the main difference between Class A/B and Class C compliance is that a Class C clock should meet the frequency wander tolerance from the enhanced G.8262.1 rather than the EEC specification in G.8262.

Like the SyncE noise tolerance case, the behavior of the clock can also be monitored through QL values generated by the DUT (or clockClass as well). To pass, the clock should not signal any degradation in clock quality levels.

■ **Characteristics of the transfer of phase/time noise through the clock:** This is a complicated set of tests to run, and many operators do not attempt them. G.8273.2 clause 7.3 defines several types of noise transfer, which are as follows:

■ PTP to PTP and PTP to 1PPS transfer, which measures the transfer of phase/time error from the input to output interfaces. This test resembles the noise transfer test for EEC clocks as it tests the characteristics and performance of the pass-band filter. The criteria to pass is that in the passband, the phase gain of the DUT should be smaller than 0.1 dB.

■ SyncE to PTP and SyncE to 1PPS transfer, which measures the transfer of phase wander from the physical frequency interface to the phase/time outputs. This test also resembles the previous test as it tests the passband filter. The tester generates a series of input wander at different frequencies, both outside and inside the range of the passband. There are different criteria for Class A/B versus Class C/D clocks because of the difference in the upper limits of the bandpass filter (1–10 Hz versus 1–3 Hz). But the result is the same; the phase gain inside the passband should be smaller than 0.2 dB.

■ **Transient response (response in the packet layer to a physical layer transient):** This (short) test involves adding a transient to the SyncE frequency signal and monitoring the response on the packet layer. The tester generates a transient that includes a degraded QL value (for about 13 s) to indicate loss of traceability with a return to a normal signal after that. There is a mask from G.8273.2 that defines the limits on the TE measured. Again, the clock requires some settle time before the test is executed.

■ **Holdover, or long-term transient:** Measures phase error under two conditions. One is when the PTP input has been removed but the SyncE is still available. The second is when both signals are removed. For G.8273.2 compliance testing, only the first case applies, as the behavior with loss of both signals is for further study.

The tester can perform this test standalone with the DUT, although many operators prefer to do it when the DUT is getting its signal from a network. The metrics consist of the dTE with low-pass filter applied and an MTIE mask for both constant and variable temperature indicating pass/fail. The values are 62–123 ns MTIE after 1000 s with the constant temperature and 63–173 ns MTIE after 1000 s with the variable temperature.

Of course, holdover tests should only be initiated following a period that allows the clock to settle down following setup. For accurate long-term tests, that should be up to 24 hours.

Of all these tests, the most popular performance feature to test is the noise generation, with the cTE being the metric most operators are interested in. Table 12-2 outlines the parameters that define which class of clock the T-BC/T-TSC fits into, based on the noise generation results.

Table 12-2 *G.8273.2 Timing Classifications for Class of Boundary/Slave Clock*

Parameter	Conditions	Class A	Class B	Class C	Class D
max|TE|	Unfiltered TE for 1000 s	100 ns	70 ns	30 ns	Unspecified
max |TE_L|	0.1-Hz LPF for 1000 s	—	—	—	5 ns
cTE	1000 s averaged	50 ns	20 ns	10 ns	Unspecified
dTE_L MTIE	0.1-Hz LPF, 1000 s constant temperature	40 ns	40 ns	10 ns	Unspecified
	0.1-Hz LPF, 1000 s variable temp	40 ns	40 ns	Unspecified	Unspecified
dTE_L TDEV	0.1-Hz LPF, 1000 s constant temperature	4 ns	4 ns	2 ns	Unspecified
dTE_H	Peak-to-peak, 1000 s constant temperature	70 ns	70 ns	Unspecified	Unspecified

Some additional tests that operators like to perform include testing the behavior of the clockClass variable, to confirm that the values being transmitted in the Announce message accurately reflect the state of the T-BC. Also popular is checking the PTP 2wayTE measurement during an event like a switchover in the control plane processors or a rearrangement in the network that changes the preferred GM.

This section covered the case for clocks supporting the FTS in G.8273.2, so the next section looks at the PTS cases and the partial and assisted-partial versions of the T-BC and T-TSC clocks.

Testing T-BC-A and T-TSC-A to G.8273.4

In the preceding section, G.8273.2 covered the cases of the PTP-aware clock with full on-path support. This section, based on G.8273.4, covers the cases where there is only partial PTP support from the network, but a local time source is present to *assist* the clock. These test cases confirm the performance of Telecom Boundary Clock with assisted partial support (T-BC-A) and Telecom Time Slave Clock with assisted partial support (T-TSC-A) clocks while running G.8275.2 PTPoIP. There is no frequency assistance available across the network in the assisted (T-BC-A and T-TSC-A) cases. Of course, frequency is available from the local source, but G.8273.4 has no test cases for that.

The testing methodology is equivalent to the methodology described in the preceding section for testing according to the requirements on G.8273.2, so the preceding Figure 12-19 is still mostly valid for the test setup. The main difference between these tests and the previous FTS case is that the assisted clocks have two possible reference input signals: the local time source (via 1PPS, for example) as a primary, and a PTP slave port aligned to a remote T-GM over a PTS network.

Because the PTP is remote and the 1PPS is local, the PTP input limits are based on the network limits at the end of a chain based on G.8271.2, and the 1PPS limits are based on the limits at the beginning of a chain. As one example, for the PTP input, the clock must tolerate the noise limit that might appear at the end of the timing chain, whereas it needs to only tolerate 1PPS noise that is generated by a PRTC (at the start of the chain).

Figure 12-20 shows possible test topologies for testing T-BC-A clocks.

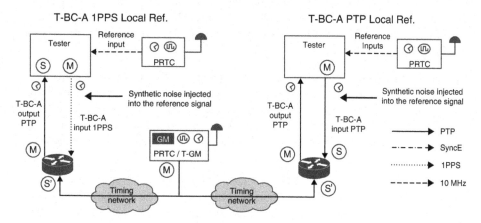

Figure 12-20   *Testing T-BC-A Clocks with 1PPS or PTP Inputs*

As shown in the figure, the new factor is that the T-BC-A (and T-TSC-A) clocks receive a local reference signal from either a local PTP clock (right-hand side) or a PRTC clock via a 1PPS signal (left-hand side). Even though this is a valid setup, G.8273.4 has not yet specified limits for cases where the local clock is a PTP clock, but only when it uses 1PPS.

Figure 12-20 shows that the DUT is also equipped with a PTP port (marked S')—the slave port aligned with a remote T-GM over a partially aware network. This port is to be used as a backup source during outage of the local clock. For the purposes of this testing, this input port is like the T-BC case, although there are no test cases that include frequency input on that port.

In the center of Figure 12-20, a PRTC+T-GM is shown as the reference source for the PTP slave port, but for some test cases, the tester would have to be configured as the master input to that PTP slave port on the DUT, so that it can inject noise into the signal.

A problem arises where the T-BC-A/T-TSC-A has an embedded GNSS receiver as the local time source. In this case, the tester does not have a port to inject the generated test because that connector is internal to the DUT. The T-TSC-A may be even more difficult to test because it may be equipped only with a single 1PPS port, and if that 1PPS is needed to inject the generated signal from the tester, then there is no 1PPS available to use as a measurement port. Of course, the T-TSC-A cannot be measured using a PTP master port, whereas that would be an option if it was a T-BC-A.

The G.8273.4 recommendation has not yet specified holdover limits for the clock relying on the remote PTP GM; it specifies only the case where both the local time source is lost and the backup PTP source is unavailable (the oscillator-assisted case). For that reason, these components do not need to be there unless the engineer wants a network phase alignment test during failure of the local source but with the PTP slave clock providing a secondary signal (see the section "Testing Networks for Phase Alignment (Packet)" earlier in the chapter).

Regarding the equipment limits, a good amount of reference information is available in the recommendations, specifically the clock limits in G.8273.4 (in clause 7) and the network limits from G.8271.2. The following represents a synopsis of the performance tests of T-BC-A/T-TSC-A clocks from G.8273.4:

- **Frequency accuracy:** Confirm that when running free, the clock's oscillator is accurate to within 4.6 ppm of the nominal frequency (of the Ethernet signal). There is no mandatory need for a frequency reference, and this is the same requirement on frequency accuracy as the standalone EEC clock. For further details, refer to the preceding section "Testing EEC and eEEC SyncE Clocks" for testing frequency accuracy.

- **Noise generation:** The tester applies an ideal signal and measures the phase error on the signal returning from the DUT. So, like the T-BC/T-TSC case, the tester generates a time signal, sends it out one port, receives the return signal on another port, and measures the amount of TE introduced by the time signal passing through the DUT. G.8273.4 currently defines noise generation limits only where 1PPS is the input signal to the DUT, not when it is delivered by PTP.

  The output of the T-BC-A is measured at both the 1PPS output and the PTP master port; the output of the T-TSC-A is measured at the 1PPS output (if available).

  Unlike the G.8273.2 case, the value for the limits of max|TE| is currently for further study. The tester calculates the cTE (averaged over 10,000 s) and the MTIE and TDEV for the (filtered) dTE. For cTE, the limit applicable for Class A is ±50 ns, and for Class B, it is ±20 ns (the same as the T-BC/T-TSC case). The dTE_L (after 0.1-Hz filtering) should be less than 50 ns peak-to-peak (again, tested over 10,000 s), while the dTE_H limit is for further study.

  See section 7.2 of G.8273.4 for more information on the noise generation limits and performance of T-BC-A/T-TSC-A clocks.

- **Tolerance to phase/time noise from the input (PTP or 1PPS) signal:** The tester applies a noise pattern to the timing input signal while the engineer monitors the clock behavior. A pass result is that the DUT should maintain its lock to the reference signal without switchover or moving into holdover.

  The PTP noise is the peak-to-peak pktSelected2wayTE network limit (1100 ns from clause 7.3.1.1 in G.8271.2) applied as an input (it is only defined for option

1 networks). This is the maximum limit allowed at the end of a chain of clocks, so a clock must be able to tolerate it as input.

For the 1PPS input, the limit is defined by a max|TE| of 100 ns, which is the max|TE| that can occur at the output of a PRTC. When the DUT has an embedded PRTC, this 1PPS case is not applicable because there might not be a 1PPS port to give an input.

In addition, a T-TSC-A must maintain the $max|TE_L|$ (0.1-Hz LPF) at its 1PPS output no greater than 1350 ns (only in the case where the T-TSC-A is external to the end application). This value is 250 ns more than the normal 1100 ns because the usual 400 ns of TE (where the T-TSC is embedded in the end application) is split into two: 250 ns for the holdover in the T-TSC-A and 150 ns for TE in the final application. So, the 1350-ns limit applies at the output of the T-TSC-A going into the end application (presumably with a 1PPS signal).

- **Characteristics of the transfer of phase/time noise through the clock:** G.8273.4 defines only one type of noise transfer, the 1PPS input to PTP output or 1PPS to 1PPS noise transfer. As in the T-BC/T-TSC case, the criteria to pass is that the phase gain of the DUT should be smaller than 0.1 dB.

- **Transient response (response in the PTP/1PPS output during an interruption in the 1PPS input):** This test involves a loss of the local 1PPS time signal, and then a subsequent return of the signal, all while monitoring the response at the output. The transient response needs to be less than 22 ns MTIE during the loss and again less than 22 ns MTIE following restoration. During the loss, it shall meet the holdover requirement defined in the next test case.

- **Holdover, or long-term transient:** Measures the phase error under two conditions for the assisted clocks. One is when the local time input (1PPS) has been removed and the remote PTP signal is still available, and the second is where both signals are absent and the clock goes into holdover based on its oscillator.

  For the oscillator holdover case, the tester measures the phase error over 1000 s and plots it against a mask that rises to 1 µs after 1000 s. Once again, holdover tests should only be initiated following a period that allows the clock to settle down following setup. For accurate long-term tests, that should be up to 24 hours.

Frankly, the testing of assisted clocks is not frequently executed. In some ways, these tests are simpler than the G.8273.2 case, but quite a few combinations of possible cases are not yet specified. This is because the assisted scenario has more contributing factors that are difficult to model and control. Another reason is that G.8273.4 is a much later document, so the definition of more limits and cases may come with time. That is the case with the assisted clocks; now to look at the case of unassisted PTS clocks, which is probably a more realistic scenario to test against.

Testing T-BC-P and T-TSC-P to G.8273.4

In the preceding section, G.8273.4 covered the cases of the PTP-aware clock with local assistance. This section, also based on G.8273.4, covers the case where there is only partial support from the network with no local time source. The Telecom Boundary Clock with partial support (T-BC-P) and Telecom Time Slave Clock with partial support (T-TSC-P) clocks also run G.8275.2 PTPoIP, and using a physical frequency source, such as SyncE or eSyncE, is optional.

If SyncE is used as a frequency source, then the T-BC-P/T-TSC-P clock must implement an assisting clock that meets the clock requirements of G.8262 and the ESMC capabilities of G.8264. If a TDM signal is used as a frequency source, then it must meet the clock requirements of G.813 and the SSM requirements of G.781. This is as you would expect.

The testing methodology is equivalent to the methodology described in the preceding section for testing according to the G.8273.2 recommendation for T-BC/T-TSC, so the preceding Figure 12-19 is still valid for the test setup.

Regarding the equipment limits, the same source of reference information is available, specifically the clock limits in G.8273.4 (in clause 8) and the network limits in clause 7.3.2 in G.8271.2. The following represents a synopsis of the performance tests of T-BC-P/T-TSC-P clocks from G.8273.4:

- **Frequency accuracy:** Confirm that when running free, the clock's oscillator is accurate to within 4.6 ppm of the nominal frequency.

- **Noise generation:** The tester applies an ideal signal and measures the phase error on the signal returning from the DUT. For this case, the tester generates a PTP signal, passes it to the measurement port via the DUT, and measures the amount of TE introduced.

  The output of the T-BC-P is measured at both the 1PPS output and the PTP master port; the output of the T-TSC-P is measured at the 1PPS output (if available).

  As in the assisted case, the value for the limits of max|TE| is currently for further study. The tester calculates the cTE (averaged over 10,000 s) and the MTIE and TDEV for the (filtered) dTE.

  For cTE, the limit applicable for Class A is ±50 ns, and for Class B it is ±20 ns (the same as the T-BC-A/T-TSC-A case). The dTE_L (after 0.1-Hz filtering) should be less than 200 ns peak-to-peak (again, tested over 10,000 s), while the dTE_H limit is for further study.

  See section 8.2 of G.8273.4 for more information on the noise generation limits and performance of T-BC-P/T-TSC-P clocks.

- **Tolerance to phase/time and frequency noise from the input (PTP and SyncE) signals:** The tester applies a noise pattern to the timing input signal while the engineer monitors the clock behavior. A pass result is that the DUT should maintain its lock to the reference signal without switchover or moving into holdover.

  The PTP noise is the max|pktSelected2wayTE| network limit (1100 ns from clause 7.3.2.1 of G.8271.2) applied as an input (it is only defined for option 1 networks), which is the maximum allowed TE at the end of a chain of clocks. The tolerance to frequency wander must be the same as an EEC from G.8262, which specifies MTIE (option 1) and TDEV (option 1, 2) masks. See the section "Testing EEC and eEEC SyncE Clocks" for details.

In addition, a T-TSC-P must maintain the max|TE$_L$| (0.1-Hz LPF) at its 1PPS output no greater than 1350 ns (only where the T-TSC-P is external to the end application). This value is 250 ns because the normal 400 ns of TE (where the T-TSC is embedded in the end application) is split into two: 250 ns for the holdover in the T-TSC-P and 150 ns for the final application.

■ **Characteristics of the transfer of phase/time noise through the clock:** Measures the transfer of TE from the input to output interfaces from PTP to PTP and PTP to 1PPS. This tests the characteristics and performance of the passband filter. As in the other cases, for the test to pass, the phase gain of the DUT should be smaller than 0.1 dB.

The transfer of physical frequency to PTP or 1PPS is not currently specified.

■ **Transient response (response in the PTP/1PPS output during an interruption in the PTP input):** This test involves a loss of the PTP time signal, and then a subsequent return of the signal, all while monitoring the response at the output. The transient response needs to be less than 22 ns MTIE during the loss and again less than 22 ns MTIE following restoration. During the loss, it shall meet the holdover requirement defined in the next test case.

The transient response of the PTP/1PPS outputs during a switchover of the source of physical frequency or during the switchover of both frequency and PTP signals is not currently specified.

■ **Holdover, or long-term transient:** Measures the phase error under two conditions for the assisted clocks. One is when the PTP input has been removed and the physical frequency signal is still available, and the second is where both signals are absent, and the clock goes into holdover based on its oscillator.

For the oscillator holdover case, the tester measures the phase error over 1000 s and plots it against a mask that rises to 1 µs after 1000 s (the same as the assisted case).

In the case that a frequency source such as SyncE is still available, the limits are the same as the T-BC/T-TSC case. The test calculates the dTE with an LPF applied and there is an MTIE mask for both constant and variable temperature indicating pass/ fail. The values are 62–123 ns MTIE after 1000 s with the constant temperature and 63–173 ns MTIE after 1000 s with the variable temperature.

Once again, holdover tests should only be initiated following a period that allows the clock to settle down following setup. For accurate long-term tests, that should be up to 24 hours.

In the authors' experience, much like the test cases for the T-BC-A and T-TSC-A assisted clocks, these tests are not widely executed. Generally, engineers are more interested in the T-BC cases from G.8273.2.

Testing T-TC to G.8273.3

One slightly different form of testing involves that of T-TCs. The testing layout is very much the same as the case for the T-BC with PTP and SyncE delivered to one port on the T-TC and returned out another port to be measured by the tester. The major difference is that the tester directly confirms the accuracy of the values written into the correction-Field of the Sync and Delay_Req messages. Figure 12-21 shows the topology needed to support testing.

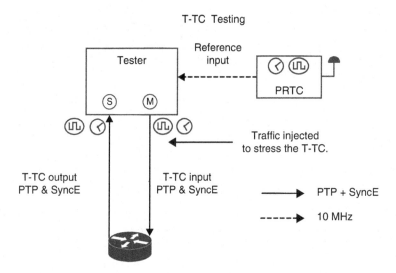

Figure 12-21 *Testing T-TC Transparent Clocks*

The tester can determine the actual latency by timestamping the packets on egress and again on ingress. It will then be able to compare the difference in value between the latency and the values written into the CF by the T-TC (the propagation time for the cable is entered into the tester and calibrated out). The engineer can configure the tester to generate both one-step and two-step PTP packet streams to ensure proper behavior for both schemes. It is also possible to run these tests under varying traffic patterns and priorities to confirm whether the T-TC keeps its accuracy under high-stress traffic situations.

For reference, there is more information on testing these clocks in Appendixes A.4 and B.4 of ITU-T G.8273 as well as the equipment limits that are specified in G.8273.3.

Note that to provide the best performance, the T-TC needs connection to a form of physical frequency, and of course, the most likely choice to provide that signal is SyncE or eSyncE. The reason this signal is needed is the same reason that a tester needs it. The T-TC needs to measure the transit time that the PTP message spends resident in the clock, and to do that its oscillator needs to be accurate. Consequently, G.8273.3 recommends that the T-TC should be frequency synchronized to an external frequency reference to remove errors due to oscillator inaccuracy.

As with the T-BC recommendations, G.8273.3 specifies that a T-TC can perform at either Class A or Class B level of performance, with the latest revision adding a Class C. To achieve the level of Class C, T-TC clocks can only be used in combination with eSyncE clocks (from G.8262.1). Many of the values used to define the classes of performance are the same as those for the T-BC case.

The following represents a synopsis of the performance tests of a T-TC clock from G.8273.3:

- **Noise generation:** The tester applies an ideal signal (PTP with SyncE), sends messages out a master port, receives the return signal on a slave port, and measures the TE introduced by the time signal passing through the DUT.

  The tester measures the 2wayTE (see Chapter 9) and confirms a pass/fail against the specified max|TE| T-TC clock limit, such as 100 ns for Class A and 70 ns for Class B. Currently, the value for Class C is not specified. The tester also calculates the cTE (averaged over 1000 s) and the MTIE for the (filtered) dTE.

  Each of these calculated values is used to measure compliance against class (A, B, or C) for noise generation. For cTE, Class A is ±50 ns, Class B is ±20 ns, and Class C is ±10 ns when averaged over 1000 s (the same as for the T-BC case).

  The dTE_L (after filtering) is compared to the relevant MTIE value for each class of performance. For dTE_L (0.1-Hz LPF) at constant temperature over 1000 s, the limit for Class A is 40 ns, for Class B is 40 ns, and for Class C is 10 ns. For dTE_L (0.1-Hz LPF) at variable temperature over 10,000 s, the limit for Class A is 40 ns and for Class B is 40 ns, while Class C is not yet specified. The TDEV results for dTE_L are not yet specified.

  The dTE_H (after filtering with a 0.1-Hz high-pass filter [HPF]) over 1000 s must be less than 70 ns for Classes A and B, while the value for Class C is not yet specified.

- **Tolerance to phase/time and frequency noise from the input (reference) signals:** Currently the ability to syntonize (frequency synchronize) the T-TC is only specified using the physical method and not the PTP method, so there is no specification for tolerance of phase/time noise via the PTP message stream.

  The T-TC must tolerate frequency noise (wander tolerance) that is the same as the SyncE clock (G.8262) for Classes A and B, and the same as the eSyncE clock (G.8262.1) for Class C. See the earlier section "Testing EEC and eEEC SyncE Clocks."

- **Characteristics of the transfer of phase/time noise through the clock:** For similar reasons to the noise tolerance test case, there is no specification for PTP noise transfer to PTP or PTP to 1PPS.

  SyncE to PTP noise transfer measures the transfer of noise from the physical frequency interface to the phase/time outputs. For the case of a T-TC, this noise affects the residence time measurements (wander in the physical signal shows up as inaccuracies in the CF measurements). There are different criteria for Classes A and B versus a Class C clock because of the difference in the limits of the bandpass filter

(1–10 Hz for SyncE versus 1–3 Hz for eSyncE). But the result is the same; the phase gain inside the passband should be smaller than 0.2 dB.

- **Transient response specifications:** Not yet specified (the term used in ITU-T recommendations is "for further study").

- **Holdover, or long-term transient:** Does not apply to the T-TC because it is not phase aligned to the PTP message stream.

The testing specification for the T-TC mirrors, as much as possible, the same values as the T-BC. This is very handy for budgeting purposes because it allows the T-BC and T-TC to be treated much the same, certainly from the point of view of noise generation performance. Over the past few years, the authors have not seen many operators performing T-TC testing, but that is because the T-TC has not been a popular option for deployment. Most effort goes into testing the T-BC.

Testing (Cascaded) Media Converters

There is a slightly different consideration when testing T-BC clocks that consist of two network elements connected to each other in a *back to back* configuration. In PTP terms, these devices are known as "media converters," and there are sections of the ITU-T recommendations (such as Annex A in G.8273) describing their performance.

This state may arise in a lab where a T-BC does not have interfaces that can be connected directly into the timing tester. Perhaps the DUT needs testing with an Ethernet variant that is not supported on a tester (such as 400 GE at the time of writing) or some specialized optic that the tester does not support or recognize.

A similar problem arises when the DUT is part of a transport link using technology such as microwave, passive optical networking (PON), or cable that almost certainly comes as a pair. It might have an Ethernet port that can accept a timing signal from the tester but lack a port to return an output signal for measurement. The return signal comes out of a second device.

In this situation, the way to test it is as a back to back pair, as shown in Figure 12-22.

The tester introduces the generated test signal to one device that does the conversion to the specialized media. Another device at the other end of that link recovers the clock and returns the timing signal to the tester with PTP and SyncE. The tester then measures the TE characteristics of the pair, including the media.

Note that we are not testing the transport of PTP over these transports. In fact, for some of these transport schemes, it is known that they produce unacceptable amounts of PDV and asymmetry. That is what would happen if these devices were completely PTP-unaware, which is not the case here.

In this scenario, these pairs of devices act as a two-device distributed T-BC; they recover the clock from the PTP on the slave port, carry the frequency and phase using some native mechanism, and regenerate the clock signals at the master port of the second device.

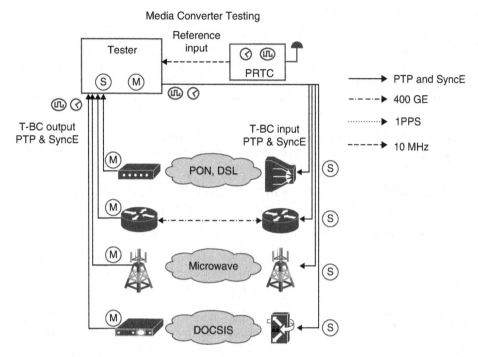

Figure 12-22 *Testing T-BC Performance in Media Converters*

The problem arises because the tester is now testing two nodes as a T-BC and not one, so a new method is needed to compare the equipment limits in the recommendations. Appendix V of G.8273.2 gives details on addressing this scenario when testing T-BC performance. The following shows how the various test cases are treated in the case of cascaded media:

- **Noise generation:** In summary, the cTE adds linearly, the dTE_L accumulates as a root mean square (RMS), while the dTE_H stays at the same value as for one clock.

  When testing for cTE, the requirement is ±50 ns to meet Class A in the case of a single T-BC clock. If the clock is made up of two nodes, then they would still meet Class A if the cTE was less than ±100 ns (adding ±50 ns from each).

  The dTE_L accumulation is treated differently because of the results of filtering. Normally, the requirement for Class A is 40 ns, but when testing two devices back to back the RMS of the accumulated errors is used. When combining two Class A devices, the RMS of 40 ns + 40 ns is used, which comes to almost 57 ns. This can be rounded up to 60 ns, so testing a pair of clocks in this method would meet Class A if the measured dTE_L is under 60 ns.

  The dTE_H does not accumulate because the noise from the first device is substantially filtered out by the second device. So, the peak-to-peak limit is the same as for a single device.

The max|TE| is a maximum of the absolute value of all the above TE values, accumulating the cTE, dTE_L, and dTE_H values for both devices. But there is a complication in that the symmetry of the dTE_H must be considered in its contribution to the calculation of max|TE|. Appendix V of G.8273.2 proposes two assumptions/methods for this asymmetry to calculate max|TE| and averages the results of both cases to arrive at a value. For Classes A, B, and C, the max|TE| values are 160 ns, 100 ns, and 45 ns, respectively. See Appendix V for details.

■ **Noise tolerance:** The noise tolerance is the same as a single clock; just because there is a different form of media downstream does not mean that this clock should be less tolerant to noise.

■ **Noise transfer:** Noise transfer for PTP to PTP or PTP to 1PPS at the output of the pair of clocks should have a maximum phase gain of 0.2 dB at a bandwidth of 0.1 Hz.

■ **Transient response:** This is for further study, and no requirement has been specified.

■ **Holdover requirement:** This should not change, although the performance depends on whether there is still a frequency signal available during a failure of the media converter interface. If it is not available, then the case is the same as the oscillator holdover.

Note that for noise generation, a similar method can be used to calculate the combined error from a chain of single-node T-BC clocks across a timing network. Values derived from MTIE and TDEV (such as dTE_L) accumulate as the square root of the sum of the squared values and the cTE values simply add up.

Overwhelmed! What Is Important?

At this point, it would be quite reasonable to expect most readers to be exhausted with all the details about the myriad of possible test cases against the performance limits in the clock specifications. For vendors, and for engineers concerned with all factors of timing performance, many of these tests are very important and are performed carefully and methodically. But for operators who are just interested in getting their 5G radios deployed, some might suspect obvious amounts of overkill. For their sake, the following list highlights the tests that are most important for everyday deployments:

■ **End-to-end network frequency wander:** Frequency wander tests are easy to set up and execute, so it makes sense to perform them. The performance is rarely an issue with SyncE, but frequency testing may be necessary if, for example, you use it to detect a timing loop in the network. The other possible (troubleshooting) case is where every node in the chain appears to be faithfully reproducing the frequency signal, but the hardware is not doing what it claims (in the ESMC values). See the following section "Field Testing."

■ **End-to-end network phase alignment:** Getting your network in phase and keeping it there is really the goal of this whole book, so it makes sense to measure phase alignment. Every operator wants to perform these tests to see what the phase alignment is at the end of the timing distribution network and what happens to it under various conditions and events.

- **Holdover:** It is vital to understand that when the PTP and SyncE (or both) signals are interrupted in the timing network, what happens to the clocks downstream? Typically, most operators test this by removing the timing signal(s) from the last clock in the chain and monitoring the phase alignment over time. It is also a good test to investigate the difference between holdover assisted by SyncE and oscillator holdover.

- **Network rearrangements:** This is testing the case when the T-GM for the timing chain fails and the network selects another source, or there is a network arrangement upstream of the last clock. Similar cases involve monitoring TE during events like control plane processor switchovers in those same network elements.

- **Frequency clock (EEC/eEEC) tests:** The frequency tests executed most often would be wander noise generation, but even then, most engineers are not that interested in testing SyncE clock performance on a single node. The exception is operators who are interested in circuit emulation of TDM links—they love to measure frequency!

- **G.8273.2 T-BC/T-TSC clock performance:** This is the clear winner for the most popular test set to perform. Just about every operator interested in phase/time deployments wants to test the performance of the T-BC. The very clear favorite is the noise generation tests, specifically to measure max|TE| and cTE. For platforms with redundant control plane processors, measuring PTP 2wayTE (and wander generation) during switchover is also very popular.

- **Functional testing:** The most useful of these tests is the behavior of the clock quality variables (in the Announce message) during the various changes of state in the T-BC or T-GM. The same applies to the ESMC QL messages. It is important to check that the values of these variables reflect your expectations during events like switchover, acquiring, T-GM failover, and holdover.

The reason why operators execute the preceding test cases more often than any others is that they are the important ones when all you want to do is to get the design certified and the solution deployed. The authors would agree that these points cover most of what you should need. Also remember that if you are not concerned with phase/time, then you only need to perform the preceding test cases relating to frequency.

This list is a good place to start, but you should be prepared to move into other areas as you see the results unfold. A final piece of advice: Never walk out of a lab with an unanswered question. One day, that question will return and demand an immediate answer, and at the worst possible time. Test cases have a finite runtime, but the mobile network must run 24×365.

Automation and Assurance

Network providers and operators are always challenged with turning up new services in an error-free, flexible, and agile manner. Of course, all the operator wants to do is to provide a service to its end customers, but it needs to map that goal into configurations on a multitude of network elements. This gets more difficult when the networks are multi-vendor, and each vendor has its own management interface for configuration, monitoring, and control.

The same issue arises with provisioning a path for a timing signal, which to the network is just another service. In theory, the operator would like an end-to-end configuration and visibility across all PTP network elements—from the GM through BCs to the remote application. To allow the network to be operated well and with high efficiency, there must also be a robust set of PTP performance metrics and monitoring tools.

IEEE 1588-2008 specifies a management node (a device that configures and monitors clocks) for configuration and performance management of PTP clocks. The management node can read and write parameters in the other PTP nodes via PTP management messages. The management node itself does not take part in the synchronization or BMCA and is only used for the management of PTP clocks. Refer to section "PTP Clocks" in Chapter 7, "Precision Time Protocol," for some more details on management nodes. Note, however, that management nodes and the PTP management messages are not supported in any of the telecom profiles.

The goal is to provide a mechanism where one or more nodes can perform actions such as GET and SET on a range of parameters (known in 1588-2008 as datasets) on other PTP nodes via management messages. The target PTP node will reply to each GET or SET with either an *OK* or *not OK* status to indicate the result of the operation. Refer to Chapter 15 of 1588-2008 for details on management nodes and management messages.

1588-2008 did not always specify datasets for optional features. IEEE 1588-2019 improves on the optional features from 1588-2008 and adds several new ones (for details, see clauses 16 and 17 of 1588-2019). Additionally, 1588-2019 specifies datasets for all these optional features as well as the standard features. If a feature can be managed remotely, then a dataset is specified in 1588-2019 and, with that, datasets clearly become an information model for management.

However, this approach of using management messages without any comprehensive system of built-in security can expose PTP clocks to numerous types of attacks from within the network. And so, most of the equipment vendors either do not recommend these management messages or have disabled them. As previously mentioned, they are not supported in many profiles, including the telecom profiles. However, designing an accurate timing network is not enough if we cannot monitor and manage it; hence, IEEE and IETF have defined other mechanisms that provide detail state and performance statistics of the timing device.

SNMP

The history of network operations has shown that SNMP is a good mechanism for fault handling (sending SNMP traps to a centralized collector for fault isolation) and monitoring of the devices (for example, enabling the retrieval of performance data). With the increasing adoption of PTP for timing distribution, clearly an SNMP Management Information Base (MIB) was needed to apply similar management techniques to timing.

The IETF proposed a standard, "Precision Time Protocol Version 2 (PTPv2) Management Information Base," in RFC 8173. To be fair, this proposed standard is dated June 2017 and has not been updated since then, and it doesn't apply to PTPv1. From the point of view of the telecom profiles, RFC 8173 only considers the G.8265.1 telecom profile for frequency, and the rest are out of scope.

Much like the Yet Another Next Generation (YANG) models, this SNMP MIB focused on managing the standard PTP data elements, which has some shortcoming in the 1588-2008 edition. There is also a project at the IEEE to draft an amendment to 1588-2019 to define a MIB to support the new version. See the following section on the YANG developments for more details on both these topics. Note that there is not a currently defined MIB for the ITU-T G.8275.1 profile, which makes SNMP wholly unsuited for management of most WAN-based timing solutions, including those for mobile.

On the other hand, there are proprietary, vendor-specific MIBs that are available for management of PTP clocks. CISCO-PTP-MIB is one such example, which captures many of the elements of PTP configuration and performance monitoring.

Although widely adopted, SNMP has some deficiencies that make it an unsuitable tool for configuration management. One of the main reasons is that SNMP handles a mix of configuration and non-configuration data. This mix makes it difficult to implement even basic network automation functions such as configuration backup and restore because it is not clear what elements of a MIB are for configuration.

An additional problem is that, architecturally, SNMP is a device view rather than a network and services view. And operators no longer wish to manage devices but want to configure and manage networks and/or services. As an example, if an interface were to start seeing errors, SNMP would alert the network management system (NMS) that interface GE 10/1 on Router "Fred" is showing a very high error count. But SNMP does not tell the operator that this link, servicing ACME Incorporated, has an extremely high service level, and the operator is contractually committed to inform the customer of its degraded state within a defined time limit.

Such gaps, along with several shortcomings of other management protocols (such as CORBA, SOAP, and others) gave rise to the development of NETCONF and YANG.

NETCONF and YANG

In 2006, the IETF published the Network Configuration (NETCONF) specification in RFC 4741 (now updated by RFC 6241) to provide mechanisms to install, manipulate, and delete the configuration of network devices. NETCONF uses an eXtensible Markup Language (XML)-based data encoding for the configuration data as well as the protocol messages. NETCONF has been adopted by major network equipment providers and has already gained strong industry support. Also, vendors of other (non-network) equipment are starting to support NETCONF on their devices as well.

Yet Another Next Generation (YANG) is a data modeling language (DML) that is used to model configuration, state data, and administrative actions to be used by network configuration protocols such as NETCONF. YANG was originally published in September 2010 as RFC 6020, and it was updated in August 2016 by RFC 7950 to YANG 1.1. YANG allows the definition of a unified model to describe a service as well as device configurations and is referred to as a YANG model.

A central network orchestrator (or network configuration server) utilizes the YANG model(s) to configure devices in the network using the NETCONF protocol (this is different from a PTP management node). Thus, the NETCONF protocol is a generic implementation that acts on YANG models, conceptually the same as how SNMP acts on MIBs. In theory, a new network service should only require a new YANG model to enable network-wide configurations.

While YANG models have already been standardized for many protocols, at the time of writing, a standardized YANG model for PTP is still not complete. IETF RFC 8575, published in May 2019, is a proposed standard for a YANG data model for PTP. This proposed model targets only the IEEE 1588-2008 default profiles, which means the data model only includes the standard dataset members specified by IEEE 1588-2008.

The optional features (as specified in clauses 16 and 17) of IEEE 1588-2008 did not always specify datasets, so it is not clear how to manage those features using a data model. Several shortcomings in the 1588-2008 version were rectified in the 2019 version, which will simplify the development of a model—such as making all features to be managed available as a dataset. Similarly, the telecom profiles are not covered by the model in this RFC, as there are features in those profiles that are not captured in 1588-2008.

Such omissions (albeit not deliberate) and the continuing evolution of network protocols are anticipated by YANG, and the standard has a provision to allow data models to be extended. For any update to the data model resulting from a later version of 1588 or some specific profile, the original 1588 model is imported, and additional elements added with the *augment* YANG DML keyword. It is the authors' expectation that as RFC 8575 becomes established as a standard, it will be improved to include elements and datasets required by IEEE 1588-2019 as well as by different profiles specified by ITU-T recommendations (such as G.8275.1, G.8275.2, and so on).

So, for now, RFC 8575 proposes the hierarchy of a YANG module for IEEE 1588-2008, which encompasses query and configuration of device, timing port, and clock datasets. Query and configuration of clock information includes the following:

- Clock dataset attributes in a clock node, such as *Current dataset*, *Parent dataset*, *Default dataset*, *Time Properties dataset*, and *Transparent Clock Default dataset* (see Chapter 7 for details of these datasets).

- Port-specific dataset attributes, such as *Port dataset* and *Transparent Clock Port dataset*. Again, see Chapter 7 for details of these datasets.

In IEEE 1588-2008, each member of the various PTP datasets is classified as one of the following:

- **Configurable:** The value can be read and written by the management node (an example is the domain number of the clock).

- **Dynamic:** The value can only be read, but the value is updated by normal PTP protocol operations (an example is the clockClass of the Default dataset).

- **Static:** The value can only be read, and the value typically does not change (an example is the two-step flag).

For details on the classification of each PTP dataset, refer to 1588-2008 or the overview of the various datasets in Chapter 7.

To handle the situation where an element can change between categories, RFC 7950 specifies a *deviation* mechanism whereby the category of the field can be updated. In a future implementation, if a read-only element of a dataset becomes read-write and the management node can update it, then the YANG model can be adjusted to allow that.

Besides the proposed model in RFC 8575, some equipment vendors are also designing and implementing their own proprietary YANG models that describe both configuration and performance aspects of PTP clocks. The YANG model for Cisco XR–based PTP clocks is one such example.

Figure 12-23 illustrates a simplified mechanism of using NETCONF and YANG to manage PTP clocks from a central management node. Note that the figure illustrates a theoretical model, which, at the time of writing, is possible only by using proprietary YANG models and when all the devices in the topology support NETCONF and YANG.

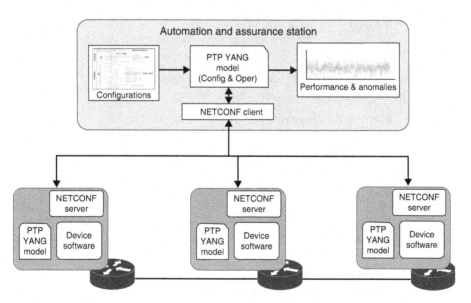

Figure 12-23 *Management of PTP Clocks Using NETCONF and YANG*

To configure and provide full performance monitoring, these proprietary YANG models go beyond the standardized model that has been proposed. The Cisco proprietary YANG models for Cisco XR–based PTP clocks is one such example. Their implementation of this YANG model has two main categories and multiple subcategories:

- **YANG models for configuration:** Different subcategories under this include generic configuration and equipment-specific configurations. For example, Cisco has *Cisco-IOS-XR-ptp-cfg.yang* and *Cisco-IOS-XR-ptp-pd-cfg.yang*.

- **YANG models for monitoring of operational parameters:** *Cisco-IOS-XR-ptp-oper.yang*, *Cisco-IOS-XR-ptp-pd-oper.yang*, and *Cisco-IOS-XR-ptp-pd-oper-sub1.yang* are models that capture the data elements needed to give an operational view of the status of the PTP clocks.

These YANG models can be inherited and can be used by network management solutions (using either NETCONF or some other protocol) for configuring and managing PTP clocks in the network. At the time of writing, there are very few solutions that exist to manage SyncE and PTP clocks exclusively via NETCONF/YANG, at least for the telecom profiles.

Almost all operators use a combination of different methods to manage the network devices. The following are some of these methods:

- A *command-line interface* (CLI) with a direct connection (via SSH/Telnet) for configuration as well as collecting statistical information and performing troubleshooting.

- Using proprietary SNMP MIBs, primarily for performance monitoring, as part of an automated management system. The Cisco Evolved Programmable Network Manager (EPNM) is an example of one such management solution, which provides automation and assurance of SyncE and PTP clocks for Cisco devices.

- Using the standard NETCONF protocol and proprietary YANG models for configurations, performance monitoring, and troubleshooting purposes.

- Streaming telemetry data from devices for performance monitoring.

Other SDOs, such as the IEEE, are also working on YANG data models to supplement their existing specifications for PTP. Two projects are currently underway:

- **P1588e:** The organization responsible for 1588 is working on several amendments to the 1588-2019 version. The focus of this specific project, P1588e, is to specify MIB and YANG data models for all PTP datasets included in the 1588 standard. The goal is to produce module files that specify the MIB and YANG data models and make them publicly accessible.

- **P802.1ASdn:** The organization responsible for the 802.1AS PTP profile has a project to define a YANG data model for the 2020 versions of that PTP profile. The goal is to allow configuration and reporting of all managed objects in the base standard. As many of the managed objects in 802.1AS originate from 1588, it makes sense that this model will augment the model from P1588e.

For further information on both these projects, refer to their references in the section "Further Reading" at the end of the chapter.

It is the authors' opinion that the automation and assurance aspects of both SyncE and PTP are beginning to evolve and will proceed rather quickly. Standardization activities on this front are only now gaining traction, such as projects to extend the YANG models to 1588-2019; the telecom profiles; and frequency synchronization. Once such standards become stable, they will be quickly adopted. It will soon become commonplace to configure and monitor SyncE and PTP clocks with a single, unified approach and in a completely vendor-agnostic manner.

Troubleshooting and Field Testing

Congratulations, the timing network has passed all the acceptance criteria and is in production, so a job well done by all. The celebrations have died down and the project is coming to an end. But now the question is, "What happens when something goes wrong?"

This section covers the highlights of how the operations staff can help keep the network aligned and the application people happy. Obviously, some of what can go wrong depends on the behavior and quality of the implementation on the devices you have, but the authors feel that there are some lessons to pass on that apply generally.

Common Problems

The most obvious problems are those relating to configuration. Much of the configuration of PTP is very straightforward; the problem is that many people configure it without knowing what every feature and flag does. The following section on troubleshooting covers a few of the mismatches between master and slave that cause problems, and many of them come down to configuration. So, check the basics first, such as to confirm the values of domain numbers and message rates (if they are configurable) and to ensure that the configured interface is the one passing the PTP traffic.

The next lesson is not to overcomplicate the solution. It is not uncommon to see people configure a cross-linked grid of nodes and every interface is configured as a possible source of SyncE and PTP. The thinking is that if a backup link is good, then three or four of them must be better. This is not the case, and doing so is just asking for never-ending instability that is usually traced back to timing loops caused by unfathomable topologies. Design the timing flow for normal and reasonable failure conditions. Do not just turn it on every link and expect it to work reliably. Note that this lesson applies especially to SyncE, PTP profiles such as G.8275.1 are more forgiving, just do not go overboard with large numbers of input ports.

Be careful of the correct configuration of TDM and SyncE frequency inputs and their quality levels. For the North American markets, the SyncE for any network equipment should be configured as an option 2 EEC clock, and for other markets, it is usually option 1 (ITU-T/ETSI). This difference is a mirror of the distinction between the E1 and T1 systems. Remember that option 1 (the E1 option) might be the default; it certainly is on Cisco devices.

The SyncE value for the configured option may take a default value that aligns with any configuration for TDM, so setting option 2 for TDM might make option 2 the default for ESMC. Either way, especially if you are working on systems in North America, double-check that the configuration of SSM and ESMC is set to be option 2.

Another trap is the so-called *revertive mode*, which can be specified as an option for the frequency selection process on some devices. Normally, when the primary source of frequency (say, SyncE from Ethernet interface numbered as port 1) goes down and port 2 is valid (and configured) as a secondary source, the frequency system will select port 2.

When port 1 signals that it has regained traceability to a QL-PRC/QL-PRS source of frequency, most operators expect the system to select the higher-priority port 1 again as the source of frequency for the system. This is the case only when revertive mode is set (and, of course, the QL received on port 1 is the same as on port 2), because otherwise it will continue to use port 2 until there is another rearrangement in the sources of frequency. This catches out a lot of engineers.

A similar problem arises with a timer associated with the frequency source. If revertive mode is configured, then port 1 will become the source for the system because it is configured as a higher-priority input than port 2. However, it will revert to port 1 only after a *wait to restore* timer has expired, and the default for this timer is 300 s or 5 minutes. The idea behind this timer is that the interface should demonstrate a period of fault-free operation before being relied upon as a source for the whole system.

Engineers testing in a lab frequently tear their hair out wondering why the device is selecting frequency sources seemingly at random and never switching to the correct port when it should. This type of behavior is specified in G.781, and it is working exactly according to the recommendations. The cure for this is to configure revertive behavior and reduce the time to restore timer down to a reasonable minimum (for example 10 s), especially for lab and testing work.

Troubleshooting

The preceding "Common Problems" section mentions that trying to write a detailed troubleshooting guide for a diverse range of implementations is a difficult task, as much of the behavior of a clock is affected by the design of its features. The approach to solving a timing problem can also be influenced by what sort of information is available by the software.

Although these lessons learned are mostly based on the authors' substantial experience with Cisco routers (lesser so for Linux running ptp4l), some general principles apply universally. The following is the basic fault-finding methodology that Cisco suggests when helping customers who are trying to isolate a timing issue in their network:

Step 1.   Start with the PRTC source (likely a GNSS receiver) and confirm it is receiving good satellite signals (see the section "GNSS Receivers and Signal Strength" later in the chapter for detailed coverage of this topic).

Step 2.   Confirm that the T-GM is getting valid frequency, phase, and time signals from the PRTC. This is more likely to be an issue if the two functions are not combined into a single device—if they are separate devices, double-check the cables for continuity and signal integrity and the connectors for compatibility.

Step 3.   Follow the chain of timing through the network downstream from the T-GM. At each clock, start by ensuring that the node is receiving a good frequency signal and then check phase/PTP. Remember that using a physical frequency source is mandatory for G.8275.1 and optional for G.8275.2. For frequency:

   a. Check that the proper frequency source is selected and that the device receives ESMC packets with QL-PRC/QL-PRS (if the QL value looks weird or is unexpected, check that the SSM/ESMC is configured with the correct option). If the clock indicates that the frequency source is internal (meaning internal oscillator), then something is wrong (there is no valid reference).

   b. Frequency selection follows the ESMC, so check that first. If no ESMC packets are being received, then you might need to ensure that untagged traffic is allowed on the source interface. Make sure that the ESMC packets reflect the capability of the link on the link they arrive on (such as the ESMC packets have not been "tunneled" and that there is no copper SFP in the path).

   c. Check that the device is properly frequency aligned and stable (no "flapping").

Step 4.   Now that the node is in the state FREQUENCY-LOCKED, perform the following steps to discover why the clock is not PHASE-ALIGNED:

   a. Confirm the intended interface is configured with the correct profile, domain number, and transport. For G.8275.1, ensure that untagged traffic is allowed, and for G.8275.2, confirm that the IP path to the master is via the interface expected (for timestamping support). Your implementation of G.8275.2 might support PTP only on a physical or layer 3 (L3) interface versus some logical device (such as a loopback device).

   b. Obtain a list of the *foreign masters* discovered by the Announce message and check what values the potential masters are advertising. Specifically, check the clock quality fields, domain number, and priority fields. Also ensure that the GM is setting bits to indicate that it has a traceable source of time and frequency. Make sure the clock is selecting the correct master based on those Announce messages and that the port facing that foreign master goes into slave state.

   c. Ensure the packet counters are increasing steadily and in proportion to the expected packet rates. For G.8275.1, confirm that the packet rates are set according to the profile and are flowing in the correct direction. As a quick check, the count of Delay_Req, Delay_Resp, and Sync (and optional Follow_Up) should be twice the number of Announce messages. Only Delay_Req should be transmitted from the slave port; the rest should only be received by it (the opposite is true for a master port).

d. For G.8275.2, ensure that the clock is completing negotiation with the master and confirm that the configured packet rates match the master configuration. A mismatch of packet rates can cause Announce timeouts, which interrupts the whole acquiring process. A packet trace helps considerably because it allows you to see what is happening if you suspect that the negotiation process is failing to complete.

Announce timeouts happen when the master is sending (for example) only one Announce message per second and the slave is expecting more, such as eight per second. If the announce timeout is set to three messages on the slave, then the announce timeout event will trigger after the slave thinks it has missed three Announce messages, or after 0.375 s. This will restart the whole negotiation process, and it will never lock.

Step 5. If there is no alignment, check the one-step versus two-step interoperability. Ensure that the two-step bit is set in the Sync packet if the master is a two-step clock and that a Follow_Up arrives with a matching sequence number. The slave port should accept either, but if you know the slave device cannot accept what it is being offered, then adjust the master.

Here are some additional troubleshooting steps based on some commonly observed symptoms:

An engineer might observe flapping between different sources of frequency. The most likely reason that this occurs is because of a timing loop affecting the ESMC and SyncE. The fix is to check the topology for any timing loops. One hint would be to reduce the number of frequency inputs to each node to only a single primary and backup interface. Turning on every possible interface as a possible source of SyncE frequency complicates the troubleshooting and can easily cause trouble. Of course, the node can output SyncE on every port if the node supports it, but just be careful to limit the number of inputs, especially in complex topologies.

Another common issue is where a clock flaps between the states of FREQUENCY-LOCKED and PHASE-ALIGNED. This is one of the most common problems that the authors see customers run into, and there is really very little that a vendor can do because it is normally the result of a deployment decision (such as the network has excessive PDV). Some of the reasons that this could be happening include the following:

■ There are active timing loops. Go back to the frequency steps to ensure there is not a frequency loop with the SyncE configuration. One (subtle) way you might see a timing loop is where the SyncE from node A is being given to node B and node B is giving PTP (phase) to node A. Because the frequency signal flows into the time clock on node B, this can produce loop-like behavior between them.

- The servo, upon calculating a very high variation in phase (taking the calculated phase offset outside some preset limit like 1.5 µs), may go back to FREQUENCY-LOCKED state and try to reacquire phase. The most common cause of this would be very high fluctuations in TE caused by excessive PDV and asymmetry in the path between the master and the slave.

- Unaware nodes in the path (especially for the PTS case with G.8275.2) or behavior from the packet transport (for example, a timing-unaware microwave system in the path) might allow the accumulation of PDV. For the FTS case with G.8275.1 (because it has full on-path support and resets PDV at every T-BC), this would strongly suggest some issue with the underlying transport system or optical devices rather than the PTP clocks.

- There is traceability to different clocks for frequency and phase that causes issues because of a slight misalignment between the frequency signal and the movement of the phase signal.

Troubleshooting is a learned skill. So, the more experience that you have performing the task, the better you become at doing it. Obviously, you do not want to have so many issues that you become an expert. Thankfully, in just about all cases, this technology simply works when using equipment that is designed to properly support it. Most issues come from flaws in design, non-timing-aware nodes, or shortcomings from the transport and optical systems.

The next task is to understand what the slave clock can tell the operator about the results of the clock recovery mechanism and its alignment with the master. For that, operations staff like to monitor the values calculated from the PTP timestamps, the offset from master (OFM) and the mean path delay (MPD).

Monitoring Offset from Master and Mean Path Delay

Remember that four PTP timestamps are used by the slave to calculate two values, the OFM and MPD. Many operators know how to extract these values from the servo, and some monitor it quite closely. A common fallacy is that somehow the OFM value provides a reliable indication of how accurate the slave clock is, when in fact it simply tells you what values the calculations produced for the last set of timestamps. And obviously, because of factors covered in preceding sections, these calculated values can vary quite markedly for every discrete set of timestamps.

This is what you can learn from monitoring these two values:

- If the value of OFM is varying widely, that would be caused by both PDV and varying asymmetry. The same is true for MPD, but a real example might make it easier to comprehend. Assume a situation where the path delay between a master and a slave is 100 µs in both directions, the slave has arrived at this value over many steps, and the clocks are perfectly aligned.

  So, assume that a Sync timestamp leaving the master would arrive 100 µs later under ideal conditions. The slave would calculate OFM and MPD and arrive at the OFM value, which should equal 0 ns (as the clocks are perfectly aligned) and MPD of 100 µs. Even without the mathematics, you can see that the slave is basically adding the 100-µs path delay to the received T1 timestamp. When comparing that adjusted value to its local time, it should show that they agree because the clocks are aligned and the path delay on those timestamps was exactly as expected.

Assume that the next Sync timestamp took 400 µs in the forward path (master to slave) and still only 100 µs (Delay_Req) in the reverse direction. The slave would calculate that the round-trip time is 500 µs and therefore assume that the MPD is 250 µs (half the round trip). Then the slave would interpret the T1 timestamp as indicating the slave clock is 150 µs too fast (it took 400 µs to arrive, but the MPD is 250 µs).

So, the OFM would show the slave to be 150 µs fast. In fact, it is perfectly aligned; the problem is that this series of timestamps has 300 µs of asymmetry and therefore shows 150 µs of error (half of 300).

- The OFM should normally average around the zero value, which is to be expected because that is the goal of the servo—it wants to converge on a state where the slave clock is aligned with the master clock. However, the OFM will not be zero for every set of timestamps, unless you have a perfect set of clocks and a perfect network. So, the value of OFM for every timestamp set is not so important, but how it varies might tell you a lot about the network stability and the timestamp accuracy.

- When the slave clock is acquiring a master signal for the first time and is trying to align to it (to achieve PHASE-ALIGNED state), the OFM could start with very large values. If the slave has just gone into holdover and is switching to a new master, then the values should not be very large, unless there is a large phase offset between the two masters (which is a different problem). In both cases, the OFM value should quickly converge toward zero. If it does not, then the PDV is very bad, and under worst-case conditions, the slave clock may never be able to lock at all.

- For the MPD, ideally it should remain constant. If it is varying, then there is PDV in the network. The slave will have no idea whether this is coming from the forward path or reverse path because the calculated mean is an average of the round-trip time.

Some servos display a separate value for both directions, but to calculate these values, the servo assumes the local clock is correct. In the previous example, if the servo assumes the local clock is correct, then it would be able to detect that the forward path took 400 µs (because assuming this would allow the OFM calculation to equal 0 µs).

Taking the preceding example, if the clock assumes that its time is accurate, it could see that the forward path was 400 µs and the reverse path was 100 µs and so may ignore that data because it knows it to be in error. Of course, it cannot continue to treat the data as wrong because it is basically not believing anything coming from the master unless it agrees with the local clock. And not to believe the signals from the master is the opposite of what the slave is supposed to do. If the slave continues to receive sets of timestamps telling it that its clock is too fast, then it probably is!

There is valuable information to be learned from the timestamp calculations, but just be careful to understand what these pieces of data are showing you. As has been reiterated several times, there is almost no way to determine if a standalone slave clock is truly aligned to the master clock. To ensure that, the slave clock must be compared to another reference clock, and for that, the engineer needs to make use of probes or a local reference.

Probes: Active and Passive

Making sure synchronization is working as designed is not a trivial task. As mentioned previously, the synchronization quality depends on many factors including network PDV, asymmetry, and environmental conditions. Besides monitoring the behavior and metrics of the transport nodes, one of the methods that can be employed is to integrate measurement and monitoring devices into timing distribution network itself.

These devices act as probes (they are also referred to as sensors) in the live network, providing continuous real-time monitoring information on the state of the time distribution. Operators deploy these probes at strategic points in their operational network; therefore, they differ from lab test systems, certainly in terms of cost and form factor. Compared with lab timing testers, probes are usually much cheaper and are capable of being managed remotely.

Because these probes are used for measurement purposes, they support some or all the following interfaces:

■ **GNSS receiver:** The probes have a GNSS receiver that is connected to a GNSS antenna, and that becomes their source to use as a measurement reference. Of course, the probes can carry out meaningful measurements only while a signal is available at the GNSS receiver.

■ **Frequency test input signal (such as SyncE):** Using a frequency test input signal, the probes can measure against the primary reference (such as that from the GNSS receiver). Generally, this is an Ethernet interface connected to a SyncE output on a T-BC/EEC router that is part of the timing distribution network.

■ **1PPS phase/time test signal from a node in the timing network:** The 1PPS signal can be used by probes to measure the phase of that node. Note that this might require some additional configuration on the T-BC/T-TSC timing transport node (allowing the 1PPS port to output a phase/time signal for measurement). The 1PPS signal might be used in the case where a network element does not feature a PTP master port (such as a T-TSC clock).

■ **PTP-enabled Ethernet interface:** A PTP-enabled Ethernet interface allows the probe to establish a PTP session with the test router and receive a PTP message flow. The probe uses the timestamps obtained from this PTP session to monitor and measure the phase/time alignment error of the network T-BC node.

With these input sources, the probes can measure and monitor the performance of a clock node or clock distribution network in real time. The exact type of measurement and monitoring is dependent on the type of deployment. For example, for PTS deployments, a probe can measure the PDV introduced by the unaware nodes. For FTS deployments, a probe might monitor the 2wayTE, and for frequency-only deployments, it might generate MTIE/TDEV graphs of frequency wander.

There are generally two categories of probes, which are as follows:

- **Active probes:** These devices act and behave like just another slave clock node in the network. The network element in the timing network treats the active probe connected to it as a real slave clock.

  In an FTS network deployment, the engineer would connect these probes to a T-BC node, and the probe is treated just like any other T-TSC in network. The active probe can then either export synchronization *key performance indicators* (KPIs, such as TE and/or TIE) periodically or trigger notifications based on certain criteria (such as cTE or dTE exceeding predefined thresholds) to a centralized NMS. This helps to pinpoint the type and location of a problem or allows early detection of developing issues in specific regions of the timing network.

- **Passive probes:** These probes do not become part of the synchronization network, but they are able to monitor and measure synchronization quality in real time if provided with timing information from active network elements.

  Because these probes do not directly connect to timing network elements, they are mostly used for monitoring PTP sessions, based only on time data shared from active PTP clocks. For this to work, the passive probe needs to capture timestamp data (t_1 and t_4 timestamps) from a PTP master port to allow measurement and analysis.

  PTP messages containing timestamps are either mirrored from a PTP master port or wire-tapped by the probes. The passive probe then uses its own slave-based timestamps (t_2 and t_3) to carry out the analysis. When the passive probes are installed, the fiber distance from network element (or from the wire-tapping point) is measured, which allows the probes to calibrate their measurements.

  Note that these probes can only use t_1 and t_4 timestamps from the PTP master for measurements and should *never* use the value of t_3 in the Delay_Req message from the slave. The t_2 timestamp never gets transmitted (from slave to master), so it cannot be captured from any PTP messages.

  Meanwhile, the t_3 timestamp, which is returned to the PTP master from the PTP slave (in the Delay_Req message), does not need to be accurate; it is only an approximation. In fact, a slave can even set the timestamp in the Delay_Req to be zero. So, you need to make sure that the passive probes do not rely on t_3 timestamps for any measurement because any results will almost certainly be wrong.

Because these probes are placed in live production networks, they become part of the network itself. And just like any other network devices, the operator needs to be able to manage these devices remotely and automatically. Like many other devices in the network, these devices come with their own individual management characteristics, which may present a challenge to integrate with the existing management infrastructure.

Probes are normally sold by different vendors than the vendors who sell network devices, so the probe management approach might not align cleanly with that of network devices. For example, a network device might support streaming telemetry and automation, whereas these probes might not support streaming telemetry at all. Or even if they do support it, the format of the telemetry data might be different or proprietary, which becomes an additional burden on operations staff. On the other hand, the specialized management of these devices could be more relevant for timing staff than a general-purpose NMS.

So, although the probes are useful in measuring and monitoring synchronization aspects of a network, understand that there is additional cost and effort needed to integrate these devices into the overall network management. Perhaps this is the reason, at the time of writing, such probes are not yet as prevalent as you might think.

Field Testing

One aspect often overlooked in the deployment of a timing network is the issue of field testing. The need for field testing arises when a segment of the network seems to be exhibiting clocking issues, or the application using the clocking appears to have problems. An example might be if the mobile radio people are noticing substandard radio performance or unexpected interference in some location. To eliminate clocking issues as a potential cause, the operator might decide to undertake some on-site field testing.

Obviously, this step should be taken only after all other likely causes have been eliminated. By this stage, the operations staff should understand that the timing nodes are operating correctly: all nodes are traceable to sources of frequency and phase, and the clocks are aligned and stable. See the preceding section on troubleshooting for details of what to check on the clocks. If all that seems normal, and any probe deployed in that location does not detect any issues, then it is time to conduct field testing.

The idea of field testing is to use a reference clock to confirm that the suspected nodes (such as a cell site router) connected to the application (such as a mobile radio) are frequency locked and phase aligned. For this reason, the DUT should have *measurement ports* that allow a quick check of the frequency and phase of the node. There are two measurement ports commonly used by operators:

- BITS port that transmits a frequency signal such as 2048 kHz or E1/T1. An alternative might be a 10-MHz timing port that has been configured for output.

- 1PPS signal, either from a standalone connector (such as BNC or DIN 1.0/2.3) or via a signal on a pair of pins output from an RJ-48c port.

Operators can configure their nodes to produce both these signals for output, which would allow an engineer to quickly test the synchronization situation on that network element. Obviously, for this to work, the equipment needs to support some way to output these signals, but some types of equipment may not.

For example, it is unlikely that a mobile radio has a 1PPS output jack that can be used to measure phase/time error. In that case, the only alternative available might be to test the next-to-last router in the chain to see how good the TE is immediately before the last hop. Another (developing) alternative would be to test the time alignment of the radio signal itself—in what is often called *over the air* testing.

> **Note**   It is also possible to use SyncE to measure frequency and PTP to measure the phase. However, there are negative security and operational complications if you do that. No operator wants to leave equipment on remote locations with active Ethernet ports ready for a test because of the risk to security. The alternative is that the field engineer must coordinate with the network operations to have them configure a test port whenever a field test is scheduled and close it afterwards.

So, if measurement ports are available on the DUT, the next requirement is a piece of test equipment that has a frequency and phase reference to measure the DUT signals. One piece of equipment that the authors commonly encounter for this role is the Sentinel from Calnex Solutions. Figure 12-24 illustrates one of those test devices—as you can see, it is designed to be handy and portable.

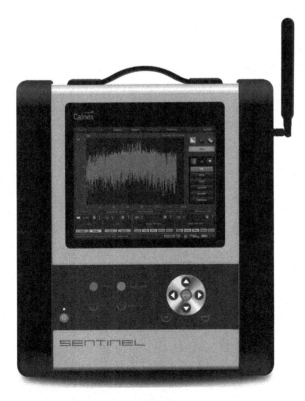

Figure 12-24   *Sentinel Field Tester from Calnex Solutions*

These devices allow the connection of frequency sources (such as SyncE, E1/T1, 2048 kHz, 10 MHz) and phase signals (1PPS, PTP) to enable time error and frequency measurements. See the preceding sections on testing networks for frequency wander and phase alignment for details on the specifics of these tests. The next issue is to find a frequency/phase reference—they are often unavailable at the remote location where the field engineer needs to test the equipment.

So, devices like the Sentinel have two components that combine to provide a stable timing reference. They are equipped with both a GPS receiver and a rubidium (Rb) oscillator in such a way that the GPS receiver can be used to condition and align the Rb oscillator. The engineer connects the Sentinel to a portable GPS antenna and leaves it with a sky view for an extended period of learning (which could be up to 12 hours, depending how long since the Rb oscillator last received a valid GPS signal).

To conduct field testing, the engineer disconnects the antenna, and the Rb oscillator goes into (battery powered) holdover. When onsite, the engineer carries the device to the suspect network clock, plugs in the BITS and 1PPS ports, and measures the phase alignment and frequency wander using the Rb oscillator (that is in holdover). Obviously, the longer the Rb oscillator is in holdover, the less accurate the measurement is likely to be.

Performing this test gives a definitive answer to the question of whether or not the network clock is properly phase aligned. One situation in which this might be valuable is where there is a large amount of asymmetry in the PTP link to the DUT that has remained undiscovered, and nobody knew that compensation was needed to correct it.

GNSS Receivers and Signal Strength

GNSS receivers, no matter what the application, all have a method for indicating the signal strength of the different satellites they are tracking. Some receivers display the signal strength in visual form such as vertical bars, but most can express signal strength in terms of actual metrics such as carrier-to-noise-density ratio (C/N_0) or signal-to-noise ratio (SNR). These two terms are so regularly used interchangeably that the differences between them are often overlooked. A full understanding of the differences is beyond the scope of this book, but you should know that they are not the same measurement, and their respective acceptable range of values is very different from each other.

SNR is usually expressed in terms of decibels. It refers to the ratio of the signal power and noise power at a given bandwidth. Those power values are expressed either as decibel-milliwatts (dBm, sometimes dB_{mW}) or decibel-watts (dBW). Note that for satellite reception, the SNR is negative, which means that the signal power is much less than the noise, which is why acquiring the signal is a complex and sensitive process.

C/N_0, on the other hand, is usually expressed in decibel-hertz (dB-Hz) and refers to the ratio of the carrier power and the noise power per unit bandwidth. Confusingly, some GNSS receivers call the signal strength values "SNR" but report C/N_0 numbers.

There are some typical ranges of expected values for reception of GPS L1 coarse/acquisition (C/A) signals. A reasonable receiver would expect to receive signals in the range of ~ 37 to 45 dB-Hz C/N$_0$, which is roughly equivalent to an SNR of ~ −29 dB to −21 dB (a higher negative dB number is a weaker signal). The amount of data that a receiver can decode from signals begins to drop very sharply when it is less than 35 dB-Hz.

There are multiple stages in the state of a GPS receiver. In the acquisition phase, the receiver is trying to detect and acquire the satellite signal from among the noise. After the receiver has acquired the satellite signal, it moves to the tracking loop. Once the tracking has begun, the receiver can (slowly) receive the time and navigation data and start to calculate a solution for location (in three dimensions) and time. Most receivers determine the location and average the solution for position over time (to arrive at a more accurate position). Finally, when the receiver decides that it has determined the position to sufficient accuracy, it switches to a timing mode (a more accurate value for location improves the accuracy of the time).

A problem arises if the antenna has only a limited view of the sky or is suffering some form of interference or multipath reflection. The survey process might struggle to complete because the receiver never arrives at a solution for location that is consistent enough for accurate timing. Even worse, if the signal from the satellite becomes too weak or disappears, then the receiver will lose that signal and discard its data from the solution.

As a rule of thumb, for a typical GPS receiver, acquisition is limited to signals above 35 dB-Hz, but it is feasible to continue tracking with a signal strength down to about 25 dB-Hz. This means that acquisition requires a higher signal strength compared to tracking. Below about 28 dB-Hz, the receiver risks losing its ability to continue receiving the satellite signal.

Table 12-3 is one example of a set of received C/N$_0$ values from a GPS receiver owned by one of the authors. The pseudorandom noise (PRN) is the satellite's (GPS) code number, the signal is the C/N$_0$ reading, the health bit indicates the signal data is valid, and the azimuth and elevation indicate the direction of the satellite and the angle above the horizon, respectively.

Table 12-3   *Example of Satellites Visible to GPS Receiver*

PRN No.	Signal	Healthy	Azimuth	Elevation
10	44	Yes	247	51
13	37	Yes	40	13
15	38	Yes	63	38
16	36	Yes	277	18
18	39	Yes	50	62
20	42	Yes	320	78
23	44	Yes	300	79
26	37	Yes	241	19
29	44	Yes	148	27

The satellites in the table with a low signal strength (36–37) mostly have low elevation levels, which means they are both further away (up to 5000 km further) and their signals must pass through more atmosphere (because of the angle). For these reasons, those signals encounter another few decibels of power reduction caused by loss from the increased path length. You can see in Table 12-3 that the satellites with the lowest three signal strengths are from those with elevations less than 20 degrees above the horizon.

The low elevation means the signal has crossed through more of the atmosphere and is likely to suffer more from any propagation anomalies and errors introduced by the atmosphere. This is one reason why receivers tend to exclude satellites from the solution if they are low in the sky (say, less than 15 degrees). But whatever the elevations, if most of the signal strengths are too low, then the place to begin investigation is with the antenna and cabling for the receiver.

The next aspect to consider is the number of satellites that the receiver can receive with good signal strength. It is necessary to have at least four satellite signals to be able to determine an accurate location and timing solution. It is also advantageous to have the visible satellites widely dispersed around the receiver (in azimuth and elevation) to improve the geometry of the data reference points.

Once the survey is completed, receivers can recover a satisfactory timing solution even with only one satellite visible. For that mode to be feasible, the antenna must be fixed in a location known by the receiver (this can come from the survey mode). This is basic mathematics; the receiver needs four satellites to solve the equations for four unknowns (x, y, z, and t), and if the location (x, y, z) is known, then a single satellite can solve for a single unknown, t.

Once you are satisfied that the receiver is successfully receiving good signals, the next steps are to ensure that the receiver is showing the expected location and height and to ensure that there is a valid value for the cable compensation. If the location is wrong or incompletely surveyed, then the antenna installation and view of the sky is the logical place to start investigating.

In summary, when investigating problems with a timing distribution network, do not forget to first check the sources of time to ensure they are accurately receiving a good timing signal and are traceable to UTC.

For further information on GNSS systems in general, and their use as sources of time for telecommunications networks, a good reference source is the ITU-T technical report on GNSS systems, "GSTR-GNSS Considerations on the use of GNSS as a primary time reference in telecommunications."

GPS Rollover

The last issue to be aware of is the problem with GPS week rollovers and how that is handled in receivers that are already in the field. This is because the original GPS signals outputting GPS calendar information (using the GPS timescale) carry the GPS week number in a 10-bit field. This field counts the week number since the epoch of the GPS timescale (6 January 1980).

Of course, a 10-bit value overflows, or *rolls over*, after 1024 weeks (approximately 19.6 years). This has already happened twice in the history of GPS, first in August 1999 and second in April 2019. Modernized signals available on new generations of satellite use 13-bit representations for the GPS week that will largely avoid this problem in the future, but it is still a problem for receivers of the widely used L1 signals.

The most common approach to solve this problem was to use the idea of a *base date* in the receiver software. Take an example of a date before the last rollover, say 22 June 2015. In the GPS calendar, this is Julian day 173 in GPS week 1850 (the Julian date is simply the day number of the year with January 1 starting as day 1). But notice that the week number is already greater than 1024, which is because the week number has already rolled over once. So, 22 June 2015 is in week 826 of the second set of weeks from 1024–2047. The GPS L1 legacy signal will only show that it is week 826, since they only have 10 bits to represent the value 1850.

So, what a receiver manufacturer does is to release their software that is hard coded using a base date—a date such as the compilation date of the software. So, software compiled on 22 June 2015 assumes that the date it receives from GPS can never be before that date. If the week number the receiver receives from GPS is 826 or greater, then it assumes that the date is in the range of 22 June 2015 to 6 April 2019 (it adds 1024 to it, giving a range of 1850–2047). If the week number is less than 826, then the current date must be after another rollover, and be in the date range of the next set of 1024 weeks (7 April 2019 onward) and so it adds 2048 to the received week number (giving a week number of 2048 and upward).

This means the receiver will correctly report the date from the start of GPS week 1850 (1024 + 826) to the end of week 2873 (1850 + 1023). But what happens when the week number exceeds 2873 on 28 January 2035? The receiver will revert to the original base date in June 2015! The thing to understand is that the GPS rollover problem might arise in your receivers at a date not aligned with the well-publicized end of a 1024-week GPS era.

The solution to this internal rollover problem is to simply release a new version of the software with a later base date, so it is not a difficult problem to fix. Many receivers, even smart phones and tablets, have used this approach and fix the base date problem through automatic updates to their software. More concerning to operators though, is that there are models of a very widely deployed GPS PRTC/T-GM that will suffer this problem in September 2022, and they do not have an automatic update feature like a smart phone.

The lesson is to make sure the software in your receivers is kept up to date.

Summary

The goal of this chapter was to cover the "life cycle" of steps that an operator might take to decide upon, roll out, and operate a timing solution. An additional bonus was to tie up some loose ends and include additional material that did not seem to belong elsewhere in the book. To begin with, the chapter provided a broad overview of the features and components that an engineer should look for in a network element that performs well as a clock.

The section on hints for designing a timing vendor selection process and then the testing of timing was meant to share with operators the authors' experience in these important areas. This experience is gained from answering literally hundreds of technical RFPs and RFIs and helping just as many customers perform timing testing on network equipment and PTP/SyncE clocks. Testing is a very specialized and detail-oriented activity that a lot of engineers really struggle with, so this chapter made a special effort to include comprehensive coverage of the topic.

Finally, the focus switched to the current state of play with network management and automation, an area where you can see that PTP still has some work to do. And the chapter would not be complete without some additional details of the day-to-day operations and management of a timing solution, including some details on difficulties that can arise with GNSS receivers and antennas.

Conclusion

Congratulations for making it to the end of the book, although there is no expectation that readers could have (or should have) read every section. The goal of the book was to cover a great breadth of material—more than most readers would need—and in enough depth to suit just about everybody. Of course, the idea was to introduce the topic from a very low level of understanding and point you in the right direction should you need to go further.

This book has a been both a labor of love and a welcome distraction for the authors during a very difficult year for many people, both locally and globally. First and foremost, we wanted to make something available to the timing community that is chiefly educational and informative. Our greatest reward would be that it helps you both in your learning and when undertaking future projects.

References in This Chapter

3GPP. 36.104, "Evolved Universal Terrestrial Radio Access (E-UTRA); Base Station (BS) radio transmission and reception." *3GPP*, Release 16, 36.104, 2021. https://www.3gpp.org/DynaReport/36104.htm

Arul Elango, G., Sudha, G., and B. Francis. "Weak signal acquisition enhancement in software GPS receivers – Pre-filtering combined post-correlation detection approach." *Applied Computing and Informatics* 13, no. 1 (2017). https://www.sciencedirect.com/science/article/pii/S2210832714000271

Calnex Solutions

"Considering Cables." *Calnex Solutions*, Application Note CX5009, 2013. https://www.calnexsol.com/en/docman/techlib/timing-and-sync-lab/137-managing-the-impact-of-cable-delays/file

"G.8262 SyncE Conformance Testing." *Calnex Solutions*, Test Guide CX5001, Version 6.1, 2018. https://info.calnexsol.com/acton/attachment/28343/f-c92c1cbe-20a7-4b32-b238-0116abcb7bd7/1/-/-/-/-/CX5001%20G.8262%20SyncE%20conformance%20testing%20app%20note%20v6.1.pdf

"G.8262.1/G.8262 EECs Conformance Test." *Calnex Solutions*, Test Guide CX3010, Version 1.0, 2018. https://info.calnexsol.com/acton/attachment/28343/ f-66bffe95-9a26-4955-9c64-2b3bbc953b7b/1/-/-/-/-/CX3010_G.8262.1_ G.8262%20EECs%20Conformance%20Test.pdf

"G.8273.2 BC Conformance Test." *Calnex Solutions*, Test Guide CX3009, Version 1.0, 2018. https://info.calnexsol.com/acton/attachment/28343/f-75fc9b8f-6884- 4c7e-bfa8-0f108a5cea0a/1/-/-/-/-/CX3009_G.8273.2%20BC%20Conformance% 20Test.pdf

"G.8273.2 T-TSC Conformance Test." *Calnex Solutions*, Test Guide CX3008, Version 1.0, 2018. https://info.calnexsol.com/acton/attachment/28343/ f-7d718c02-16a1-4899-9ad9-584e6a8243f3/1/-/-/-/-/CX3008_G.8273.2%20T- TSC%20Conformance%20Test.pdf

"Measuring Time Error Transfer of G.8273.2 T-BCs." *Calnex Solutions*, Application Note CX5034, 2017. https://info.calnexsol.com/acton/attachment/28343/ f-305478ee-b527-46d6-a20a-b57f34882104/1/-/-/-/-/Measuring%20TE% 20Transfer%20of%20T-BCs.pdf

"T-BC Time Error." *Calnex Solutions*, Test Guide CX5008, Version 6.1, 2018. https://info.calnexsol.com/acton/attachment/28343/f-16531778-3816-4524- 9834-586a8616ea17/1/-/-/-/-/CX5008_G.8273.2%20Conformance%20Tests.pdf

"T-TSC Time Error." *Calnex Solutions*, Test Guide CX5020, Version 1.4, 2018. https://info.calnexsol.com/acton/attachment/28343/f-9be6a7a2-b3c0-49f8- 99ea-37db1403ba3c/1/-/-/-/-/CX5020%20G.8273.2%20T-TSC%20Conformance %20Test%20Application%20Note.pdf

IEEE Standards Association

IEEE Conformity Assessment Program (ICAP). *IEEE*. https://standards.ieee.org/ products-services/icap/index.html

"IEEE Standard for a Precision Clock Synchronization Protocol for Networked Measurement and Control Systems." *IEEE Std 1588-2008*, 2002. https:// standards.ieee.org/standard/1588-2008.html

"IEEE Standard for a Precision Clock Synchronization Protocol for Networked Measurement and Control Systems." *IEEE Std 1588:2019*, 2019. https:// standards.ieee.org/standard/1588-2019.html

International Telecommunication Union Telecommunication Standardization Sector (ITU-T)

"G.703: Physical/electrical characteristics of hierarchical digital interfaces." *ITU-T Recommendation*, 2016. https://handle.itu.int/11.1002/1000/12788

"G.781: Synchronization layer functions for frequency synchronization based on the physical layer." *ITU-T Recommendation*, 2020. https://handle.itu.int/11.1002/ 1000/14240

"G.811: Timing characteristics of primary reference clocks." *ITU-T Recommendation*, Amendment 1, 2016. https://handle.itu.int/11.1002/1000/12792

"G.811.1: Timing characteristics of enhanced primary reference clocks." *ITU-T Recommendation*, 2017. https://handle.itu.int/11.1002/1000/13301

"G.812: Timing requirements of slave clocks suitable for use as node clocks in synchronization networks." *ITU-T Recommendation*, 2004. https://handle.itu.int/11.1002/1000/7335

"G.813: Timing characteristics of SDH equipment slave clocks (SEC)." *ITU-T Recommendation*, Corrigendum 2, 2016. https://handle.itu.int/11.1002/1000/13084

"G.823: The control of jitter and wander within digital networks which are based on the 2048 kbit/s hierarchy." *ITU-T Recommendation*, 2000. https://www.itu.int/rec/T-REC-G.823/en

"G.824: The control of jitter and wander within digital networks which are based on the 1544 kbit/s hierarchy." *ITU-T Recommendation*, Corrigendum 1, 2000. https://www.itu.int/rec/T-REC-G.824/en

"G.8260: Definitions and terminology for synchronization in packet networks." *ITU-T Recommendation*, 2020. https://handle.itu.int/11.1002/1000/14206

"G.8261: Timing and synchronization aspects in packet networks." *ITU-T Recommendation*, Amendment 2, 2020. http://handle.itu.int/11.1002/1000/14526

"G.8261.1: Packet delay variation network limits applicable to packet-based methods (Frequency synchronization)." *ITU-T Recommendation*, Amendment 1, 2014. https://handle.itu.int/11.1002/1000/12190

"G.8262: Timing characteristics of synchronous equipment slave clock." *ITU-T Recommendation*, Amendment 1, 2020. https://handle.itu.int/11.1002/1000/14208

"G.8262.1: Timing characteristics of enhanced synchronous equipment slave clock." *ITU-T Recommendation*, Amendment 1, 2019. https://handle.itu.int/11.1002/1000/14011

"G.8263: Timing characteristics of packet-based equipment clocks." *ITU-T Recommendation*, 2017. https://handle.itu.int/11.1002/1000/13320

"G.8265.1: Precision time protocol telecom profile for frequency synchronization." *ITU-T Recommendation*, Amendment 1, 2019. https://handle.itu.int/11.1002/1000/14012

"G.8271: Time and phase synchronization aspects of telecommunication networks." *ITU-T Recommendation*, 2020. https://handle.itu.int/11.1002/1000/14209

"G.8271.1: Network limits for packet time synchronization in packet networks with full timing support from the network." *ITU-T Recommendation*, Amendment 1, 2020. https://handle.itu.int/11.1002/1000/14527

"G.8271.2: Network limits for time synchronization in packet networks with partial timing support from the network." *ITU-T Recommendation*, Amendment 2, 2018. https://handle.itu.int/11.1002/1000/13768

"G.8272: Timing characteristics of primary reference time clocks." *ITU-T Recommendation*, Amendment 1, 2020. https://handle.itu.int/11.1002/1000/14211

"G.8272.1: Timing characteristics of enhanced primary reference time clocks." *ITU-T Recommendation*, Amendment 2, 2019. https://handle.itu.int/11.1002/1000/14014

"G.8273: Framework of phase and time clocks." *ITU-T Recommendation*, Corrigendum 1, 2020. https://handle.itu.int/11.1002/1000/14528

"G.8273.2: Timing characteristics of telecom boundary clocks and telecom time slave clocks for use with full timing support from the network." *ITU-T Recommendation*, 2020. https://handle.itu.int/11.1002/1000/14507

"G.8273.3: Timing characteristics of telecom transparent clocks for use with full timing support from the network." *ITU-T Recommendation*, 2020. https://handle.itu.int/11.1002/1000/14508

"G.8273.4: Timing characteristics of telecom boundary clocks and telecom time slave clocks for use with partial timing support from the network." *ITU-T Recommendation*, 2020. https://handle.itu.int/11.1002/1000/14214

"G.8275: Architecture and requirements for packet-based time and phase distribution." *ITU-T Recommendation*, 2020. https://handle.itu.int/11.1002/1000/14509

"G.8275.1: Precision time protocol telecom profile for phase/time synchronization with full timing support from the network." *ITU-T Recommendation*, Amendment 1, 2020. https://handle.itu.int/11.1002/1000/14543

"G.8275.2: Precision time protocol telecom profile for time/phase synchronization with partial timing support from the network." *ITU-T Recommendation*, Amendment 1, 2020. https://handle.itu.int/11.1002/1000/14544

"GSTR-GNSS: Considerations on the Use of GNSS as a Primary Time Reference in Telecommunications." *ITU-T Technical Report*, 2020. http://handle.itu.int/11.1002/pub/815052de-en

"Q13/15 – Network synchronization and time distribution performance." *ITU-T Study Groups*, Study Period 2017–2020. https://www.itu.int/en/ITU-T/studygroups/2017-2020/15/Pages/q13.aspx

Internet Engineering Task Force (IETF)

Bjorklund, M. "The YANG 1.1 Data Modeling Language." *IETF*, RFC 7950, 2016. https://tools.ietf.org/html/rfc7950

Bjorklund, M. "YANG – A Data Modeling Language for the Network Configuration Protocol (NETCONF)." *IETF*, RFC 6020, 2010. https://tools.ietf.org/html/rfc6020

Enns, R. "NETCONF Configuration Protocol." *IETF*, RFC 4741, 2006. https://tools.ietf.org/html/rfc4741

Enns, R. "Network Configuration Protocol (NETCONF)." *IETF*, RFC 6241, 2011. https://tools.ietf.org/html/rfc6241

Jiang, Y., X. Liu, J. Xu, and R. Cummings. "YANG Data Model for the Precision Time Protocol (PTP)." *IETF*, RFC 8575, 2019. https://tools.ietf.org/html/rfc8575

Shankarkumar, V., L. Montini, T. Frost, and G. Dowd. "Precision Time Protocol Version 2 (PTPv2) Management Information Base." *IETF*, RFC 8173, 2017. https://tools.ietf.org/html/rfc8173

Navstar GPS Directorate. "NAVSTAR GPS Space Segment/Navigation User Segment Interfaces." *GPS Interface Specification*, Revision L, 2020. https://www.gps.gov/technical/icwg/IS-GPS-200L.pdf

Chapter 12 Acronyms Key

The following table expands the key acronyms used in this chapter.

Term	Value
1PPS	1 pulse per second
2wayTE	two-way time error
3GPP	3rd Generation Partnership Project (SDO)
5G	5th generation (mobile telecommunications system)
APTS	assisted partial timing support
ATIS	Alliance for Telecommunications Industry Solutions
BC	boundary clock
BERT	bit error rate tester
BITS	building integrated timing supply (SONET)
BMCA	best master clock algorithm
BNC	bayonet Neill-Concelman (connector)
C/A	coarse/acquisition (codes for GPS)
CAT	Calnex Analysis Tool
CF	correctionField
CFP	C form factor pluggable (100GE)
CLI	command-line interface
C/N$_0$	carrier-to-noise-density ratio
CPU	central processing unit
cTE	constant time error
dB	decibel
dB-Hz	decibel-hertz
dBm, dB$_{mW}$	decibel-milliwatts

Term	Value
dBW	decibel-watts
DIN	Deutsches Institut für Normung (German Institute for Standardization)
DML	data modelling language
dTE	dynamic time error
dTE_L	dynamic time error with low-pass filter
dTE_H	dynamic time error with high-pass filter
DUT	device under test
E1	2-Mbps (2048-kbps) signal (SDH)
EEC	Ethernet equipment clock
eEEC	enhanced Ethernet equipment clock
EPNM	Evolved Programmable Network Manager
ePRC	enhanced primary reference clock
ePRTC	enhanced primary reference time clock
ePRTC-A	enhanced primary reference time clock—class A
ePRTC-B	enhanced primary reference time clock—class B
ESMC	Ethernet synchronization messaging channel
eSyncE	enhanced synchronous Ethernet
FTS	full timing support (for PTP from the network)
FPGA	field-programmable gate array
FPP	floor packet percent
GE	gigabit Ethernet
GM	grandmaster
GNSS	global navigation satellite system
GPS	Global Positioning System
HPF	high-pass filter
ICAP	IEEE Conformity Assessment Program
IEEE	Institute of Electrical and Electronics Engineers
IETF	Internet Engineering Task Force
IoT	Internet of things
IP	Internet protocol

Term	Value
ITSF	International Timing & Synchronization Forum
ITU	International Telecommunication Union
ITU-T	ITU Telecommunication Standardization Sector
KPI	key performance indicator
L1, L2, L5	bands of radio spectrum between 1–2 GHz
L2	layer 2 (of the OSI model)
L3	layer 3 (of the OSI model)
LPF	low-pass filter
MAC	media access control
max|TE|	maximum absolute time error
max|TE$_L$|	maximum absolute time error with low-pass filter
MIB	Management Information Base
MPD	mean path delay
MRTIE	maximum relative time interval error
MTIE	maximum time interval error
MUX	multiplexer
NCO	numerically-controlled oscillator
NE	network element
NETCONF	Network Configuration
NIST	National Institute of Standards and Technology (USA)
NMS	network management system
OC	ordinary clock
OCXO	oven-controlled crystal oscillator
OFM	offset from master
PDU	protocol data unit
PDV	packet delay variation
PEC-S-F	packet-based equipment clock—slave—frequency
PHY	PHYsical Layer (of OSI Reference Model)—an electronic device
pktSelected2wayTE	packet-selected two-way time error
PLL	phase-locked loop
POC	proof of concept

Term	Value
PON	passive optical networking (optical broadband)
ppm	parts per million
PRC	primary reference clock (SDH)
PRN	pseudorandom noise
PRS	primary reference source (SONET)
PRTC	primary reference time clock
PRTC-A	primary reference time clock—class A
PRTC-B	primary reference time clock—class B
PTP	precision time protocol
PTPoIP	PTP over Internet protocol
PTS	partial timing support
QL	quality level
QSFP+	quad small form-factor pluggable-plus (transceiver)
Rb	rubidium
RF	radio frequency/function
RFC	Request for Comments (IETF document)
RFI	request for information
RFP	request for proposal
RJ-45, RJ-48c	registered jack 45, 48c
RMS	root mean square
SDH	synchronous digital hierarchy (optical transport technology)
SDO	standards development organization
SEC	synchronous equipment clock
SEC	SDH equipment clock
SETS	synchronous equipment timing source
SFP	small form-factor pluggable (transceiver)
SFP+	small form-factor pluggable-plus (transceiver)
SONET	synchronous optical network (optical transport technology)
SNMP	Simple Network Management Protocol
SNR	signal-to-noise ratio
SSM	synchronization status message

Term	Value
SSU	synchronization supply unit (SDH)
SyncE	synchronous Ethernet—a set of ITU-T standards
T1	1.544-Mbps signal (SONET)
TAI	Temps Atomique International
T-BC	Telecom Boundary Clock
T-BC-A	Telecom Boundary Clock—Assisted
T-BC-P	Telecom Boundary Clock—Partial support
TCXO	temperature-compensated crystal oscillator
TDEV	time deviation
TDM	time-division multiplexing
TE	time error
T-GM	Telecom Grandmaster
TIE	time interval error
TLV	type, length, value
ToD	time of day
TSU	timestamping unit
T-TC	Telecom Transparent Clock
T-TSC	Telecom Time Slave Clock
T-TSC-A	Telecom Time Slave Clock—Assisted
T-TSC-P	Telecom Time Slave Clock—Partial support
UI	unit interval
Uipp	unit interval, peak-to-peak
UTC	Coordinated Universal Time
VCO	voltage-controlled oscillator
WAN	wide-area network
WSTS	Workshop on Synchronization and Timing Systems
XFP	ten gigabit small form factor pluggable (transceiver)
XML	eXtensible Markup Language
YANG	Yet Another Next Generation (data modelling language)

Further Reading

Refer to the following recommended sources for further information about the topics covered in this chapter.

Calnex Solutions: https://www.calnexsol.com/

Calnex Solutions, Technical Library: https://www.calnexsol.com/en/solutions-en/education/techlib/timing-and-sync-lab

Cisco MIB, repository of Cisco SNMP MIB definitions: ftp://ftp.cisco.com/pub/mibs/v2/

Cisco YANG, repository of Cisco YANG data models: https://github.com/YangModels/yang/tree/master/vendor/cisco

IEEE Standards Association

"IEEE Approved Draft Standard for Local and Metropolitan Area Networks – Timing and Synchronization for Time-Sensitive Applications, Amendment: YANG Data Model." *IEEE P802.1ASdn*, Amendment to IEEE Standard 802.1AS-2020, 2020. https://www.ieee802.org/1/files/public/docs2020/dn-draft-PAR-0520-v01.pdf

"Standard for a Precision Clock Synchronization Protocol for Networked Measurement and Control Systems, Amendment: MIB and YANG Data Models." *IEEE P1588e*, Amendment to IEEE Standard 1588-2019, 2020. https://standards.ieee.org/project/1588e.html

Index

Numbers

A

F

H

I

N

S

Register Your Product at ciscopress.com/register

Access additional benefits and **save 35%** on your next purchase

- Automatically receive a coupon for 35% off your next purchase, valid for 30 days. Look for your code in your Cisco Press cart or the Manage Codes section of your account page.
- Download available product updates.
- Access bonus material if available.*
- Check the box to hear from us and receive exclusive offers on new editions and related products.

Registration benefits vary by product. Benefits will be listed on your account page under Registered Products.

Learning Solutions for Self-Paced Study, Enterprise, and the Classroom

Cisco Press is the Cisco Systems authorized book publisher of Cisco networking technology, Cisco certification self-study, and Cisco Networking Academy Program materials.

At ciscopress.com, you can:

- Shop our books, eBooks, practice tests, software, and video courses
- Sign up to receive special offers
- Access thousands of free chapters and video lessons

Visit ciscopress.com/community to connect with Cisco Press

 Pearson

Addison-Wesley • Adobe Press • Cisco Press • Microsoft Press • Pearson IT Certification • Que • Sams • Peachpit Press